NEW MEDIA *THE DIGITAL SOURCE BOOK*

SHOWCASE

NEW MEDIA THE DIGITAL SOURCE BOOK **SHOWCASE**

President and Publisher
IRA SHAPIRO

Senior Vice President
WENDL KORNFELD

Vice President Marketing
ANN MIDDLEBROOK

Director of Production
KAREN M. BOCHOW

Director of Sales
ROB DRASIN

MARKETING
Promotion Manager
EMILY RUBIN
Marketing Coordinator
LISA WILKER
Labels-To-Go Coordinator
CAROLYN FINGER
Book Sales Administrator
MICHAEL SHUSTIN

ADVERTISING SALES
Sales Administrator
SHANA SHUA
Sales Representatives:
MITCH HOCHMAN
DAVE TABLER

ADMINISTRATION
Controller
BRENDA MASSY
Office Manager
ELAINE MORRELL
Accounts Manager
CONNIE MACLEOD
Assistant to the Controller
MILA LIVSHATZ
Bookkeeper
MICHELLE ROBERTS
Administrative Assistant
PAULA COHEN

PUBLISHED BY:
AMERICAN SHOWCASE, INC.
915 Broadway, 14th Floor
New York, New York 10010
(212) 673-6600
FAX: (212) 673-9795

PRODUCTION
Project Managers
CHUCK ROSENOW
PAMELA SCHECHTER
Production Coordinators
ZULEMA RODRIGUEZ
TRACY RUSSEK
Production Administrator
DENISE BURKA
Traffic Assistant
GEORGE PHILLIPS
CREATIVE
Art Director/Studio Manager
MARJORIE FINER
Assistant Art Director
MARY BETH TWOMEY
Studio Assistant
HARDY HYPPOLITE

Data Systems Coordinator
JULIA CURRY
Research
JOHN RUSSELL

Technical Consultant
MIHAI NADIN
Viewpoints Editor
JEANNE BROWNE

Special Thanks to:
ELIZABETH ATKESON
ED BAKER
BOB CARLSON
HELENE GRESSER
WENDY GROSSMAN
CHRISTIANE MCKENNA
ANNE NEWHALL
OMAR PORTIELES
SEAN RUNNETTE
JOY VITALE

U.S. Book Trade Distribution:
WATSON-GUPTILL
PUBLICATIONS
1515 Broadway
New York, New York 10036
(212) 764-7300

For Sales outside the U.S.:
ROTOVISION S.A.
9 Route Suisse
1295 Mies, Switzerland
Telephone: 022-755-3055
Telex: 419246 ROVI
FAX: 022-755-4072

Listings Production:
GOODMAN/ORLICK
DESIGN, INC.

Color Separation:
UNIVERSAL COLOUR
SCANNING LTD.
Printing and Binding:
EVERBEST PRINTING CO., LTD.

New Media Showcase 4
ISBN 0931144-89-2 – Paperback
ISBN 0931144-90-6 – Hardback
ISSN 1063-6471

COVER CREDITS:
Front Cover:
LANCE JACKSON
Title Page:
THOMAS UPTON

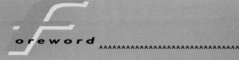

contents

Foreword ^^^^^^^^^^^^^^^^^^^^^^^^^^^^^ 4

Ira Shapiro
Publisher

Visuals—Ads & Portfolios

Index to Visuals ~~~~~~~~~~~~~~~~~~~~~~~~~~~~~~ 5

Electronic Imaging ~~~~~~~~~~~~~~~~~~~~~~~~~~~~ 7

Interactive Multimedia/Production Services ~~~~~~~~~~~~~ 93

Hardware/Software ~~~~~~~~~~~~~~~~~~~~~~~~~ 105

Viewpoints

Creative Property Rights Affect Us All +++++ 6
DONNA A. DEMAC, ESQ.

Print—The Final Frontier ++++++++++++ 39
FRANK J. ROMANO

CD-ROM: The Medium Is Still The Message ++++ 46
CHARLIE MAGEE

Neo-Classical New Media Education:
Preparing For Tomorrow
++++++++++++++++ 80
HARVEY FLAXMAN

Contemplating The Digital Future
++++++++++++++ 118
DAVID BIEDNY

Professional
Directory

Regional Categories and
Contents ▪ 121
Alphabetical Index ▪ 210

Electronic Imaging

Photographers ▪ 122
Holographers ▪ 128
Retouchers ▪ 129
Illustrators ▪ 131
Designers ▪ 141
Representatives ▪ 151

Presentation Graphics ▪ 152
Desktop Publishing Hardware ▪ 154
Desktop Publishing Software ▪ 155

**Interactive
Multimedia**

Presentation ▪ 158
Animation ▪ 162
Education/Training ▪ 165
Media Integration ▪ 166
Visualization ▪ 167
Interactive Screen Design ▪ 168
Consultants ▪ 168

Production Services

Optical ▪ 170
Video ▪ 171
Video Independent ▪ 178
Video Animation ▪ 179
Video Sound Integration ▪ 184
Prepress 188
Service Bureaus ▪ 189

Integrated Printing ▪ 194
CD-ROM Publishers ▪ 194

Hardware

CD Interactive ▪ 196
CD-ROM ▪ 196
Computer Platforms ▪ 196
DAT ▪ 196
Digital Editing ▪ 197
Digitizers ▪ 197
Display Systems ▪ 197
Electronic Photography ▪ 197
Graphic Input Devices ▪ 197
HD TV ▪ 197
Optical Disks and Drives ▪ 198
Printers/Output Devices ▪ 198
Scanners ▪ 198
Video-Digital Interactive ▪ 198
Video Production ▪ 198
Virtual Reality ▪ 199

Software

3D Modeling ▪ 200
Animation ▪ 200
Authoring ▪ 200
Digital Editing ▪ 201
Imaging/Digital Effects ▪ 201
Raytracing/Rendering ▪ 201
Video Production ▪ 201

**Resources &
Information**

Publications (Magazines/Newsletters) ▪ 203
Professional Organizations ▪ 203
Trade Shows/Conferences ▪ 204
Professional Development ▪ 205
Consultants ▪ 207

foreword
foreword
foreword foreword
foreword foreword

Change. Some people thrive on it, others fear it. Sometimes change comes slowly; at other times, it moves like a high-speed train. Change has been the only constant in the world of computers and new media — to the delight of some, and the consternation of others! Innovations in hardware and software have been presented at every turn, and new outlets for creative expression have been numerous. But even in this exciting, ever-changing arena, the substance and volume of change that we're now experiencing is virtually unprecedented.

Three important trends comprise the *revolution* that is currently occurring within the digital *evolution*. The first is *convergence,* a combination of technological breakthroughs that will, among other things, make possible the greatly-anticipated information super-highway.

Convergence is the synergy of computer graphics, video, animation, sound and data. It goes beyond multimedia in that it integrates other means of communication. Convergence is why the film industry, television, publishers and telephone companies are merging into mega-businesses. Under many new names, they are charging into the marketplace, pledging new and refined delivery systems. By no accident, AT&T has been running a series of TV commercials with glimpses into the future: a good-night kiss to baby via video-telephone; a hand-written fax sent from the beach; an interactive TV class on the history of music. These illustrate the powerful potential of convergence. But more is yet to come: video delivery on demand, digital filmmaking, and new forms of entertainment.

The second trend could be called a chip off the new block. Literally, *new computer chips* are being developed that are the first significant challenges to the IBM and Apple platforms. The old-style mainframe computer and the contemporary desktop may be rendered passé in just a few years, as new machines emerge.

These new machines are more powerful and much faster, and they integrate many of the functions that are essential to the creative work of professionals working in new media. Clearly, the microprocessor industry had convergence in mind when it designed its new Pentium (*Intel inside!,* goes the ad), Power PC, Alpha and MIPS chips. These chips have a level of speed that will allow faster processing of images, the integration of video and sound, and much more. Their performance equals that of mainframes. The Quadra 840 AV is one example. It can record QuickTime movies from a video source, "write" to tape, play audio, and more. This is only a suggestion of what the future will bring through amazing hybrid machines: a PC-based Mac, or whatever it might be called.

Simultaneous with Apple's Quadra, companies such as Silicon Graphics and SUN are using new chips to offer similar functions on their INDY and SPARC stations, respectively. A digital video camera and an analog microphone are standard equipment here, and testify to the focus of the fast-growing multimedia market. At the high end of the technology spectrum, 24-bit color, and the ability to render three-dimensional, solid graphics with full-motion video and compression, further suggest where the future lies.

Right now, your computer undoubtedly contains separate processors for audio and video; the new chips will combine them. Progress has already been made in better 3D programs, as well as in virtual reality applications. Another area of progress that is benefiting creative uses is networking. It's now possible to let different machines truly talk to each other, to the advantage of everyone using them. As a result, computer networking – within companies, and from one system to another, anywhere in the world – will accommodate more natural and complex styles of communication and cooperative work.

The final trend is, in truth, a movement – the *Open Systems Movement.* IBM's already-open system for the personal computer is what allowed for the production of IBM clones (compatibles). Apple's proprietary (or "closed") system is the reason that only a Mac is a Mac, and only Apple has it. But both giants are now being challenged by a marketplace that increasingly rejects having to choose between platforms. New operating systems currently in development by both leaders, as well as by formidable competitors – such as Microsoft, Silicon Graphics, Hewlett-Packard, 3DO and SUN – will allow consumers to build systems to their own specifications, selecting equipment from a variety of vendors.

The fruits of this opening will also facilitate a quantum leap in creative, cooperative work and communication through what is called the "client-server model," which is a decentralized use of machines. This ability to share information is much closer to what artists and communications professionals do than is the centralized use of machines. The client-server model allows for fast downloading and exchange of information. It has the potential for extraordinary sophistication. The result is greater mobility and less reliance on office equipment.

Some of these innovations, such as the integration of applications (programs "talking" to each other), are with us right now. Others, like full video, are but months away. And something such as interactive planning and design is a few years ahead. But all this change is having a dramatic impact on computer strategies for both big and small businesses, as well as individual artists.

Since its inception in 1990, *New Media Showcase: The Digital Sourcebook* has been dedicated to the explosion of creative opportunity and technological change in new media: charting, reflecting and participating in it – and helping our advertisers and our readers to do the same. From the beginning, we have given creative professionals a place to exhibit their unusual talents, and to give art and creative directors, corporate communications specialists, and graphic design firms a simple way to find these experts.

The current era of change is helping us to fulfill the total scope of our original mission. For example, many people working in new media are beginning to reject the term "computer artist," preferring instead to be regarded simply as artists working with a range of technologies and techniques. Indeed, a significant number of the artists in *New Media Showcase* who used to work exclusively for print media are now including multimedia. This year's book reflects this trend. In addition to beginning its own evolution into a sourcebook that serves more than print, *New Media Showcase* reveals other changes, as well. The book has grown by 25% and its graphic redesign is intended to reflect cutting edge design trends: colorful, asymmetrical, offbeat and daring. We've also re-designed our logo with an original typeface.

All of us involved with this book are intrigued and excited by the opportunities presented in this fast-paced, far-sighted industry. But we know that, as with any kind of change, it can be scary, confusing and uncomfortable. However, with information, preparation and a good guide, the trip to tomorrow can be an exhilarating journey. We look forward to continuing to help you along your way.

IRA SHAPIRO
Publisher

index to Visuals

ACCORNERO, FRANCO ========= 28
ACE GROUP, THE ============ 101
AGFA ====================== 109
ANTHONY, MITCHELL ========== 74
ARTCO ==================== 50-53
BAKER, KOLEA ============= 54,55
BAKER, DON ================= 55
BELDING, PAM ============== 77
BERMAN, HOWARD ============ 8,9
BERNSTEIN & ANDRIULLI REPRESENTATIVES = 64
BIG PIXEL, THE ================= 70
BLACKMAN, BARRY========== 14,15
BLANK, JERRY ================ 59
BRICE, JEFF ================= 54
BRONSTEIN, STEVE ============ 8,9
BUCHMAN, DOUGLAS ========= 70
BURKEY, J.W. ================= 35
CARROLL, DON ============= 18,19
CHROMA STUDIOS, INC. ========= 92
CONRAD & ASSOCIATES, JON ====== 68
COPYTONE ================== 103
COVEY, MIKEL =============== 10
CUSTOM COLOR CORPORATION = 102
DAMAN STUDO ============== 25
DEAN DIGITAL IMAGING ========= 38
DIGICHROME IMAGING ======== 92
DIGITAL ART ================ 40,41
DIGITAL IMAGE ============= 90
DIGITAL STOCK ============== 99
DREAMTIME SYSTEMS ========== 29
FARR PHOTOGRAPHY ========= 36,37
FEARLESS DESIGNS, INC. ======== 58
GLASGOW & ASSOCIATES, DALE === 67
GROSSMAN, MYRON ========== 75
GROSSMAN, WENDY =========== 49
GUDYNAS, PETER ============ 24
HERBERT, JONATHAN =========== 63
H-GUN LABS ============== 43
HOWE, PHILIP ================ 32,33
HUNT, STEVEN ============== 22
ICON GRAPHICS INC. =========== 76
IKEGAMI ================= 108
JACKSON, LANCE ============= 61
LAX SYNTEX DESIGN =========== 61
LETRASET ================= 113
LVT A KODAK COMPANY ====== 117
LYNCH, ALAN / ARENA ARTISTS === 24
MacNEILL, SCOTT ============= 69

MAD WORKS ================= 74
MATSURI CORPORATION ======= 72
McELROY, OLIVIA ============= 64
MC2 ===================== 48
MERSCHER, HEIDI ============= 48
META 4 DIGITAL DESIGN ======== 95
MICROCOLOR ================= 97
MILLET, CÉCILE ============= 73
MORRELL, PAUL =============== 17
NEITZEL, JOHN ============== 34
NEW MEDIA PRODUCTIONS, INC.== 96
OLDROYD DIGITAL =========== 100
OSTROFF, SANDY ============= 84
OUTERSPACE =============== 91
O'VERY COVEY, TRACI =========== 10
PACE, JULIE ================= 79
PALMS HIGH RES DIGITAL IMAGERY, THE =85
PELIKAN PICTURES ========== 88,89
PETERSON, BRUCE ============== 11
PHOTOEFFECTS ============== 87
PIXAR ===================== 116
PUNIN, NIKOLAI============== 60
RAPHAËLE/DIGITAL TRANSPARENCIES,INC. === 82,83
RENARD REPRESENTS=========== 63
R/GA PRINT ================ 44,45
ROMERO, JAVIER============== 57
SANDBOX DIGITAL PLAYGROUND === 84
SCHLOWSKY, BOB & LOIS ==== 20,21
SILICON GRAPHICS ==== 106,107, BACK COVER
SKYLITE PHOTO PRODUCTIONS, INC. 34
SMITH, MARTY================== 65
SPECULAR INTERNATIONAL= 114,115
STOKES RETOUCHING, LEE ======== 86
STRATA INC. ============= 110,111

STUDIO MacBETH =============== 50,51
STRUTHERS, DOUG ============= 52,53
TAYLOR - PALMER, DOROTHEA========= 62
TEICH, DAVID ============= 71
TRACER DESIGN, INC. =========== 42
TUCKER, MARK ============== 23
TULL, JEFF ================= 58
UPTON, THOMAS ============ 12,13
VAN NOSTRAND REINHOLD ========= 120
VARIS PHOTOMEDIA =========== 26,27
WACOM ==================== 112
WEISS, MICHAEL ============== 16
WESTLIGHT ================= 94
WILEY, PAUL ================= 56
WOJNAR DRAKE PHOTOGRAPHY, INC. == 30,31
ZAP ART ================== 24
ZUBER-MALLISON, CAROL =========== 66

viewpoints

THE DILEMMA

When David Letterman made his highly-publicized move from NBC to CBS last year, the question of creative property rights was given considerable prominence. NBC claimed that it owned the rights to Letterman's famous Top Ten List on the grounds that anything he and his writers dreamed up while they were working for the network was NBC's intellectual property. Letterman – all on-air smiles but genuinely outraged – disregarded an injunction against his use of the Top Ten List and devoted the one he recited on the night of his CBS debut to the NBC brass. With audiences chuckling in support, he scathingly portrayed NBC as a ridiculous poor loser trying lamely to compensate for its faulty strategy in the late-night TV wars.

This seemingly-silly dispute was in fact serious business. And it rang a familiar bell and pushed some important buttons for writers, artists and others who feel as if they're in constant battle to protect their work. It pushed similar buttons for publishers, ad agencies and other corporations that see themselves – like NBC – as powerless to control the creative work they frequently commission and expensively support.

Letterman is a TV superstar, not a digital graphic designer; NBC is a powerful TV network, not a mid-size ad agency. But there is a significant parallel to the concerns and rights of these disparate parties. For the most part, the law regards the creator of an image or body of language as its owner. These ownership rights *can* be forfeited in a "Work for Hire" or similar agreement, in which an artist, writer or other "creative provider" agrees that anything he or she develops – an article, photograph, video, software program – or comedy routine – becomes the property of the company or individual commissioning the work. Therefore, depending on the contracts David Letterman and his writers had with NBC and their legal status with the network, the Top Ten dispute could be resolved in favor of Letterman *or* NBC.

If an existing work is being purchased, the terms of the sale can outline ownership and control. In a neat, orderly universe, this would seem to solve the matter of creative and intellectual property rights. But our universe is neither neat nor orderly. And the explosion of technological opportunity and potential in the digital marketplace has created an urgent need for renewed consideration of the legal, economic, ethical and social aspects bearing on property rights.

Never before have entertainment and information been transmitted in such a variety of ways. But regardless of the medium, the material being dis-

CREATIVE PROPERTY RIGHTS AFFECT US ALL

DONNA A. DEMAC, ESQ.

played or transmitted consists of someone's intellectual property. Today's media products may contain hundreds or thousands of images for a single project, be they photographs, film or video images, computer graphics or animation. They may be used to create a conventional book, an interactive reference, a video game, a shopping experience, an educational tool, an on-line computer information network, etc. Whatever the product or service created, the brave new digital world requires that copyright privileges and restrictions be observed at every stage of production and distribution by all concerned.

Why "all concerned"? Because these considerations are not just for the benefit of the artist creating or providing images to commercial users. Remember the public awareness campaigns waged by Xerox and Kleenex years ago? Xerox didn't want all photocopies to be called "xeroxes" and Kleenex didn't want all tissues to be tossed off as "kleenexes." They were guarding the integrity of their products by enforcing the protection of trademark law against having their brand names turned into generic terms.

Like individuals, companies have something they want to protect when they enter the commercial arena – and the broader that arena, the greater the need for protection. For example, the 1993 GATT accord (General Agreement on Tariffs and Trade) was considered both a victory and a disappointment because of how it handled property rights protection. Previous GATT agreements dealt exclusively with industrial goods; this recent deal was a breakthrough because the U.S. succeeded in gaining agreement that intellectual property laws will be enforced internationally. This will benefit many kinds of companies, since the language of the agreement is necessarily hazy. The crystal-clear disappointment was that film, television and music companies were left out in the cold - and you can be sure that the issue of their protection will be raised again until it's resolved.

THE FRAMEWORK

Only a mutual understanding of each other's rights and concerns can help create a flexible and congenial marketplace – and it must begin with an understanding of how the law currently works in the digital marketplace. Media products are protected in several different ways, each of which has its own function. The primary legal doctrines include copyright and patent law, which are rooted in the U.S. Constitution. Trademark law, which covers names, phrases and symbols that identify product developers, as well as unfair competition law, are also available to media producers. Furthermore, there are state laws that can be used to resolve disputes related to privacy and the right of publicity. Publicity law protects an individual's control over the commercial exploitation of his or her identity and is of increasing importance to those working in the advertising and entertainment fields, where personalities reign.

Until recently, three basic goals had gained tacit support among creators and merchants alike in evaluating the effectiveness of these laws:

1. Are the needs and rights of intellectual creators and suppliers well served on several levels: creatively, financially, and in terms of the personal rights of individuals as well as operating rights of companies?
2. Is the public well served, possessing easy access to the widest possible range of works and items, readily available in the market at reasonable prices?
3. Are the remedies for infringements fair and effective?

When these goals were applied to conventional work distributed by traditional means – a story published in a book, a photograph printed in a magazine, an encyclopedia studied in a library – this process of evaluation covered most bases. However, during the last decade, these very questions have planted seeds of anxiety among a range of creative profess-

continued on page 78

ELECTRONIC
IMAGING

Steve Bronstein
Howard Berman
Photography

Howard Berman

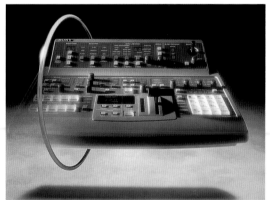

Steve Bronstein

Steve Bronstein

Landscape photo by Jennie Madder

Howard Berman

Howard Berman

Steve Bronstein

Steve Bronstein

ABSOLUT MIAMI.

Steve Bronstein

**Represented by
Gary Hurewitz
& Stephanie Coté**
38 Greene St
NYC 10013
Tel 212 925 2999
Fax 212 925 3799

9

Photographer Mikel Covey and designer/illustrator Traci O'Very Covey's

combined talents offer innovative digital illustrations

that bring an idea into the realm of the senses.

O'VERY/COVEY   PHONE 801-582-8505 FAX 801-582-8545   1577 SHERMAN AVENUE, SALT LAKE CITY, UTAH 84105

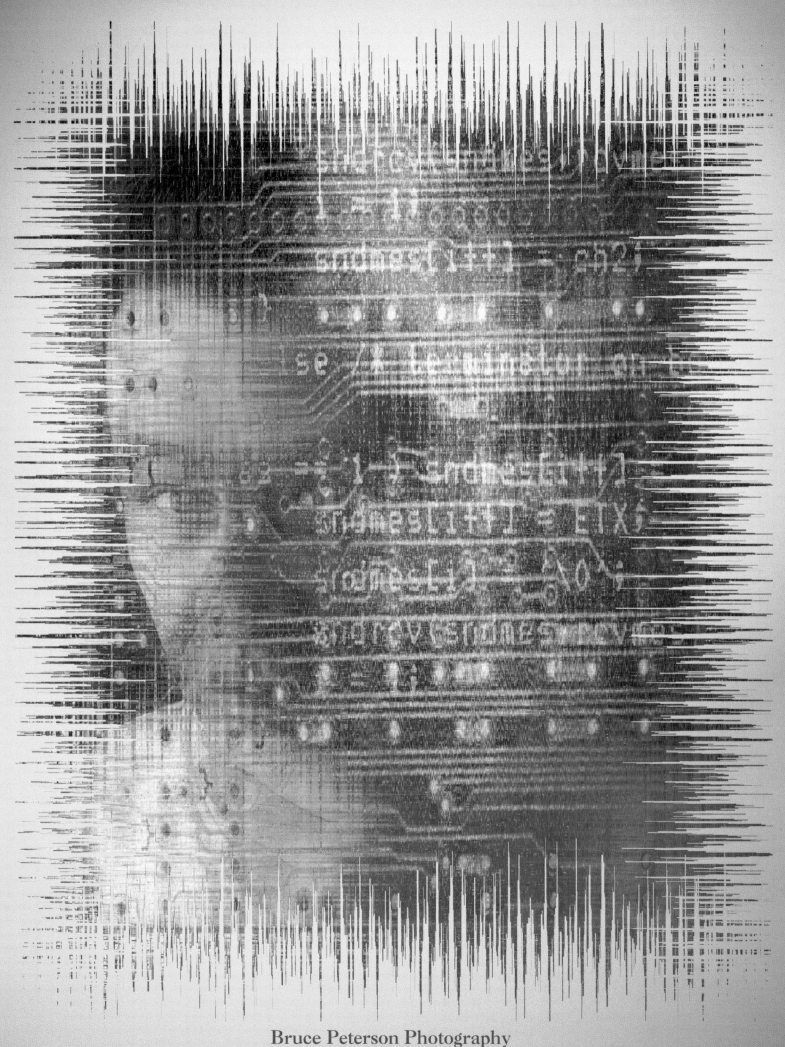

Bruce Peterson Photography
Digital Visions
602.820.8965.602.820.8650 fax

UNEXPECTED IMAGINARY NARRATIVE LYRICAL ETHEREAL DREAM GUT

THOMAS UPTON
Vocal: 415.325.8120
Faxual: 415.327.9224
E.Mail: photons@netcom.com

1. *Direct TV, New Media Magazine* **2**. *"That giant suckin' sound,"* **3**. *Traditional Family Values* **4**. *The Whalers' Graves, Nantucket.* *All Photos © T.U. 1993*

NUTS AND BOLTS APPROPRIATE PRACTICAL STRAIGHT FORWARD DIRECT

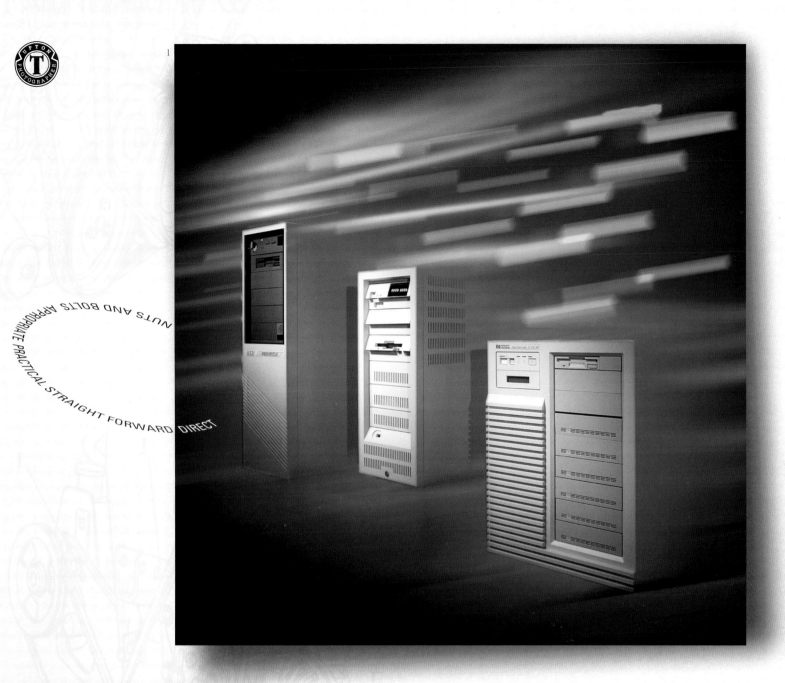

1. *Fast Pentium Systems*, Windows Source Magazine **2 & 3**. *Read/Writable CD's*, New Media Magazine **4**. *Breaking Creative Blocks*, MacUser Magazine. Photoshop by a photographer! THOMAS UPTON 415.325.8120

B L A C K M A N

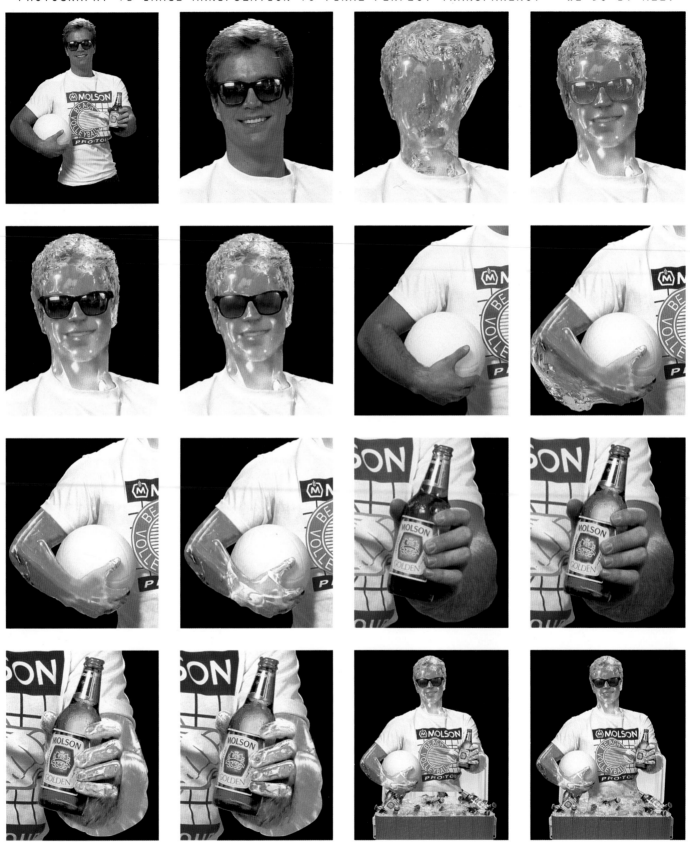
14

S T U D I O

Clients include:

- Sony
- Duracell
- General Electric
- Panasonic
- Savin
- Gillette
- NFL
- Best Foods
- Sprint
- NYNEX
- UPS
- Showtime

Michael Weiss Studio, Inc
251 Guard Hill
Mt. Kisco, NY 10549
Studio (914) 241-3456 Fax (914) 241-3488
Represented by Mary DeVlieger

Michael Weiss Studio

Photography,
Computer Retouching
& 2D/3D Special Effects

3D EfX

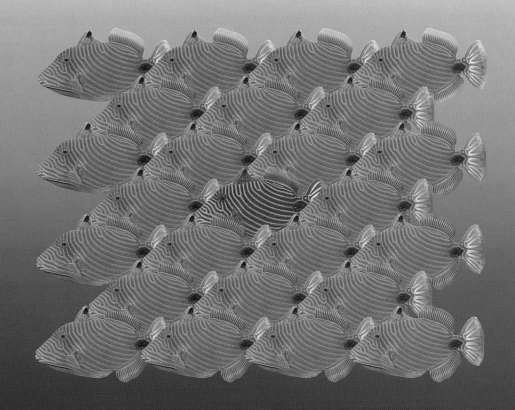

PAUL MORRELL

PHOTOGRAPHY & IMAGING

composites • concepts • murals
tinted black & white images
ultra-sharp 8 × 10 transparencies

300 brannan street, suite 207
San Francisco, California 94107
415-543-5887   Fax 543-5848

DON CARROLL'S
MULTI DIMENSIONAL IMAGES, LTD.

For assignment call us direct, 1-800-219-2825. Fax your layout to, 212-274-9349

489112

489118

48117

489115

200083

489119

Computer Artist : NORIKO IIZUKA

489114

489113

489120

Represented by

THE IMAGE BANK®

For information on usage of these works, refer to image number. To see additional select stock, call The Image Bank nearest you, and request,

DON CARROLL'S MULTI-DIMENSIONAL IMAGES CD vol.1, or The Image Bank stock catalogs.

489124

489121

489122

489123

PHOTOGRAPHY

SCHLOWSKY

COMPUTER IMAGERY

73 OLD ROAD
WESTON . MA . 02193
617.899.5110 . FAX: 617.647.1608

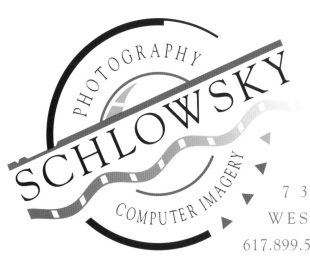

PHOTOGRAPHY

SCHLOWSKY

COMPUTER IMAGERY

7 3 O L D R O A D
W E S T O N . M A . 0 2 1 9 3
617.899.5110 . FAX: 617.647.1608

Bob Schlowsky - *Photography* • *Lois Schlowsky* - *Computer Imagery*
E S T A B L I S H E D I N 1 9 7 4

© 1994

STEVEN HUNT

Photo/Digital Illustration

801/544-4900

Fax/Modem 801/544-8826

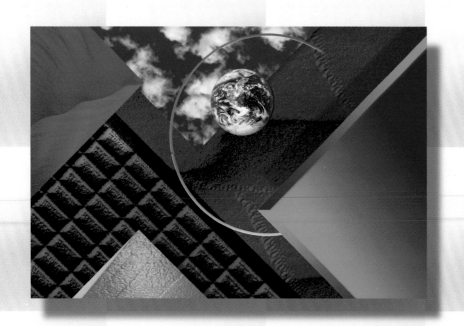

MARK TUCKER 615·254·6802

nutrients

CALCIUM	MILK, DAIRY, SALAD
SODIUM	SALT
POTASSIUM	FRUITS, MILK, CEREALS
IRON	LEGUMES, GRAINS, MEAT
ZINC	MILK, SHELLFISH, BRAN
IODINE	SEAFOOD, IODIZED SALT
FLUORIDE	RICE, SOYBEANS, ONIONS
CHROMIUM	CORN OIL, GRAINS, MEAT

PHOTO IllUSTRATION

Fusion of Illustration & Photography

VARIS

PHOTOMEDIA

(213) 874 0129

Fax (213) 874 8357

Pager (213) 303 8381

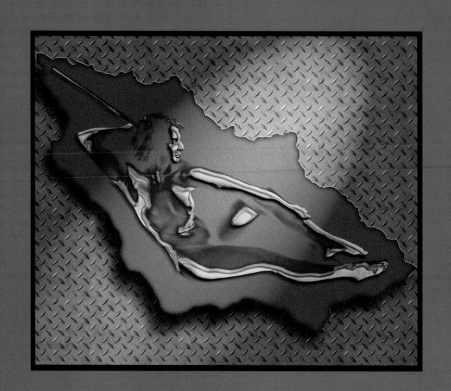

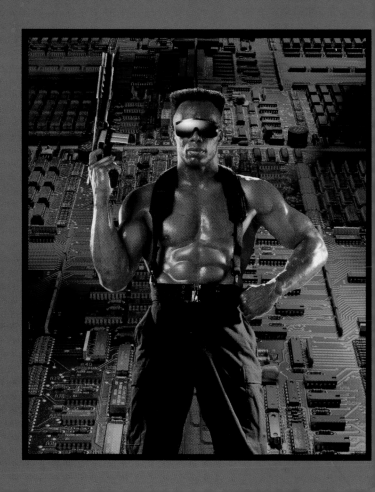

620 BROADWAY, NY. NY 10012   FRANCO ACCORNERO   212-6740068

NO LIMITS

COMPLETE DIGITAL PHOTO-ILLUSTRATION & EFFECTS

DREAMTIME SYSTEMS™

404 421 8800

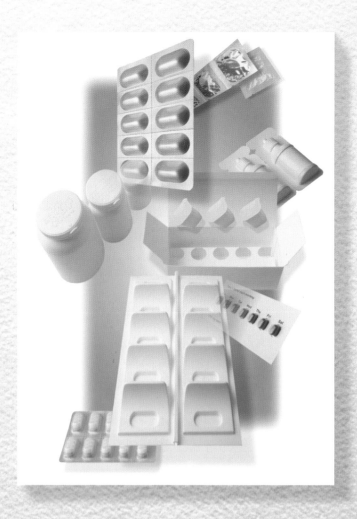

CLIENTS
Air Products
Aldus
Bell & Howell
Beta Products
Binney & Smith
Breyers Ice Cream
Campbell Soup
CibaVision
Cigna Corporation
Crayola
Endust
Frusen Glädjé
Health Infusions
Historical Documents
ICI America
Jefferson Hospital
Kiwi Brands

Linear
Maalox
Meridian Bank
Mobil Oil
Nova Nordisk Pharmaceuticals
PCI Services
Rhône-Poulenc Rorer
Roy Rogers
Schering-Plough
Scott Paper
Seaboard Marine
Squibb Pharmaceuticals
Sweartheart Cups
SmithKline Beecham
Telerx
Trump Castle
Trump Taj Mahal
Wyeth-Ayerst International

Wojnar Drake Photography Inc.

326 Kater Street, Philadelphia, PA 19147   215-922-5266   FAX 215-922-6024

FAST

FLEXIBLE

DIGITAL PAINTING

FINI

MICROSOFT
HEWLETT PACKARD
APPLIED MICROSYSTEMS
READERS DIGEST
NINTENDO
KIWANIS
AVON BOOKS
DELACORTE PRESS
CATERPILLAR
JOHN FLUKE

Photography: Ed Lowe, Don Mason
Calligraphy: Glenn Yoshiyama

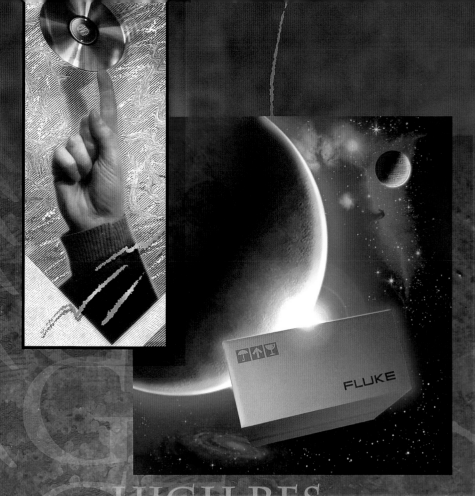

HIGH RES
DIGITAL
PAINTING

PHILIP HOWE
I MERGE
TRADITIONAL
ILLUSTRATION
AND PHOTOGRAPHY
WITH HI-END
DIGITAL
MANIPULATION

STUDIO
206 682 3453
FAX
206 623 2554

John Neitzel

Skylite Photo Productions, Inc.
152 West 25th Street, New York, N.Y. 10001   212-807-9754 FAX: 212-627-5181

A photographer who can bring your concepts to life with creative photography, combined with the mastery of contemporary creative image manipulation, utilizing the latest technology in applied digital photographic imaging. If you can conceive it, he can make it so. To review the complete portfolio call 212-807-9754

34

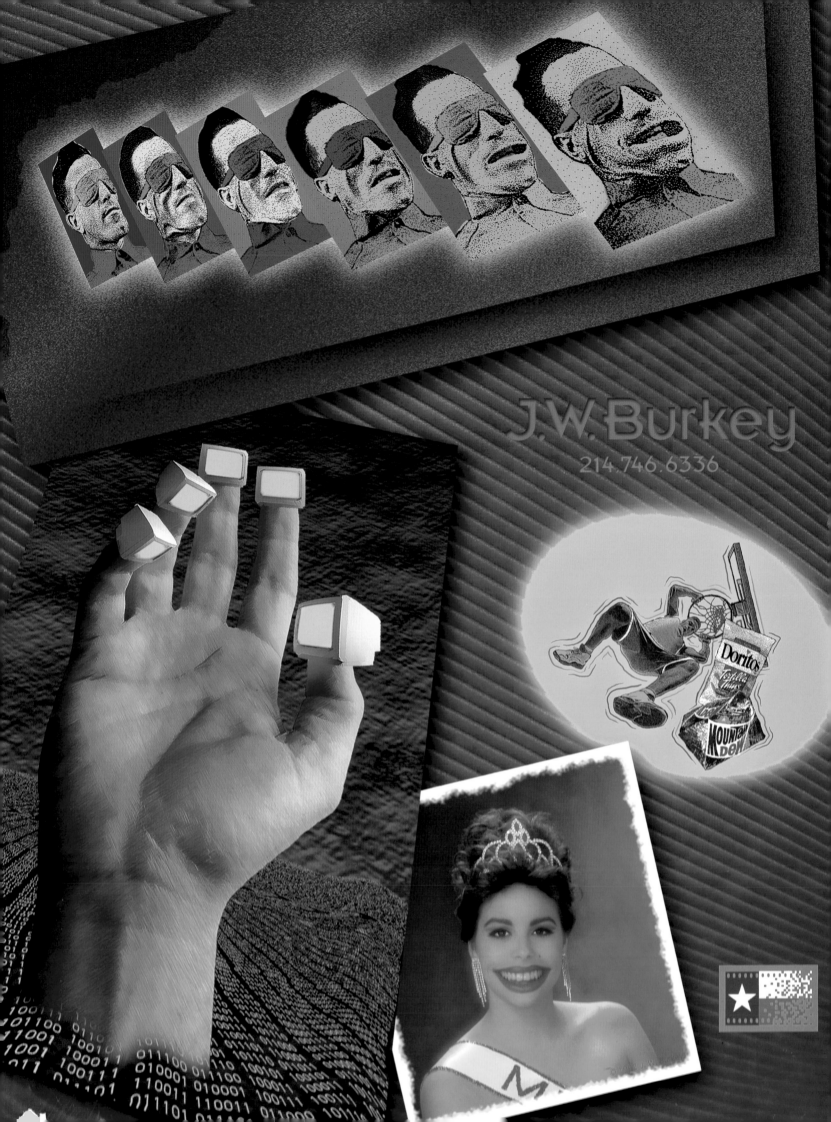

J.W. Burkey

214.746.6336

Farr Photography   12 Walnut Street   Rochester, New York 14611   Voice 716 235 1479   Fax 716 235 8176

F A ^R ^R

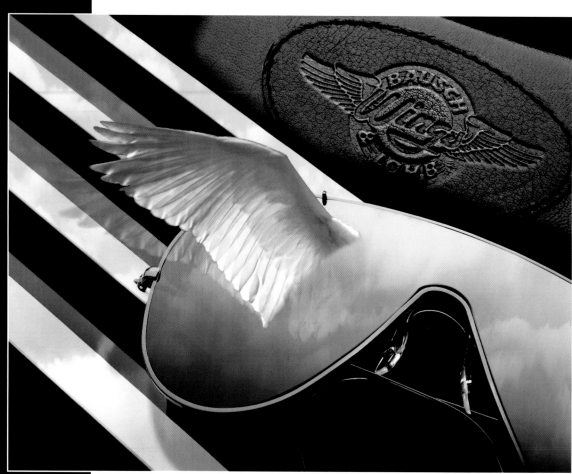

F A R R

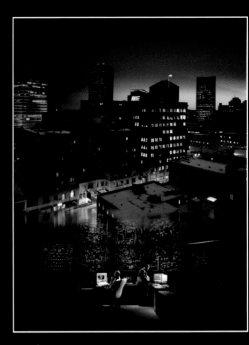

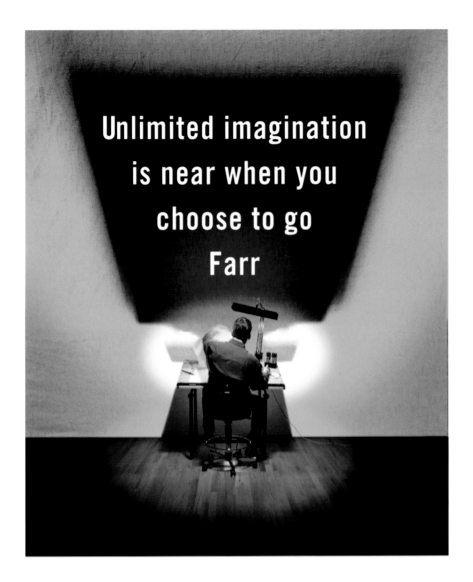

Unlimited imagination is near when you choose to go Farr

High Resolution Digital Imaging

Unix Workstation

Digital and Film Photography

Output Files to RGB or CMYK

DEAN DIGITAL IMAGING

IMAGINALITY
The art of making an idea real.

2B South Poplar Street
Wilmington, Delaware 19801
(302) 655-6992
(800) 969-7198
(302) 655-6770 facsimile
(302) 655-7248 modem

What has a gigabyte of immediately accessible random access memory, a user-friendly, high resolution graphical user interface and the ultimate in portability? It could be the latest laptop or palmtop computer — but more likely, it's the latest issue of a magazine, newspaper, or catalog.

You have heard that print will change over the next decade. Actually, print itself will *not* change — but the information that is disseminated by print will be different. Print may be produced with unconventional or improved technology, but it will remain print on paper, and that paper will have to be mailed or delivered to you by one means or another.

Right now, there is a growing group of electronic media competing with print as a vehicle for information delivery. Catalogs and reference works are being routinely released on computer disks, displaying color pages with graphics, sound and even video. This is called multimedia, because it combines visual and audio information in new and exciting ways. The only limitation is that you need a computer to see and use it.

The necessary computer disks can be floppy disks, CD-ROMs, flopticals, removables and non-removables of all kinds. There is no standardization, and someday there may be data archaeologists whose only job is to identify strange electronic storage media from the last decade. If you don't believe me, go into your basement and identify the old 78 rpm turntable, 8-track tape cartridge and Philips cassette player that have been replaced by your new compact disc player upstairs. Will you be able to play the Beta video tapes of your wedding in the year 2034? Recorded media keeps changing with the production of smaller units that hold more data. In contrast, the other day I read a book that was printed in the year 1501.

The great data highway will have us sitting at computers connecting into on-line services and universal e-mail or voice mail (which I guess is v-mail). Think of the data highway functioning like CB radio, only you have to type on your computer keyboard to announce your "handle" and communicate with your "good buddies" on the information road.

Broadcast media will give us 500 channels of cable TV, complete with downloaded movies and entire channels devoted to *Gilligan's Island*. The telephone and cable companies are fighting for the link into your home — which is your on-ramp to the data highway — and there will probably be a toll booth. Most everyone will have a computer, whether they want one or not. And the remote control that you now use to surf through the current roster of 80 TV channels will activate a new control box that will do what a computer does. Your *Yellow Pages* will be electronic, and so will your shopping.

Print will be part of the traffic on this information highway, but in a new way. Earlier this year, Hewlett-Packard announced a color printer that hooks into a television, allowing you to print screens full of information on paper. In the future, rather than printing one million units on a printing press, one million users will print one unit on their own home printer. Thus, the print industries as we know them are under attack. New media proponents bombard us with statistics: half of all magazines displayed on the newsstand are thrown away, unsold; 40% of what goes into landfills is paper. A book only needs to sell 100,000 copies to be a bestseller, but 42 million people watch *Roseanne*. Many of these proponents say that magazines should now be electronic, allowing readers to download only the information they want or need. Catalogs could also go this route with multimedia presentations of products.

But in both cases, there is still a need for print. The electronic magazine will find some of its information on paper — not from a large printing press, but from the office or home printer. The electronic catalog will let you find what you want, but you will also still want to print the page out for reference. Let's be honest: are we all going to carry around computers? Are we going to spend

PRINT - THE FINAL FRONTIER

FRANK J. ROMANO

our days at the office in front of a computer screen and our time at home huddled around a TV screen? Will everyone on the 7:42 commuter train be reading their computer?

Probably not — but the information super-highway will unavoidably eclipse the country-road of print. Remember that famous scene in *The Graduate* where Dustin Hoffman, as a young man confused about his future, is cornered by a well-meaning friend of his parents who is full of career and investment advice. "I have one word to say to you," he says, "plastics!" I also have one word to say to you: batteries! The daily tools that we will use to do our work in the future will be electronic and they will require lots of energy.

And telling you how wonderful print is will not change the inevitability of change, but it is a reminder of how we have come to be who we are. Our entire culture is derived from the printed word. Our powers of concentration are based on the impedance factor between the eye/brain and the printed page. The ephemeral images of the video screen alter that. Yet to come is the Nintendo Generation with the attention span of non-vertebrates — but we are powerless to stop the onward rush of technology.

Nonetheless, there are undeniable benefits. In a class I teach on electronic publishing, I take one day's worth of the *Congressional Record* (some 500 to 900 pages) and throw it to the floor. I ask some poor unfortunate student in the front row to find every occurrence of Senator Kennedy speaking about an amendment to a bill on comparable worth. Invariably, the student goes through it for awhile and gives up. I ask what they were looking for and after a while they say it: "an index." So, I helpfully throw down the bi-weekly cumulative index to the *Congressional Record* (about 400 pages). *Then* I call up the CD-ROM version on a computer and search to retrieve the information on screen. Finally, we print out the pages we wanted. The value of electronic publishing in such an instance is clear.

By law, there are depository libraries in the United States that must keep the *Congressional Record*. Some of them now get the print version, others get it on microfilm. The electronic version could arguably eliminate all printing of the *Record*, but not all of these libraries have computers. They continue to accumulate the printed *Record* and at some later point, discard older editions. The Government Printing Office will still be printing the *Record* long after the electronic version has taken hold, but as time goes by, it will be in smaller and smaller quantities.

However, in many areas, print is fighting back, adapting in order to

continued on page 98

DIGITAL ART ®

COMPUTER GRAPHICS FOR PRINT SINCE 1984

3166 EAST PALMDALE BLVD. SUITE 120 PALMDALE, CA 93550

805.265.8092   FAX 265.8095

DIGITAL ART™ ®

COMPUTER GRAPHICS FOR PRINT SINCE 1984

T R A C E R

Computer-Generated Art • Animation • Multimedia • Video • Full-Service Marketing Communications

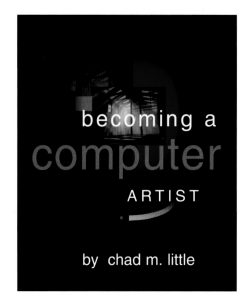

becoming a
computer
ARTIST

by chad m. little

We wrote the book on computer art. Really.
Look for *Becoming a Computer Artist*
by TRACER Design's Chad Little, available at bookstores everywhere.

TRACER DESIGN, INC. • 4206 N. CENTRAL AVENUE • PHOENIX, ARIZONA 85012
PHONE 602-265-9030 • FAX 602-263-0405

PHOTO: NOEL HAPGOOD
COMPUTER-GENERATED FILMSTRIP AND DIGITAL COMPOSITE: R/GA
AD: RYSZARD HOROWITZ; RON SULLIVAN, RUMRILL-HOYT
CLIENT: EASTMAN KODAK COMPANY

PHOTO: DAN WILBY
COMPUTER-GENERATED PANELS AND DIGITAL COMPOSITE: R/GA
AD: JOSEPH FRANCIS, R/GA
CLIENT: HOW MAGAZINE

DIGITAL EFFECTS

PHOTOS: HOWARD BERMAN
DIGITAL COMPOSITE: R/GA
AD: MICHAEL SCHELL, SCHELL MULLANEY
CLIENT: COMPUTER ASSOCIATES

Newsweek
May 31, 1993 : $2.95

in'ter·ac'tive

1. new technology that will change the way you **shop**, **play** and **learn**
2. a zillion-dollar **industry** (maybe)

PHOTO: MYKO
COMPUTER-GENERATED EGG: JOSEPH FRANCIS, R/GA
AD: PETER COMITINI, NEWSWEEK
CLIENT: NEWSWEEK

PHOTOS: RYSZARD HOROWITZ
COMPUTER-GENERATED TORNADO AND DIGITAL COMPOSITE: R/GA
AD: ANESTOS TRITCHONIS, CME
CLIENT: 3M

R/G A
PRINT

212 239 6767
Please call Jimm Burris for a complete portfolio.
Fax 212 947 3769   350 West 39th Street   New York City 10018
A Division of R/GA Digital Studios Inc.

*i*t's hard to believe that little more than 20 years ago, the activity of publishing was exclusively associated with books, newspapers and magazines produced on conventional presses. Desktop publishing was hardly conceived; no one had ever heard of fax publishing; and CD-ROM would have been mistaken for the robot in *The Day the Earth Stood Still*.

Today, new digital technologies are changing the face of publishing faster than you can say "offset press." One of the most important of the new technologies is CD-ROM. The acronym stands for Compact Disc Read-Only Memory, and the medium consists of a shiny metallic disc, much like the ones you now use to play Beethoven and the Beatles on your home stereo. (Who would have thought 20 years ago that vinyl records would become extinct!) A CD-ROM disc can hold huge quantities of information — over 600 megabytes, the equivalent of more than 200,000 pages. It is accessed through a CD-ROM player connected to a computer, and its contents are displayed on the monitor screen.

The lively, versatile combination of text, graphics, animation and sound that CD-ROM features is the multimedia everyone's excited about. Its interactive qualities are radically altering the ways publications are "read." Its unprecedented memory capacity, and the ability to program non-sequential retrieval of images, sounds, and text are driving CD-ROM's impact on the publishing marketplace. And as a whole, this technology is inspiring new ideas and understanding about communicating, teaching, learning and playing.

Presently, the CD-ROM market is shared by publications based on PC/Windows (or DOS) and Macintosh computer platforms, Nintendo and Sega video game platforms, and others. The growing number of available titles includes reference publications, games, educational programs, and promotional materials. The market is flourishing at a tremendous rate, with the pace expected to continue for the next five years. At the Macromedia International Users Conference held in San Francisco in October, 1993, speakers from IBM, Sega, 3DO and Apple presented consistent and impressive growth calculations. By the end of 1993, approximately 5 million CD-ROM players had been sold. Conservative estimates are that another 3 million will be sold in 1994. By 1995, the CD-ROM market will exceed 9 billion dollars! If platform trends continue, 60-70% of these machines will be connected to PC-compatibles, 30% to Macs, and 10% divided among the other machines.

The expansion and importance of CD-ROM technology can be summed up in three words: *volume, cost, and variety*. Each disc contains massive amounts of information in one compact unit; is cheap to produce and duplicate; and can accommodate an assortment of communication styles. Compared to the high costs of traditional publishing (including printing, shipping and storage), CD-ROM is a high-tech bargain miracle. Additionally, CD-ROM is appealing because it allows the viewer/user to work very quickly — at least compared to reading and research with print. Data retrieval and cross-referencing are fast, and can be customized to the information involved.

However, insufficient speed in other areas is a major CD-ROM issue, affecting both access time and CD-ROM's capability to support full-motion video — which is defined as full-screen images moving at 30 frames per second. This limitation is expressed through two characteristics: the data transfer rate (from disc to computer), and average access time (to the data stored). Until recently, CD-ROM players couldn't spin the discs fast enough to satisfy computer users. The swiftest players are still more than 20 times *slower* than the average computer hard drive. But users are increasingly willing to accept this limitation — which has led to an increase in the number of titles available.

CD-ROM's inability to support full-motion video stems from limits in the technology's storage characteristics and software limitations (compression, in particular). A semblance of video — in short sequences and programmed through QuickTime animation — is all that is currently possible. When the technology improves to that point that true video can be contained on a CD-ROM disc, it will be possible to generate products that communicate with the magic and immediacy of a *Star Trek* computer diary — but that's a ways off.

There is also the matter of shelf-life. I have reservations about how long individual CD-ROM titles will be available and/or in use, particularly as the technology grows and changes. The basic elements of CD-ROM as we know it now will no doubt be applied to new and perhaps more efficient formats, even within the next ten years. A change of formats is both an advantage and a disadvantage. Having one format, such as the CD used for music, ensures uniformity and compatibility. But it also locks in possible limitations. The new formats — including Photo CD and CD-I — are attempts to better adapt the medium to specific applications. However, the major characteristic of CD-ROM, regardless of format, is *the digital storage of information*. Once information is digitized, it can be published in any digital medium — past, present or future. So, whether CD-ROM is *the* digital medium of the future, or just the first seminal step towards it, is really a moot point.

No one understands this better than those who are resisting the ride on the CD-ROM bandwagon, for fear that its popularity will mean the death of print. It's not a concern I share, because I trust the flexibility that historically accompanies change. Did TV kill radio? Did video kill film? No. Every time a new medium comes along, the old media shift direction to accommodate the new. Radio drama, comedy and quiz shows are a thing of the past, having given way to television in these areas. But radio is still the mass medium of choice for music, and as a national speaker's platform through call-in talk shows. It is also an important crisis medium; in the hours that followed the Los Angeles earthquake, while the rest of the nation watched TV reports of the devastation, L.A. residents themselves (many of whom were without electrical power) turned to battery-operated radio for vital information.

However, as that disaster also demonstrated, when the power *is* on, the texture and technology of video is vastly superior to film in delivering the impact of real-time events into our homes. And in general, video is best at carrying the messages of short, special-interest stories — as evidenced by the growing body of work by independent video-makers, who use this less-expensive medium

CD-ROM: THE MEDIUM IS STILL THE MESSAGE

CHARLIE MAGEE

to state their creative or political cases. In sharp contrast (all puns intended), the resolution and screen size of film still have no rival when it comes to communicating emotional content, and at presenting stories that have greater depth. Steven Spielberg could never have made either *Jurassic Park* or *Schindler's List* on video; he didn't even trust the latter to *color* film, preferring instead the stark shapes and shadows of black & white for his exploration of the Holocaust.

These are tangible examples of how the advent of new technologies added to, and changed, the overall communications mix. In response to one another, radio specialized and personalized; video united and empowered; and film broadened and matured. A similar dynamic is occurring with the introduction of CD-ROM technology into the etched-in-steel arena of publishing. We can only imagine how print, with its flexibility and accessibility, will shift and grow as CD-ROM seeks its niche in the communications mix.

Many projects or publications that will benefit from the addition of sound, animation and viewer participation will be published on CD-ROM; so will projects of considerable size, taking advantage of the technology's extraordinary storage capacity and meaningful cost effectiveness. For example, the fast-access, non-sequential reading and cross-referencing associated with encyclopedias, telephone directories and catalogs (for products *and* art) will find an ideal medium in CD-ROM. Even certain publications that cry out for portability, such as maps that will assist us in navigating through city traffic, will be more and more frequently imprinted on CD-ROM. (Publications which, by design, leave a lot of room for the reader's imagination, will still be published on paper. Novels would fall into this category.)

And, as I mentioned earlier, CD-ROM is prompting new ideas, as well as creating new opportunities. Scientists and psychologists have long understood that there is a strong visual component to the way people think

(more in some people than in others). For many people, the things that are learned and remembered most effectively are those things that are seen (as opposed to read or heard), or which can otherwise be translated into visual terms. Much of digital technology is rooted in this principle.

CD-ROM is elaborating on this understanding, creating an important new library of images, ranging from art to science. While I am by no means anti-language or anti-print, and believe in the importance of literacy, I also recognize the quantum leap that CD-ROM may produce in the area of communication and education of all kinds — and there's an information-hungry public eagerly awaiting it.

However, this technology, like most everything else, is market-driven, and not everything being released on CD-ROM is a boon to the common good. Many titles published in the past few years have been called *shovelware*. This refers to what's created when a company indiscriminately takes a massive amount of print or visual information, digitizes it, throws a few buttons on a main menu screen (called an interface), adds a rudimentary "interactive" search and retrieval mechanism, and calls it a finished product. It is quantity, not by any means quality — and it is deemed justified by its producer on the basis of volume alone.

Obviously, profit potential is the motivator for schlock like shovelware. Fortunately, there is ample opportunity for profit and quality to co-exist in CD-ROM, and I believe high-quality products and publications will eventually dominate the market, as long as the market demands them. The masterpieces of CD-ROM publishing have yet to be produced — but we can look forward to thrilling adventure games, engrossing stories, warehouses of easy-to-access information, more helpful ways of learning, and puzzling new projects we don't yet have names for.

We are, at heart, a creative industry. Many artists steeped in print tradition want to learn and do something new. The

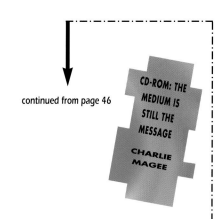

continued from page 46

rising acceptance of CD-ROM is prompting dramatic change in the design industry: yesterday's art director, illustrator or animator is today's interactive multimedia artist. Thousands of new artists are entering the work force with considerable skill in digital media, and many kids are learning digital art techniques on computers at home, or on the machine at mom or dad's studio. A few smart, brave publishers are leading the parade by taking upfront losses in order to establish themselves, or to maintain their positions as leaders in the publishing industry. Their production staffs are learning a whole new language, and their public relations and marketing experts are learning new markets. It's a nervous and exciting time.

So, what does all this mean? Where will it lead? I suggest that the real issue raised by CD-ROM is communication and how to achieve it more successfully and creatively than ever before. I further suggest that the process of human evolution has been one long Age of Communication, that evolution is essentially the *survival of the fittest communicators*.

Throughout human history, the most successful groups of people — whether tribe, city-state, kingdom, nation or corporation — have had one thing in common: a system of communication superior to that of their competition. In the battle for survival, the fittest communicators have better access to necessary information, the intelligence to transform that information into knowledge and action, and the freedom to communicate that knowledge to other members of the group, thereby ensuring continued survival.

CD-ROM technology is successful because the criteria of volume, cost and variety parallel the standards for fittest communication. CD-ROM currently works better than anything else at increasing the volume of information we can obtain and transmit. In general, quantity can't help but enhance quality, and competition will make sure that quality wins out. The low cost per unit of a CD-ROM disc increases our ability to share the information it contains. It is easy to distribute, and therefore enhances decentralization and the freedom to communicate. And CD-ROM's colorful, animated, visual approach to information enables us to learn and remember more and different kinds of information, even complex ideas.

When Marshall McLuhan told us back in 1967 that "the medium is the message," and that electronic connection and interdependence would make of our world a global village, he was writing about the power of television and telephones. He hadn't even dreamed of the information super-highway that will bring the global village out of the backland and into the bright city lights of tomorrow.

We don't yet know if CD-ROM will help society transform the knowledge that the new technology provides into positive action and a major step in communication evolution. Whether the writing is on the wall, the page or the screen, it must still be read. And whether or not we will ultimately survive as the fittest communicators remains to be seen. But CD-ROM offers the most exciting potential for a new form of mass communication since the invention of television. It should be welcomed, not feared; and it should be used wisely, not squandered.

Charlie Magee, president of One World Interactives in Eugene, OR, is a title developer who, with his partner, Vicky Ayers, is currently designing an alien treasure hunt game for the Mac and Windows CD-ROM market.

HEIDI MERSCHER

Digital Illustration • Photo Manipulation • Retouching

505 *776•1333*

Represented in Philadelphia by Terry Putscher 215•569•8890

WENDY GROSSMAN

2 1 2 • 2 6 2 • 4 4 9 7

more work can be seen in American Showcase #17

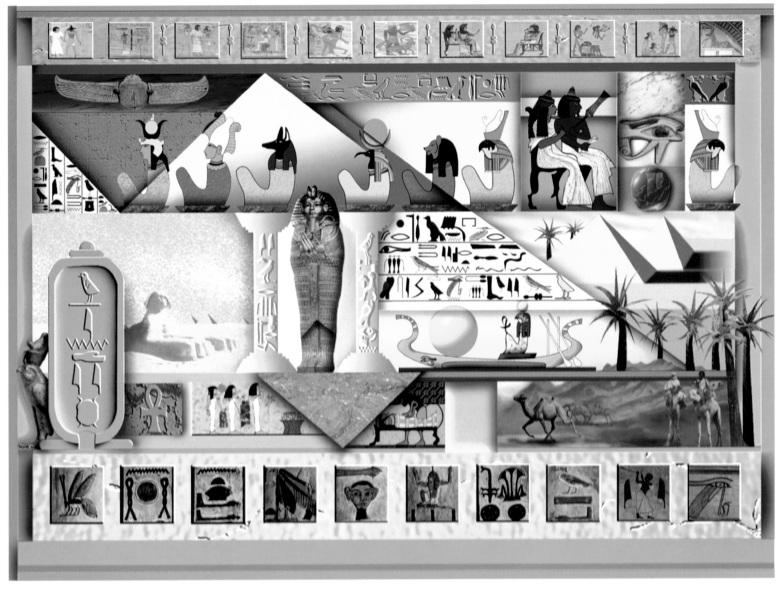

ARTCO

STUDIO MACBETH

Digital Illustration, Retouching, Photography, Electric Image Manipulation

ARTCO L.L.C. • **Gail Thurm and Jeff Palmer** • **Tammy Shannon, Associate**

Serving clients within New York City:
Serving clients outside New York City:

232 Madison Ave • Room 402 • New York, NY 10016
227 Godfrey Road • Weston, CT 06883

(212) 889-8777 • Fax: (212) 447-1475
(203) 222-8777 • Fax: (203) 454-9940

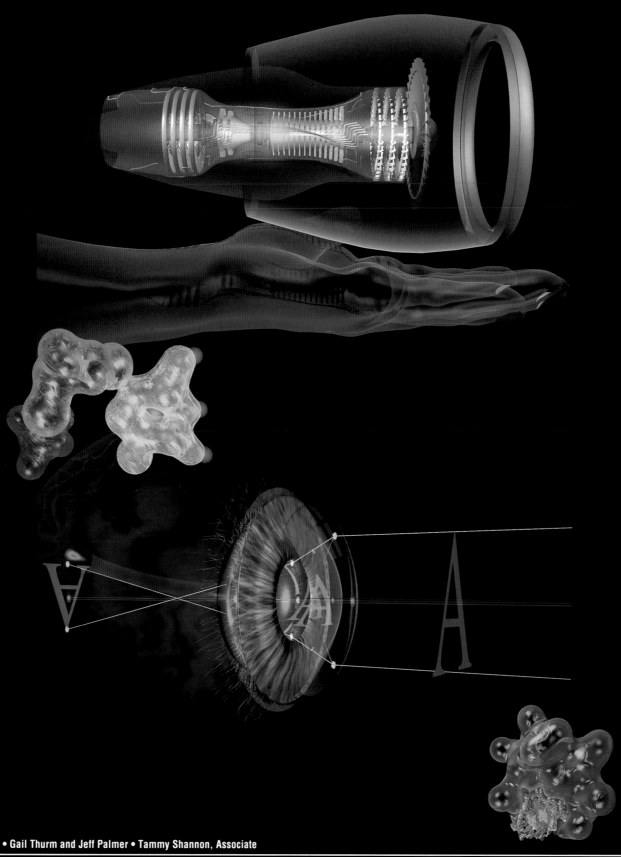

DOUG STRUTHERS

Computer Illustration, Animation

KOLEA BAKER
Artists Representative Inc

2814 - NW 72nd Street   Seattle WA   98117   206.784.1136   fax 206.784.1171

DON BAKER

KOLEA BAKER
Artists Representative Inc

2814 - NW 72nd Street   Seattle WA   98117   206.784.1136   fax 206.784.1171

PAUL
WILEY
DIGITAL ILLUSTRATION
410 W24th Street #12i
New York City 10011
212
800 627.8071
fax/modem 212.633.9392

JAVIER ROMERO
DESIGN GROUP

24 East 23rd Street, New York, NY 10010 [212] 420 0656 Fax 420 1168

Paseo Castellana 179, 4D1, 28046 Madrid, España 571 8183 Fax 571 8055

PACKAGING • GAMETEK

IBM • SPAIN

LOGO • FILM WORKS

ANIMATION • TELEMADRID

MACUSER • SPAIN

POINT OF PURCHASE •
ROUNDHEADS WORLDWIDE INC.

JEFF TULL

Fearless Designs, Inc

622 E. Main St., Suite 206

Louisville, KY 40202

502 584-1333

FAX 502 584-1332

Jerry Blank

Telephone: 408 289 9095
Facsimile: 408 289 8532

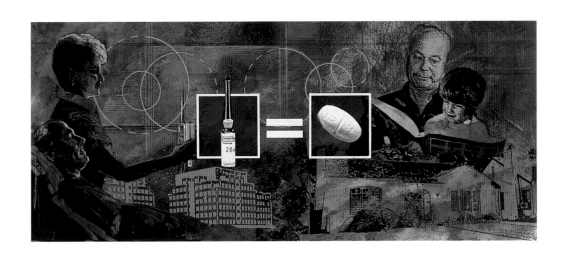

NIKOLAI PUNIN

161 WEST 16
STREET 18E
NYC, NY 10011
TEL.: 212 727
7237
FAX: 212 620
4056

LOGOS

COMPS

REPRESENTED
BY:

TONAL
VALUES

TEL.: 305 576
0142
FAX; 305 576
0138

ANNUAL REPORTS

EDITORIAL

CHIP
CATON

TEL.: 203 523
4562
FAX : 203 231
9313

Lance Jackson © Lax Syntax Design 510.849.4313 fax: 510.849.0574 1790 Fifth Street Berkeley, Ca. 94710

DOROTHEA TAYLOR-PALMER

T.P. DESIGN
ATLANTA, GEORGIA
FON: (404) 413-8276
FAX: (404) 413-9856

MACINTOSH GENERATED DESIGN AND ILLUSTRATION

CRANKING THE Creative

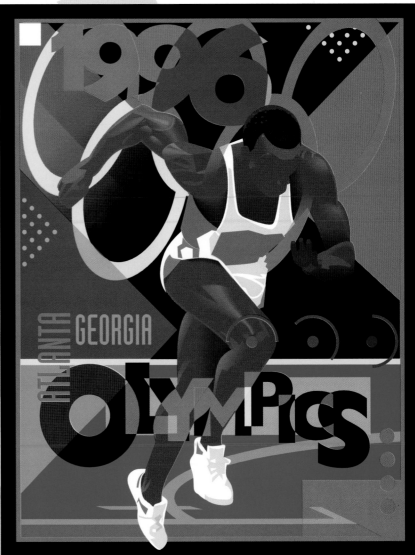

1996

ATLANTA GEORGIA

OLYMPICS

Jonathan Herbert COMPUTER Illustration

STATE–OF–THE–ART
3D MODELING
FOR PRINT AND
ANIMATION.
CALL FOR BOOK
OR REEL.

RENARD

REPRESENTS
∎
TEL: 212•490•2450
FAX: 212•697•6828

BERNSTEIN & ANDRIULLI REPRESENTATIVES 60 EAST 42ND STREET NEW YORK NY 10165 FAX (212)286-1890 PHONE (212)682-1490

SiliconGraphics

IRIS INDIGO

colorbus
CYCLONE

CPU

GRAPHICS

PULL

A B

AXLE SHAFT

BOOT

SNAP
RINGS

BOOT
CLAMP

INNER
RACE

THREADS

SPLINE

OUTWARD
JOINT
HOUSING

BALLS

CAGE

BOOT
CLAMP

MARTY SMITH
COMPUTER GENERATED TECHNICAL ART

PHONE: (714) 962-0461   FAX: (714) 851-9238

CAROL ZUBER-MALLISON

Maps
Infographics
Charts
in Adobe Illustrator

Rep:
Brooke & Co.
4323 Bluffview Blvd.
Dallas, TX 75209

(214) 352-9192

Studio:
(214) 906-4162

Studio fax/modem:
(817) 924-7784

Clients include:
Alcatel
Burlington Northern
Dallas Morning News
Dryden Press
Exxon
GTE
Harcourt Brace
Holiday Inn
Host Marriott
Chaparral Steel
Texas Rangers
Southwestern Bell
Pier 1
Playboy
Tenneco
Whittle
Communications

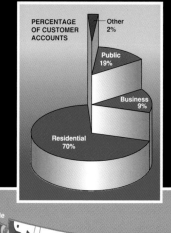

PERCENTAGE OF CUSTOMER ACCOUNTS
Other 2%
Public 19%
Business 9%
Residential 70%

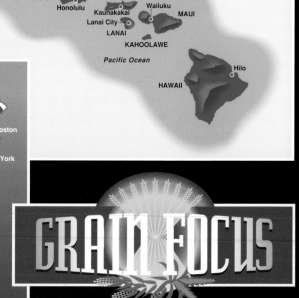

KAUAI
Puuwai
Lihue
NIIHAU
OAHU
Honolulu
MOLOKAI
Kaunakakai
Wailuku
Lanai City
MAUI
LANAI
KAHOOLAWE
Pacific Ocean
HAWAII
Hilo

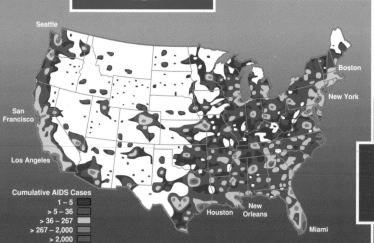

Seattle
Boston
New York
San Francisco
Los Angeles
Houston
New Orleans
Miami

Cumulative AIDS Cases
1 – 5
> 5 – 36
> 36 – 267
> 267 – 2,000
> 2,000

GRAIN FOCUS

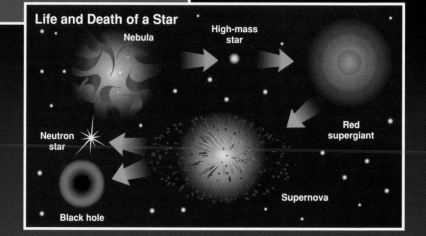

Life and Death of a Star
Nebula
High-mass star
Neutron star
Red supergiant
Supernova
Black hole

AMOCO OIL CO—Trial graphics, the plasma cell shooting antibodies

ASTD—Troubleshooting the puzzle

The White House—The Health Security Card

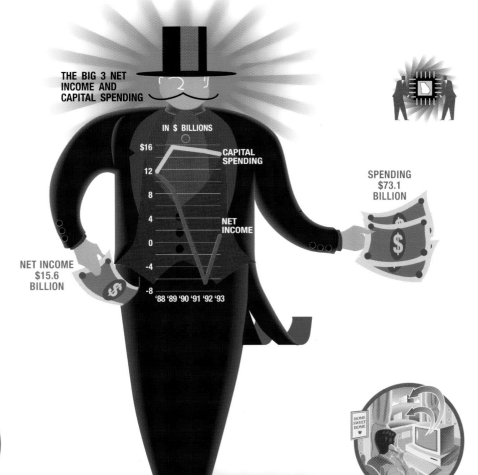

THE BIG 3 NET
INCOME AND
CAPITAL SPENDING

IN $ BILLIONS

$16

CAPITAL
SPENDING

12

SPENDING
$73.1
BILLION

8

4

NET
INCOME

0

-4

NET INCOME
$15.6
BILLION

-8

'88 '89 '90 '91 '92 '93

Pacific Telesis—Digital data circles the globe

Pacific Telesis—Telecommuting from home

Eisner & Associates—The big 3

Radon—Animation of how it gets in and how to get it out

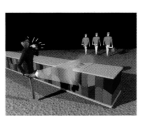

AMOCO OIL CO—Trial graphics, animation of cellular repair team

Georgia Trend Magazine—Top 100 companies in Georgia

   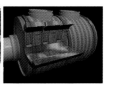 

Nasa—Animation the shuttle & spacelab

MOST OF THE ART WE CREATE COMES FROM MACINTOSH COMPUTERS AND CAN BE MADE COMPATIBLE IN MOST FILE FORMATS INCLUDING PC, NEXT, SCITEX, SILICON GRAPHICS, AND SUN

STUDIO PHONE 818 301 9662 FAX 818 303 1123

Outside Magazine : jetlag

Dugan Farley : Catapress, TTS

Big Step Market Place : Osaka Japan, summer fair

for custom, digital or reflective illustration: **JON CONRAD & ASSOCIATES 221 W MAPLE AVE MONROVIA CALIFORNIA 91016**

MAC
NEILL
and
MAC
INTOSH

74 YORK STREET
LAMBERTVILLE NJ 08530
609·397·4631
CALL FOR
FLOPPY PORTFOLIO

Published in USAir Magazine. Subject: Dallas, Texas Trolley service.

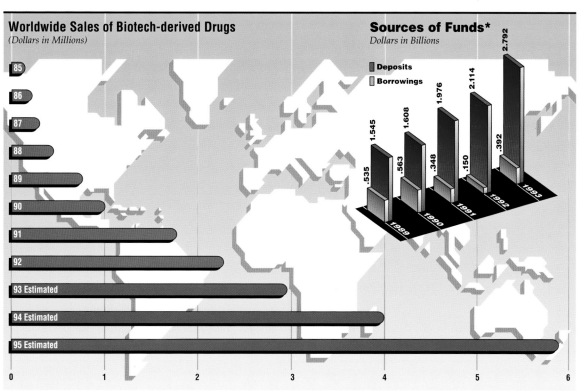

Worldwide Sales of Biotech-derived Drugs
(Dollars in Millions)

85
86
87
88
89
90
91
92
93 Estimated
94 Estimated
95 Estimated

0   1   2   3   4   5   6

Sources of Funds*
Dollars in Billions

■ Deposits
□ Borrowings

.535   1.545
.563   1.608
.348   1.976
.150   2.114
.392   2.792

1989   1990   1991   1992   1993

Published in Enzon's 1993 Annual Report. Source: Medical Advertising News

*Published in Collective Bancorp's 1993 Annual Report

All artwork created with Adobe Illustrator 3.2 & 5.0

DOUGLAS BUCHMAN
DIGITAL ILLUSTRATION & GRAPHICS

▲ MEN'S JOURNAL MAGAZINE

▲ MULTIMEDIA WORLD MAGAZINE

▲ REGISTERED REPRESENTATIVE MAGAZINE

▲ PRISCOMM INC.

▲ PICKWORLD MAGAZINE

▲ THE BLACK DIAMOND GROUP

DAVID TEICH
41 Tamara Drive, POB 246
Roosevelt NJ 08555-0246
609.448.5036   FAX 443.3228
AOL: DaveTeich

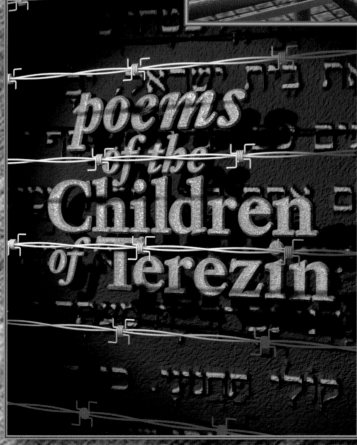

Computer Illustration

matsuri

MATSURI Corporation
150 West 56th Street
New York, N.Y.10019

Tel: 212-582-6001
Fax: 212-582-6002

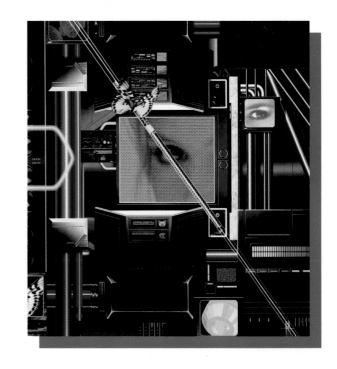

Cécile MILLET Computer Artist

Computer Graphic Design
High Res. Illustration, Logo, Publication
PH. : (612) 587 0897
FAX : (612) 231 3279

December
1992

Minnesota Milestones

2020
PLANNING

A report card for the future

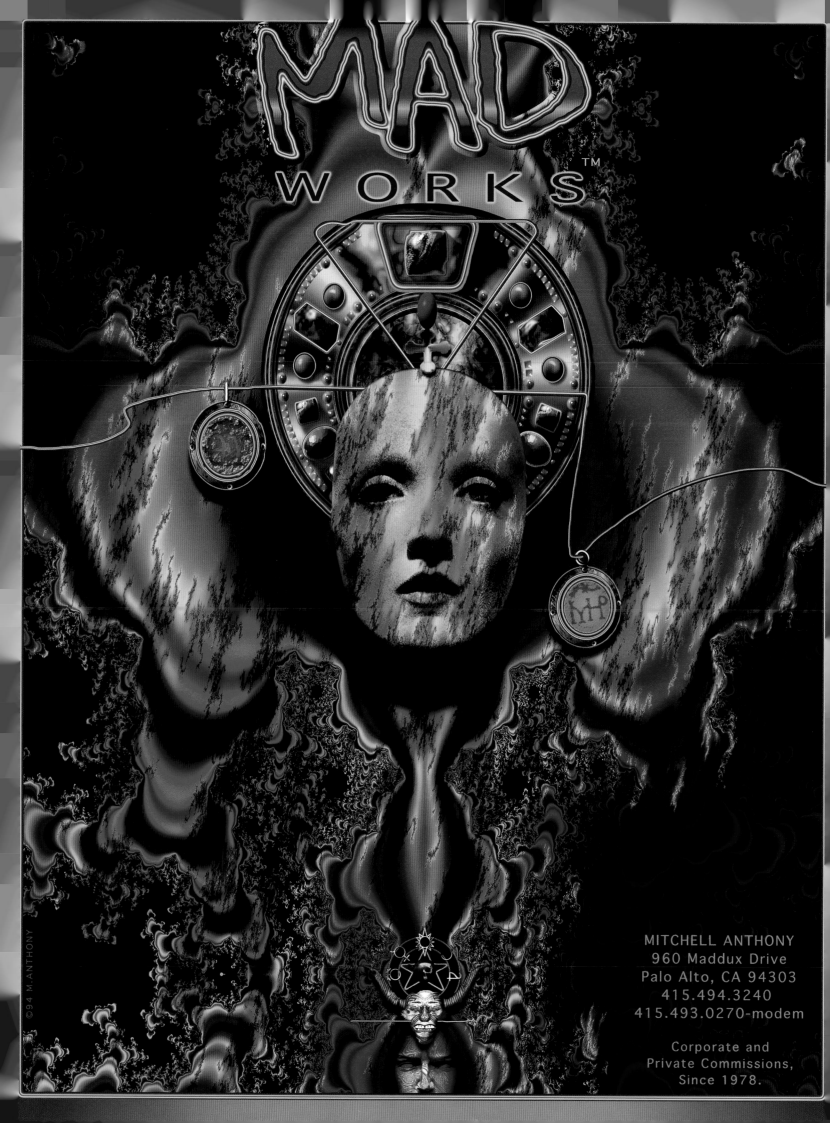

MAD WORKS™

MITCHELL ANTHONY
960 Maddux Drive
Palo Alto, CA 94303
415.494.3240
415.493.0270-modem

Corporate and
Private Commissions,
Since 1978.

Myron Grossman

Studio: (818) 797-8422
Fax in Studio

National Health Enhancement Systems

National Health Enhancement Systems

TGI Fridays

Embassy Suites Hotels

Advertising Production Assoc. of LA

Goodyear

ICON GRAPHICS INC. • **34 ELTON STREET** • **ROCHESTER, NEW YORK 14607** • **716 271-7020** • **FAX 716 271-7029**
All illustrations are produced electronically. Call for more illustration samples.

Pam Belding
Digital Illustration

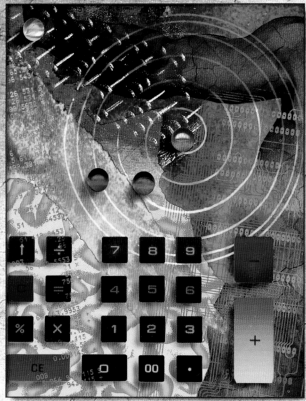

USING CALCULATORS
For Business Problems
Third Edition

Gary Berg, Ph.D. ◆ Leo Gafney, Ed.D.

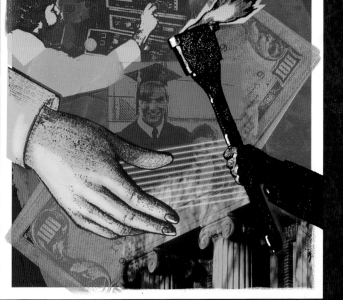

Voice 612.476.1338

Fax 612.476.0024

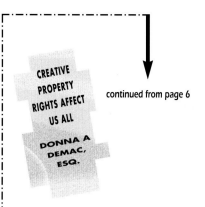

CREATIVE PROPERTY RIGHTS AFFECT US ALL

DONNA A DEMAC, ESQ.

continued from page 6

ionals and corporate investors who are concerned about the control of digital art and products. These digital creations are so dynamic in their nature and so prone to manipulation, they are far less easy to control and are much harder to keep track of.

Representative of this concern is an organization called The Writers Rights Coalition – comprised of the National Writers Union, Writers Guild of America/East, Volunteer Lawyers for the Arts, Arts Management International, International Association of Art Critics, and American Society of Journalists and Authors – which recently banded together to address the issue of on-line publishing. Many of the nation's top newspapers, news magazines and wire services are now on-line in digital computer networks. Any subscriber in good standing can down-load information from these publishers, which previously were secure and discreet sources. Writers are greatly concerned about the unauthorized use, alteration and uncompensated resale of their work – and this unprecedented coalition of groups demonstrates the depth of their distress.

Similarly, this dilemma, which affects artists, designers and other image makers just as it does writers, has prompted stock houses and individual photographers to embark upon significant financial and organizational initiatives. Archive New Media in New York has in recent years made large investments in new digital transmission capabilities and has acquired more images to support its work with multimedia developers, such as Microsoft. In 1993, a new licensing agency, Picture Network International in Arlington, Virginia, began providing users with an on-line image resource for searching, acquisition and licensing of premium images from diverse sources. This approach allows users to have the speedy access to images they now want, while protecting image creators, because it is they who hold the digital access purse strings.

Artists and writers are not alone in wanting protection and fair compensation. Ad agencies and publishers, for example, survive on a combination of creativity and exclusivity. If an ad campaign is built around a stylish image – and that image shows up in someone else's ad – the only thing created is an unhappy client (and an unemployed agency). The rush to speedy access has produced something of a loose cannon of information and images.

A LEGACY OF PROTECTION

Which brings us back to the need and importance of copyright, designed to protect the works of writers, photographers and many others, including publishers and other kinds of companies. It's interesting to note that copyright began in Europe in the 17th century, and the U.S. Constitution, adopted in 1787, has contained a provision authorizing Congress to pass copyright and patent laws since its inception. Over nearly four centuries, copyright protection has been further expanded and now encompasses works created using cameras, film, computer software and more.

The extension of copyright to photography came about in a dispute over a photographic portrait of Oscar Wilde taken by Napoleon Sarony. When a lithographer reproduced the photo without permission, Sarony went to court charging his rights had been violated, and the Supreme Court agreed. The last comprehensive copyright statute enacted by Congress was the 1976 Copyright Act. This law makes an author or any other artist the principal rights-holder of his or her work. Specifically, this Act states that creators have the exclusive right to authorize the publication, reproduction, alteration and manipulation of their work. It was also in 1976 that songwriter Ronald Mack won his case against ex-Beatle George Harrison; Mack contended that Harrison plagiarized the melody line of his song, *He's So Fine*, in Harrison's composition of his hit, *My Sweet Lord*. While the court conceded that Harrison's act was unintentional, he nonetheless had to compensate Mack. The value of good law and adequate protection is that it applies to all.

But in daily practice, the effectiveness of copyright law depends on vigilance by both the creator (owner) and the person or company desiring access to the work. Photographers and other image makers are often reluctant to spend the time necessary to copyright each of their images, yet cite the value of copyright law in establishing their ownership rights. While it is an admittedly difficult clerical task to copyright hundreds if not thousands of images, they really have little recourse if they wish to address their concern about delivering their images to a multiplicity of new users. Not everyone understands (or wants to understand) copyright and licensing protocols. But if image makers of all kinds want to take advantage of the protection of the law, while also capturing new royalty-generating markets, the paperwork must be done. In so doing, rights-holders reinforce the objectives of self-expression and the existence of an uncensored media marketplace. Copyright is a commercial right – but it is also a democratic responsibility.

FAIR USE

Copyright, however, is not inflexible. The doctrine called Fair Use makes it lawful to use a copyrighted work without permission, based on specific criteria. It's a good example of how copyright protection strives for equitable resolutions, based largely upon the user's common sense. The criteria for determining fair use takes into consideration a variety of factors, including:

1. The purpose and character (commercial or not-for-profit) of the use;
2. The nature of the copyrighted work;
3. The amount and substance of the portion used;
4. The effect of the unauthorized use upon the potential market for the copyright holder's work.

In essence, if a project or publication makes "reasonable" use of copyrighted work in a manner that does not take advantage of the rights-holder, and which serves the public good, Fair Use gives fair users a break.

In addition to Fair Use is the legal doctrine of Public Domain, which applies to creative works or inventions on which copyrights or patents have expired, or on which they never existed. According to the 1976 statute, copyright exists for the length of a creator's life, plus 50 additional years after his or her death. This does not mean that all works become Public Domain after that; other individuals or companies may hold additional licensing rights based on ownership or as granted by the image creator's estate. But a good deal of existing work is in Public Domain and provides image users with creative opportunities that do not impinge on the rights of living artists.

CONCLUSION

The changing world of digital media is creating an environment in which creative property owners, users and the public are confronted with new rights issues. But if copyright protection is to remain meaningful, owners and users of digital images must be attentive to the changing ways of valuing both ownership rights and the diverse applications of the digital image. The protection of one's rights requires informed awareness of how new technologies challenge one's basic rights under copyright law. Accordingly, I offer this Top Five List of safeguard suggestions to rights-holders:

1. Recognize that rights enforcement is often a matter of common sense and always a matter of self-initiative. Litigation should be viewed as a last resort.
2. Know your eligibility for pro-

continued on page 104.

JULIE PACE
678 WELLSLEY DRIVE, LAKE ARROWHEAD, CA 92352 909-337-0731 PHONE 909-337-5703 FAX-MODEM

The philosophers cautioned us. *Those who cannot remember the past are condemned to repeat it,* said George Santayana. Aristotle said, *Education is the best provision for old age.* Those of us who are educators, artists, or communicators of any kind, would do well to remember these two sage proverbs as we contemplate the growth of new media, its impact on society, and our ability to use it wisely.

The checkered, undirected, market-driven history of television stands as a critical example of what can happen when technological progress is placed above social values, and what can result when the quality of programming fails to keep pace with the miraculous technology that delivers it. Today, with new media playing an ever-greater role in all areas of business, culture, politics and communication, it is incumbent upon new media professionals to prepare themselves with a thorough and balanced education.

We must study new media in all its aspects: technological, creative and sociological. And we must reinforce this study with a foundation in all of the arts, history and sciences – the components that used to comprise a well-rounded higher education (in the days before education itself was reduced to the specialization necessary to get a job). Nothing less will prevent us from repeating the past. Nothing more can best prepare us for the future.

These ideas reached a high point of clarity for me in the spring of 1993, when I listened to computer pioneer John Sculley's keynote address at the National Association of Broadcasters convention. He compellingly told those assembled about the profound shifts and changes in American society, and the ways that advances in media are helping to shape them. Sculley reminded us that the American media industry is the largest in the world – to the tune of six *trillion* dollars. The audience – almost equally divided among computer nerds, techies, creative producers and directors – applauded both the financial implications of Sculley's information, as well as his description of the future of communication: micro-processing, digital compression, and fiber optic information highways.

I was attending this event in my capacity as chairman of the Department of Visual and Performing Arts at Fairleigh Dickinson University (Madison Campus). For me, Sculley's speech was a call to action, prompting me to acknowledge and respond to the economic and social metamorphosis around us. As an educator, I knew I had to make a new and different contribution to shaping the future. As an artist who has never given up hands-on experience in film and video, I knew that these were the disciplines I should use as a springboard.

NEO-CLASSICAL NEW MEDIA EDUCATION: PREPARING FOR TOMORROW

HARVEY FLAXMAN

But wait, I anticipate your questions and confusion: why the urgency, and what's it all about? Haven't art, design and photography schools increasingly responded to the growth of new media with special courses? Hasn't the establishment of the Center for Creative Imaging in Maine provided an oasis for study and development in new media? Aren't there networks of artists, designers and others working and experimenting together, teaching themselves and each other, often in exciting on-the-job training situations? Aren't the hardware and software manufacturers offering seminars and crash courses that help users get started? Don't artists already know all about speaking to, and reflecting, the human condition? Isn't it just the technical training they need, and isn't that already available?

Allow me to try to provide some answers, as I see them. The urgency stems from the fact that visual forms of art and communication have become the dominant paradigm in society. Their role will definitely increase, because they are the most efficient form of communicating the types of messages that society is exposed to, from news to entertainment. But new media technology is fast outpacing society's ability to digest and manage it, hurtling towards the next stage even before the current advance has been absorbed and understood. Yet, understood or not, technology is changing the way people communicate and learn. And those who control what is *seen* will shape and control the way millions of others think and learn.

Some 10 to 12 years ago, computer graphics triggered the first attempts in education to cope with the change brought about by computers and their use in visual communication. When desktop publishing came along, the need to re-evaluate the education of graphic designers, advertisers and illustrators became so obvious, these efforts increased, particularly in specialized educational settings. More recently,

multimedia captured everyone's imagination, and opportunities for learning this mixed-media abounded – even before educators fully understood what multimedia is.

But even though educational opportunities for learning about the technical side of new media have sprung up in many places, from artists' studios to sophisticated, specialized facilities, it still falls to mainstream institutions of higher learning to recognize and respond to the totality of today's new media challenges. Hardware and software manufacturers are, understandably, primarily concerned with selling products. Employers, however avant-garde they may be, are focused on getting a job done and serving whatever part of the marketplace they occupy. It is essential that colleges and universities do what they have historically done best: provide a thorough and unbiased foundation in the study of many disciplines; provide a parallel groundwork in significantly related disciplines; and provide integral study of the social and creative impact of all these disciplines.

Accordingly, my answer to John Sculley's call was to help establish Fairleigh Dickinson's new program in electronic filmmaking and digital video design – the first university program of its kind in the country. Determining its mission and scope was an education in itself for my colleagues and me. Our discussions led us to conclude that we had to help our students think and see in new ways. We didn't want to reinforce the traditional concepts that separate filmmaker from video maker, computer graphic artist from computer scientist, or even computer artist from "conventional" artist. We wanted to stress a larger vision of technical expertise and a commitment to having content drive the use of technology – not the other way around.

For these reasons, the program we created is interdisciplinary, preparing graduates to fully function in the post-modern world. He or she must be able to draw not only from past culture, but also from contemporary cultures around the world. This is necessary if we want the films, ads and video on the interactive channels that everyone talks about to reflect quality, civility, and ethics.

A truly proficient electronic filmmaker or digital video designer will emerge from this convergence of creativity and technology. Courses cover the tradition, chronology and theoretical practices of film and video. But we also offer classes that help students think conceptually: to understand color and form, to see vision, light and sound as a unique mix that tells a story. Technically, students get hands-on training in a variety of platforms, and work in our Visualization Lab helps students master non-linear editing and digitized graphics.

Something else, equally important, has the potential to come from this integrated approach to teaching new media skills. When university educators fully understand the impact of computers on the way people think and learn – about computers themselves, as well as everything else – this knowledge will filter down to elementary and secondary levels of education. Right now, parents who are concerned about their children being "computer literate," and therefore prepared to compete for employment, are satisfied when they see a well-equipped computer lab in their children's school. But computers should not be confined to that lab, standing as formidable machine monsters to be vanquished with specialized study. Ideally, computers should be in every classroom, and integral to the study of every subject.

New computer programs, simulations and other products must be created to help people learn at every level. Those products can and will be developed by the kind of professional turned out by a far-sighted, interdisciplinary university program. Additionally, people in business and government can benefit from new, improved and different uses of computers – not just for data storage, text processing and speedy publishing, but for analyzing, planning and strategizing. Again, these innovations will be made possible by artists, designers and technicians who value the opportunities for communication

that new media can provide, rather than being hung up on the whiz-bang of technology for its own sake.

Finally, I began this piece with a less-than-flattering description of television's rocky course. Television's early pioneers must ultimately be forgiven for not fully understanding the scope of what they were dealing with. They especially deserve this dispensation because they instinctively brought into the fledgling industry the best and brightest minds working in journalism and the arts. They *were* concerned, first and foremost, with content quality. How could they have known that television would become something far more than an ingenious device for bringing the theater stage, concert stage and world stage into the homes of average citizens?

But *we* know. We're fully aware that the simulation of reality that television creates has become a form of reality unto itself. We know that people can be desensitized by this technology just as easily as they can be enlightened by it. We know that market concerns, when imposed on this kind of parallel reality, can truncate information, putting what is saleable above what is important. We know that the lasting power of what is seen poses a serious challenge to what can be learned by other means; that the subtlety, detail and privacy of the printed page has been forever usurped by the strength of the moving image.

Educators can no longer afford to divorce themselves from a society steeped in and controlled by technology. There are intellectuals among us who take pride in rarely watching television, and who think it's clever to dismiss new media technology as the high-tech plaything of a younger generation. If the New Age is to embrace and reflect the best of humanity, then the best of humanity must embrace and reflect the highest potential of the powerful machines it has created. Today, as in Aristotle's time, education in the forces that shape our lives is the best way to provide for a civilized tomorrow.

continued from page 80

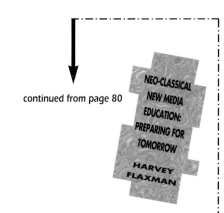

NEO-CLASSICAL NEW MEDIA EDUCATION: PREPARING FOR TOMORROW

HARVEY FLAXMAN

Professor Harvey Flaxman is the chairman of the Department of Visual and Performing Arts at Fairleigh Dickinson University (Madison Campus). He is also an established film and video director.

Photography by Dennis Manarchy

Our clients come to us again and again. And that says more about our abilities than anything else. Our client list for the past year includes:

Acer	AmeriTech	Canada Dry	Dr. Pepper	Jeep
AIG	AT&T	Caterpillar	Ernest & Julio Gallo	Keds
Alfa Romeo	Beechnut	CenTel	Ford	Kikkoman
American Airlines	Bell Helicopter	Chrysler	Frito-Lay	Kodak
American Express	BizMart	Champion International	Greyhound	L&M Cigarettes
	Blue Cross/Blue Shield	Clarion Cosmetics	Hanes	L.A. Gear
	BMW	Clorox	Herman Miller	La-Z-Boy
	Brown & Root Braun	Cole-Haan	Hilton	Levi's
	Budweiser	Colgate	Infiniti	M&M/Mars
	Busch	Costa Cruises	Intel	Marlboro
	Calvin Klein	Dow Chemical	Isuzu	Mazda

Showcase: 1984-1993   Black Book: 1984-1993

SANDBOX DIGITAL PLAYGROUND

SPECIAL EFX / 3D IMAGING / MORPHING / ILLUSTRATION / RETOUCHING

SANDY OSTROFF / 312 / 372 / 1170   BOB WOLTER / REP / 312 / 951 / 6161

photographer Richard Hamilton Smith
art director Doug Trapp
writer Christopher Wilson
agency Martin Williams
client Target/Minneapolis Institute of Arts

THE PALMS HIGH RES DIGITAL IMAGERY   612-339-2000

Jan. 18, 1994

Lee Stokes Retouching
TCF Tower
121 S. 8th Street
#825
Minneapolis MN 55402
612-339-5770

Dear Lee,

It seems I shot a brick when in fact the client needs a picture
of Cindy Crawford. Can you help?

Hopefully,

Steve Mitchell
Art director

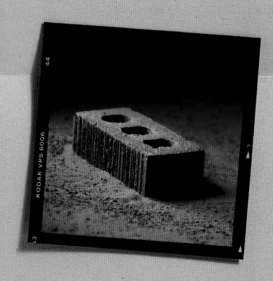

3350 Peachtree Road NE
Suite 1130
Atlanta, Georgia 30326

800-962-9993
404-816-8000

PHOTO*effects*

**Digital Photo
Illustration,
Retouching,
2D & 3D Special
Effects**

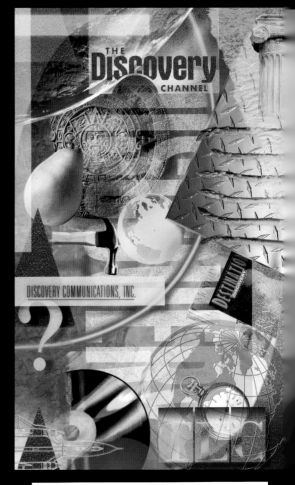

ELIKAN PICTURES

8 1 0 / 3 5 6 / 2 4 7 0

LECTRONIC ART

and Retouching

Credits: John Dudek, Kathleen Martin, Dean Armstrong, Bernie Solo, Ed Dye, Laura Lybeer-Hilbert, John Latin, Chuck White, Petra Pepellashi. Film Separations by Precision Color Inc. • 20500 Civic Center Drive, Suite 2800 • Southfield, Michigan 48076 •

ELECTRONIC ART
and Retouching

PELIKAN PICTURES
810 / 356 / 2470

Pelikan Pictures Ltd. • 20500 Civic Center Drive, Suite 2800 • Southfield, Michigan 48076 • Credits: Peggy Day, Dean Armstrong, Bernie Solo, Ed Dye, Laura Lybeer-Hilpert, John Latin, Chuck White, Petra Pepellashi, Film Separations by Precision Color, Inc.

Computer Illustration by Tom Janousek, Artist, Digital Image, Inc.

DIGITAL
I M A G E

Providing complete digital service that includes:

- •High resolution Isomet drum scans

- •Photographic quality LVT film outputs

- •Shima Seiki & Dicomed photo retouching

- •Large format digital prints

- •High resolution MAC outputs

Our staff is composed of computer science
engineers working hand in hand with
talented artists, photographers and
illustrators to create a professional digital
imaging team. This team strikes a perfect
balance, providing us with the artistic talent
and technical knowledge to provide you with
the perfect product.

Photorealistic Illustration by Tom Janousek, Artist, Digital Image, Inc.

Composition & Photography by Ray Parrott, President, Digital Image, Inc.

DIGITAL IMAGE
...WHERE ART MEETS
TECHNOLOGY

1758 Tullie Circle
Atlanta, GA 30329
(404) 325-6955
fax (404) 325-5828

1302B E. Walnut Ave.
Dalton, GA 30721
(706) 275-0683
fax (706) 275-0127

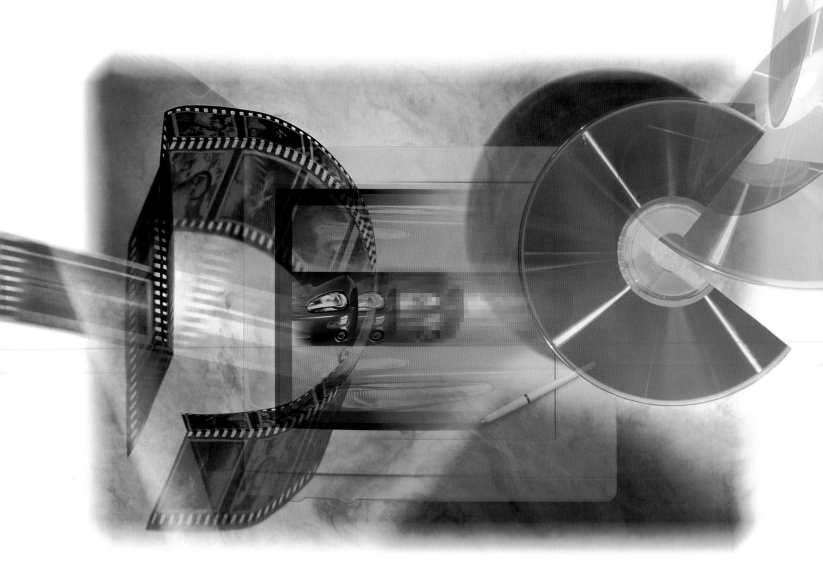

DigiChrome Imaging
"IMAGING THE FUTURE WITH TECHNOLOGY & VISION"

UNRIVALED RESOLUTION ▪ *Electronic image capture* ▪ *Kodak Pro Photo CD* ▪ *hi-speed,*

hi-res scanning ▪ **ACCESSIBLE ARTISTS** ▪ *image manipulation and enhancement* ▪ *digital*

photo-composition ▪ *electronic retouching* ▪ **UNPARALLELED OUTPUTS** ▪ *original quality*

continuous tone Kodak negatives and transparencies ▪ *large format digital color prints*

▪ *color laser or dye sublimation prints & overheads* ▪ **COMPREHENSIVE PRO LAB SERVICES**

1·800·471·1118

DIGICHROME IMAGING, *A SERVICE OF CHROMA STUDIOS, INC.*
2300 MARILYN LANE   COLUMBUS, OHIO   43219

Photo CD symbol is a trademark of Kodak used under license.

INTERACTIVE
MULTIMEDIA

PRODUCTION
SERVICES

Congratulations. You've just gone interactive. Take a gigabyte ride on the wave of the future - a wave we've been surfing since the beginning. We're Meta4 Digital Design, and we can show you a dazzling mix of computer graphics, video, 3D animation, voice over & music that can deliver ideas more powerfully than you've ever dreamed - and will forever change what you think is possible. But to see it a little human interaction is needed. So call us. At the touch of a button we'll come show you how we're pushing the outer edge of imagination.

For a free demo, brochure & portfolio call: Paul Lemberg at 1 800 638 -2414.

Or write Meta4 Digital Design, 245 West Norwalk Rd., Norwalk CT 06850-0432

MicroColor provides computer illustration, animation and interactive media services to the country's most demanding designers. Clients range from Pentagram and Frankfurt Gips Balkind to small independent design firms.

Recent projects include a digital version of the American Express® Card, the Chicago Transit Authority's system map, a 3-D rendering of ASI's Infinity signage system, directory signs with isometric plans for the TVA's offices, ergonomic drawings for a reference book for industrial designers by Henry Dreyfuss Associates, Sony's internal CD-ROM drive packaging, an animation of Samsung's neon sign to be erected in Times Square and an interactive simulation of AT&T Bell Laboratories' Touch Tone II user interface.

For a complimentary portfolio on Macintosh disk, please call 212.787.0500 or send a fax to 212.787.6740.

Micro Color

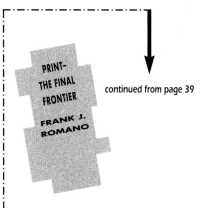

**PRINT-
THE FINAL
FRONTIER**

**FRANK J.
ROMANO**

continued from page 39

lower costs, increase quality and ensure publishing's place in communications. The traditional printing press is geared for long runs (5,000 copies or more) of large sheets, called signatures, and is not generally cost-effective for shorter runs. This is changing as new plate and press technologies emerge.

For example, some of this new plate technology allows offset lithographic printing without water (just ink), capable of a 400 line screen (133 is the standard). Stochastic (or frequency modulated) screening eliminates half-tone dots, as we know them, to provide higher levels of print quality than have ever been achieved before. New output devices generate multi-page imposed plates directly from electronic data. Some new presses automatically load plates and substantially cut make-ready times. Other presses make plates right *on* the press from digital data. And high-fidelity color printing is striving to achieve photographic quality.

At the same time, digital printers are evolving for print runs well under 5,000 copies, which will give rise to new forms of publishing. This brings us to Demand Printing. In academia, it is now possible to have a textbook customized with chapters from a number of books; McGraw-Hill has been a pioneer in this area. The ultimate in on-demand publishing is the fax newsletter, with information magically appearing in print in one's fax machine.

In addition, Just in Time delivery is based on printing what you need when you need it. Instead of printing a million units of a brochure with information that may change and require many of them to be discarded, the Just in Time approach allows you to produce just those quantities required. As a business procedure, this will affect customary print run lengths, as well.

There is also the link between traditional photography and print with PhotoCD, and there are other technologies that are moving print toward continuous-tone reproduction. These innovations are called *high-end printing* to indicate that print is changing for the better and determined to hold its own — as well as a significant share of the communications marketplace. Additionally, economies and improvements in print will encourage *more* print. Individuals, companies, schools and other organizations that now feel certain "slick" printed communications are beyond their budgets will discover they can afford to promote, market, sell, fund-raise and inform with new sophistication and style.

All together, these changes are likely to work in print's favor, because print is so handy for a number of reasons. You can make comments in the margins and emphasize information. You can pass it around to almost anyone. And there is the matter of permanence. Paper may degrade with time, and potential fire might destroy — but barring disaster, print lasts.

Even if we value and utilize the new electronic approaches to communications in many areas, I just don't see our entire society becoming computerized across the board. One of the chief reasons is the book, books that are read for pleasure: mysteries, romances, westerns, thrillers, biographies, poetry, philosophy and more. There is an ease, portability and sense of privacy provided by print that will always be desirable — not to mention the practical ability to share books with others, use them without energy or machinery (remember the generations who read by candlelight?), and recycle their machinery.

Print is not dying; the role and process of print are changing — and print will be around for a long time to come.

Frank J. Romano is the Melbert B. Cary, Jr. Professor of Graphic Arts at the Rochester Institute of Technology School of Printing.

THE ONLY SERIOUS STOCK PHOTOGRAPHY ON PHOTO CD.

Before today, if your work demanded the highest quality stock photography, you endured research fees, usage fees, and time delays. Now you have a choice.

Digital Stock Professional offers you an extensive library of extremely high resolution stock photography on Photo CD.

These exceptional images, created by renowned assignment and stock photographers, can be yours to use, royalty-free, in your advertising, brochures, and reports — forever. *A disc can pay for itself the first time you use it.*

Each Digital Stock Photo CD contains 100 photographs from a specific category, and comes with a color contact sheet so you can immediately select exactly the right image.

Our original transparencies have been meticulously scanned using Kodak's state-of-the-art Photo CD technology. These 2048 x 3072, 24-bit color scans will produce *stunning full-page color separations* at 150 lines per inch. You can use the images as they are, or transform them into original works using programs like Photoshop and Picture Publisher.

Digital Stock discs are *Mac and PC* compatible. Our constantly expanding library includes the sixteen titles shown above and others. Call for latest categories.

To order, call
1-800-545-4514

The Photo CD logo is a trademark of the Eastman Kodak Company, and is used under license. Photoshop is a trademark of Adobe Systems, Inc. Picture Publisher is a trademark of Micrografx. Digital Stock Professional and the Digital Stock Professional logo are trademarks of Digital Stock, Inc.

DIGITAL STOCK PROFESSIONAL

DIGITAL STOCK INC., 400 South Sierra Avenue, Suite 100, Solana Beach, CA 92075, Phone 619-794-4040, Fax 619-794-4041

WE'RE A TOUGH ACT TO FOLLOW.

THE ACE GROUP SM

THE PRE-PRESS SOLUTION FOR ADVERTISING PROFESSIONALS FEATURING MACINTOSH AND PC PLATFORMS · HIGH-END IMAGE SCANNING AND RETOUCHING · TYPESETTING GRAPHICS · ELECTRONIC IMAGING · MATCHPRINTS · DIGITAL COLOR PROOFING · ENGRAVING · DISK CONVERSIONS · PACKAGING DESIGN AND ASSEMBLY · SIGNAGE CAMERA-READY MECHANICALS · ILLUSTRATION · HAND LETTERING · LOGO DESIGN · SCREENPRINTING · ULTRACOMP TRANSFERS AND FOILS · DIRECT IMAGING

149 WEST 27TH STREET, NEW YORK CITY 10001 212.255.7846/243.2929 FAX: 212.989.1028 MODEM: 212.243.1625

EXPANDING THE HORIZON OF PHOTO IMAGING

Copytone is a complete and versatile photographic service that handles everything from the simplest photostat to even your most complicated color photo comp mural, including all the latest computer design and imaging.

We are centrally located in the heart of New York City, around the corner from Times Square. We offer a convenient and complimentary messenger service.*

Our computerized state-of-the-art equipment and top notch professionals enable us to guarantee your job will be completed to your satisfaction time after time.

Our owners work side-by-side with our technicians, insuring we deliver the best quality to all our clients.

*Available for all our local clients. We utilize shipping services for those of you who are out of town.

We are a customer-driven service organization.

COPYTONE NYC
COMPLETE PHOTOGRAPHIC SERVICES

115 WEST 45th • NYC, NY 10036 • (212) 575-0235

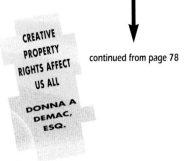

CREATIVE PROPERTY RIGHTS AFFECT US ALL

DONNA A DEMAC, ESQ.

continued from page 78

tection under copyright law, including how Fair Use and Public Domain apply to you.

3. Focus on the specifics of a situation to determine when your rights should be reserved, and when they might be shared.

4. Look at options for licensing work to different categories of users.

5. Look for opportunities to team up with other individuals and groups to enhance your protection and supervise the use of your work.

To rights users, I encourage familiarity with the law at least sufficient to comply with it – and a spirit of fair play, as well. To facilitate your knowledge and understanding of the law, contact the U.S. Copyright Office in Washington, D.C., at 202-707-3000 and request a copy of the 1976 Copyright Act. The integrated issues of ethical responsibility, good business procedure and creative integrity comprise another good list of reasons to honor the proverbial Golden Rule. When it comes to copyright law, *do unto others as you would have them do unto you* is the best protection for all.

Donna A. Demac is an attorney specializing in telecommunications and intellectual property, presently based at Columbia University. She is the author of Intellectual Property in the Multimedia Age, *jointly published in 1994 by the Interactive Multimedia Association and Prentice Hall.*

Hardware

Software

IRIS Showcase digital interactive media presentation package, Digital Media Clip Library, Soundscheme and Prosonus sound library, IRIS Explorer™ visualization application, IRIS Inventor™ 3-D Viewer, IRISInsight™ Viewer and complete on-line documentation, Getting Started Guide); Hot Mix Application Trial CD (trial copies of Insignia SoftPC 3.1 DOS/Windows emulator and Quorum software application adaptor for Macintosh MSWord™ and MSExcel™), ability to read Kodak Photo CD™; and, IRIX™

ANY PORT IN

STEREOVIEW
Stereoscopic glasses for an even more virtually realistic 3D look.

MONITOR
Get the best graphics on the planet with up to 19," 1280 x 1024 multi-scan monitors.

GIO EXPANDABILITY
Options! Indy-Video™ for I/O and special effects—CosmoCompress™ for JPEG compression.

COMPOSITE VIDEO IN
Pipe in video from VHS, Laser-disc, etc.

S-VIDEO IN
Higher quality? You got it. Including Hi-8. Record projects in American or European standards.

ETHERNET AUI
Zoom story-boards, clips, or graphics around your facility at high-speed.

MOUSE
The way this puppy reacts to a Silicon Graphics mouse is startling.

DUAL SERIAL
MAC-compatible. Talk to any-thing from MIDI to touchscreens to VTRs.

PARALLEL
Let 'er RIP! This thing is a Post-Script printing marvel.

HEADPHONES
Slip 'em on and play to your neighbor's content.

MICROPHONE
Add a voice-track to your animatic.

ANALOG STEREO IN
Music beds, voiceovers, and sound effects—the analog access pipe.

ANALOG STEREO OUT
Out to VTRs, audio decks, or external speakers. Play it. Play it loud.

DIGITAL STEREO IN/OUT
Record and playback CD-quality sound. Sweet.

INDYCAM™ PORT
For the Indy-Cam color digital video camera. Meet face-to-face without using your feet. Or send videomail.

ISDN
Network with clients around the corner, or around the planet.

ETHERNET 10BASE-T
A twisted-pair connection to share, deliver, or just play with ideas.

KEYBOARD
The 101-Keyboard is PS2™ compatible, just like the mouse.

FAST SCSI-II
External disk and tape drives, CD-ROM, scanners, Photo CD™, printers, plotters, you name it!

East Coast
37 Brook Ave.
Maywood, NJ 07607
(201)368-9171

West Coast
23105 Kashiwa Ct.
Torrance, CA 90505
(310) 534 - 0050

Ikegami
®

Ikegami, the leader in television broadcast cameras brings a new standard of performance to computer displays. Ikegami displays provide more value and performance in digitally controlled Trinitron and Invar mask displays. Ikegami displays meet all VLF emissions and saftey standards and provide outstanding ergonomics for professional use. Our displays even come with software to allow control of all monitor functions right from your keyboard! So call Ikegami for the dealer nearest you and get the best displays!

" Trinitron is a registered trademark of Sony Corporation "

BUYING OUR NEW SCANNER IS LIKE TRAVELING FIRST CLASS ON AN ECONOMY TICKET.

AGFA INTRODUCES STUDIOSCAN.™ SO MUCH SCANNER FOR SO LITTLE MONEY.

It used to be that unless you had a lot of money to spend, a scanner was simply out of the picture. But now there's Agfa StudioScan, the new feature-rich scanner that's wonderfully affordable.

StudioScan offers the speed and efficiency of one-pass scanning for both black-and-white and color images. Its flatbed design accommodates sizes up to 8.5″ x 14″, and an optional transparency module scans 35mm slides up to 8″ x 10″.

What's more, StudioScan comes with a complete software package, including Agfa's proven labor-saving FotoLook,™ FotoTune LE,™ FotoSnap™ programs, and Adobe PhotoShop LE. Compatible with both Macintosh and PC systems, StudioScan actually guides an entry-level user through the entire scanning process.

To find out more about StudioScan's high-quality imaging for your layouts, in-house publications, illustrated reports, and more, call 1-800-685-4271 ext. 1164 today. And discover the Agfa scanner that offers first-class features at an economy price.

AGFA ◆ *Agfa*

The complete picture.

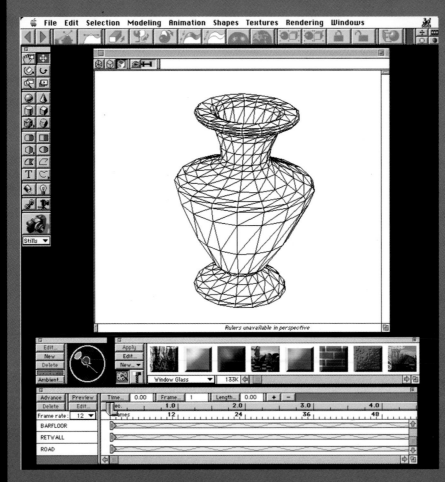

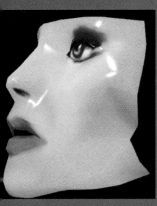

DON'T CALL ORKIN.*

When the Mac first came around, everybody loved its *little rodent.* No cursor up. No cursor down. Man, what a handy device. Then people got into graphic arts. Whoops. *Mousetrap!* INTRODUCING ARTZ.™ The flashy, jammin', way cool ADB graphics tablet from Wacom. ArtZ simply *plugs into your ADB port* and lets you create graphics using a cordless, batteryless, pressure-sensitive stylus, about the size and weight of a fine pen. Now, creating art on a computer is as *natural* as creating art on a note pad, a canvas, or even a subway. To see just how flashy and way cool the ArtZ is, TAKE THIS SIMPLE TEST: *(1)* Cut out the mouse at the top of the page. Blow it up if you like. Now *(2) trace it* with the mouse  on your computer. Then *(3)* print it, and *(4)* scurry on down with both to your local Wacom dealer. Don't know where one is? Just call *1-800-922-6613.* When you get there, *(5)* trace the mouse on an ArtZ. *Feels good,* doesn't it? Now *(6)* print it and *(7)* compare the two. OK? Now *(8)* pull out your pocketbook. You've been doing *mickey mouse art* long enough.

*For the mice that run on a Mac, you can't call for pest control. But for the mice that run on top of a Mac, and under a Mac, and in the cupboards and stuff, call these guys: Orkin Exterminating.

WACOM®

Putting technology in its place.™

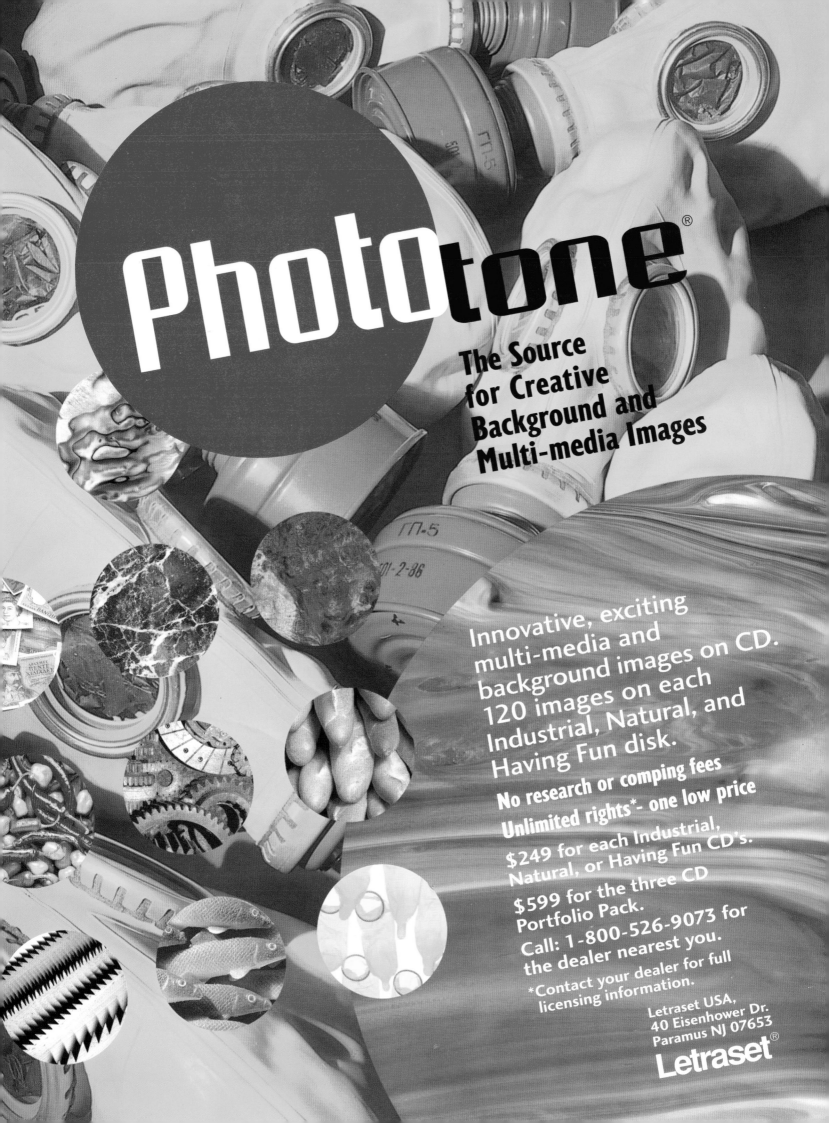

Compose High-Resolution Photoshop Artwork. Fast.

1 Use Adobe Photoshop to capture and edit individual images on your Macintosh. Retouch the images and create masks. Now you're ready to compose in Collage.

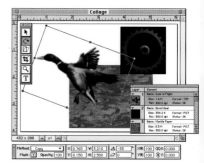

2 In Collage, you work with screen-resolution 'proxies,' so changes are easy, flexible and *fast*. Use familiar page-layout tools to move, layer, scale, and rotate images. Dynamically turn masks on or off. Adjust transparency and feathering. Add or remove effects at any time. Even edit or move text after it has been placed!

3 Instantly create sophisticated effects like soft drop shadows. Use Collage's built-in effects or third-party Photoshop filters. When you move an image its effects move with it! At your command, Collage accesses the original images and renders a high-resolution composite. Don't wait—*create!*

SPECULAR
COLLAGE™
FOR MACINTOSH®

The Image Composition Tool for Adobe Photoshop™
CALL SPECULAR INTERNATIONAL AT 1-800-433-SPEC

Specular International

INFINI-D™
doesn't just
win awards

1992 WINNER

"Best 3D Program"

Five Mice, Highest Rating
MacUser Product Review, April 1993

MACWORLD
MW
EDITORS' CHOICE

"Best All-Around 3D Package"
Winner of Two Editors' Choice Awards

R A T E D
NewMedia
MAGAZINE

AWESOME

NOVEMBER 1992

C. David Piña drew the Academy Awards logo using Fontographer® and Aldus FreeHand.® He used Infini-D to instantly turn the FreeHand document into a 3D model, adding bevels and reflections. Infini-D's automatic alpha channel allowed him to composite the logo with an image in Photoshop for the final artwork.

It helps make them.

Infini-D is the powerful and easy-to-use 3D graphics tool that has swept virtually every award in the industry. So it comes as no surprise that professionals such as C. David Piña rely on Infini-D to produce graphics and animations for their most demanding clients.

Specular's Infini-D integrates seamlessly with programs like Adobe Premiere™, Photoshop™ or Illustrator;™ allowing you to quickly add the impact of 3D to any project.

Infini-D's interface provides the easiest-to-use 3D environment—period. And the **PowerPC version of Infini-D**, which will ship at the same time as Apple's Macintosh on PowerPC, will offer blazingly fast workstation-class speed!

If you're involved in desktop video, multimedia, graphic design, or illustration on the Macintosh,® call today to learn how Infini-D can put the amazing world of 3D graphics at your fingertips.

1-800-433-SPEC

479 WEST STREET / AMHERST, MA 01002 / 413-253-3100

Specular International

P·I·X·A·R

Award-Winning

Animation Production

...*specializing in inspiring characters and stories.*

Image rendered with RenderMan.
J. Walter Thompson/Warner-
Lambert, Inc. ©1994.

World Class

Graphics Products

...*for creating stunning images and animations.*

Image created with Pixar Typestry. "Celebrate the Spirit."
Main title loop design by C. David Piña.
©1992 The Walt Disney Company .

Image created with Pixar Showplace, Pixar One Twenty Eight
and Glimpse. Inspired by the work of M.C. Esher. ©1993 Pixar.

The Industry Leader in Computer Graphics Products & Services.

P·I·X·A·R

1001 West Cutting Boulevard Richmond, CA 94804 510.236.4000 Fax 510.236.0388

Create the Impossible . . . .

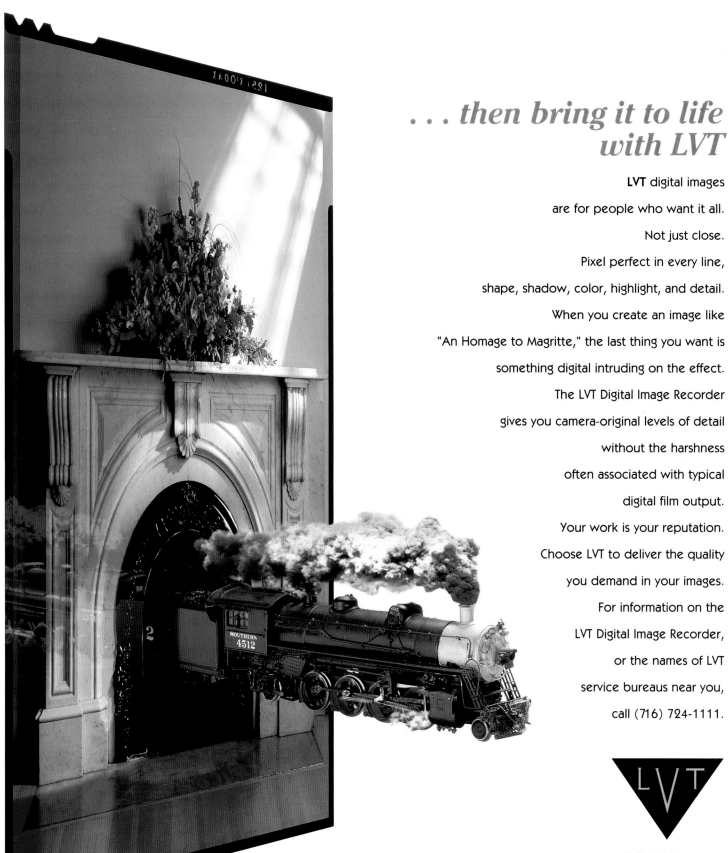

. . . then bring it to life with LVT

LVT digital images
are for people who want it all.
Not just close.
Pixel perfect in every line,
shape, shadow, color, highlight, and detail.
When you create an image like
"An Homage to Magritte," the last thing you want is
something digital intruding on the effect.
The LVT Digital Image Recorder
gives you camera-original levels of detail
without the harshness
often associated with typical
digital film output.
Your work is your reputation.
Choose LVT to deliver the quality
you demand in your images.
For information on the
LVT Digital Image Recorder,
or the names of LVT
service bureaus near you,
call (716) 724-1111.

LVT

A **Kodak** Company

100 Kings Highway, Suite 1400
Rochester, NY 14617

You never know how a revolution will begin – or where it will end. On the night of November 8, 1972, HBO sent forth its first program from a tiny transmission facility in Wilkes-Barre, Pennsylvania. It was a polka contest, filmed in black and white. HBO had 400 subscribers back then, and leaders in the fledgling cable industry didn't believe that consumers, who were already suspicious of "pay TV," would shell out for basic cable *and* premium channels on top of that. And behind closed doors, everyone in the industry worried about where all the programming would come from to fill up their roster of new channels. Out in American living rooms, most people couldn't even grasp the concept of cable itself. Their hazy idea of pay TV envisioned the same old shows, on the same old channels, on the same old set – that might soon have a coin slot attached to it. The technology was new, the programming was sparse, and the foundation of consumer demand had not yet been laid.

Little more than 20 years later, cable television is an institution, providing a sizable line-up of entertainment, sports, news, and specialty channels – as well as doing ongoing battle against the intrusion of home video. HBO, with 18 million subscribers in the U.S. alone, has become the largest movie producer in Hollywood, attempting to preserve its dominance over a new selection on the cable menu: pay-per-view. And out in almost every room in American homes, consumers sit with TV remote controls in hand, blithely surfing through 75 channels.

Now, people are hearing about another innovation: the information super-highway, a brave new road to tomorrow that will offer them a host of technological miracles, from video telephone, to 500 TV channels, many of which will feature "interactive" programming. The technology is new, the programming is sparse, and the foundation of consumer demand has not yet been laid.

But you would think tomorrow was here right now, judging by the "feeding frenzy" of hype being conducted by the news media, business and government. Although the reality of the super-highway defies simplistic description and analysis, attempts to do just that are abounding. Many different and, in some cases, contradictory, technological and social phenomena have been crowded together under this highway heading. That's because it's easier to talk about the companies that will own some aspect of the highway (the telephone, cable and entertainment giants, among them), rather than address the social and economic ramifications of democratic access. However, if a book such as this one is important to your business, it is equally important for this book to make sense of the confusion created by the hype, and to attempt to outline the potential impact on the business community of such revolutionary technology.

In trying to separate the high-tech wheat from the marketing chaff, certain questions must be posed, and answers speculated. What, in reality, will be the predominant forms of communications media covered by the highway? Will it be a single-lane "highway," or is this metaphor limiting? In keeping with the analogy, who will build this road, collect the tolls, and pay for its maintenance? Who will be the state troopers, guarding against recklessness and coordinating aid after an accident (and what will constitute an accident)? Most important, who will be using this highway? How will they get there – and why will they want to take the ride?

Conventional analysis of this evolving communications system rarely makes a clear distinction between the actual physical structures of the information highway (the technology that is the equivalent of access roads, bridges and tunnels), and the vehicles on the highway (information, new services and new forms of TV programming). Last to be fully and appropriately considered are drivers – the businesses and consumers whose use of the highway will justify all the ambitious construction. This is an issue of some concern for designers and other new media professionals, who stand to be among the greatest beneficiaries of the super-highway.

Part of the faulty analysis is rooted in confusion between *content* (the vehicles) and the *distribution method* (the road itself). For example, multimedia has come to be understood as a combination of diverse media (graphics, text, sound, animation, video, etc.). But what makes multimedia a major breakthrough is that all types of media can now be represented in digital format. This is important, because it means that anyone who has a microcomputer and other appropriate (and increasingly inexpensive) hardware can convert all forms of communications media into a stream of digital bits. This can then be transmitted by numerous distribution methods to any location with the appropriate receiving gear. Disks, telephone lines, switched packet services, radio and TV satellite transmissions are all distribution methods for sending digital data down the street or across the globe.

Therefore, when we talk about the super-highway, we are actually referring to what is being presented to the public by the most powerful players in the communications industry as the *primary distribution method* for all forms of information and activity. However, this distribution method is at least ten years away, and even once it's in place in select, largely urban, areas, huge portions of the country will be excluded. Additionally, the super-companies that build the super-highway are going to want super-control, if only to recoup their super-financial investment. So, an important question arises: if only a handful of vehicles (information, etc.) are covered by the super-highway (primary distribution method), and only a limited number of drivers (businesses and consumers) can get onto it, then what will make it so super? The reason that actual modern roadways are important to our lives is that anyone with a car and a driver's license in good standing gets to use them. *Anyone.* If the same is not true of the information highway, nothing much will have changed, except that some of us will be moving a lot faster.

In an ideal and truly democratic technological future, distribution methods will not be the sole province of the corporate giants. But all we're hearing about is what *they* can deliver to us. They don't want us to think about who else can get into the game, let alone what we can deliver to each other! To be sure, powerful distribution methods in unscrupulous hands – political terrorists, for example – is a sobering prospect, and distribution standards and safeguards will have to be established. But by definition, democracy is dangerous and difficult. If the information future is

CONTEMPLATING THE DIGITAL FUTURE

DAVID BIEDNY

established. But by definition, democracy is dangerous and difficult. If the information future is to contain more than shopping and sitcoms on demand, a visionary re-evaluation of what it means to have free and open communications media must be made.

However, once construction is completed on the super-highway, millions of people will want to use it for just as many reasons, regardless of its imperfections and limitations. That's because this highway will truly be greater than the sum of its parts, representing far more than the vehicles riding on it. Fully realized, the highway will provide many new services (not just entertainment). It will offer new forms of social interaction and new cultural forms that will reflect the global scope of human endeavor. In business, it will offer entirely new ways for people to work together from remote and mobile locations, and provide new avenues for communicating with clients and the marketplace. Perhaps most important, it will create new ways for teachers to teach and learners to learn.

Interactive television, the most highly touted feature of the future, will, separate and apart from all other delivery methods, offer many of the new ways to go to school, shop, play, borrow material from libraries, even search out new friends and experiences. Anyone who isn't computer-phobic and has mastered the art of programming a VCR (a yardstick, not a requirement) is likely to view the prospect of interactivity with enthusiasm and interest, and will probably not be daunted by a mega-menu of 500 channels.

Current trends support the anticipated interest in many features of the super-highway. In today's communications market-place, there is already a well established information highway that closely mimics the country's actual road system. Instead of a single central highway, there is a dense network of access roads: phone lines, satellite transmission systems and cable TV, all of which serve different purposes and markets. The principal difference among them is *cost*. But any resourceful computer owner with a

modem can participate in a diverse, globally connected digital world and enjoy a smorgasbord of information just waiting to be accessed via public services such as Prodigy, CompuServe, America Online and Internet. For the most part, these services have done good work in the area of user interfaces for data access, making it relatively easy for non-technical participants to get online and be productive. As a result, they've attracted a wide following and will continue to do so.

Therefore, what is the real future of the information super-highway and interactive television programming likely to be? For starters, those responsible for programming will have to take into consideration that not everyone is a convert to the digital age. Distrust, ignorance and established behavior patterns will pose great obstacles to reaching the "average" consumer, especially those over the age of 35. In order for interactivity to truly gain ground, it will be necessary to turn to traditional reward mechanisms to motivate it. For this reason, interactive game shows offering real prizes, shopping networks, and movies-on-demand will likely be the leading initial applications.

Generally, entertainment and humor, proven approaches to making many types of information easier to understand, will be primary ingredients in future programming, and will help provide incentives for interactivity. In addition, video game structures, which are already an important component of the most successful education software products, will continue to have an important place in programming.

In the next century, when today's "computer babies" have come of age, the interactive television world will blow wide open. And eventually, education will be a major force in interactive media and could significantly address the problems of teaching and learning, bringing revitalized and individualized training to masses of people.

From a commercial perspective, brand/image awareness and loyalty will be the advertising industry's greatest marketing imperative. This will

continued from page 118

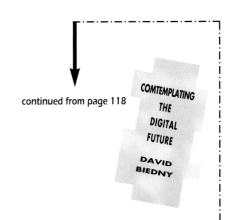

COMTEMPLATING THE DIGITAL FUTURE

DAVID BIEDNY

apply to products being sold, as well as to the channels themselves. The nature and form of advertising will change radically; some of the trends are evident now. Program-length "infomercials" are becoming an advertising staple in a TV climate that has viewers zapping away from short spots faster than they honk their car horns when a traffic light changes. The infomercial trend will continue.

Innovative forms of advertising will undoubtedly be developed, just as the music industry initially created music videos to sell records. And while education will be an important part of the television future, public television as we know it might be one of the first major crack-ups on the information super-highway. Advertisers who used to benefit from the social cachet of public TV underwriting may find new opportunities in the sponsorship of educational niche channels; a televised "GM University" or "IBM Institute" is not far-fetched.

Ultimately, as in any major capitalist undertaking, money is the bottom line, and the cost of wiring the country for the super-highway is likely to become the biggest obstacle to its potential success. It's interesting to note that the highway hype that began in late 1993 was sparked by the October 13th announcement of the proposed merger between TCI, the nation's largest cable system, and Bell Atlantic, the country's third-largest phone company. When that proposed mega-marriage fell through last February, Wall Street lost 42 points in one day and hopes for the highway lapsed into temporary but serious uncertainty. As you read this now, important new deals may have already taken its place. Certainly, the formidable merger between Viacom and Paramount, long preceded by the Time-Warner merger, signals that major players will continue to join forces for the move into the communications future.

But the TCI/Bell Atlantic faltering at the altar is revealing – of business' fiscal unwillingness to foot the entire highway bill; of government's expectations that they *should* do so; and of consumers' refusal to have the costs passed on to them. (That the failed merger came on the heels of the FCC's mandated decrease in consumer cable fees is no accident.) In the final analysis, if consumers are presented with new communications utility bills that rival their supermarket budgets, the message to those offering the information super-highway will be simple: no thanks, we'll walk. But surely, the promise of this communications wonder will motivate all concerned to work it out. Look how far HBO went once they mastered the polka.

David Biedny is a leading multimedia designer, as well as a new media educator and writer whose work appears frequently in top magazines serving the industry.

CLICK. CLICK. CLICK.

An evolutionary guide to the revolutionary world of color electronic prepress

◄ •

The 1st comprehensive guide to designing, storing, transmitting, correcting, and manipulating digital images for print and non-print uses. The work of 30 leading digital artists is showcased as well as numerous other full-color composite images with step-by-step photos of exactly how each was created.

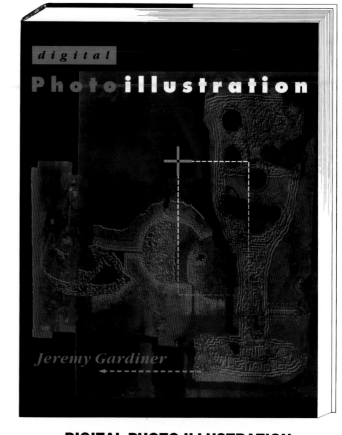

DIGITAL PHOTO ILLUSTRATION
By Jeremy Gardiner
200 color plates/160 pages/0-442-01167-9/$39.95

 VAN NOSTRAND REINHOLD
Books with Style + Substance
115 Fifth Avenue, New York 10003

Contents

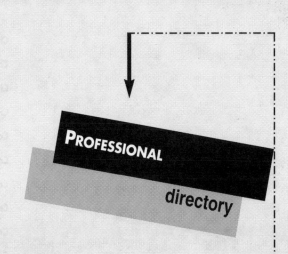

PROFESSIONAL directory

Electronic Imaging — 122
Photographers — 128
Holographers — 129
Retouchers — 131
Illustrators — 141
Designers — 151
Representatives — 152
Presentation Graphics — 154
Desktop Publishing - Hardware — 155
Desktop Publishing - Software — 155

Interactive Multimedia — 158
Presentation — 162
Animation — 165
Education - Training — 166
Media Integration — 167
Visualization — 168
Interactive Screen Design — 168
Consultants —

Production Services — 170
Optical — 171
Video — 178
Video Independent — 179
Video Animation — 184
Video Sound Integration — 188
Prepress — 189
Service Bureaus — 194
Integrated Printing — 194
CD-ROM Publishers —

Hardware — 196
CD Interactive — 196
CD-ROM — 196
Computer Platforms — 196
DAT — 197
Digital Editing — 197
Digitizers — 197
Display Systems — 197
Electronic Photography — 197
Graphic Input Devices — 197
HDTV — 198
Optical Disks and Drives — 198
Printers/Output Devices — 198
Scanners — 198
Video-Digital Interactive — 198
Video Production — 199
Virtual Reality —

Software — 200
3-D Modeling — 200
Animation — 200
Authoring — 201
Digital Editing — 201
Imaging/Digital Effects — 201
Raytracing/Rendering — 201
Video Production —

Resources & Information — 203
Publications (Magazines/Newsletters) — 203
Professional Organizations — 204
Trade Shows/Conferences — 205
Professional Development — 207
Consultants —

Regional Categories

New York City

East
Connecticut
Delaware
Maine
Maryland
Massachusetts
New Hampshire
New Jersey
New York State
Pennsylvania
Rhode Island
Vermont
Washington, D.C.

South
Alabama
Arkansas
Florida
Georgia
Kentucky
Louisiana
Mississippi
North Carolina
Oklahoma
South Carolina
Tennessee
Texas
Virginia
West Virginia

Midwest
Illinois
Indiana
Iowa
Kansas
Michigan
Minnesota
Missouri
Nebraska
North Dakota
Ohio
South Dakota
Wisconsin

West
Alaska
Arizona
California
Colorado
Hawaii
Idaho
Montana
Nevada
New Mexico
Oregon
Utah
Washington
Wyoming

Canada

International

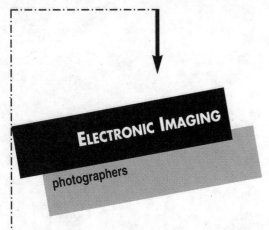

ELECTRONIC IMAGING

photographers

Photographers

NEW YORK CITY

Abbott, Waring
78 Franklin St, New York, NY 212-925-6082

Abolafia, Oscar
215 W 98th St, New York, NY.............................212-662-8472

● **ACCORNERO, FRANCO (P 28)**
620 Broadway #6R, New York, NY.....................212-674-0068

Aich, Clara
218 E 25th St, New York, NY..............................212-686-4220

Akavia, Eden
PO Box 545, New York, NY.................................212-533-0633

Barba, Jac
247 Mulberry St #14, New York, NY 212-431-3294

Barbero, Teri
134 Spring St #502, New York, NY 212-925-1380

Bechtold, John
117 E 31st St, New York, NY...............................212-679-7630

Beebe, Rod
127 W 26th St #800, New York, NY 212-727-9727

Beechler, Greg
200 W 79th St PH #A, New York, NY..................212-580-8649

Behr, Sheri Lynn
1675 York Ave #26F, New York, NY 212-922-9020

Beidler, Barbara
648 Broadway #506, New York, NY....................212-979-6996

Berkley, Miriam
353 W 51st St #1A, New York, NY 212-246-7929

Berkwit Studio/Lane Berkwit
262 Fifth Ave 5th Fl, New York, NY 212-889-5911

Betz, Charles
108 E 16th St 6th Fl, New York, NY 212-228-3080

Black Lamb Studios Ltd/Ariel Jones
31-33 Walker St 2nd Fl, New York, NY 212-334-1209

● **BLACKMAN STUDIOS INC,
BARRY (P 14,15)**
40 W 25th St 6th Fl, New York, NY 212-627-9777

Blell, Dianne
125 Cedar St #11N, New York, NY.....................212-732-2990

Blink Inc/Stan Shaffer
2211 Broadway, New York, NY...........................212-580-5522

Borderud, Mark
151 W 19th St 12th Fl, New York, NY 212-242-5425

Bordnick, Barbara
39 E 19th St, New York, NY................................212-533-1180

Bourke, Dennis
37 W 20th St #606, New York, NY 212-741-3211

Breskin, Michael
135 W 29th St 5th Fl, New York, NY 212-330-7619

Brett, Clifton
242 W 30th St #3R, New York, NY.....................212-947-0139

● **BRONSTEIN/BERMAN/WILLS/**
Howard Berman & Steve Bronstein **(P 8,9)**
38 Greene St, New York, NY...............................212-925-2999

Bryan-Brown Photo, Marc
534 W 35th St, New York, NY.............................212-594-1360

Buhl Studios/Carol Shadford
114 Green St 5th Fl, New York, NY 212-274-0100

Callis, Chris
91 5th Ave 7th Fl, New York, NY 212-243-0231

Campus, Peter
110 Bleeker St, New York, NY............................212-529-4856

Carroll, James
382 Central Park West, New York, NY.................212-865-5321

Chalkmark Graphics/David Chalk
348 W 38th St #7C, New York, NY....................212-279-2433

Chiba, Kaz
303 Park Ave S #412, New York, NY..................212-674-7575

Chotas, James
11 E 87th St, New York, NY................................212-410-3917

Cirone, Bettina
240 Central Park #4H, New York, NY.................212-262-3062

Cobi Productions/Jacob Babchuk
132 W 22nd St 3rd Fl, New York, NY.................212-929-8811

Collas, Pito
425 Broome St #3M, New York, NY...................212-966-8450

Computer Solutions/Jeff Blechman
593 Broadway, New York, NY.............................212-431-5239

Creative Media Concepts/Regina Tierney
8 W 38th St, New York, NY.................................212-391-0202

Critical Image/Michael Berger
4 W 37th St 2nd Fl, New York, NY.....................212-594-2586

Cunningham, Peter
53 Gansevoort St, New York, NY........................212-633-1077

Cutler Studio Inc/Craig Cutler
628-30 Broadway #403, New York, NY..............212-473-2892

Davidian, Peter
340 W 22nd St, New York, NY............................212-675-8077

Davis, Harold
250 W 104th St #81, New York, NY...................212-316-5903

Denbo Multimedia/Robert E Denbo
135 W 29th St, New York, NY.............................212-465-2297

Depthography Inc/R Anthony Munn
122 E 27th St 2nd Fl, New York, NY..................212-679-8101

Dervis, Aris
214 Riverside Dr #207, New York, NY................212-666-5967

Diadul, Robert
111 E 14th St # 394, New York, NY...................212-982-7109

Diamant Noir/Philippe Louis Houze
519 Broadway, New York, NY.............................212-226-9299

Diaz, Jorge
340 E 80th St #16D, New York, NY...................212-439-6324

Digital Photo Studio/Vince Aiosa
5 W 31st St, New York, NY.................................212-563-1859

Dodge, Jeff
133 Eighth Ave, New York, NY...........................212-620-9652

Dole, Jody (Mr)
95 Horatio St, New York, NY..............................212-691-9888

Dorf Studios/Myron Jay Dorf
205 W 19th St 3rd Fl, New York, NY.................212-255-2020

Dressler, Marjory
4 E 2nd St, New York, NY...................................212-254-9758

Drivas, Joseph
15 Beacon Ave, Staten Island, NY.....................718-667-0696

Electronic Publishing Ctr/Irving S Berman
110 Leroy St 2nd Fl, New York, NY....................212-727-2655

Feldman, Simon
45 E 20th St 5th Fl, New York, NY 212-260-7890

Finzi, Jerry
36 W 20th St 6th Fl, New York, NY...................212-255-2110

Fischer Photography, Ken & Carl
666 Greenwich St #656, New York, NY.............212-675-8655

Fishman, Miriam
12 E 14th St #2E, New York, NY.......................212-633-9126

Flatow, Carl
20 E 30th St, New York, NY...............................212-683-8688

FPG International/Gary Elsner
32 Union Sq E 6th Fl, New York, NY..................212-777-4210

Frasser, Oscar
555 Main St, New York, NY................................212-486-0596

Freed, Gary
130 W 25th St 4th Fl, New York, NY.................212-218-5029

Galante, Dennis
133 W 22nd St 3rd Fl, New York, NY.................212-463-0938

Gallucci Studio Inc/Ed Gallucci
147 W 25th St 11th Fl, New York, NY................212-675-0097

● **GARDINER, JEREMY**/Pratt University **(P 120)**
Computer Graphics Ctr ARC Bldg,
Brooklyn, NY...718-636-3600

Gartel, Laurence
270-16 B Grand Central Pkwy, Floral Park, NY....718-229-8540

Gittler, Barbara
157 W 57th St #700, New York, NY...................212-246-8282

Gloffke, Barry
99-32 66 Rd, Queens, NY.................................718-897-4372

Goldberg, Lenore
155 6th Ave, New York, NY................................212-627-1770

Goldman, David
22 E 17th St #2F, New York, NY........................212-807-6627

Goll, Charles R
404 E 83rd St, New York, NY.............................212-628-4881

Goossens, Jan
518 East 11th St #2F, New York, NY.................212-228-4142

Gordon, Joel
112 Fourth Ave 4th Fl, New York, NY 212-254-1688

Gove, Geoffrey
117 Waverly Pl #5E, New York, NY...................212-260-6051

Greenberg Associates, R
350 W 39th St, New York, NY............................212-239-6767

Greenberg, Jill
322 Seventh Ave, New York, NY........................212-594-5624

Gregg, Rene
1233 York Ave, New York, NY...........................212-631-1080

Gross, Jonathon
20 West 20th St, 10th Fl, New York , NY...........212-255-5752

Grundy, Jane
150 W 82nd St, New York, NY...........................212-787-8454

Halsband, Michael
1200 Broadway #3A, New York, NY..................212-889-2994

Hamann, Horst
73 Warren St #6, New York, NY.........................212-677-5509

Han, Calla & Peter
284 Mott St #6C, New York, NY........................212-274-1571

Heal, Patricia
229 E 12th St #65, New York, NY.....................212-995-2505

Heiberg Studio/Milton Heiberg
71 W 23rd St #614, New York, NY....................212-741-6405

Heisler, Gregory
568 Broadway #701, New York, NY...................212-777-8100

Heyl, Fran
230 Park Ave #2525, New York, NY 800-327-0333

High-Res Solutions Inc/Jim Casey
23 E 20th St, New York, NY...............................212-475-6107

Hodgson, David
550 Riverside Dr #1, New York, NY...................212-864-6941

Hoffer, Jane
509 W 110th St #5D, New York, NY.................212-663-6404

Hoover, Duane
22 Watt St #4, New York, NY.............................212-334-4134

Hornbacher, Sara
40 Harrison St #9B, New York, NY 212-964-9582

Horowitz, Ryszard
137 W 25th St, New York, NY............................212-243-6440

Hovde , Nob
14638 Third Ave, New York, NY 212-753-0462

Huang, Eric
5 W 19th St 5th Fl, New York, NY 212-924-4545

Huba, John
168 Ludlow St, New York, NY............................212-533-8648

Image Bank, The
111 Fifth Ave, New York, NY..............................212-529-6700

Imagery Center Inc/Barbara Ellen Adelman
5 E 22nd St #25A, New York, NY.......................212-353-8931

Index Stock International Inc
126 Fifth Ave 7th Fl, New York, NY....................212-929-4644

Jarpotawanich, Ampon
238 E 58th St #7, New York, NY........................212-371-9244

Joel, Seth
32 Union Square E, New York, NY.....................212-674-0852

Jordan, G Steve
116 W 72nd St #15A, New York, NY.................212-724-7309

Kahler, Charlotte
165 E 72nd St, New York, NY............................212-988-6945

Kana, Titus
876 Broadway, New York, NY............................212-473-5550

KanImage Division/Barbara Leven
Cosmos Comm 11-05 44th Dr,
Long Island City, NY..718-482-1800

Kaplan, Alan
7 E 20th St 2nd Fl, New York, NY......................212-982-9500

Kelly, Beth
321 W 88th St #1B, New York, NY 212-580-4940

Kilkelly Associates/Jim Kilkelly
40 W 73rd St #2R, New York, NY.......................212-873-6820

Knight, Kevin
116 Lexington Ave, 3rd Fl, New York, NY...........212-532-9482

Knight, William J
405 W 23rd St #7A, New York, NY....................212-727-0066

Koopman, Mary Ann
33-34 28th St 2nd Fl, Long Island City, NY.........718-932-2720

Koudis Studio/Nick Koudis
30 W 22nd St, New York, NY.............................212-475-2802

Krongard, Steve
116 Mercer St, New York, NY............................212-334-1001

L Squared Studios/Linda Lauro
20 W 20th St #1000, New York, NY 212-675-4681
Lane, Morris
212-A E 26th St, New York, NY 212-696-0498
Leduc, Lyle
320 E 42nd St #1014, New York, NY 212-697-9216
Leinwand, Freda
463 West St #229G, New York, NY 212-691-0997
Lesinski Photography, Martin
35 Pierrepont St #2B, Brooklyn Heights, NY 718-463-7857
Levin, Bruce
231 E 58th St, New York, NY 212-832-4053
Lightpath Imageworks/Barry Pribula
59 First Ave, New York, NY 212-777-7612
Lopez, Nola
684 Broadway #4A , New York, NY 212-254-2582
Marr, Julie
121 Macdougal St #7, New York , NY 212-477-3352
Maurio
258 W 22nd St #4C, New York, NY 212-229-1479
McCarthy, Tom
115 E 90th St #1B, New York, NY 800-344-2149
McDonough, Patricia
28 Hubert St, New York, NY 212-226-9017
Mendez, Raymond A
220 W 98th St #12B, New York, NY 212-864-4689
Meola, Eric
535 Greenwich St, New York, NY 212-255-5150
Merrill, Lizanne
333 Rector Pl #5N, New York, NY 212-945-6115
Miller, Marilyn
36 W 20th St 10th Fl, New York, NY 212-206-1000
Mills, Robert
211 E 4th St #2, New York, NY 212-228-3589
Milne, Bill
140 W 22nd St 8th Fl, New York, NY 212-255-0710
Mitchell, Diane
175 W 73rd St, New York, NY 212-877-7624
Modricker, Darren
50 Lexington Ave #159, New York, NY 212-978-8877
Morita, Tatsuya
224 Ave B #25, New York , NY 212-982-8742
Morse, Michael
6 W 87th St #6, New York, NY 212-721-2638
● **MULTIDIMENSIONAL IMAGES LTD**/Don Carroll
(P 18,19)
188 Grand St, New York, NY 800-219-2825
Nakamura, Ikuo
864 President St, Brooklyn, NY 718-399-6786
Neumann, William
119 W 23rd St #206, New York, NY 212-691-7405
Nexvisions/Guy Fery
150 Fifth Ave #202, New York, NY 212-255-4646
Novak, Dennis
232 Elizabeth St #6E, New York, NY 212-334-5129
Ogrudek Ltd
17 W 17th St, New York, NY 212-645-8008
Ommer, Uwe
84 Riverside Dr #5, New York, NY 212-496-0268
Orcutt, Bill
190 E 2nd St #18, New York, NY 212-254-1312
Page, Lee
68 Thomas St #5R, New York, NY 212-233-2227
Pearle, Eric
752 West End Ave, New York, NY 212-662-3022
Pereg, Larry
40 W 17th St, New York, NY 212-645-9136
Pernambuco, Fernando
261 Broadway #12E, New York, NY 212-608-3913
Peterson, Elsa
250 W 100th St #1200, New York, NY 212-678-7768
Peterson, Grant
568 Broadway #1003, New York, NY 212-219-0004
Photographers/Aspen/Nicholas Devore III
40 W 12th St, New York, NY 212-255-8815
Pittman, Dustin
210 W 16th St #2H, New York, NY 212-242-3702
Porto, James
480 Canal St, New York, NY 212-966-4407
Poster Studio, James
210 Fifth Ave #402, New York, NY 212-689-3820
Price, Clayton J
205 W 19th St 2nd Fl, New York, NY 212-929-7721
● **R/GA PRINT**/Jimm Burris **(P 44,45)**
350 W 39th St, New York, NY 212-239-6767

Raabe, Dan
80 Montague St, Brooklyn, NY 718-260-9666
Rapoport, David
55 Perry St #2D, New York, NY 212-691-5528
Richards, Tamarra
537 Underhill Ave, New York, NY 718-328-7440
Roaman, Brad
125 E 87th St, New York , NY 212-996-9498
Roberts, Mathieu
122 E 13th St #3A, New York, NY 212-460-5160
Roche, Alan
270 W 22nd St #4, New York, NY 212-741-1861
Rosenthal, Marshal M
231 W 18th St, New York, NY 212-807-1247
Rosenthal, Mel
740 West End Ave, New York, NY 212-666-8521
Rothstein, Jeff
95 Christopher St #3D, New York, NY 212-924-5898
Rubenstein, Raeanne
8 Thomas St, New York, NY 212-964-8426
Rubin, Susan
4700 Broadway #4A, New York, NY 212-569-4860
Russell, John
144 Sullivan St #23, New York , NY 212-473-6562
Schierlitz, Tom
242 W 38th St 15th Fl, New York, NY 212-764-4832
Scruggs, Jim
32 W 31st St #3, New York, NY 212-629-6495
Sednaoui, Stephane
325 Lafayette St, New York, NY 212-431-1321
Shahidi, Behrooz
551 W 22nd St, New York, NY 212-206-8724
Simon, Peter Angelo
520 Broadway #702, New York, NY 212-925-0890
Skogsbergh, Ulf
5 E 16th St, New York, NY 212-255-7536
Skolnik, Lewis
135 W 29th St 4th Fl, New York, NY 212-239-1455
● **SKYLITE PRODUCTIONS**/John Neitzel
(P 34)
152 W 25th St 12th Fl, New York, NY 212-807-9754
Smith, Stephen C
101 W 12th St #174, New York, NY 212-691-8420
Sorensen, Chris
PO Box 1760 Murray Hill Sta., New York, NY.....212-684-0551
Spaeth, Dana
167 Crosby St, New York, NY 212-941-1892
Spellman Photomontage, Naomi
60 Watts St, New York, NY 212-274-8654
Stockton, Cheryl
257 W 12th St, New York, NY 212-255-4109
● **STUDIO MACBETH (P 50,51)**
232 Madison Ave #402, New York, NY 212-889-8777
Studio Q/Carolyn Quan
285 W Broadway #620, New York, NY 212-343-1108
Taylor, John B
12 E 86th St, New York, NY 212-988-8900
Tcherevkoff, Michel
15 W 24th St 11th Fl, New York, NY 212-229-1733
Tompkins, Bill
25 E 67th St #11D, New York , NY 212-249-2067
Tsutsumi, Adi
301 W 29th St, New York , NY 212-564-8865
Tugboat/James Levin
45 W 21st St, New York, NY 212-242-5337
Urbane USA Inc/Aiko Otake
141 Fifth Ave #8S, New York, NY 212-505-9000
Veldenzer, Alan
160 Bleecker St 10th Fl, New York, NY 212-420-8189
Vickers, Camille
200 W 79th St PH #A, New York, NY 212-580-8649
Viesti, Joe
627 West End Ave, New York, NY 212-787-6500
Wagner Studio/David A Wagner
126 Eleventh Ave 5th Fl, New York, NY 212-243-4800
Wagner, Daniel
231 W 29th St, New York, NY 212-279-4042
Waltzer Digital Service, Carl/Carl Waltzer
873 Broadway #412, New York, NY 212-475-8748
Wang Studio Inc, Tony
18 W 21st St 6th Fl, New York, NY 212-229-1900
Weil, Christopher
438 W 37th St, New York, NY 212-564-0660
Weiser, Barry
336 W 11th St, New York, NY 212-620-4525

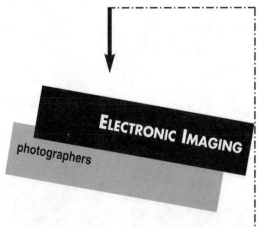

ELECTRONIC IMAGING

photographers

Weissman, Walter
463 West St #B332, New York, NY 212-989-9694
Westheimer, Bill
167 Spring St #2W, New York, NY 212-431-6360
Whitby, Michelle
117 E 57th St, New York, NY 212-223-1063
Williamson Janoff Aerographics/Dean Janoff
514 W 24th St 3rd Fl, New York, NY 212-807-0816
Williamson Photography/Richie Williamson
514 W 24th St, New York, NY 212-807-0816
Witkin, Christian
297 Church St 2nd Fl, New York , NY 212-334-0060
Wong, Leslie
110 W 74th St, New York, NY 212-595-0434

E A S T

Abbey Photographers
416 E Central Blvd, Palisades Park, NJ...............201-947-1221
Abt, Patricia A
PO Box 196, Stuyvesant Falls, NY 518-828-7615
Addis, Kory
144 Lincoln St #4, Boston, MA 617-451-5142
Agelopas, Mike
2510 N Charles St, Baltimore, MD 410-235-2823
Akis, Emanuel
102 Railroad Ave, Hackensack, NJ...................201-342-8070
Andersen-Bruce, Sally
19 Old Mill Rd, New Milford, CT 203-354-0640
Avid Productions Inc
10 Terhune Pl, Hackensack, NJ 201-343-1060
Bakhtiar Photo, Sherry
7927 Inverness Ridge Rd, Potomac, MD...........301-299-2671
Bancroft, Monty
46 Doeview Ln, Pound Ridge, NY 914-764-1554
Bates, Carolyn
20 Caroline St, Burlington, VT 802-865-0445
Bedford Photo Graphic/Doug Abdelnour
Rte 22/PO Box 64, Bedford Village, NY 914-234-3123
Berinstein, Martin
249 A St #31, Boston, MA.............................617-261-4777
Berle Cherney/Visual Productions
2121 Wisconsin Ave NW #470, Wash, DC 202-337-7332
Bernsau, W Marc
PO Box 520, Springvale, ME 207-324-1741
Blakeslee Group
916 N Charles St, Baltimore, MD 410-727-8800
Blate, Samuel R
10331 Watkins Mill Dr, Gaithersburg, MD 301-840-2248
Bodin, Fredrik
5 Wily St, Gloucester, MA 508-281-3771
Bolduc, Damon
PO Box 2168 Main St, Conway, NH 603-539-5101
Bond, Kendra
PO Box 552, Buzzards Bay, MA 508-759-7359
Borkoski Photography, Matthew
912 Loxford Terrace, Silver Spring, MD.............301-681-3051
Bornstein, Myer
13 Green St, Berkley, MA...............................508-824-9365
Boston Photographers
56 Creighton St, Boston, MA..........................617-491-7474
Boudreau, Janice
HC2 Box 32 , Charlemont, MA 413-337-4033
Brown, Jim
PO Box 936, Salem, MA 508-741-2783
Bruno, Patricia J
PO Box 8042, Ward Hill, MA...........................508-374-8037

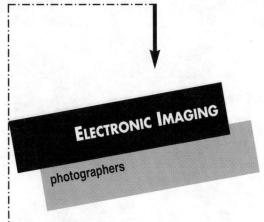

ELECTRONIC IMAGING

photographers

Bubriski, Kevin
PO Box 559, Shaftsbury, VT................802-442-4516
Burdick, Gary
9 Parker Hill, Brookfield, CT................203-775-2894
Buschner Studios
450 W Metro Park, Rochester, NY................716-475-1170
Bush Associates, Diane
328 Ashland Ave, Buffalo, NY................716-882-7371
Business Presentation Services/Mary Clupper
40 Cameron Ave, Somerville, MA................617-666-1161
Callaghan, Charles
214 S 4th St, Perkasie, PA................215-453-9952
Cantor, Phil
209 Montclair Ave, Upper Montclair, NJ................201-783-1065
Castle, Ed
4242 E West Highway #509, Chevy Chase, MD.301-585-9300
Chey, Dong Cheol
233 12th St#1B, Palisades Park, NJ................201-944-0539
Collins & Collins/James Collins
PO Box 10736, State College, PA................814-234-2916
Comb, David
107 South St 2nd Fl, Boston, MA................617-426-3644
Corkum, Paul
75 Donald St #41, Weymouth, MA................617-337-8139
CR 2 Studio Inc
36 St Paul St #601, Rochester, NY................716-232-5140
Crossley, Dorothy
Mittersill Rd, Franconia, NH................603-823-8177
Cunningham, Jerry
RR 2, Box 55, E Corinth, ME................207-228-5708
Daniels, Mark
8413 Piney Branch Rd, Silver Spring, MD................301-587-1727
Davidson, Peter
9R Conant St, Acton, MA................508-635-9780
● **DEAN DIGITAL IMAGING**/Floyd Dean (P 38)
2B South Poplar St, Wilmington, DE................302-655-6992
Delbert, Christian
19 Linell Circle, Billerica, MA................617-273-3138
DiChello, Joseph
Box 432, North Haven, CT................203-562-9981
Diebold, George
416 Bloomfield Ave, Montclair, NJ................201-744-5789
DiGena, Louis
356 First St, Hoboken, NJ................201-963-5339
Dillon Jr, Emile
PO Box 39, Orange, NJ................201-675-5668
Edgerton, Brian
10 Old Rte 28/PO Box 364, Whitehouse, NJ................908-534-9400
El-Darwish, Mahmoud
9112 Kittery Ln, Bethesda, MD................301-469-6383
Electro/Grafiks/Paul Wilson
1238 Callow Hill St #405, Philadelphia, PA................215-923-6440
Ennis, Phillip H
98 Smith St, Freeport, NY................516-379-4273
● **FARR PHOTOGRAPHY**/Lori Farr (P 36,37)
12 Walnut St, Rochester, NY................716-235-1479
Farr Photography/Guy Miller
12 Walnut St, Rochester, NY................716-235-1479
Fastforward Communications Inc/Jaime Martovamo
401 Columbus Ave, Valhalla, NY................914-741-0555
Fatone, Bob
166 W Main St, Niantic, CT................203-739-2427
Feidler, Anita
25 C Patten St, Watertown, MA................617-926-6211
Feiling, David
7804 Ravenswood Ln, Manlius, NY................315-682-7937

Ferreira, Al
237 Naubuc Ave, East Hartford, CT................203-569-8812
Fisher, Elaine
7 McTernan St, Cambridge, MA................617-492-0510
Flynn, Bryce
14 Perry Dr Unit A, Foxboro, MA................508-543-3020
Foremost Comm's/David Fox
360 Woodland St, Holliston, MA................508-820-1130
Fox Hollow Enterprises/Rosalind Morgan
RR 2/Box 2670, Houlton, ME................207-532-7286
Francois, Emmett W
208 Hillcrest Ave, Wycoff, NJ................201-652-5775
Ganson, John
14 Lincoln Rd, Wayland, MA................617-444-5101
Gates, Jeff
2000 Hermitage Ave, Silver Spring, MD................301-949-0436
Gerardi Studios
27 Pippen Pl, New City, NY................914-639-9809
Gidley, EF
43 Tokeneke Rd, Darien, CT................203-655-1321
Gilbert, Thom
31 Farrell, Weston, CT................203-454-0219
Gold, Gary D
454 N Pearl St, Albany, NY................518-434-4887
Hahn, Bob
3522 Skyline Dr, Bethlehem, PA................215-868-0339
Harholdt, Peter
7525-Q Connelley Dr, Hanover, MD................410-766-9809
Helmar, Dennis
416 W Broadway, Boston, MA................617-269-7410
Hemmeon, Bruce
267 Beacon St, Framingham, MA................508-872-3043
Heppert, David
563 Gleasondale Rd, Stow, MA................508-562-7812
Hewitt, Scott
501 Tatnall St, Wilmington, DE................302-654-1397
Hickman, Louis
Box 5358, Plainfield, NJ................908-561-2696
Holmes Agency/Greg Holmes
9153 Brookville Rd, Silver Spring, MD................301-589-1251
Houck, Julie
46 Oakland St, Melrose, MA................617-247-2240
Image Foundry/Tim Fields
3100 Elm Ave, Baltimore, MD................410-889-6660
Impact Studios/Lisa Smart
1084 N Delaware Ave, Philadelphia, PA................215-426-3988
Impact Studios/Scott Fibauer
1084 N Delaware Ave, Philadelphia, PA................215-426-3988
Insight Photo
55 Gill Ln #5, Iselin, NJ................908-283-4727
Iverson Photomicrography, Bruce
7 Tucker St #65, Pepperell, MA................508-433-8429
Johnson, Paul
84 Shepton St, Boston, MA................617-282-0223
Keeley, Chris
4000 Tunlaw Rd NW #1119,
Washington, DC................202-337-0022
Kothari, Sanjay
531 Main St #813, New York, NY................212-319-0807
Kozlowski Productions/Mark Kozlowski
48 Fourth St, Troy, NY................518-274-8512
Kress, Michael
7847 Old Georgetown Rd, Bethesda, MD................301-654-0909
Kristofik, Bob
23 Colonial Rd, Westport, CT................203-454-2541
Kruza, Jay
PO Box M, Franklin, MA................508-528-6211
Landwehrle, Don
9 Hother Ln, Bayshore, NY................516-665-8221
Larsen, Ernie
139 Highland Ave, Meriden, CT................203-982-2003
Leibowitz, David Scott
333 Vandelinda Ave, Teaneck, NJ................201-836-6712
Leonard Still Video/Barney Leonard
134 Cherry Ln, Wynnewood, PA................610-649-5588
Leung, Jook
35 S Van Brunt St, Englewood, NJ................201-894-5881
Lieberman, Fred
2426 Linden Ln, Silver Spring, MD................301-565-0644
Lightworks/George Dodson
692 Genessee St, Annapolis, MD................301-261-2099
Lincon, Denise
19 Newtown Turnpike, Westport, CT................203-226-3724
Lynn, Jenny
18 S Letitia St, Philadelphia, PA................215-925-8967

Mahoney, Bob
347 Cameco Cir, Liverpool, NY................315-652-7870
Mandelkorn, Richard
309 Waltham St, W Newton, MA................617-332-3246
Manning, Ed
875 E Broadway, Stratford, CT................203-375-3384
Marsico, Dennis
110 Fahnestock, Pittsburgh, PA................412-781-6349
Mauro, George
9 Fairfield Ave, Little Falls, NJ................201-890-0880
McCarthy, Tom
73 Glen Cove Drive, Glen Head, NY................516-671-1156
McCormack, Richard
459 Fairmount Ave, Jersey City, NJ................201-435-5836
McCoy, Dan
PO Box 573/1079 Main St, Housatonic, MA................413-274-6211
McGrath, Bonnie
7 Adams St, Acton, MA................508-897-8043
McNamara, Casey
109 Broad St, Boston, MA................617-542-5337
McWilliams, Jack
15 Progress Ave, Chelmsford, MA................508-256-9615
Medvec, Emily
151 Kentucky Ave SE, Washington, DC................202-546-1220
Meehan, Joseph
360 Between the Lakes Rd, Salesbury, CT................203-824-9848
Melton, Janice Munnings
692 Walkhill St, Boston, MA................617-298-1443
Michael Furman Photography Ltd/Michael Furman
115 Arch St, Philadelphia, PA................215-925-4233
● **MICHAEL WEISS PHOTO**/
Michael Weiss Studio Inc (P 16)
251 Guard Hill, Mt Kisco, NY................914-241-3456
Michaud, Brian
2320 Main St, Bridgeport, CT................203-335-0632
Migliozzi, Kathy
7 Wainwright Rd #15, Winchester , MA................617-721-0101
Millenium Group, The/Jose R Garcia
65 Adams Rd, Haydenville, MA................413-268-9391
Miller, J T
12 Forest Edge Dr, Titusville, NJ................609-737-0803
Moser/Media, Michael
2000 P St NW #500, Washington, DC................202-293-1780
Muskie, Stephen O
23 Lookout Hill, Peterborough, NH................603-924-6541
O'Neil, Julie
5 Craigie Cir #7, Cambridge, MA................617-547-5168
Olsen, Dan
280 Summer St, Boston, MA................617-951-0765
Oxberry/Div of Cybernetics Products/Steve Hallett
180 Broad St, Carlstadt, NJ................201-935-3000
Page & Page Slidemarket/Allen Lee Page
703 Maple Hill Dr, Woodbridge, NJ................908-750-4171
Palette Studios
399 Market St, Philadelphia, PA................215-440-0500
Palmieri, Jorge
516½ 8th St SE, Washington, DC................202-543-3326
Parker, Charles A
PO Box 750, Woodstock, VT................802-672-3388
Paskevich, John
2215 Arch St, Philadelphia, PA................215-587-0287
Peirce, George E
133 Ramapo Ave, Pompton Lakes, NJ................201-831-8418
Petronio, Frank
113 Commonwealth, Rochester, NY................716-461-5583
Phaneuf, Arthur P
168 Stage Rd, Plainfield, NH................603-675-9268
Phase One Graphics/Donald Rea
315 Market St, Sudbury, PA................717-286-1111
Pixcel Eyes/Reza Estakhrian
600 Taft Ave, Morrisville, PA................215-295-9141
Pixel Factory/Sergio Levin
25 Grace Terrace, Passaic, NJ................201-472-0650
Polatty, Bo
4309 Wendy Ct, Monrovia, MD................301-865-5728
Pollack, Steven
7th & Ranstead Sts, Philadelphia, PA................215-649-1514
PRAXIS Digital Solutions/Katrin Eismann
PO Box 831, Camden, ME................207-236-7400
Pulling, Nathaniel H
11 Manito Rd Box 608, East Orleans, MA................508-255-0293
Raphael, Dick
239 Causeway St, Boston, MA................617-523-4664
Rezendes, Paul
Bearsden Rd Star Route, S Royalston, MA................508-249-8810

Rhoades, Dean
Highway 30 North, Lake Clear, NY 518-327-5000
Riley, George
PO Box 1025, Yarmouth, ME 207-846-5787
Rivera, Tony
28 Dutch Ln, Ringoes, NJ 908-788-3991
Roberts, Terence
1909 N Market St, Wilmington, DE 302-658-8854
Rowin, Stanley
791 Tremont St #W515, Boston, MA 617-437-0641
Ruggeri, Lawrence
10 Old Post Office Rd, Silver Spring, MD 301-588-3131
● **SCHLOWSKY PHOTOGRAPHY AND COMPUTER IMAGERY**/Bob & Lois Schlowsky (P 20,21)
73 Old Rd, Weston, MA...............617-899-5110
Silver Shadow Images/Stephanie Strain
4 Atwood St #AA1, Hartford, CT 203-246-3062
Simons, Stuart
71 Highland St, Paterson, NJ 201-278-5050
Smith, Brian
7 Glenley Terrace, Brighton, MA...............617-926-8311
Smith, Dick
Box 300, N Conway, NH 603-356-2814
Smith, Rusty
Box 8169 Buckland Station, Manchester, CT........203-644-9733
Smith, Thayer
PO Box 1794 , Waterville, ME...............207-872-6383
Smyth, Kevin
31 Arbor Circle, Basking Ridge, NJ 908-580-1150
Snitzer, Herb
64 Roseland St, Cambridge, MA...............617-497-0251
Sonneville, Dane
PO Box 155, Passaic, NJ 201-472-1225
St Onge, Cheryle
PO Box 67, Durham, NH 603-659-6528
Star Connection/Jim Pickerell
110 Frederick Ave #A, Rockville, MD...............301-251-0720
Steve Greenberg Photography/Steve Greenberg
368 Congress St 5th Fl, Boston, MA 617-423-7646
Sukolsky-Brunelle Inc/David Kull
908 Penn Ave 2nd Fl, Pittsburgh, PA 412-391-6440
Swan Engraving/Stuart Swan
31 Wordin Ave, Bridgeport , CT...............203-366-4308
Swenson, Barbara
376 Wells Rd, Wethersfield, CT...............203-563-2557
The Real Design/Seth Resnick
28 Seaverns Ave #8, Boston, MA 617-983-0291
Turner Photography/Pete Turner Photo
5 Longhouse Westwood Rd, Wainscott, NY 516-537-2434
Tyava Productions/Mary Ross
259 Oak St, Binghamton, NY 607-722-1457
Ulsaker Studio/Michael Ulsaker
275 Park Ave, E Hartford, CT...............203-282-0341
Vanderwarker, Peter
28 Prince St, W Newton, MA 617-964-2728
Ventura, Michael
11722 Highview Ave, Silver Spring, MD...........301-933-5066
Visions/Jim Flanigan
1325 N 5th St #F4, Philadelphia, PA 215-236-4448
Wagner, Marilyn M
91 N Third St, Meriden, CT 203-238-1194
Wasserman, Cary
6 Porter Rd, Cambridge, MA...............617-492-5621
Weems, Al
18 Imperial Pl #5B, Providence, RI 401-455-0484
Weidlein, Peter
19 Vanderbilt Rd, W Hartford, CT 203-231-9009
Weiland, Juliette
67 Signal Hill Rd, Wilton, CT 203-762-8958
Weinrebe, Steve
755 N Taylor, Philadelphia, PA...............215-232-8080
Weston Productions/John Weston
6140 E Molloy Rd, E Syracuse, NY 315-463-9140
Wexler, Jerome
13 Langshire Dr, Madison, CT...............203-245-2396
● **WOJNAR DRAKE**/Lee Wojnar (P 30,31)
326 Kater St, Philadelphia, PA 215-922-5266
Zubkoff Photography, Earl
2426 Linden Ln, Silver Spring, MD 301-585-7393

S O U T H

Aker/Zvonkovic Photography LLP
4710 Lillian St, Houston, TX 713-862-6343
Allen, Don
1787 Shawn Dr, Baton Rouge, LA 504-925-0251

Arday Illus/Don Arday
616 Arbor Creek Dr, De Soto, TX 214-223-6235
Axiom Inc/Craig Guyon
120 S Brook St , Louisville, KY...............502-584-7666
Bachmann Photography/Bill Bachmann
PO Drawer 568248, Orlando, FL...............407-333-9988
Baird, Darryl
5916 Anita Dr, Dallas, TX 214-824-0911
Baker Photography/I Wilson Baker
1094 Morrison Dr, Charleston, SC...............803-577-0828
Bateman, John H
3500 Aloma Ave #D-10, Winter Park, FL 407-671-2516
Berman, Bruce
140 N Stevens #301, El Paso, TX...............915-544-0352
Bilby, Glade
606 Esplanade Ave, New Orleans, LA...............504-949-6700
Britt, Ben
2007 S Ervay St #301, Dallas, TX...............214-421-9846
Brooks, Charles
800 Luttrell St, Knoxville, TN 615-525-4501
Buffington, David
2401 S Ervay #105, Dallas, TX...............214-428-8221
● **BURKEY, J W (P 35)**
1526 Edison St, Dallas, TX...............214-746-6336
Cannedy, Carl
3333 Elm St, Dallas, TX...............214-748-1048
Carney Photography/Dennis Carney
2804 Elm Hill Pike, Nashville, TN...............615-889-5000
Chamowitz, Mel
3931 N Glebe Rd, Arlington, VA...............703-536-8356
Chenn, Steve
6301 Ashcroft, Houston, TX...............713-271-0631
Chisholm Rich & Assoc
6813 Northampton Way, Houston, TX...............713-957-1250
Clear Light Studio
1818-102 St Albans Dr, Raleigh, NC...............919-876-5544
Cook, Jamie
1740 Defoor Pl, Atlanta, GA...............404-351-1883
Crocker, Will
1806 Magazine St, New Orleans, LA...............504-522-2651
Dana Industries
PO Box 641, Louisburg, NC...............919-496-3262
Darshan Associates/Alex Jamison
3501 N 21st Ave, Arlington, VA...............703-536-8980
Debold, Bill
4501 Swiss Ave, Dallas, TX...............214-553-5406
DFW Photography Inc/Donald Wristen
2025 Levee St, Dallas, TX...............214-748-5317
Dietrich, David
115-11 Sweetwater Rd, Horseshoe, NC...............704-891-5235
Digital Imaging Group/Ray Burnette
1120 W 11 St #200, Houston, TX...............713-802-9466
DiVitale Photo Inc/Sandy DiVitale
420 Armour Circle N E , Atlanta, GA...............404-892-7973
● **DREAMTIME SYSTEMS**/
Arni Katz (P 29)
108 Ayers Ave, Marietta, GA...............404-421-8800
Dressler, Brian
111 Northway Rd #A, Columbia, SC...............803-254-7171
Elliot, Tom
19756 Bel Aire Dr, Miami, FL 305-251-4315
Ewasko Studios/Tom Ewasko
5250 Gulfton/Bldg 4/Studio 4, Houston, TX 713-668-2202
Figura, Paul
1555 San Marco Blvd #2A, Jacksonville, FL...............904-398-7285
Garrison, Gary
4921 Jefferson Hwy, New Orleans, LA...............504-734-0916
Gefter, Judith
1725 Clemson Rd, Jacksonville, FL...............904-733-5498
Gomez, Rick
2545 Tigertail Ave, Miami, FL 305-856-8338
Grigg, Roger Allen
PO Box 52851, Atlanta, GA...............404-876-4748
Hand, Ray
10921 Shady Trail #100, Dallas, TX 214-351-2488
Hitchcock, Paul
12865 Kingsway Rd, Wellington, FL 407-795-3228
Hixson, George
PO Box 131322, Houston, TX...............713-225-1421
Horowitz, Jason
2531 N Lexington St, Arlington, VA...............703-532-2531
Impossible Images Inc/JC Bourque
12153 SW 131 Ave, Miami, FL...............305-256-1013
Isgett, Neil
4303-D South Blvd, Charlotte, NC...............704-376-7172

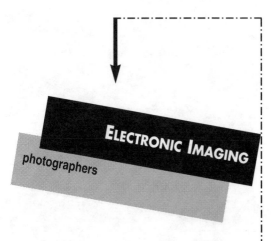

ELECTRONIC IMAGING
photographers

Jamison, Chipp
2131 Liddell Dr NE, Atlanta, GA...............404-873-3636
Joachim Studio Photography, Bruno
1577 Spring Hill Rd #105, Tysons Corner, VA 703-448-9501
Kearney, Mitchell
301 E 7th St, Charlotte, NC...............704-377-7662
Kern, Geof
1355 Conant, Dallas, TX...............214-630-0856
Kohanim, Parish
1130 W Peachtree NW, Atlanta, GA 404-892-0099
Lipson, Stephen
409 Miller Rd, Coral Gables, FL 305-662-8657
Lowery, Ron
409 Spears Ave, Chattanooga, TN...............615-344-3701
Luckett, Julie L
810 Bellevue Rd #280, Nashville, TN 615-646-9581
Marcus, Joel
2250 Ulmerton Rd E #15, Clearwater, FL...............813-573-0575
Marotte, Frank
1028 Northgate Dr W, Irving , TX...............214-650-0811
McDonald, Keith
Box 1157, Alexandria, VA...............703-578-0165
Meredith, Diane
6203 Westcott, Houston, TX...............713-862-8775
Meyler, Dennis
1903 Portsmouth #25, Houston, TX...............713-520-1800
Millington, Rod
PO Box 49286, Sarasota, FL...............813-388-1420
Minardi, Mike
PO Box 14247, Tampa, FL...............813-251-1696
Molina, Jose R
1861 SW 37th Ave, Miami, FL 305-443-1617
Moore Photo, Rick
323 De la Mare Ave, Fairhope, AL...............205-990-8622
Morrissey Photography/Timothy Morrissey
1315 53rd St #3, W Palm Beach, FL...............407-863-6625
Murphy, Michael
4322 El Prado Blvd, Tampa, FL...............813-831-1210
Myers, Jeff
5250 Gulfton/Bldg 4/Studio A, Houston, TX 713-661-9532
NuLight Studio/Rick Bostick
497 Semoran Blvd #105, Casselberry, FL...............407-331-5717
O'Dell, Dale
2040 Bissonnet, Houston, TX...............713-521-2611
Olive, Tim
754 Piedmont Ave NE, Atlanta, GA...............404-872-0500
Olson, Carl
3325 Laura Way, Winston, GA...............404-949-1532
Pearlstone & Assoc
PO Box 6528, San Antonio, TX...............210-826-1897
Phelps, Greg
1245 North Expressway, Brownsville, TX...............210-541-4909
PhotoSonics/Audio Visual
1116 N Hudson St, Arlington, VA...............703-522-1116
Ponzoni Photographic Inc/Bob Ponzoni
703 Westchester Dr, High Point, NC...............910-885-8733
Possenti, Peter
314 W Cary St, Richmond, VA 804-788-1766
Powerhouse Productions/Bob Shaw
1723 Kelly Ave, Dallas, TX...............214-428-1757
Puryear, Jack
11606 DK Ranch Rd, Austin, TX...............512-345-9903
Rodgers, Ted
544 Plasters Ave, Atlanta, GA...............404-892-0967
Roolaart, Harry
2401 Randolph Rd, Charlotte, NC...............704-347-4280

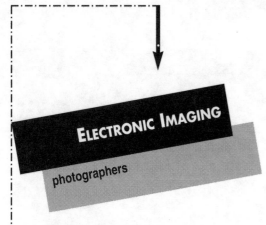

ELECTRONIC IMAGING

photographers

S A & A Inc/Suzanne Anderson & Assocs
17 Executive Park #100, Atlanta, GA 404-636-0134
SANA/Edgehill Studio
17 Executive Park Dr NE #100, Atlanta, GA 404-636-0134
Scott, Ron
1000 Jackson Blvd, Houston, TX 713-529-5868
Seeger, Stephen
2931 Irving Blvd #101, Dallas, TX 214-634-1309
Seitz, Arthur
1905 N Atlantic Blvd, Ft Lauderdale, FL..............305-563-0060
Sense Interactive/Dave Crossley
1412 W Alabama, Houston, TX 713-523-5757
Sherman, Ron
PO Box 28656, Atlanta, GA................................404-993-7197
Smith/Bailey Photography
PO Box 565046, Dallas, TX..............................214-941-4611
Sochurek, Howard
25 Seabreeze Ave #400, Delray Beach, FL 407-243-3691
Staartjes, Hans
34 Lana Ln, Houston, TX 713-621-8503
Stuck, Jon D
6138 Edsall Rd #203, Alexandria, VA 703-354-5158
The Image Bank/The Image Bank
Wms Sq #700/5221 N O'Connor, Irving, TX 214-432-3900
Touchton, Ken
1100 NE 28th Terrace, Pompano Beach, FL.........305-785-0104
Trinity Photo Graphic/Timothy Cracchiolo
8625 Tourmaline Blvd, Boynton Beach, FL 407-736-3916
Truitt Photographics
3201 Stuart, Ft Worth, TX.................................817-924-8783
● **TUCKER, MARK (P 23)**
508 Lea Ave, Nashville, TN 615-254-6802
Uni-Graphix/Monica Harrion
2715 Kingsroad Ave, Jackson, MS 601-355-4008
Washington House Photography/Sharon Swab
5520 Cherokee Ave, Alexandria, VA 703-642-8000
Weeks, Christopher
1260 E 31st Pl, Tulsa, OK................................918-749-1917
Whitehead, Dennis R
1410 N Nelson St, Arlington, VA 703-524-6814
Wiley & Flynn
379 W Michigan St #200, Orlando, FL 407-843-6767

MIDWEST

Altman, Ben
1420 W Dickens St, Chicago, IL 312-935-9007
Amenta Jr, Joseph L
555 W Madison #3802, Chicago, IL 312-248-2488
Ayala Computer Imaging
631 W Stratford Pl, Chicago, IL.........................312-327-1156
Barlow Productions Inc/Ron Barlow Sr
1115 Olivette Executive Pkwy, St Louis, MO 314-994-9990
Bartholomew Enterprises/Gary Bartholomew
433 Golf Rd, Des Plaines, IL.............................708-635-0799
BDG Production Studio/Julie Mikos
213 W Institute Place #702, Chicago , IL............312-440-0240
Bender + Bender
281 Klingel Rd/PO Box 201, Waldo, OH.............614-726-2470
Besser, Doug
4947 N Seeley Ave, Chicago, IL.........................312-334-1521
BKR Studio/Mike Prout
55685 Currant Rd, Mishawaka, IN 219-259-9576
Braun Photography
1245 S Cleve-Mass Rd #216, Akron, OH 216-666-4540
Bruton, Jon
3842 W Pine Blvd, St Louis, MO.......................314-533-6665

Cocchiarella Design/Nino Cocchiarella
22 E Powell Ave, Evansville, IN 812-422-6250
Color Image/Jim Trotter
12342 Conway Rd, St Louis, MO.......................314-878-0777
Conrad, Larry
2009 S 19th St, Milwaukee, WI.........................414-671-1424
Copeland, Burns
6651 N Artesian, Chicago, IL............................312-465-3240
Cossette, Diane I
3516 Emerson Ave S #1, Minneapolis, MN 612-827-0511
Creative Concepts/Doug Crane
94-D Westpark Rd, Centerville, OH 513-436-2020
CSA Archive/Rice Davis
30 N First St, Minneapolis, MN..........................612-339-1263
Dale Photo, Larry
4865 Gallagher Rd, Rochester, MI......................313-656-5151
DeBolt Photography, Dale
120 W Kinzie St, Chicago, IL.............................312-644-6264
Digital Ink/David Garon
409 W Maryland St, Duluth, MN 218-724-3020
Digital Knowledge Corp/Duane Franke
8100 Wayzata Blvd, Golden Valley, MN 612-595-0801
Eclipse/Michael Spaw
8117 Rosewood, Prairie Village, KS 913-864-2515
Elledge, Paul
1808 W Grand Ave, Chicago, IL........................312-733-8021
Ewert, Steve
17 N Elizabeth St, Chicago, IL...........................312-733-5762
Frank Miller Illus & Photo/Frank Miller
6016 Blue Circle Dr, Minnetonka, MN 612-935-8888
Fritz, Tom
2930 W Clybourn, Milwaukee, WI.....................414-344-8300
Gilo, Dave
2009 S 19th St, Milwaukee, WI.........................414-671-1910
Glausen, Judy
213½ N Washington Ave,
Minneapolis, MN..612-332-5009
Graham-Henry, Diane
442 W Belden, Chicago, IL...............................312-327-4493
Griffith, Sam
345 N Canal, Chicago, IL.................................312-648-1900
Halsey Creative Svcs Inc, Dan
12277 Nicollet Ave S, Minneapolis, MN 612-894-2722
Hamill, Larry
77 Deshler, Columbus, OH................................614-444-2798
Hammarlund, Vern
135 Park St, Troy, MI.......................................313-588-5533
Husom, David
1988 Stanford Ave, St Paul, MN 612-699-1858
Image Productions
115 W Church, Libertyville, IL 708-680-7100
Imagine That/Geof Payne
817 Prairie Ln, Columbia, MO 314-443-1995
Imaging, Inc./Harry Przekop Jr
950 W Lake, Chicago, IL...................................312-829-8201
Izquierdo, Abe
213 W Institute #208, Chicago, IL.....................312-787-9784
Izui, Richard
315 W Walton, Chicago, IL...............................312-266-8029
Jacobs Digital Arts, Todd
2437 W Wilson, Chicago, IL.............................312-769-0383
Jilling, Helmut
2735 Eighth St, Cuyahoga Falls, OH 216-928-1330
Jones, Mark
5529 Concord Ave, Minneapolis, MN 612-929-8597
Jung, Mike
87-31 Hamlin, Skokie, IL..................................708-677-2914
K Squared Inc/David Kogan
1242 W Washington Blvd, Chicago, IL...............312-421-7345
K&S Photographics/Ron Caldeleugh
222 N Canal St, Chicago, IL.............................312-207-1212
Kaltman, Len
3708 York Ln, Cincinnati, OH 513-761-6104
Kauffman, Kim
444 Lentz Ct, Lansing, MI.................................517-371-3036
Kean, Christopher
3 S Prospect/Pickwick Bldg, Park Ridge, IL........708-292-1144
Kildow, William
1743 W Cornelia, Chicago, IL...........................312-248-9159
Lacy, John
800 N Kenwood, Royal Oak, MI........................313-548-3842
Lauth Photography/Lyal Lauth
456 N Leavitt, Chicago, IL................................312-829-9800
Lehn & Assocs, John
2601 E Franklin Ave, Minneapolis, MN 612-338-0257

MacDonald, Al
1221 Jarvis Ave, Elk Grove, IL..........................708-437-8850
Mankus, Gary
835 N Wood St #101, Chicago, IL 312-421-7747
Marshall Photo, Don
415 W Huron, Chicago, IL................................312-944-0720
Matrix Photo Labs/Mike Tuttle
118-120 W North St, Indianapolis, IN...............317-635-4756
McCann, Michael K
15416 Village Woods Dr, Eden Prairie, MN........612-949-2407
Meteor Photo & Imaging /Steve Grundner
1099 Chicago Rd, Troy, MI...............................313-583-3090
Meyer & Assocs, Aaron D
1302 W Randolph, Chicago, IL 312-243-1458
Mofoto Graphics/Bob Moore
3651-A Shenandoah, St Louis, MO 314-231-1430
Moy, Willie
364 W Erie, Chicago, IL...................................312-943-1863
Muresan, Jon
6018 Horger St, Dearborn, MI..........................313-581-5445
Nelson, Tom
800 Washington Ave N #301, Minneapolis, MN 612-339-3579
Nielsen, Ron
1313 W Randolph #326, Chicago, IL................312-226-2661
O'Barski, Don
17239 Parkside Ave, S Holland, IL 708-596-0606
Obata Design
1610 Menard, St Louis, MO..............................314-241-1710
Olausen, Judy
213 1/2 N Washington Ave, Minneapolis, MN ..612-332-5009
Perry, Eric
2616 Industrial Row, Troy, MI............................313-280-0640
Petroff, Tom
9 W Hubbard St, Chicago, IL............................312-836-0411
Philpott, Keith
13736 W 82nd St, Lenexa, KS 913-492-0715
Photo Concepts/Jack Forbes
957 Phillips Rd, Cincinnati, OH 513-232-5999
Photonics Graphics/Alan J Brown
700 W Pete Rose Way Lobby B-360,
Cincinnati, OH...513-723-4440
Photos Ink/Medical & Legal Visuals
1819 Minnehaha, Minneapolis, MN 612-371-0140
Popa, Joseph
727 West Briar Pl Ste A3, Chicago, IL................312-248-6464
Prism Studios/Tom Farmer
2412 Nicollet Ave, Minneapolis, MN 800-659-2001
Puza, Greg
PO Box 1522, Milwaukee, WI..........................414-671-1960
Quicksilver Assoc Inc/Kevin Johnson
18 W Ontario St, Chicago, IL............................312-943-7622
Radencich, Mike
Rt 2/Box 185B, Holt, MO.................................816-264-2542
Ray C Photographics
2241 S Michigan Ave, Chicago, IL....................312-842-3835
RE Miller Communications/Roger E Miller
3520 W 21st St, Minneapolis, MN.....................612-925-0781
● **SANDBOX DIGITAL PLAYGROUND**/
Sandy Ostroff (P 84)
203 N Wabash #1602, Chicago, IL...................312-372-1170
Sanderson, Glenn
1000 Pine St, Green Bay, WI............................414-437-6200
Sladcek Studio/William K Sladcek
215 W Illinois, Chicago, IL................................312-644-7108
Solomon, Paul
354 Buttles Ave, Columbus, OH 614-421-7448
Spectra Action/Lewis Portnoy
5 Carole Ln, St Louis, MO.................................314-567-5700
Systems/Will Bannister
849 W Lill Ave #K, Chicago, IL..........................312-327-2143
Thomas, Tony
676 N Lasalle St 6th Fl, Chicago, IL...................312-337-2274
Tillis, Harvey S
1050 W Kinzie St, Chicago, IL...........................312-733-7336
Univ of Ill at Champaign/Theodore Zernich
Dept of Art/408 E Peabody, Champaign, IL.........217-333-7713
Vera, Clara Claudia
1446 W Polk St 2nd Fl, Chicago, IL...................312-227-9372
Wooden, John
219 N 2nd St #306, Minneapolis, MN 612-339-3032
Yaworski, Don
600 White Oak Ln, Kansas City, MO.................816-454-4011
Young McKenna & Assocs Inc/Christine Young
2106 Maple Ave, Evanston, IL..........................708-866-9388
Zamiar Photography/Thomas Zamiar
210 W Chicago Ave, Chicago, IL 312-787-4976

21st Century Media/Denise Schwend
2013 Fourth Ave, Seattle, WA............................800-528-3472
Abaci Gallery of Computer Art/Daria Barclay
312 NW Tenth Ave, Portland, OR..................503-228-8642
Andrew Rodney Photography/Andrew Rodney
30 Gavilan Rd, Santa Fe, NM.........................505-983-1933
Arens & Associates/William Arens
7915 Silverton Ave #301, San Diego, CA 619-536-9051
ARticulate Imaging/Dovida
849 Mission Way, Sacramento, CA...................916-736-2108
ASA Darkrooms
905 N Cole Ave #2, Los Angeles, CA................213-463-7513
Ashley, Bruce
Box 2955, Santa Cruz, CA...............................408-429-8300
Baldwin, Doug
216 S Central Ave, Glendale, CA 818-547-9268
Balfour Walker Photography/Balfour Walker
1838 E 6th St, Tucson, AZ...............................602-624-1121
Base Arts/John Sappington
1060 Malone Rd, San Jose, CA 415-821-4989
Beals, Steven K
807 N Russell St, Portland, OR.......................503-288-0550
Biggs, Ken
1147 N Hudson Ave, Hollywood, CA................213-462-7739
Blair Photography, Richard
2207 Fourth St, Berkeley, CA..........................510-548-8350
Blaustein, Alan
154 Harvard Ave, Mill Valley, CA 415-383-1511
Brookhouse, Winthrop
473 Jefferies/PO Box 3502, Big Bear Lake, CA...909-866-7020
Brown, Ron
33 East 2750 South, Bountiful, UT................800-992-3801
Burke/Triolo Productions
8755 Washington Blvd, Culver City, CA............310-837-9900
Burns & Assocs Inc
2700 Sutter St, San Francisco, CA..................415-567-4404
Burt, Pat
1412 SE Stark, Portland, OR...........................503-284-9989
Busher, Dick
7042 20th Pl NE, Seattle, WA 206-523-1426
Bybee Digital Studios, Gerald
1811 Folsom St, San Francisco, CA..................415-863-6346
Caesar Photo Design Inc/Lima Caesar
21350 Nordhoff St #103, Chatsworth, CA 818-718-0878
Cal Poly State Univ/Eric R Johnson
Art Department, San Luis Obispo, CA 805-544-5121
Campbell Studio, Thomas /Thomas Campbell
6442 Santa Monica Blvd #204, Hollywood, CA.800-400-1070
CDS Design/Colin Sims
645 E Speedway, Tucson, AZ...........................602-770-1193
Cinque, Michael
PO Box 1322, Aptos, CA..................................408-423-8342
Collector, Stephen
4209 N 26th St, Boulder, CO 303-442-1386
CPC Studios/George Aiello
PO Box 6424, Concord, CA..............................510-627-9350
Crosier, Dave
1201 First Ave S #327, Seattle, WA..................206-682-6445
Cruz, Frank
1855 Blake St #2E, Denver, CO.......................303-297-1007
D'Hamer, M
PO Box 5488, Berkeley, CA..............................510-530-8433
● **DAMAN STUDIO, TODD (P 25)**
2019 Third Ave #200, Seattle, WA..................206-448-9318
Daniels, Charles
905 N Cole Ave #2120, Hollywood, CA............213-461-8659
Dazzleland Digital
611 Fifth Ave, Salt Lake City, UT.....................801-355-8555
Digit Eyes/Rik Henderson
11231 Otsego, N Hollywood, CA.......................818-769-3021
Digital Foto Arts/John Elgin
20812 Vose St, Canoga Park, CA......................818-347-7719
● **DIGITAL STOCK PROFESSIONAL**/Charles Smith **(P 99)**
400 S Sierra Ave #100, Solana Beach, CA.........619-794-4040
Digital Vision
55 Francisco St, PH, San Francisco, CA 415-397-2700
Direct Images/Bill Knowland
PO Box 29392, Oakland, CA.............................510-614-9783
Dow, Carter
1041 25th St, San Francisco, CA......................415-824-5660
Dressler, Rick
1322 Bell Ave #M, Tustin, CA..........................714-259-9113
Duke, William
90 Hillview Ave, Los Altos, CA.........................415-949-1344

Dyna Pac
7926 Convoy Ct, San Diego, CA......................619-560-0117
Electronic Images/Nick Fain
300 Broadway #32, San Francisco, CA............415-398-3434
Familian, David
1031 25th St, Santa Monica, CA....................310-829-9636
Farace Photography & Comm/Joe Farace
14 Inverness Dr E #B-104, Englewood, CO 303-799-6606
Farmer, Roscoe
PO Box 422833, San Francisco, CA 415-474-8457
Felzman Photography, Joe
4504 SW Corbett Ave #120, Portland, OR 503-224-7983
Fox, John
211 E Carrillo #205, Santa Barbara, CA............805-564-8499
Frank, Brooks
229 Duncan St #1, San Francisco, CA..............415-641-5904
Franklin, Charly
3352 20th St, San Francisco, CA......................415-824-4000
Fry III, George B
PO Box 2465, Menlo Park, CA.........................415-323-7663
Gage, Hal
2008 E Northern Lights Blvd, Anchorage, AK 907-272-4356
Garber, Helen Kolikow
2607 Second St #1, Santa Monica, CA............310-396-4982
Gerry McIntyre Photography/
Gerry McIntyre
849 Mission Way, Sacramento, CA....................916-736-2108
Ginsberg, Robin
2475 Third St #257, San Francisco, CA 415-621-1215
Goavec Photography/Pierre Yves Goavec
340 Harriet St, San Francisco, CA 415-552-7331
Graves, Tom
296 Randall St, San Francisco, CA...................415-550-7241
Greene Productions, Jim
PO Box 2150, Del Mar, CA..............................619-454-4133
Haislip, Kevin
PO Box 1862, Portland, OR.............................503-254-8859
Hamilton, Jeffrey Muir
6719 N Quartzite Canyon Pl, Tucson, AZ 602-299-3624
Hart Productions Inc, Geb
8659 Hayden Pl, Culver City, CA......................310-280-0650
Haynes, Barry
820 Memory Ln, Boulder Creek, CA 408-338-4569
Highton, Scott
996 McCue Ave, San Carlos, CA......................415-592-5277
Hollenbeck, Cliff
Box 4247 Pioneer Sq, Seattle, WA 206-443-1333
Honowitz, Ed
2265 E Foothills Blvd, Pasadena, CA...............818-405-1425
Hopkins, Paul
747 Wilcox Suite 201, Los Angeles, CA............213-469-3336
● **HUNT, STEVEN (P 22)**
2098 E 25th St S, Layton, UT 801-544-4900
Image Bank, The
2400 Broadway #220, Santa Monica, CA 310-264-4850
Imagic
1545 N Wilcox Ave, Hollywood, CA 213-461-7766
Ingham Photo Inc, Stephen
2717 NW St Helens Rd, Portland, OR 503-274-9788
Johnson, Stephen
607 Carmel Ave, Pacifica, CA 415-355-7507
Jones, Aaron
3 Vaquero Rd, Santa Fe, NM 505-988-5730
Kaake, Phillip
1987 El Dorado Ave, Berkeley, CA...................510-525-6297
Katzenberger, George
211-D E Columbine St, Santa Ana, CA 714-545-3055
KAZ
215 Adobe Canyon Rd, Kenwood, CA.............707-833-2536
Keenan Assocs/Larry Keenan
7101 Saroni Dr, Oakland, CA..........................510-339-9733
Kelley Studios Inc, Tom
2472 Eastman Ave #35-36, Ventura, CA 805-658-9908
Kemper, Lewis
3251 Lassen Way, Sacramento, CA 800-925-2596
Kerns, Ben
1201 First Ave S #321, Seattle, WA.................206-621-7636
Kimball, Ron
1960 Colony, Mt View, CA..............................415-969-0682
Kirkland, Douglas
9060 Wonderland Park Ave, Los Angeles, CA.....213-656-8511
Koosh, Dan
PO Box 6038, Westlake Village, CA..................818-991-2105
Krasner, Carin
3239 Helms Ave, Los Angeles, CA 310-280-0082

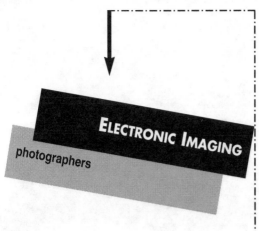

ELECTRONIC IMAGING
photographers

Lamotte Studios Inc/Michael Lamotte
828 Mission St, San Francisco, CA......................415-777-1443
Land, Fred
12612 Rose Ave #17, Los Angeles, CA 310-398-6250
Landreth Studios
2020 Airport Way S, Seattle, WA 206-343-7118
LaTona, Kevin
159 Western Ave W #454, Seattle, WA 206-285-5779
Lawson, Pamela
8809 Oakwild Lane, Los Angeles, CA................213-656-3886
Laxer, Jack
16952 Dulca Ynez Ln, Pacific Palisades, CA 310-459-1213
Leng, Brian
1021½ N La Brea, Los Angeles, CA.................213-850-0995
Lockwood, Scott
2109 Stoner Ave, Los Angeles, CA...................310-312-9923
Lund, John
860 Second St, San Francisco, CA....................415-957-1775
Maisel, David
180 Sunset Way , Muir Beach, CA....................415-380-8700
Maloney, Jeff
2415 De La Cruz Blvd, San Jose, CA................408-980-8004
Manning, Lawrence
2506 Huntington Ln #1, Redondo Beach, CA 310-379-0977
Marley, Stephen
1160 Industrial Way #5, San Carlos, CA...........415-595-4226
Martin Brinkerhoff Associates/
Randall Lubert
17767 Mitchell, Irvine, CA 714-660-9396
Maxon, Paul
4812 Hillsdale Dr, Los Angeles, CA 213-224-8685
● **MC2**/Heidi Merscher **(P 48)**
65 Lower Arroyo Hondo Rd,
Arroya Hondo, NM..505-776-1333
McClain, Craig
9587 Tropicana Dr, La Mesa, CA 619-469-9599
Michaels, Christion
3580 Utah St, San Diego, CA...........................619-574-7686
Michaels, Robert
6716 Zumirez Dr, Malibu, CA 310-457-5859
Michels Photo, Bob
821 Fifth Ave N, Seattle, WA 206-282-6894
Miller, Jordan
506 S San Vicente Blvd, Los Angeles, CA...........213-655-0408
Mitchell Photo, Josh
8033 Sunset Blvd #490, Los Angeles, CA..........213-227-1000
Molenhouse, Craig
PO Box 7678, Van Nuys, CA 818-901-9306
Montano, Daniel
1616 17th St #369, Denver, CO 303-628-5440
Moojedi, Kamran
900 W Sierra Madre #122, Azusa, CA...............818-969-5508
Morgan Inc., Scott
711 Hampton Dr, Venice, CA 310-392-1863
● **MORRELL, PAUL (P 17)**
300 Brannan St #207, San Francisco, CA...........415-543-5887
Mulkey, Roger
3042 Fillmore St, San Francisco, CA.................415-346-7343
Multimedia Professionals
5835 Avenida Encinas #127, Carlsbad, CA 619-929-8788
Murray, Bill
PO Box 2442, Redmond, WA 206-441-2154
Mustafa Bilal Photography/Mustafa Bilal
5429 Russell NW, Seattle, WA 206-782-4164
O'Toole, Terence
255 W 700 St, Salt Lake City, UT......................801-355-5970

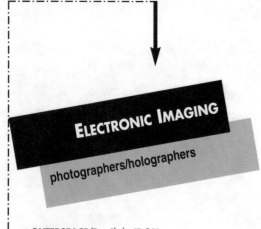

ELECTRONIC IMAGING
photographers/holographers

● **OUTERSPACE**/Tom Slatky **(P 91)**
  1119 Colorado Ave #110, Santa Monica, CA 310-394-5031
Outwater Prods, Robert
  14926 Calvert, Van Nuys, CA 818-994-5406
Parian, Levon
  5769 W Venice Blvd, Los Angeles, CA 213-934-3685
Parks, Peggy
  21 Broadview Dr, San Rafael, CA 415-457-5300
Perception Labs/Rick English
  1162 Bryant St, San Francisco, CA 415-255-0751
Percey, Roland
  626 N Hoover, Los Angeles, CA 213-660-7305
● **PETERSON PHOTOGRAPHY, BRUCE**
(P 11)
  6105 S Ash Ave #A3, Tempe, AZ....................602-820-8965
Pfleger, Mickey
  PO Box 280727, San Francisco, CA 415-355-1772
Phaedrus Productions/Dennis Dunbar
  714 Hill St #2, Santa Monica, CA 310-450-9319
Photogroup
  3500 SE 22nd Ave Bldg 41, Portland, OR 503-797-7817
Photosynthesis/Michael James
  1555 8th Ave #9, San Diego, CA 619-231-9061
Pitts, Tom
  1045 17th St #A, San Francisco, CA...............415-861-7053
Plager, Doug
  2320 First Ave 2nd Fl, Seattle, WA 206-441-5366
Pleasant, Ralph B
  1131 Alta Loma Rd # 529, W Hollywood, CA 310-289-1097
Powell, Todd
  PO Box 4097, Frisco, CO 303-668-2280
Print Expression/Phil Tekunoff
  2939 E Broadway, Tucson, AZ......................602-327-0077
Pro Photo Productions/Neil Anderson
  2921 N Birch Pl, Fullerton, CA....................714-990-0405
Randall, Robert
  15626 156th Pl SE, Renton, WA 206-228-1933
RB Photographic/Reddie Henderson
  3837 E Wier #5, Phoenix, AZ 602-437-4545
Riggs, Robin
  3785 Cahuenga W, Studio City, CA.................818-506-7753
Robinson, Dave
  138087 Cascadian, Everett, WA 206-743-9833
Roden, Bill
  670 Sinex Ave, Pacific Grove, CA..................408-649-8462
Rosen, Barry
  733 N Kings Rd, Los Angeles, CA 213-653-2043
Ross, Bill
  7056 Fletcher Bay Rd, Bainbridge Island, WA 206-842-7722
Ross, Ken
  PO Box 6168, Scottsdale, AZ.......................602-319-2974
Sacred Land Photography/C A Pavlinac
  PO Box 613, Mill Valley, CA.......................415-381-9685
The San Diego Convention & Visitors
Bureau/Jim Hance
  401 B St #1400, San Diego, CA 619-557-2852
Sanford, Ron
  PO Box 248 , Gridley, CA..........................916-846-4687
Scharf, David
  2100 Loma Vista Pl, Los Angeles, CA..............213-666-8657
Scoggins, Anthony
  701 Island Ave #3C, San Diego, CA 619-338-0543
Segal Photography, Susan
  11738 Moor Pk #B, Studio City, CA................818-763-7612
Semple, Rick
  1115 E Pike, Seattle, WA 206-325-1400

Shaner, Tom
  323 Main St 2nd Fl Box 311, Park City, UT........801-649-4499
Sholik, Stan
  1946 E Blair Ave, Santa Ana, CA 714-250-9275
Silva, Keith
  771 Clementina Alley, San Francisco, CA 415-863-5655
Sinick, Gary
  3246 Ettie St, Oakland, CA........................510-655-4538
Sokol, Mark
  6518 Wilkinson Ave, N Hollywood, CA 818-506-4910
Spaceshots Inc/Greg Lytle
  12255 Foothill Blvd, Sylmar, CA 818-890-7870
Spellman, David
  618-A Moulton Ave, Los Angeles, CA 213-223-7770
Stapleton, Kevin
  712 S Grand View St, Los Angeles, CA.............213-487-1609
Steele, Melissa
  PO Box 280727, San Francisco, CA 415-355-1772
Stone Images, Tony
  6100 Wilshire Blvd #1250, Los Angeles, CA 800-234-7880
Sugar, James
  45 Midway Ave, Mill Valley, CA....................415-388-3344
Sund, Harald
  PO Box 16466, Seattle, WA 206-938-1080
Tachibana, Kenji
  1605 12th St #10, Seattle, WA 206-325-2121
Tarleton, Gary
  PO Box 183, Corvallis, OR 503-752-3759
Taylor Photo, Dan
  3131 Western Ave #513, Seattle, WA 206-285-5662
Thurlbeck, Ken
  990 Del Rey Ave, Pasadena, CA 818-797-7454
Timescape Image Library
  12700 Ventura Blvd, Studio City, CA 818-508-1444
Todd Johnson Productions/Todd Johnson
  647 Angelus Pl, Venice, CA 310-577-1767
Torrance, Scott
  1627 Camden #204, Los Angeles, CA 310-444-9033
Trailer Photo, Martin
  8615 Commerce Ave, San Diego, CA 619-549-8881
Turner & DeVries
  701 N Kalaheo Ave, Kailua, HI.....................808-261-2179
● **UPTON, THOMAS (P 12,13)**
  1879 Woodland Ave, Palo Alto, CA.................415-325-8120
Vander Houwen, Greg
  PO Box 498, Issaquah, WA 206-999-2584
VanderSchuit Studios Inc, Carl
  751 Turquoise St, San Diego, CA 619-539-7337
● **VARIS PHOTOMEDIA**/Lee Varis **(P 26,27)**
  922 N Formosa Ave, Los Angeles, CA 213-874-0129
Wahlstrom, Richard
  650 Alabama St 3rd Fl, San Francisco, CA 415-550-1400
Warren, William James
  509 S Gramercy Place, Los Angeles, CA 213-383-0500
Weinberg & Clark
  160 E Dana St, Mountain View, CA 415-962-1752
● **WESTLIGHT INC**/WestLight New Media
(P 94)
  2223 S Carmelina Ave, Los Angeles, CA 800-622-2028
Wexler, Glen
  736 N Highland Ave, Los Angeles, CA 213-465-0268
White, Lee
  14005 Sylvan St, Van Nuys, CA....................818-779-1002
Widstrand, Russ
  5423 Marburn Ave, Los Angeles, CA...............213-298-7470
William Stetz Design/William Stetz
  1108 Seward St, Los Angeles, CA..................213-461-4267
Williams, Wayne
  15423 Sutton St, Sherman Oaks, CA 818-905-8097
York Photo Imagery
  2867-1/2 W 7th St, Los Angeles, CA...............213-387-2489

CANADA

Aartvark Photography/Jean-Luc Denat
  134 St Paul St, Ottawa Canada, ON...............613-748-5357
Allen, Jim
  245 Carlaw Ave #505, Toronto, ON...............416-778-0910
Cooper, David
  950 Powell St #202, Vancouver Canada, BC 604-255-4576
Cooper, Ken
  1730 Vernon Dr, Vancouver, BC....................604-253-3315
Dojc, Yuri
  74 Bathurst St, Toronto Canada, ON 416-366-8081
Etkin, Rick
  21 E 5th Ave, Vancouver Canada, BC 604-875-0535

Fine Art/Andrew Wysotski
  1224 Cedar St, Oshawa, ON 416-721-9488
French, Graham
  388 Carlaw Ave #303, Toronto, ON 416-778-7922
Habal, Jan
  1020 Hamiton St #203, N Vancouver, BC 604-988-6970
Kent Kallberg Studios/Kent Kallberg
  1138 Homer, Vancouver , BC.......................604-876-5116
Wiley, Matthew
  483 Eastern Ave, Toronto, ON 416-462-0112

INTERNATIONAL

Cordero, Felix L
  1259 Ponce DeKeon #6-5, Santurce, PR............809-725-3465

Holographers

NEW YORK CITY

Aust, Hale
  521 E 12th St #11, New York, NY 212-982-3713
Berkhout, Rudie
  223 W 21st St, New York, NY......................212-255-7569
Dimension 3 Inc/Craig Diamond
  11 Maiden Ln 2nd Fl, New York, NY 212-732-2176
Fisch, Amy
  718 Broadway, New York, NY......................212-677-3509
Flatow, Carl
  20 E 30th St, New York, NY 212-683-8688
Fornari, Arthur
  813 Eighth Ave, Brooklyn, NY 718-965-3956
Holographic Studios
  240 E 26th St, New York, NY 212-686-9397
Holographics Inc/Ana Maria Nicholson
  44-01 11th St, Long Island City, NY 718-784-3435
Laser Light Ltd
  28 Old Fulton, Brooklyn Hts, NY 718-226-7747
Maxedon, Terry
  718 Broadway, New York, NY......................212-677-3509
Mediainterface, Ltd./Ronald Erickson
  215 Berkeley Pl, Brooklyn, NY 718-398-1136
New York Holographic Labs
  PO Box 20391/Tompkins Sq, New York, NY.......212-254-9774
Vila, Doris
  157 E 33rd St, New York, NY 212-686-5387

EAST

Acme Holography
  12 Sunset Rd, W Somerville, MA...................617-623-0578
American Bank Note
  399 Executive Blvd, Elmsford, NY 914-592-2355
Art Science & Tech Inst/
Holography Collection/Laurent Bussaut
  2018 R St NW, Washington, DC...................202-667-6322
Blank, Pat
  337 Vesta Ct, Ridgewood, NJ 201-447-2931
Bush, Edward
  403 Highland Ave, Clifton, NJ 201-473-6718
Casdin-Silver, Harriet
  51 Melcher St #501, Boston, MA...................617-423-4717
Clark University
  Dept of Visual/Performing Arts, Worcester, MA....508-793-7113
Defreitas, Frank
  PO Box 9035, Allentown, PA 800-458-3525
Diffraction Ltd/Julie Parker
  PO Box 1115, Waitsfield, VT 802-496-6642
Dimensional Foods Corp
  8 Faneuil Hall, Boston, MA 617-973-6465
Direct Holographics
  P.O. Box 295, Strasburg, PA 717-687-9422
Hologram Research Inc
  25 E Loop Rd, Stony Brook, NY....................516-444-8839
Holographic Applications Inc
  21 Woodland Way, Greenbelt, MD.................301-345-4652
Holographics North Inc
  444 S Union St, Burlington, VT.....................802-658-2275
Jurewicz, Arlene
  RR 1/Box 4235, Lincolnville, ME 207-763-3182
Kurzen, Aaron
  PO Box 3233, Stony Creek, CT 203-488-4711
Law Holographics/Linda E Law
  234 Main st, Huntington Station, NY 516-673-3138
Nemtzow, Scott
  242 E Highland Ave, Philadelphia, PA 215-242-2848
NG Visual Art/Geza Nemeth
  185 Summit Ave, Upper Montclair, NJ 201-744-0806

Oxberry/Div of Cybernetics Products/
Steve Hallett
180 Broad St, Carlstadt, NJ 201-935-3000
Rothaus, Valli
PO Box 72, Carversville, PA 215-297-8490
Stockler, Len
7227 Eastwood St, Philadelphia, PA 215-331-5067
Toppan Printing Co
1100 Randolph Rd, Somerset, NJ 908-469-8400
Yeager, Carol
5761 Cauterskill Rd, Catskill, NY 518-943-2007

S O U T H

Holographic Images Inc
1301 Dade Blvd, Miami Beach, FL 305-531-5465
Kline, Daniel A
1810 Airport Blvd, Cayce, SC 803-739-0592
Laser Co/James H Bowman
1900 Gore Dr, Haymarket, VA 703-754-2526
Zimmerman, Sharon
1704 Lookout Point Ct, Raleigh, NC 919-848-0452

M I D W E S T

Berglund, Lawrence
3840 10th Ave SW, Rochester, MN 507-284-5784
Chicago Schl of Holography/Loren Billings
1134 W Washington Blvd, Chicago, IL 312-226-1007
Chromagem Inc
573 S Schenley, Youngstown, OH 216-793-3515
Croydon, Michael
950 Benson Ln, Green Oaks, IL 708-367-8248
Dimensional Imaging Cons Inc/Douglas Tyler
111 N 2nd St, Niles, MI 616-683-0934
Fine Arts Research & Holographic Ctr
1134 W Washington Blvd, Chicago, IL 312-226-1007
Holomat/Matthew E Hansen
741 E Gorham St, Madison, WI 608-255-3580
Lasersmith Inc/Steve Smith
1000 W Monroe St, Chicago, IL 312-733-5462
MacShane Laser Art/Holography
512 Braeside Dr, Arlington Hts, IL 708-398-4983
Vance, Bill
1277 Roxbury Rd, Rockford, IL 815-399-3314

W E S T

Alexander
2311 4th St #306, Santa Monica, CA 310-393-9846
Briones Engineering/Robert Briones
6270 Galway Dr, Colorado Springs, CO 719-598-6942
CFC Applied Holographics
1721 Fiske Pl, Oxnard, CA 800-438-4656
Cherry Optical Holography
2047 Blucher Valley Rd, Sebastopol, CA 707-823-7171
COMDISC
1508 Cotner Ave, Los Angeles, CA 310-479-0899
Conner Holographic Services/Arlie Conner
5777 SW Calusa Loop, Tualatin, OR 503-691-9452
Cross, Lloyd
PO Box 672, Gualala, CA 707-884-9139
Deem, Rebecca
709-1/2 Glen Oaks Blvd, Glendale, CA 818-549-0534
Feroe Holographic Consulting /James R Feroe
1420 45th St #33, Emeryville, CA 510-658-9788
Freund, Art
124 Brookwood Dr, Santa Cruz, CA 408-426-8382
Gorglione Holography, Nancy
2047 Blucher Valley Rd, Sebastopol, CA 707-823-7171
Holography Institute
PO Box 24-153, San Francisco, CA 415-822-7123
Iylem: Artists Using Sci & Technology
PO Box 749, Orinda, CA 510-482-2483
James, Randy
503 Caledonia St, Santa Cruz, CA 408-458-4213
Jex Fx/Gary Platek
47 Paul Dr #9, San Rafael, CA 415-499-9477
Kaufman, John
PO Box 477, Pt Reyes Station, CA 415-663-1216
Lasart Ltd
PO Box 703, Norwood, CO 303-327-4701
Man/Environment Inc
2251 Federal Ave, Los Angeles, CA 310-477-7922
Muskovitz, Aaron
S Lake/PO Box 1022, Tahoe, CA 916-544-5989
Neovision-Visions/Bill Hilliard
PO Box 74277, Los Angeles, CA 213-387-0461

Northern Lightworks
2148 N 86th St, Seattle, WA 206-526-5752
Pink, Patty
PO Box 24153, San Francisco, CA 415-822-7123
Ralcon Development Lab/Richard Rallison
8501 S 400 W/Box 142, Paradise, UT..............801-245-4623
Red Beam Inc
9011 Skyline Blvd, Oakland, CA 510-482-3309
Stephens, Anait Arutunoff
1685 Fernald Point Ln, Santa Barbara, CA.........805-969-5666

C A N A D A

Dyens, Georges M
5982 Ave Durocher, Autremont, QB 514-278-4593
Holographic Studios Ltd
2525 York, Vancouver, BC..............................604-734-1614

Retouchers

N E W   Y O R K   C I T Y

Beckman, Melissa
95-02 72nd Ave #2, Forest Hills, NY 718-793-7286
Boucek, Al
37 W 20th St #902, New York, NY 212-929-5100
Color Wheel Inc
227 E 45th St 6th Fl, New York, NY 212-697-2434
Designers Atelier/Dave Perskie
45 W 45th St #808, New York, NY 212-221-8585
Digital Photo Works/Marc Silverman
62 W 70th St, New York, NY 212-595-5464
EMR Systems Comm/Frank Santisi
134 W 26th St, New York, NY 212-989-4442
FCL/Colorspace/Terry Chamberlain
10 E 38th St 3rd Fl, New York, NY 212-679-9064
FCL/Colorspace/Terry Chamberlain
10 E 38th St 3rd Fl, New York, NY 212-889-7787
Graphic Systems/Bill Hufstader
33 W 17th St, New York, NY 212-242-8787
Graphics for Industry/Mark Pahmer
8 W 30th St 7th Fl, New York, NY 212-889-6202
HBO Studio Productions/Judy Glassman
120A E 23rd St, New York, NY 212-512-7800
Image Axis Inc/Kevin O'Neill
38 W 21st St, New York, NY 212-989-5000
Image Network/Chuck Dorris
532 Broadway 5th Fl, New York, NY.................212-274-1621
Imaging Consortium Inc/Jack Yeh
1200 Broadway #2E, New York, NY 212-689-2008
Magic Graphics/Carmine Corinello
424 W 33rd St 12th Fl, New York, NY 212-947-1942
Norkin Digital Art/Scott Norkin
43 W 24th St #11A, New York, NY 212-229-0112
Show & Tell
39 W 38th St, New York, NY 212-840-2912
Site One/Oya Demirli
44 W 28th St 4th Fl, New York, NY 212-447-1516
State of the Arts Graphic Svc/Marlene Cohen
540 E 20th St, New York, NY 212-777-2599
● **STUDIO MACBETH (P 50,51)**
232 Madison Ave #402, New York, NY 212-889-8777
Tony Giordano/
Potomac Graphic Industries Inc
508 W 26th St, New York, NY 212-924-4880
Waltzer Digital Service, Carl/Carl Waltzer
873 Broadway #412, New York, NY 212-475-8748
Wood, Joan
160 W 24th St #6E, New York, NY 212-929-5309
ZER/Tony Giordano
508 W 26th St, New York, NY 212-463-0494

E A S T

Baquero, George
4 Westlay Ln, New Milford, NJ 201-261-6011
Chromagen Digital Imaging/Philip Bryan
1100 University Ave #141, Rochester, NY..........716-256-6218
Cicchetti, John
33 Stratford Rd, Scarsdale, NY 914-723-6305
Color Logic/Jim Longo
299 Webro Rd, Parsippany, NJ 201-515-0099
Daniels Printing Co/Esther Fisher
40 Commercial St, Everett, MA 617-389-7900
Diette, Paul
281 Summer St, Boston, MA 617-542-1300
Dunning, Chris
33 Stratford Rd, Scarsdale, NY 914-934-0517

Fumi Color Engineers/Elaine Segal
337 Summer St 3rd Fl, Boston, MA....................617-338-5409
Graphic Consortium/Dominic Gazzara
924 Cherry St, Philadelphia, PA 215-923-3200
Gundlach, Elisabeth
17 Ferndale St, Albany, NY 518-482-4375
Hess, Robert
63 Littlefield Rd, E Greenwich, RI 401-885-0331
Janesko, Lou
10400 Connecticut Ave #PH, Kensington, MD 301-933-5533
Phase One Graphics/Donald Rea
315 Market St, Sudbury, PA 717-286-1111
Rivera, Tony
28 Dutch Ln, Ringoes, NJ 908-788-3991
Santo, Vincent
2 Skibo Ln, Mamaroneck, NY 914-698-4667
● **SCHLOWSKY PHOTOGRAPHY AND
COMPUTER IMAGERY**/Bob & Lois Schlowsky
(P 20,21)
73 Old Rd, Weston, MA.................................617-899-5110
Steuer, Sharon
205 Valley Rd, Bethany, CT............................203-393-3981
Young Assoc, Robert
78 N Union St, Rochester, NY 716-546-1973

S O U T H

Arenella Inc/Joe Arenella
2970 Brandywine Rd #121, Atlanta, GA............404-936-9200
Art for Advertising Inc/Michael Block
4011 W Plano Parkway #102, Plano , TX...........214-596-1060
Blackhawk Color Corp/Leann Sanderson
13900 49th St N, Clearwater, FL......................813-535-4641
Carney Photography/Dennis Carney
2804 Elm Hill Pike, Nashville, TN 615-889-5000
Color Place, The/Jim Moss
1330 Conant St, Dallas, TX.............................214-631-7174
Dallas Photo Imaging/Michael Foley
3942 Irving Blvd, Dallas, TX............................214-630-4351
Davidson & Co/Ken Davidson
1750 Lower Roswell Rd, Marietta, GA 404-973-9637
● **DIGITAL IMAGE**/Ray Parrott **(P 90)**
1758 Tullie Circle, Atlanta, GA 404-325-6955
● **DREAMTIME SYSTEMS**/Arni Katz **(P 29)**
108 Ayers Ave, Marietta, GA...........................404-421-8800
Eyebeam/Miles Wright
2500 Gateway Center Blvd #600,
Morrisville, NC..919-469-3859
Herndon Jr, Tom
18175 Midway Rd #353, Dallas, TX 214-306-3475
Impossible Images Inc/JC Bourque
12153 SW 131 Ave, Miami, FL 305-256-1013
Kuehl, Allan
4261 Wheaton Ln, Clarkston, GA 404-296-1900
Laser Tech Color Inc/Damien Gough
1505 Luna Rd Bldg 2 #202, Carrollton, TX.........800-365-8957
Luckett, Julie L
810 Bellevue Rd #280, Nashville, TN................615-646-9581
Meisel Photographic Corp
9645 Webb Chapel Rd, Dallas, TX....................800-527-5186
Meteor Photo & Imaging Co/Steve Carroll
680 14th St NW, Atlanta, GA 404-892-1688
NEC Inc/Roy Luckett
1504 Elm Hill Pike, Nashville, TN.....................615-367-9110
● **NEW MEDIA PRODUCTIONS INC/**
Benjamin Nowak **(P 96)**
4209 Ivy Chase Way NE, Atlanta, GA 404-257-9220

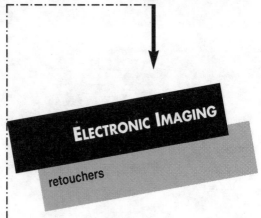

ELECTRONIC IMAGING

retouchers

Norling Studios Inc/Glenn McVicker
221 Swathmore Ave, High Point, NC..................919-434-3151
● **PHOTOEFFECTS**/Trevor Donovan **(P 87)**
3350 Peachtree NE #1130, Atlanta, GA............404-816-8000
● **RAPHAELE/DIGITAL**
TRANSPARENCIES INC(P 82,83)
616 Hawthorne, Houston, TX..................713-524-2211
Washington House Photography/Sharon Swab
5520 Cherokee Ave, Alexandria, VA..................703-642-8000
WRE/Colortech Prepress/John Comerford
533 Banner Ave, Greensboro, NC..................919-275-9821

MIDWEST

Altered+Images+Inc/Kent & Jeff Mueller
3226 S 92nd St, Milwaukee, WI..................414-327-2820
Anderson Perlstein, Ltd/Steve Murawski
1351 Barclay Blvd, Buffalo Grove, IL..................708-537-0100
Art Bunch, The/Mike Fisher
230 N Michigan Ave #618, Chicago, IL............312-368-8777
Art for Advertising Inc/Leonard Glowacki
566 W Adams St, Chicago, IL..................312-939-3393
Art in Progress
229 W Illinois St, Chicago, IL..................312-645-4500
BDG Production Studio/Julie Mikos
213 W Institute Place #702, Chicago , IL..................312-440-0240
Capps Studio/Mark Pinson
35 W Wacker Dr #3000, Chicago, IL..................312-220-0990
● **CHROMA STUDIOS INC/**
DigiChrome Imaging **(P 92)**
2300 Marilyn Ln, Columbus, OH..................800-471-1118
Color Image/Tom O'Donnell
461 N Milwaukee Ave, Chicago, IL..................312-666-2844
● **CUSTOM COLOR CORP (P 102)**/Guy Clark
300 W 19th Terrace, Kansas City, MO..................800-821-5623
Cutler-Graves/Larry Cutler
10 E Ontario #1910, Chicago, IL..................312-988-9393
Digital Imaging & Design/Sue Mann
7346 Ohms Ln, Edina, MN..................612-832-9433
Digital Wrist
900 Second Ave #930, Minneapolis , MN..................612-339-9329
Electric Art Company/Ralph Greenhow
3305 W Harrison St, Chicago, IL..................312-444-1168
GraFX Creative Imaging/Frank Defino Jr
2917 N Latoria Ln, Franklin Park, IL..................800-633-7887
Graphic Access/Rich Taylor
401 E Illinois #310, Chicago, IL..................312-222-0087
Hayes, Kevin
4808 S Union Ave, Chicago, IL..................312-227-6351
Hillstrom Stock Photo, Ray
5500 W Ardmore Ave, Chicago, IL..................312-753-2774
Jacobs Digital Arts, Todd
2437 W Wilson, Chicago, IL..................312-769-0383
Jet Litho/David Graunke
1500 Center Circle, Downer's Grove, IL..................708-932-6370
K&S Photographics/Ron Caldeleugh
222 N Canal St, Chicago, IL..................312-207-1212
Kalamazoo Color Lab
1326 Portage St, Kalamazoo, MI..................616-344-600
Kibby/Dick Kemp
25235 Dequindre, Madison Heights, MI..................313-542-1223
King Graphics Inc/Nancy Gorski
571 W Polk St, Chicago, IL..................312-943-2202
Kitzerow, Scott
3505 N Pine Grove, Chicago, IL..................312-935-9234
La Driere Studios/Bernie Potochnick
77 E Longlake Rd, Bloomfield Hills, MI..................313-644-3932

LaLiberte, Richard
211 E Delaware Pl, Chicago, IL..................312-944-1825
Licata Associates Inc/Glenn Norikane
2000 S Michigan Ave, Chicago, IL..................312-326-3600
Lubeck & Assocs, Larry
203 N Wabash #1602, Chicago, IL..................312-726-5580
Matrix Photo Labs/Mike Tuttle
118-120 W North St, Indianapolis, IN..................317-635-4756
Meteor Photo & Imaging/Steve Grundner
1099 Chicago Rd, Troy, MI..................313-583-3090
Midwest Litho Ctr/3D Imaging/Derek Frederickson
125 E Oakton, Des Plaines, IL..................708-296-2000
Modern Imaging/WACE USA/Mike DeKam
455 Grand Ave NE, Grand Rapids, MI..................616-458-1393
Palm, Brad & Maggie
2626 Second St NE, Minneapoles, MN..................612-339-2000
● **PALMS HIGH RES DIGITAL IMAGERY/**
Brad & Maggie Palm **(P 85)**
2626 Second St NE, Minneapolis, MN..................612-339-2000
● **PELIKAN PICTURES/**Petra Pepellashi
(P 88,89)
20500 Civic Center Dr #2800, Southfield, MI..................313-356-2470
Photographic Specialties Inc
1718 Washington Ave N, Minneapolis, MN..................612-332-6303
Photography Unlimited Inc/John Nate
11600 W Lincoln Ave, Milwaukee, WI..................414-321-1600
Pro-Color/Doug Meisner
909 Hennepin Ave S, Minneapolis, MN..................612-673-8900
RamPage Electronic Imaging/Tom Patterson
2233 Kemper Ln, Cincinnati, OH..................513-281-8095
Ray, Elise
4443 Bell St, Kansas City, MO..................816-756-0747
Risser Digital Studio/Leon Risser
5330 W Electric Ave/Box 19896, Milwaukee, WI..................414-545-1270
● **SANDBOX DIGITAL PLAYGROUND/**
Sandy Ostroff **(P 84)**
203 N Wabash #1602, Chicago, IL..................312-372-1170
Spectrum Image Group/Byron Sletton
10835 Midwest Ind Blvd, St Louis, MO..................314-423-8111
Spectrum Inc/Otto Grenten
200 E Ohio St 2nd Fl, Chicago, IL..................312-751-8820
● **STOKES RETOUCHING & COMPUTER**
IMAGING/Lee Stokes **(P 86)**
121 S 8th St #825, Minneapolis, MN..................612-339-5770
TDS/William McKinley
113 North May St, Chicago, IL..................312-733-4464
Techtron Imaging Center/Shelly Slimak
160 E Illinois St, Chicago, IL..................312-644-4999
Trans FX/Craig Burt
362 W Erie, Chicago, IL..................312-943-2664
Vista Color Lab/Pete Gallo
2048 Fulton Rd, Cleveland, OH..................216-651-2830
Wace USA/**Graphic Warehouse**/Bob Prange
2 N Riverside Plz #1400, Chicago, IL..................312-876-0533
Waselle Production Artists Inc/Dave Waselle
35 W Wacker Dr #3810, Chicago, IL..................312-630-0360
Willson Creative Group/William Willson Crtv Group
355 N Ashland, Chicago, IL..................312-738-3555

WEST

ABRA/Ken Kirkland/Steve Tautz
4505 W Wagontrail, Littleton, CO..................303-795-9080
Agency Images/Jack Mauck
3900 W Alameda 17th Fl, Burbank, CA..................818-972-1887
Andre's Photo Lab/Andre Schellenberg
7686 Miramar Rd, San Diego, CA..................619-549-3900
Arizona Four Color Corp/Cliff Lindstrom
2967 W Fairmont, Phoenix , AZ..................602-241-1485
ARTiculate Imaging/Dovida
849 Mission Way, Sacramento, CA..................916-736-2108
Artlab/Tony DeYoung
1603 Howard St, San Francisco, CA..................415-554-0248
Atomic Media Group/Carol Leigh
1933 Davis St #213, San Leandro, CA..................510-430-8012
Best Digital/Tony Lobue
1895 Park Ave #203, San Jose, CA..................408-261-7410
Bow-Haus Inc/Charles James
8017 Melrose Ave, Los Angeles, CA..................213-658-5129
Capitol Color Imaging Inc/John Matuze
47 W 200 South, Salt Lake City, UT..................801-363-8801
Capitol Color Imaging Inc/Greg Schriener
1463 W Alameda Ave, Denver, CO..................800-634-1564
CD Folios/CD Publishing/Randy Fugate
22647 Ventura Blvd #199,
Woodland Hills, CA..................805-527-9227

Christenson, Paul
1566 East 3080 South, Salt Lake City, UT............801-364-0919
Cies/Sexton Photo Lab/Ed Cies
1247 Santa Fe Dr, Denver, CO..................303-534-4000
Colorgraphics/Sherri Gillenwater
3707 E Broadway Rd, Phoenix, AZ..................602-437-2720
Crocker, Jim
Fullerton, CA..................714-871-5444
Cronopius
6313 Yucca St, Hollywood, CA..................213-464-6114
Dazzleland Digital
611 Fifth Ave, Salt Lake City, UT..................801-355-8555
Digital Design Inc/Jim Belderes
5835 Avenida Encinas #121, Carlsbad, CA..................619-931-2630
Dowlen Artworks/James Dowlen
2129 Grahn Dr, Santa Rosa, CA..................707-579-1535
Elite Color Inc/Tom Tivo
127 W25th 8th Fl, New York, NY..................212-989-3289
Ferrari Color Digital Imaging/Scott Powell
2574 21st St, Sacramento, CA..................800-533-6333
GP Color Inc/John Portaro
201 S Oxford Ave, Los Angeles, CA..................213-386-7901
Greene Productions, Jim
PO Box 2150, Del Mar, CA..................619-454-4133
Hallas Photo Lab/jim Davis
4532 Telephone Rd #11, Ventura, CA..................805-642-8063
Imagic/Paul Elmi
1545 Wilcox Ave, Los Angeles, CA..................213-461-7766
Intersep/Tom Pettinger
2345 S 2700 West, Salt Lake City, UT..................801-973-6720
Ivey-Seright International/David Hoffman
424 8th Ave N, Seattle, WA..................206-623-8113
Laserscan Inc/Gary Harris
10220 S 51 St, Phoenix, AZ..................602-893-7777
Littlejohn's Retouching Studio/Mike Haynes
16 Lyndon Ave, Los Gatos, CA..................408-395-7377
LunaGrafix/Gala Muench
Route 1/Box 17 28601 S Gem Rd, Harrison, ID..208-689-3877
Lund, John
860 Second St, San Francisco, CA..................415-957-1775
Lynx Digital Design/Alejandro Rubalcava
16 Technology #115, Irvine, CA..................714-727-3126
● **MC2**/Heidi Merscher **(P 48)**
65 Lower Arroya Hondo Rd,
Arroya Hondo, NM..................505-776-1333
Metafor Imaging/Judy Linden
39 E Walnut St, Pasadena, CA..................818-584-4034
Michaels, Robert
6716 Zumirez Dr, Malibu, CA..................310-457-5859
Pacific Digital Image/Jeremy Smith
1050 Sansome St #100, San Francisco, CA..................415-274-7234
Peter Green Design Studios/Chris Savage
4219 W Burbank Blvd, Burbank, CA..................818-953-2210
Phaedrus Productions/Dennis Dunbar
714 Hill St #2, Santa Monica, CA..................310-450-9319
Photo Concepts Inc/Mike Goldman
3001 E Thomas Rd, Phoenix, AZ..................602-957-8055
PIX Imaging/Jim Cox
6100 Melrose Ave, Los Angeles, CA..................213-462-2255
Pixmil Digital Imaging Service/Don Redding
6150 Lusk Blvd #B102, San Diego, CA..................619-597-1177
Pleasure Retouching, Robert
2727 Walsh Ave #200, Santa Clara, CA..................408-376-0196
Rico, Antonio
4138 Geary Blvd #1, San Francisco, CA..................415-668-9514
Stehney, Regina
25591 Via Inez, San Juan Capistrano, CA..................714-496-1652
TX Digital Imaging & Retouching/Michael Piltoff
577 Second St, San Francisco, CA..................415-905-0990
Veritel Select/Kevin Horner
350 Townsend St, San Francisco, CA..................415-597-6900
Walker Graphics/A WACE USA Com/Cindy Reese
333 Fremont St, San Francisco, CA..................415-433-7900
ZZYZX Visual Systems/Kim Kapin
1011 N Orange Dr, Los Angeles, CA..................213-883-1060

CANADA

BGM Color Labs/Lumiere Div
497 King St E, Toronto, ON..................416-947-1325
Donath, Emeric
100 Bain Ave/59 Oaks, Toronto Canada, ON....416-469-0885
French, Graham
388 Carlaw Ave #303, Toronto, ON..................416-778-7921
Linotext Group of Companies/Charles Hume
145 Front St E, Toronto, ON..................416-361-3300

Partners III/Roy France
109 Vanderhoof #202, Toronto, ON 416-425-5162
Smith & Co/Geoff Smith
2145 Avenue Rd, Toronto, ON..........................416-425-2688
Struve-Dencher, Goesta
#303 750 E 7th Ave, Voncouver, BC 604-872-3439

Illustrators

NEW YORK CITY

Albers, Louis
541 Isham St #51A, New York, NY..................212-569-6652
Alper Studio, A J
224 Elizabeth St #19, New York, NY 212-935-0039
Amendola, Steve
95 Horatio St, New York, NY............................212-989-3246
Anagram
133-48 115th St, S Ozone Park, NY................718-848-6176
Andersen, Rolf
45 W 84th St, New York, NY.............................212-787-3305
Applied Graphics/David Rickerd
50 W 23rd St, New York, NY.............................212-627-4111
Aronoff, Susan
201 E 25th St #11C, New York, NY..................212-889-0942
Art Resources & Technologies/Uri Dothan
37 W 20th St #1209-10, New York, NY............212-255-5399
Artichoke Arts/Glenn Cho
333 E 34th St, New York, NY............................212-532-2188
Artistic Perceptions/Carol Maria Weaver
70 W 95th St #21C, New York, NY..................212-864-2394
Associated Images/Brian Lee
545 5th Ave 3rd Fl, New York, NY....................212-687-2844
Bamundo, David
66 Carreau Ave, Staten Island, NY..................718-370-7726
Barrett, Anne S
300 Eighth Ave #6A, Brooklyn, NY..................718-965-3162
Batelman, Kenneth
11 Dyckman Dr, Mohegan Lk, NY 718-392-3483
Beauchamp, Jaime
99-12 62nd Rd, Rego Park, NY........................212-512-2230
Beckman, Melissa
95-02 72nd Ave #2, Forest Hills, NY 718-793-7286
Bego, Dolores
155 E 38th St, New York, NY............................212-697-6170
Behr, Sheri Lynn
1675 York Ave #26F, New York, NY..................212-922-9020
Bellamy, Gordon
129 W 22nd St, New York, NY.........................212-924-1380
Belman, Vickie
210 E 181st St #4A, Bronx, NY........................718-367-2688
Bernstein, Linda A
10-11 50th Ave, Long Island City, NY..............718-784-1599
Bigtwin/James Suhre
17 Greenwich Ave #11, New York, NY 212-647-9734
Birmingham, Barbara
133 Barrow St #4A, New York, NY..................212-691-5587
Blauweiss, Stephen
32-15 41st St, Long Island City, NY 718-204-8335
Bluming, Joel
328 E 19th St , New York, NY...........................212-673-0558
Brauer, Bruce Erik
2202 Richmond Terrace, Staten Island, NY..........718-667-8977
Breiger, Elaine
112 Greene St, New York, NY...........................212-966-3004
Brianstorm Unltd, Mark
180 Varick St 13th Fl, New York, NY...............212-647-9278
Broadway Video Graphics/Peter Rudoy
1619 Broadway 10th Fl, New York, NY.............212-265-7600
Brody, Ellen
20 Stuyvesant Oval #11C, New York, NY..........212-388-0297
Buck, Sid & Kane, Barney/Ron Victor
566 7th Ave #603, New York, NY.....................212-221-8090
Callaway Editions/Nicholas Callaway
54 Seventh Ave S, New York, NY......................212-929-5212
Castleman, Valerie
60 Gramercy Park #9A, New York, NY 212-254-5430
Center for Advanced Whimsy, The /
Rodney Alan Greenblatt
61 Crosby St, New York, NY 212-219-0342
Chada, Ritu
150 E 30th St #6A, New York, NY....................212-684-8134
Chalkmark Graphics/David Chalk
348 W 38th St #7C, New York, NY...................212-279-2433
Chattum Design Group/Shirley Chetter
10 E 23rd #200, New York, NY.........................212-673-4514

Chollick, Phyllis
8507 Homelawn St, Jamaica, NY 718-658-0670
Chung, Harry
34-29 71st St, Jackson Heights, NY 718-507-8140
Chung, Mei K
82 W 3rd St, New York, NY..............................212-228-9100
City College of NY/Computer Graphics/Prof Annette Weintraub
138th St & Convent Ave, New York, NY.............212-650-7410
Class Line Graphics/Karen Miles
368 Broadway #209, New York, NY..................212-571-3678
Cohen Design Inc, Hayes
133 Cedar Rd, E Northport, NY........................516-368-2031
Cohen, Adam
96 Greenwich Ave #3, New York, NY..............212-691-4074
Comitini, Peter
30 Waterside Plaza, New York, NY...................212-683-5120
Compugraphia
190 Avenue B, New York, NY...........................212-614-0283
Creative Freelancers
25 W 45th St #703, New York, NY....................212-398-9540
Cushwa, Tom
303 Park Ave S #511, New York, NY................212-228-2615
Daedalus Systems/Geoffrey Homan
235 Lafayette St, New York, NY.......................212-431-4365
Davis Studio/Paul Davis
14 E 4th St, New York, NY................................212-420-8789
Deak, David
354 Broome St #4I, New York, NY....................212-219-2984
Dean, Glenn
501 Fifth Ave #1407, New York, NY.................212-490-2450
Deck Design, Barry
611 Broadway, New York, NY...........................212-777-6627
Dellinger, Joseph
110 Ninth Ave, New York, NY...........................212-627-8526
Demark Keller & Gardner, Inc/Jeffrey Gardner
141 W 28th St 9th Fl, New York, NY.................212-714-1731
DeNicola, Robert
45-49 165th St, Flushing, NY...........................718-359-5336
DeSeta, Maxine
202 W 107th St #6E, New York, NY.................212-316-3563
Design Heads/Ray Sysko
270 Park Ave S #4E, New York, NY..................212-598-9710
Diaz, Jose
148 W 23rd St #1B, New York, NY...................212-255-8060
Dickson, Ellie
321 W 29th St #3D, New York, NY...................212-563-0674
Digital Photo Works/Marc Silverman
62 W 70th St, New York, NY.............................212-595-5464
DiMarco, Paula
315 E 21st St #7L, New York, NY.....................212-228-2893
Dressler, Marjory
4 E 2nd St, New York, NY.................................212-254-9758
Edwards, Andrew
395 South End Ave #2M, New York, NY 212-912-1344
Electronic Publishing Ctr/Irving S Berman
110 Leroy St 2nd Fl, New York, NY..................212-727-2655
Epure, Serban
60-11 Broadway #5L, Woodside, NY................718-335-7685
Eve Design
60 Plaza St E #6E, Brooklyn, NY......................718-398-0950
Fairchild Publications/Mark Mitchell
7 W 34th St, New York, NY...............................212-576-5916
FCL/Colorspace/Terry Chamberlain
10 E 38th St 3rd Fl, New York, NY...................212-679-9064
Feinberg, Susan
433 W 21st St, New York, NY...........................212-929-8679
Feltenstein, Keith
144-10 38th Ave, Flushing, NY.........................718-359-5140
Ferguson, Melanie
18 Gramercy Park S #706, New York, NY..........212-388-7677
Feuereisen, Fernando
885 Tenth Ave #2G, New York, NY..................212-399-3269
Filippucci, Sandra
270 Park Ave S #6E, New York, NY..................212-260-4153
Fine Print Design Co/Nick Ferrone
12 Remsen St #3, Brooklyn, NY........................718-488-0140
Fisch, Amy
718 Broadway, New York, NY...........................212-677-3509
Fishman, Miriam
12 E 14th St #2E, New York, NY......................212-633-9126
Flaherty, David
84 Forsyth St #5, New York, NY.......................212-274-9081
Fleisher, Audrey
430 W 24th St, New York, NY...........................212-463-3722

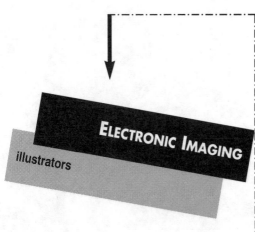

ELECTRONIC IMAGING
illustrators

Frampton, Bill
200 W 15th St, New York, NY...........................212-243-4209
Freed, Hermine
54 W 21st St #507, New York, NY....................212-229-1167
Frumkin, Peter
400 W 43rd St #30R, New York, NY.................212-694-7837
● **GARDINER, JEREMY**/Pratt University (P 120)
Computer Graphics Ctr ARC Bldg, Brooklyn, NY..718-636-3600
Gartel, Laurence
270-16 B Grand Central Pkwy, Floral Park, NY....718-229-8540
Gaz, Stan
Box 1067, Long Island City, NY.......................718-937-2766
Gellman, Rachel
192 Bleecker St, New York, NY........................212-473-7502
Gencarelli, Elizabeth
80 E 7th St, New York, NY...............................212-353-9073
Gladstone, Dale
32 Havemeyer St #2A, Brooklyn, NY...............718-782-2250
Glitch Graphics/Jessica Perry
185 E 85th St #26M, New York, NY.................212-722-0535
Gmucs, Rebecca
295 St Johns Pl #4B, Brooklyn, NY..................718-638-0903
Golden, Kenneth Sean
696 10th Ave, New York, NY............................212-246-3875
Gordon Associates, Barbara
165 E 32nd St, New York, NY...........................212-686-3514
Gosfield, Josh
682 Broadway, New York, NY...........................212-254-2582
Grace, Alexa
70 University Pl #7, New York, NY....................212-254-4424
Grace, Laurie
310 W 93rd St, New York, NY...........................212-678-6535
Graham, Thomas
408 77th St #D4, Brooklyn, NY........................718-680-2975
Graphic Chart & Map Co
236 W 26th St #801, New York, NY.................212-463-0190
Graphics in Medicine/Maura C Flynn
242-09 43rd Ave, Douglaston, NY....................718-279-1659
Grien, Anita
155 E 38th St, New York, NY............................212-697-6170
● **GROSSMAN, WENDY (P 49)**
355 W 51st St #43, New York, NY....................212-262-4497
Gussin, Jane E
243 E 18th St #22, New York, NY.....................212-777-3592
Hamann, Brad
330 Westminster Rd, Brooklyn, NY..................718-287-6086
Hamelton, Meredith
22 Remsen St #1, Brooklyn, NY.......................718-243-1244
Hammes, Alan
65 Downing St #1B, New York, NY..................212-691-6387
HBO Studio Productions/Judy Glassman
120A E 23rd St, New York, NY.........................212-512-7800
● **HERBERT, JONATHAN (P 63)**
501 Fifth Ave #1407, New York, NY.................212-490-2450
Heun, Christine
337 W 20th St #2M, New York, NY.................212-645-5536
High Priority Consulting/Bonnie Kane
45 First Ave #5-O, New York, NY.....................212-228-6000
Hong, Won-Hua
73 Monroe St, New York, NY............................212-267-4947
Honkanen, William
PO Box 20402/Tompkins Sq Stn, New York, NY ..212-982-2435
Horizon Images/Nona Abiathar McCarley
534 9th Ave #D2, New York, NY.......................212-529-5452
Hornbacher, Sara
40 Harrison St #9B, New York, NY...................212-964-9582

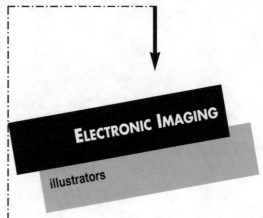

ELECTRONIC IMAGING

illustrators

Hot-tech Multimedia Inc/Lawrence Kaplan
46 Mercer St, New York, NY 212-925-3010
Hull, Cathy
165 E 66th St, New York, NY 212-772-7743
Iconink/Pia Rigby
68 Devoe St, Brooklyn, NY 718-963-2821
Iseki, Keiko
445 W 19th St #G, New York, NY.............212-255-7296
Itami, Michi
83 Grand St, New York, NY 212-925-0623
J & M Studio/Consulting/Jacqueline Skiles
236 W 27th St, New York, NY 212-675-7932
Jackson, Troi
2025 Broadway #10D, New York, NY 212-799-0936
Jiempreecha, Wichar
32-86-34th St #3A, Long Island City, NY.............718-721-6956
Kauftheil/Rothchild/Henry Kauftheil
220 W 19th St #1200, New York, NY 212-633-0222
Kelemen, Stephen
161 Henry St, Brooklyn, NY 718-855-7005
Kendrick, Dennis
99 Bank St #3G, New York, NY.................212-924-3085
Kenngott, Barbara
220-11 39th Ave, Bayside, NY 718-631-7152
Klein, Josh
235 W 71st St #3, New York, NY.................212-721-6294
Klein, Renee
164 Daniel Low Terr, Staten Island, NY 718-727-0723
Klineman, Peggy
310 W 47th St #4B, New York, NY 212-757-3460
Konkle, Kathy
313 E 9th St, New York, NY.................212-420-5970
Kursar, Ray
1 Lincoln Plaza #43R, New York, NY.................212-873-5605
Language & Graphics/Margaret Keppler
350 W 57 St #4F, New York, NY 212-315-5266
Lindgren & Smith
41 Union Sq W #1228, New York, NY 212-929-5590
Lipner, Robin
220 W 21st St #16N, New York, NY 212-929-5807
Lombardo, William
491 Broadway 12th Fl, New York, NY.........212-226-3471
Lord, Rosalind
175 W 12th St #16N, New York, NY 212-807-7959
Luce, Ben
5 E 17th St 6th Fl, New York, NY 212-255-8193
Lui, David
111 Wooster St #PH C, New York, NY 212-925-0491
Manhattan, Maria
329 W 71st St #3, New York, NY 212-799-6320
Mannes, Don
345 E 76th St #2A, New York, NY 212-288-1392
Margolin, Diane
41 Perry St, New York, NY 212-691-9537
Martino, Jacquelyn
236 Henry St, Brooklyn, NY.................718-852-6497
Maxedon, Terry
718 Broadway, New York, NY.................212-677-3509
McCormack, Geoffrey
420 Lexington Ave #PH, New York, NY.........212-986-5680
McDaniel, Jerry
155 E 38th St, New York, NY.................212-697-6170
● **MCELROY, DARLENE OLIVIA (P 64)**
60 E 42nd St #822, New York, NY.................212-682-1490
Melnick, Anil
225 First Ave, New York, NY 212-995-5176

Merrill, Lizanne
333 Rector Pl #5N, New York, NY 212-945-6115
● **MICROCOLOR INC**/Martin Haggland **(P 97)**
2345 Broadway #728, New York, NY 212-787-0500
Midnight Oil Studios/Kathryn Klein
156 Fifth Ave #623, New York, NY 212-366-9117
Mikros, Nikita
58-31 44th Ave, Woodside, NY.................718-458-6456
Mills, Elise
150 E 79th St #3, New York, NY 212-794-2042
Mison, Vesna
901 56th St, Brooklyn, NY.................718-854-7847
MLH Communications Group
51 Madison Ave #1201, New York, NY 212-576-5916
Montana Graphics/Martha Towler
34-30 78 St #1J, Jackson Heights, NY 718-651-1549
Moonlight Press Studio/Chris Spollen
362 Cromwell Ave, Ocean Breeze, NY.........718-979-9695
Moore, Tom
286 19th St #2F, Brooklyn, NY.................718-788-8474
Morawa, Amy Lynne
375 South End Ave #4K, New York, NY 212-938-5732
Morgan Associates, Vicki
194 Third Ave, New York, NY.................212-475-0440
Morris, Grey
81 Ocean Parkway #5B, Brooklyn, NY.........718-438-1476
Morrison, Joanna
322 W 57th St #10H, New York, NY.................516-536-2532
Morse, Michael
6 W 87th St #6, New York, NY.................212-721-2638
Moskovic, Stephen
268 E 10th St 4th Fl, New York, NY.................21-529-0248
Mulligan, Donald
418 Central Park W #81, New York, NY.........212-666-6079
Myers Graphics, David
228 Bleecker St #8, New York, NY.................212-989-5260
Mythic Graphics/Bonnie Robin Cohen
230 Riverside Drive #8H, New York, NY 212-866-6059
Nessim & Assocs, Barbara
63 Greene St, New York, NY.................212-677-8888
Novack, Dev
2075 Palisades Ave, Riverdale, NY.................718-884-2819
O'Kelley, Lynn
241 Bedford Ave, Brooklyn, NY.................718-599-1969
Osama Ltd/Osama Hashem
50 Lexington Ave #105, New York, NY.........212-779-1923
Pantuso, Mike
350 E 89th St, New York, NY.................212-534-3511
Paragraphics/Teresa Berasi
427 Third St, Brooklyn, NY.................718-965-2231
Paragraphics/Stanford Kay/Teresa Berasi
427 3rd St, Brooklyn, NY.................718-965-2231
Park, Chun-Sin
401 E 34th St #N20G, New York, NY.........212-725-3719
Pelavin, Daniel
80 Varick St, New York, NY.................212-941-7418
Peled, Einat
62-28 Cromwell Crescent, Rego Park, NY.........718-275-6549
Percivalle, Rosanne
132 W 21st St 12th Fl, New York, NY.............212-727-9158
Pergament Graphics
38 E 30th St, New York, NY.................212-213-8310
Pettingill, Andre
245 Bennett Ave #3G, New York, NY.........212-942-1993
Portella, Dalton
1085 Broadway Top Fl, Brooklyn, NY.........718-455-7123
Pratt Institute/Isaac Victor Kerlow
Comp Grphcs Ctr/ARC Bldg, Brooklyn, NY.........718-636-3600
Preda, Dan
598 Ninth Ave #2A, New York, NY.................212-265-2145
Principato, Salvatore
220 Sullivan St #4H, New York, NY.................212-477-8161
● **PUNIN, NIKOLAI (P 60)**
161 W 16th St #18E, New York, NY 212-727-7237
● **R/GA PRINT**/Jimm Burris **(P 44,45)**
350 W 39th St, New York, NY.................212-239-6767
Rabinovitch, William
Box 403/Canal St Station, New York, NY.........212-226-2873
Rapalee, Susan
181 8th St, Brooklyn, NY.................718-499-2301
Rick Barry Desktop Design Studio/
Rick Barry
1631 W 12th St, Brooklyn, NY 718-232-2484
Ridgeway, Ronald
530 Broadway 4th Fl, New York, NY.................212-966-9696

Rinaldi, John
26-06 212th St, Bayside, NY 718-428-6147
Rio Station Graphics/Silvio Da Silva
47 Murray St 3rd Fl, New York, NY..................212-964-9418
RJG Design/Robert J George
366 Sterling Place, Brooklyn, NY.................718-783-8514
Robinson, Lenor
201 E 69th St #6E, New York, NY 212-734-0944
Rodriguez, Claudio
304 Mulberry St #4B, New York, NY.................212-941-0573
Rohr, Dixon
155 W 68th St #26E, New York, NY 212-580-4065
● **ROMERO DESIGN, JAVIER**/Jeremy Kaplan **(PG 57)**
24 E 23rd St 3rd, New York, NY.................212-727-9445
Romney, Michael
201 E 77th St PH C, New York, NY.................212-288-0618
Rosebush Visions Corp/Judson Rosebush
154 W 57th St #826/Carnegie,
New York, NY.................212-398-6600
Rosen, Terry
101 W 81st St #508, New York, NY.................212-580-4784
Ross Culbert & Lavery
15 W 20th St 9th Fl, New York, NY.................212-206-0044
Rowland, Lauren
176 Ludlow St #1H, New York, NY.................212-254-2731
Royden, Elizabeth
309 E 18th St #4D, New York, NY.................212-228-1426
Rubyan, Robert
270 Park Ave S #7C, New York, NY.................212-460-9217
Safton, Carole
360 W 22nd St #10J, New York, NY.................212-989-1096
Saturn Productions/Ivan Katz
153 Garfield Place, Brooklyn, NY.................718-499-4379
Savard, Sister Judith
250 Riverside Dr #24A, New York, NY.................212-663-6273
Sawyer Studios Inc/Arnie Sawyer
115 W 27th St 8th Fl, New York, NY.................212-645-4455
Scallon, Ken
118 E 11th St, New York, NY.................212-979-9193
Schell, Paul
1608 E 51st St, Brooklyn, NY.................718-951-8976
Schneidman, Jared
280 Park Ave S #20L, New York, NY.................212-475-2390
Sehmi, Gagan
95-14 120th St, Richmond Hill, NY.................718-849-3882
Seibold, J Otto
38 W 21st St #1101, New York, NY.................212-366-4949
Seiffer, Alison
305 Canal St #2, New York, NY.................212-941-7076
Sela, Eliot
221 W 15th St, New York, NY.................212-627-2450
Seong, Young-Shin
235 W 22nd St #2C, New York, NY.................212-675-2019
Severtson, Jeff
221 E 12th St #2, New York, NY.................212-473-8086
Shelly, Jeffrey
55 Mercer St 4th Fl, New York, NY.................212-941-1905
Shin, Young
162-11 9th Ave #5C, Whitestone, NY.................718-767-5668
Shtern, Adele
11-21 47th Rd #3L, L I City, NY 718-706-6363
Siegel, Dink
100 W 57th St #10G, New York, NY.................212-246-9757
Silbert, Barbara Bert
40 W 24th St #7N, New York, NY.................212-741-1915
Siren Design Inc/Joan Kristensen
1204 3rd Ave #120, New York, NY.................212-580-3238
Snelson, Kenneth
140 Sullivan St, New York, NY.................212-777-0356
Spade, Sergio
295 Park Ave S #8L, New York, NY.................212-982-8862
Speer, Stephan
6105 80th St, Middle Village, NY.................718-457-7641
Stabin, Victor
8421 Midland Pkwy, Queens, NY.................212-243-7688
Stahl, Nancy
470 West End Ave #8G, New York, NY.................212-362-8779
Stankiewicz, Steven
317 E 18th St #3A, New York, NY.................212-477-4229
● **STRUTHERS, DOUG (P 52,53)**
232 Madison Ave #402, New York, NY.................212-889-8777
● **STUDIO MACBETH (P 50,51)**
232 Madison Ave #402, New York, NY.................212-889-8777
Sullivan, Steve
175 W 90th St #15B, New York, NY 212-787-7631

Sutton, Eva
239 Ninth Ave #2B, New York, NY 212-242-3599
Swann Design/Swann Smith
340 W 19th St #10, New York, NY 212-807-8261
Terezakis, Peter
50 W 22nd St, New York, NY 212-929-8978
Thornell, Ian
711 Sackett St, Brooklyn, NY 718-783-2553
Thorpe, Peter
254 Park Ave S #6D, New York, NY 212-477-0131
Tonkin, Thomas
353 W 53rd St #1W, New York, NY 212-682-2462
Tozzi, Graig
250 E Houston, New York, NY 212-477-1779
Trofimova, Marianna
105 Boerum Pl #1, Brooklyn, NY 718-625-0294
Tumble Interactive Media/Cal Vornberger
910 West End Ave #3D, New York, NY 212-316-0200
Tung, Claudia
150-21 77th Ave, Kew Garden Hills, NY 718-380-1054
Uman, Michael
1781 Riverside Dr #4C, New York, NY 212-304-0756
Von Ulrich, Mark
One Union Sq W #903, New York, NY 212-989-9325
Walker, Donna
221 E 50th St #5F, New York, NY 212-779-7158
Wallace/Church Assocs/Craig Swanson
330 E 48th St, New York, NY 212-755-2903
Waters Design/Colleen Syron
3 W 18th St 8th Fl, New York, NY 212-807-0717
Wax, Wendy
322 E 55th St #2A, New York, NY 212-371-6156
Weisser, Carl
163 Joralemon St #1500, Brooklyn, NY 718-834-0952
Whitby, Michelle
117 E 57th St, New York, NY 212-223-1063
Wiggins, Mick
58 W 15th St 6th Fl, New York, NY 212-741-2539
● **WILEY, PAUL (P 56)**
410 W 24th St #12i, New York, NY 212-627-8071
Williamson Janoff Aerographics/Dean Janoff
514 W 24th St 3rd Fl, New York, NY 212-807-0816
Wisenbaugh, Jean
41 Union Sq W #1228, New York, NY 212-929-5590
Wynn, Dan
93 Reade St, New York, NY 212-379-4719
Yankus, Marc
570 Hudson St #3, New York, NY 212-242-6334
Yourke, Oliver
525-A Sixth Ave, Brooklyn, NY 718-965-0609
Zlowodzka, Joanna
144 W 10th St #18, New York, NY 212-620-7981
Lazansky, Aaron
60 Amsterdam Ave #5A, New York, NY 212-586-8747

EAST

AB Graphics/Arnold Bombay
69 Farrington Rd, Matawan, NJ 908-566-5101
Alco Computer Graphic/Alex Guben
302 Carlton Terr, Teaneck, NJ 201-836-1294
Alsberg, Peter
33 Columbia Ave, Takoma Park, MD 301-891-3530
American Tech Systems
5 Suburban Park Dr, Billerica, MA 508-663-6755
Art on Fire/Marian Schiavo
335 38th Rd, Douglaston, NY 718-229-3660
Ascroft, Robert
287 Orchard Creek Ln, Rochester, NY 716-227-1976
Axmann, Doug
16 Schoolhouse Lane, Somerville, NJ 908-722-1925
Baldwin, Scott
125 Red Schoolhouse Rd, Chestnut Ridge, NY 914-620-0983
Baquero, George
4 Westlay Ln, New Milford, NJ 201-261-6011
Barbeau, Dan
520 Bloomfield St #2, Hoboken, NJ 201-217-1525
Barr, Kevin J
135 Lansdowne Court, Lansdowne, PA..............215-623-7403
Bartley Collection Ltd/James Webster
29060 Airpark Dr, Easton, MD 410-820-7722
Bautista, David R
50 Lillian Rd, Nesconset, NY 516-981-4092
Bellamy, Mike
70 Willow Ave, Hackensack, NJ 201-487-2342

Benante, Catherine
81 Montague Pl, Montclair, NJ 212-592-2527
Bergeron, Joe
PO Box 5864, Endicott, NY 607-786-0754
Berlin Productions Inc
Lock Lane Rd1/Box 193, Yorktown Hgts, NY 914-962-0526
Berry & Homer Inc/Marci Raneri
2035 Richmond St, Philadelphia, PA 800-522-0888
Berry, Rick
93 Warren St, Arlington, MA 617-648-6375
Blae, Ken
1089 Central Ave, Plainfield, NJ 212-869-3488
Blumrich, Christoph
149 Broadway, Greenlawn, NY 516-757-0524
Bob Clarke Illustrations/Bob Clarke
55 Brook Rd, Pittsford, NY 716-248-8683
Boone, Tim
834 Chestnut St #1212, Philadelphia, PA...........215-627-5604
Braun, Khyal
309 S Main St #C21, Middletown, CT 203-347-0434
Burwinkel, David
6055 Springhouse Pl, Bridgeville, PA 412-221-1617
Buszka, Kimberly
96 W Central St, Natick, MA 508-655-6807
Cabarga, Leslie
258 W Tulpehocken St, Philadelphia, PA 215-438-9954
Cantrell Design, David
85 Spencer Ave, Lancaster, PA 717-396-1134
Casper, Daniel S
One Devon Way, Hastings-on-Hudson, NY 914-478-7548
Centre Grafik/Arthur Bromley
900 W Valley Rd #802, Wayne, PA 215-688-2949
Chausse, Norbert
1709 Blue Spruce Dr, Sykesville, MD 410-549-1506
Chernishov, Anatoly & Irene
4 Willow Bank Ct, Mahwah, NJ 201-327-2377
Chipurnoi, Minda
1085 Warburton #508, Yonkers, NY 914-963-8959
Christopher Designs/Christopher Magalos
3308 Church Rd, Cherry Hill, NJ 609-667-7433
Cicchetti, John
33 Stratford Rd, Scarsdale, NY 914-723-6305
Ciemny, Ray
33 Bradford St, Concord, MA 508-371-2222
Clark, Gary F
823 Light Rd, Bloomsburg, PA 717-387-1689
Cliggett Design Group/Jack Cliggett
3406 Baring St, Philadelphia, PA...................215-222-8511
Collins & Collins/James Collins
PO Box 10736, State College, PA 814-234-2916
Collyer, Frank
10 Knapp Rd, Stony Point, NY 914-947-3050
Computer Graphics Design/Paul Roseneck
Box 1717, Schenectady, NY 518-381-6570
Computer Images/Steve Kirchuk
519 Grand Ave, Hackettstown, NJ..................908-850-6424
Conant, Chrissy
504 University Ave #1, Ithaca, NY 607-273-2150
Corrette, Nicholas Moses
203 College St, Burlington, VT 802-864-9241
Counts, Clinton
399 Sunset Rd, Skillman, NJ 908-359-5936
Dadabase Design/Gary Zamchick
56 Hillside Ave, Tenafly, NJ........................201-568-3727
Dahm, Bob
166 Arnold Ave, Cranston, RI 401-781-5092
Davidson, Peter
9R Conant St, Acton, MA 508-635-9780
Davison, Bill
179 Main St, Winooski, VT.........................802-655-0407
Day, Rob
10 State St #214, Newburyport, MA 508-465-1386
De Cerchio, Joe
62 Marlborough Ave, Marlton, NJ 609-596-0598
Deborah Wolfe Ltd/Bill Morse
731 N 24th St, Philadelphia, PA....................215-232-6666
Delago, Ken
706 Steamboat Rd #35, Greenwich, CT 203-661-6547
Design at Work/Steve Hoskins
220 Stony Run Ln #3D, Baltimore, MD 410-235-2168
Digital Constructs/Robert Moran
759 N Park Ave, W Redding, CT 203-452-1116
DMC & Co/Donna Chernin
169 Central St, Acton, MA 508-266-1000
Domin, Jacqueline
26 Monroe St, Honeoye Falls, NY 716-624-3318

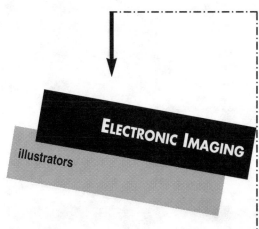

ELECTRONIC IMAGING

illustrators

Double Click Design/Susan Mason
9 Marble Terrace, Hastings-on-Hudson, NY..........914-478-4508
Drake, Patti
657 Meadow Rd, Bridgewater, NJ 908-725-4254
Dreamlight Inc/Michael Scaramozzino
50 Clifford St, Providence, RI......................401-861-8002
Drummond, Deborah
67 Concord Rd, Sudbury, MA 508-443-3160
Dunning, Chris
33 Stratford Rd, Scarsdale, NY 914-934-0517
E Fitz Art/Ellen Fitzpatrick Pinkman
22 Elm Ave, Hackensack, NJ.......................201-342-4034
Edgerton, Brian
10 Old Rte 28/PO Box 364, Whitehouse, NJ......908-534-9400
Egas, Eric
Box 600, Greenville, NY 518-966-8421
El-Darwish, Mahmoud
9112 Kittery Ln, Bethesda, MD 301-469-6383
Evans, Virginia
10 State St #214, Newburyport, MA 508-465-1386
Farrell, Richard
116 Ave C, Holbrook, NY 516-981-1807
Feigus, Jan
Box 207, Hatboro, PA 215-957-9395
Fenelon, Daniel
341 E Mt Pleasant Ave, Livingston, NJ.............201-535-5264
Ferguson, Heleman
10512 Pilla Terra Ct/Warfld Fr, Laurel, MD 301-604-4270
Flood, David Williams
43 Booraem Ave, Jersey City, NJ 201-420-7475
Ford Myers & Co/Ford Myers
326 W Lancaster Ave, Ardmore, PA 215-649-0100
Frazier, Jillian
7 Wells Rd, Lincoln, MA 617-259-9380
Fullmoon Creations/Lisa Gingras
81 S Main St, Doylestown, PA 215-345-1233
Gallagher, Matthew
1 Old Manor Rd, Holmdel, NJ 908-888-3953
Ganley Design/Mary-Anne Ganley
18B Leland Wy, Plymouth, MA 508-830-0986
Gates, Jeff
2000 Hermitage Ave, Silver Spring, MD 301-949-0436
Gavin, Bill
268 Orchard St, Millis, MA 508-376-5727
Genovese, Janell
283 Market St, Brighton, MA 617-782-3218
Gerber, Mark & Stephanie
18 Oak Grove Rd, Brookfield, CT 203-775-3658
Glenn, Mary Jane
2 Thorne Ln, Oakdale, NY 516-589-8065
Glessner, Marc
24 Evergreen Rd, Somerset, NJ 908-249-5038
Golici, Ana
225 Guy Lombardo Ave, Freeport, NY 516-623-3392
Granberg, Al
18 Mathew St, Milford, CT 203-877-2181
Graphic Images/Michele Tokach
735 Grant St, Hazelton, PA.........................717-455-2144
Green Grphc Dsgn & Adv, Mel
145 Richdale Rd, Needham, MA 617-449-6777
Green, John
2809 Boston St, Baltimore, MD 410-330-2866
Greenfield Belser Ltd/Burkey Belser
1818 N St NW #110, Washington, DC 202-775-0333
● **GUDYNAS, PETER (P 24)**
11 Kings Ridge Rd, Long Valley, NJ 908-813-8718

133

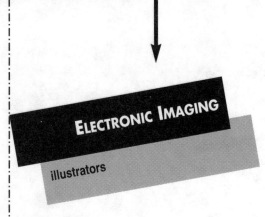

Gundlach, Elisabeth
17 Ferndale St, Albany, NY 518-482-4375
Haber-Schaim, Tamar
1870 Beacon St/Bldg 6 #B1, Brookline, MA.......617-738-8883
Haedrich, Todd
10 Byron Dr, Basking Ridge, NJ 908-204-0624
Haefele, Steve
111 Heath Rd, Mahopac, NY 914-628-1153
Harris, Ellen
45 Marion St #20, Brookline, MA 617-739-1867
Hastings, Cynthia
170 Summer St, Andover, MA 508-475-0484
Hatfield, Lee
25 Allen Ave, Newton, MA 617-527-8556
Hermine Design Group
3 Lockwood Ave, Old Greenwich, CT 203-698-1732
Hess, Mark
88 Quicks Lane, Katonah, NY 914-232-5870
Hess, Robert
63 Littlefield Rd, E Greenwich, RI 401-885-0331
Hidy, Lance
PO Box 806, Newburyport, MA 508-465-1346
Hierro Studio Inc/Claudia & Gregory Hierro
162 Tamboer Dr, N Haledon, NJ 201-427-3647
Hill-Cresson, Pat
24 Veros Ln, Franklin Park, NJ 908-422-0382
HiRes Graphics/Bob Hires
Sutton Towers #808D, Collingswood, NJ.........609-858-4770
Hoey, Peter
1715 15th St NW #2, Washington, DC 202-234-2110
Hoke, Ken
441 W Lemon St, Lancaster, PA 717-291-7227
Holewski, Jeff
11 Groendyke Circle, Robbinsville, NJ 609-443-1497
Holmes Agency/Greg Holmes
9153 Brookville Rd, Silver Spring, MD 301-589-1251
Holzer, Chris
141 Morris Turnpike, Randolph, NJ 201-895-2404
Hothouse Designs Inc/Brian Sheridan
481 Main St #100, New Rochelle, NY 914-636-0075
Hundertmark, Karen
4129 Grandview Dr, Gibsonia, PA 412-443-8527
● **ICON GRAPHICS**/Keith Meehan/
Steve Bon Durant **(P 76)**
34 Elton St, Rochester, NY 716-271-7020
Iconics Graphic Design/Jacqueline Comstock
2106 Hollow Rd, Glen Spey, NY 914-856-5544
ICONS/Glenn Johnson
76 Elm St #313, Boston, MA 617-522-0165
Inverse Media/Chris Thomas
PO Box 1072, Southport, CT 203-255-9620
Irwin, Virginia
139 S Highland St #3, W Hartford, CT............203-232-6956
Jager DiPaola Kemp Design/Michael Shea
308 Pine St, Burlington, VT 802-864-5884
James, Rick
1668 E Bishopwood Blvd, Harleysville, PA 215-361-2644
Janesko, Lou
10400 Connecticut Ave #PH, Kensington, MD 301-933-5533
Jareaux, Robin
28 Eliot St, Boston, MA.............................617-524-3099
Johnson, Craig
63 Providence Ave, Doylestown, PA 215-348-2593
Johnson, James
25 Brahms St, Boston, MA.........................617-325-7957

Jones, Maureen
PO Box 648, Amagansett, NY 516-267-6159
Jones, Ric
1149 E Court, Sunbury, PA 717-988-1701
Kaczmarkiewicz, Maryanne
3323 Beaver Dr, Yorktown Hgts, NY 914-962-7563
Kahl, David B
85 Jefferson St #13, Hoboken, NJ 201-963-7975
Kent, Nicholas
138 W Olive, Long Beach, NY 516-431-4258
Kitses, John
Longmeadow Rd, Lincoln, MA 617-259-0804
Klopp, Karyl
5209 8th Ave/Cnstitution Qtrs, Charlestown, MA 617-242-7463
Knowlton, Ken
51 Pond View Dr, Merrimack, NH 603-424-2360
Krause, Dorothy Simpson
32 Nathaniel Way, Marshfield Hills, MA............617-837-1682
Kretzschmar, Art
206 Barton Ave, Point Pleasant, NJ 908-295-5625
Lackow, Andy
7004 Boulevard E #29C, Guttenberg, NJ..........201-854-2770
Lahr, Kimberly
1019 Spruce St, Philadelphia, PA 215-887-7829
Laird Illustration & Design/Thomas L Laird
706 Scott St, Philipsburg, PA 814-342-2935
Lane, Edmund
74 Key St, Millis, MA 508-376-8752
LaRocco, Richard
8 Parnell Dr, Smithtown, NY 516-360-7796
Lavin, Arnie
23 Glenlawn Ave, Seacliff, NY 516-676-1228
Lebbad, James A
24 Independence Way, Titusville, NJ 212-645-5260
LeBlanc, Terry
65 Eustis St, Cambridge, MA 617-661-5600
Leckie, Laina
208 Bellevue Ave, Upper Montclair, NJ 201-509-2977
Leete, William W
202 Silver Lake Ave, Wakefield, RI 401-783-8055
Lehner & Whyte Design/Illustration
8-10 S Fullerton Ave, Montclair, NJ 201-746-1335
Lemelman, Martin
1286 Country Club Rd, Allentown, PA 215-395-4536
Lepine, Philip W
31 Brighton Rd, Tonawanda, NY 716-875-5490
LeVan/Barbee Studio
30 Ipswich St #211, Boston, MA 617-536-6828
Leveile Stawarz & Holl/RJ Holl
35 Old Chicopee St, Chicopee, MA 413-594-8188
LeWinter, Renee
41 Sewall St, Somerville, MA 617-628-5695
Lindroth, David
85 Broadway, W Milford, NJ 201-697-1965
Lorick, Blake
Manitou Rd RR2 Box 414, Garrison, NY 914-424-3549
Lowry, Rose
119 Little Michigan Rd , Jaffrey, NH................603-532-8433
Lussier, Robert
18 Pleasant Circle, Methuen, MA 508-670-6734
Lynch, Jeffrey
85 Roosevelt St, Garden City, NY 914-328-6709
Lynn, Jenny
18 S Letitia St, Philadelphia, PA 215-925-8967
M Scullin & Associates/Maureen A Scullin
109 W Hanover Ave, Randolph, NJ 201-907-0394
● **MACNEILL AND MACINTOSH**/
Scott A MacNeill **(P 69)**
74 York St, Lambertville, NJ 609-397-4631
Magee, Alan
Route 68 Box 132, Cushing, ME 207-354-8838
Manelis, Jessica
123 Cypress Ct, Cherry Hill, NJ 609-489-0134
Mann, Heidi
26 Custer St #2, Jamaica Plain, MA 617-522-4748
Manzione, John
7 Oak Creek Lane, Penacook, NH 603-753-6171
Marcolina Design Inc/Dan Marcolina
1100 E Hector St, Conshohocken, PA 215-940-0680
Mattingly, David B
1112 Bloomfield St, Hoboken, NJ.................201-659-7404
Mattingly, Matthew
263 Cherry, Newton, MA. 617-527-0763
Maziacyzk, Claire
834 River Rd, Schodack Landing, NY 518-732-2779

McCoy, Dan
Box 573/1079 Main St, Housatonic, MA 413-274-6211
Midnight Oil Studios
51 Melcher St, Boston, MA.........................617-350-7970
Miller, Melissa
317 High St, Milford, CT 203-389-6988
Mistretta Illustration Studio/
Andrea Mistretta
135 E Prospect St, Waldwick, NJ 201-652-7531
Montana, Leslie
35 Lexington Ave, Montclair, NJ 201-744-3407
Moran, Michael
39 Elmwood Rd, Florham Park, NJ 201-966-6229
Morecraft, Ron
97 Morris Ave, Denville, NJ 201-625-5752
Morris, Burton
400 Noble St, Pittsburgh, PA 412-682-7963
Morrow, Skip
Ware Rd/Box 123, Wilmington, VT................802-464-5523
New Edge Technologies/Nick Michael
PO Box 3016, Peterborough, NH 603-547-2263
New Media Group/Michael Endres
10018 Tenbrook Dr, Silver Spring, MD.............301-681-6100
Norton, Tom
177 Magazine St, Cambridge, MA 617-492-2609
NY Inst of Technology/Peter Voci
School of Arch & Fine Arts , Old Westbury, NY....516-876-2752
O'Connell, Francis C
468 Broadway #2, Dobbs Ferry, NY 914-693-3606
Oh, Jeffrey
6 Challenger Ct, Baltimore, MD....................410-661-6064
Otto, Jeff
161 N Keswick Ave, Glenside, PA 215-886-4690
P Beach Illustration/Peter Beach
83 Moseley Ave, Newburyport, MA 508-462-4275
P T Pie Illustrations/Peter Miserendino
33 Stonegate Dr, Southbury, CT 203-264-0908
Page-Trim, Debra
156 Putnam St, Hartford, CT 203-247-8282
Palmer, Laura Leigh
3924 Denfield Ct, Kensington, MD 301-946-5026
Panoptic Imaging/Miles Ritter
2349 St Albans St, Philadelphia, PA 215-545-4111
Paston, Herbert
28 S Silver Ln, Sunderland, MA 413-665-3366
Paul, David
264 Feronia Way, Rutherford, NJ 201-933-7157
Paul, Dayan
779 Mabie St, New Milford, NJ 201-265-3842
Pentleton, Carol
685 Chestnut Hill Rd, Chepachet, RI 401-568-0275
Perlman, David
59 Stoneham Dr, Rochester, NY...................716-381-3543
Petitto, Andrea
4309 Hosey Rd, Shortsville, NY 716-289-6004
Petrillo, Jane
22 Sunrise Dr, Williston, VT 802-434-5580
Petronio, Frank
113 Commonwealth, Rochester, NY..............716-461-5583
Phase One Graphics/Donald Rea
315 Market St, Sudbury, PA 717-286-1111
Porett, Thomas
673 Aubrey Ave, Ardmore, PA 215-896-8413
Powers, Tom
139 W Center Rd, West Stockbridge, MA..........413-232-7174
PRAXIS Digital Solutions/Katrin Eismann
PO Box 831, Camden, ME 207-236-7400
Prendergast, Michael
12 Merrill St, Newburyport, MA....................508-465-8598
Rae, William
662 Warren St, Brooklyn, NY 718-398-4423
Reuter & Associates/Bill Reuter
653 Washington Blvd, Baltimore, MD 410-385-1213
Richards, Kenn
3 Elwin Pl, E Northport, NY 516-499-7575
Rischawy, Jonathan
140 Hillcrest Rd, Flemington, NJ 908-806-3353
Robilotto, Philip
777 Stone Rd Route 9W, Glenmont, NY............518-767-3196
Rodriguez, Gisela
10 Chestnut St #3206, Springfield, MA 413-782-6870
Roland, George S
435 Sunset Dr, Meadville, PA 814-333-2006
Roman, Dianne
48 Main St #3, Somerville, MA.....................617-776-5146

Roth, Wayne
103 Cornelia St, Boonton, NJ 201-316-5411
Ruby Shoes Studio/Susan Tyrrell
124 Watertown St #1E, Watertown, MA............617-923-9965
Santo, Vincent
2 Skibo Ln, Mamaroneck, NY 914-698-4667
● **SCHLOWSKY PHOTOGRAPHY AND COMPUTER IMAGERY**/Bob & Lois Schlowsky (P 20,21)
73 Old Rd, Weston, MA.............617-899-5110
Schreiber, Dana (Mr)
36 Center St, Collinsville, CT 203-693-6688
Selman, Jan
79 Pinecrest Bch Dr, E Falmouth, MA 508-540-4586
Sesto, Carl
10 Rolfe's Ln, Newbury, MA 508-462-3783
Shaw, Barclay
170 East St, Sharon, CT 203-364-5974
Sherwood, Melanie
158 Glenwood Ave, Rochester, NY 716-458-9625
Smallwood, Bud
1505 Lewis O'Gray Dr, Saugus, MA 617-231-2075
Smith, Ellen
185 South Rd, Marlborough, CT 203-295-0004
Smith-Evers, Nancy
147 Franklin St, Stoneham, MA 617-438-5716
Smolenski, Peter
55 Olive St, Northampton, MA 413-584-5105
Sokolowski, Ted
RD #2 Box 408, Lake Ariel, PA 717-937-4527
Sonneville, Dane
PO Box 155, Passaic, NJ.............201-472-1225
Sound Image Inc/Jim Goodell
42 Richards St, Worcester, MA.............508-756-0673
Spear, Charles
456 9th St #2, Hoboken, NJ 201-798-6466
Spohn, David
614 Ellsworth Dr, Silver Spring, MD 301-589-3461
St John, Bob
PO Box 1043 , York Beach, ME 508-762-2417
Staada, Glenn
490 Schooley Mt Rd, Hackettstown, NJ 908-852-4949
Stabler, Barton
831 Willow Grove Rd, Westfield, NJ 908-789-7415
Stafford, Rod
1491 Dewey Ave, Rochester, NY.............716-647-6200
Stein, Marion
PO Box 1333, Lake Grove, NY 516-981-0685
Steuer, Sharon
205 Valley Rd, Bethany, CT.............203-393-3981
Studio Twenty Six Media/Mario Henri Chakkour
209 W Central, Natick, MA.............508-653-3132
Symington, Gary
145 Newbury St, Portland, ME 207-774-4977
Tamura, David
412 N Midland Ave, Upper Nyack, NY.............914-358-4704
Tech Graphics/Jim Sullivan
215 Salem St #3, Woburn, MA.............617-933-6988
● **TEICH, DAVID (P 71)**
41 Tamara Dr/PO Box 246, Roosevelt, NJ 609-448-5036
Thompson, Arthur
39 Prospect Ave, Pompton Plains, NJ 201-835-3534
Tiani, Alex
PO Box 4530, Greenwich, CT 203-661-3891
Tourtellott, Mark
12 Martyn St, Waltham, MA.............617-647-9615
Towler, Matthew
277 Lake Ave, Worcester, MA.............508-791-2416
Trend Multimedia/Tony Rosa
4 Holly Ave, West Keansburg, NJ 908-787-0786
Tyava Productions/Mary Ross
259 Oak St, Binghamton, NY 607-722-1457
Umansky, Steven
84 Shore Rd, Port Wahington, NY 516-621-5543
University of Vermont/Henry See
Art Dept/Williams Hall, Burlington, VT 802-656-2014
Valla, Victor R
19 Prospect St, Falls Village, CT 203-824-5014
Vella, Ray
20 N Broadway Bldg I-240, White Plains, NY 914-997-1424
Villanova, Joseph
2810 Altantic Ave, Stottville, NY 518-828-0141
Visual Graphic Communications
177 Newtown Tpke, Weston, CT 203-222-1608
Visual Logic/Jack Harris
724 Yorklyn Rd #150, Hockessin, DE 302-234-5707

Vitali, Julius
1246 Lower South Main St, Bangor, PA.............215-588-7366
Voelkl, Michael
1994 Five Mile Line Rd, Penfield, NY 716-381-4205
Wade, Renee
13 Glenforge Dr, Sicklerville, NJ.............609-875-7784
Wasserman, Cary
6 Porter Rd, Cambridge, MA.............617-492-5621
Watts, Mark
2004 Par Dr, Doylestown, PA 215-343-8490
Westinghouse CSS/John Wolowiec
11 Stanwix St #936, Pittsburgh, PA.............412-642-3246
Wilczynski, Alina
380 Barclay St, Perth Amboy, NJ 908-442-2508
Williams, Monica
418 Edgewater Rd, Pasadena , MD.............410-437-4325
Wilson, Barry
654-1/2 Morton Pl NE, Washington, DC 703-276-5386
Winston, Matthew
7 Wainwright Rd #15, Winchester, MA 617-721-0101
Wolf Design, Stephen
904 Garden St, Hoboken, NJ 201-659-2422
Wolf-Hubbard, Marcie
1507 Ballard St, Silver Spring, MD 301-585-5815
Wray, Lisa
RR1 Box 147, Springville, PA 717-836-2887
Wright, Richard
43 Potter St, Haddonfield, NJ 609-427-9369
Wuilleumier Inc, Will
607 Boylston St, Boston, MA.............617-266-0103
Yarmolinsky, Miriam
806 Kennebec Ave #2, Takoma Park, MD 301-495-6342
Young Assoc, Robert
78 N Union St, Rochester, NY 716-546-1973
Zima, Al
64 Cooke Ave, Carteret, NJ 908-969-0636

SOUTH

Alias Research/Gina Coniglio
14001 Dallas Pkwy #1200, Dallas, TX 214-934-6704
Alvarado, Leon
1212 North Post Oak #100, Houston, TX...........713-688-3636
Art Images/John Ashley Bellamy
2200 N Haskell Ave, Dallas, TX.............214-827-2032
Bartholomew, Sandra Steen
136 Bennington Rd, Charlottesville, VA.............804-979-8252
Big Hand/Craig Rispin
2226 Elm St, Dallas, TX.............214-748-2888
Blake, Juliet
4829 Lake Arjaro Dr, W Palm Beach, FL 407-478-9210
Blessen, Karen
4323 Bluffview Blvd, Dallas, TX.............214-352-9192
Boger, Claire
96 Vicksburg Cove #222, Memphis, TN.............407-487-2264
Bornstein, Stephen
2381 NE 193rd St, N Miami Beach, FL 305-933-2300
Brown, Rob
130 Springfield Ln, Madison, AL 205-772-3283
Bruner, Rick Ernest
PO Box 1469, Shepherdstown, WV.............304-876-0945
Burke Design Group/Jack Burke
301 E 7th St Suite 201, Charlotte, NC.............704-579-1545
● **BURKEY, J W (P 35)**
1526 Edison St, Dallas, TX.............214-746-6336
Carter, Greg
7704 Crown Crest Ct, Raleigh, NC.............919-676-0238
Chezem Studio/Doug Chezem
3613 Cornell Rd, Fairfax, VA.............703-591-5424
Compaq Computer Corp/Chris Purcell
PO Box 692000, Houston, TX.............713-374-4679
Corbitt, John
900 Daytona Dr NE, Palm Bay, FL 407-729-6154
Crane, Gary
1511 W Little Creek Rd, Norfolk, VA 804-523-7520
Creative Dept Inc/Douglas Bowman
303 E Fifth, Edmond, OK.............405-348-0856
Cuevas, George
4640 NW Seventh St, Miami, FL.............305-672-4142
Davick, Linda
4805 Hilldale Ln, Knoxville, TN.............615-546-1020
Davis, Stephen
365 NE 156th St, Miami, FL 305-940-9832
Dawson, Will
11004 E 11th Pl, Tulsa, OK.............918-234-1362
Debela, Acha
3547 Mayfair St #G06, Durham, NC.............919-419-0250

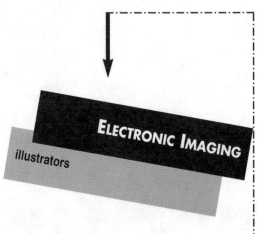

Digital Medical Images/Carrie DiLorenzo
60 Kesick Ct, N Augusta, SC 706-721-6173
Dixon, David
8 Gentry Carson Dr, Gray, TN 615-283-0484
● **DOROTHEA TAYLOR-PALMER/**
T P Design **(P 62)**
490 Stone Mtn-Lithonia Rd #124,
Stone Mountain, GA.............404-413-8276
Downtown Design/Larry Goode
2414 A South Lamar, Austin, TX 512-707-1192
Drew, Ned
1642 N Grace St, Richmond, VA 804-367-6250
Ebersol, Rob
734 Clairemont Ave, Decatur, GA.............404-371-9136
Electric Pencil Studio
PO Box 7016, Greenville, SC 803-244-1369
Encompassed Graphics/Patrick McKeon
101 South Jennings Suite 5304, Ft Worth , TX 817-336-1118
● **FEARLESS DESIGNS**/Jeff Tull **(P 58)**
622 E Main St #206, Louisville, KY 502-584-1333
Findley, John
729 Edgewook Ave NE Ste G, Atlanta, GA.........404-623-6370
Foge, Kurt
8422 Singapore Ct, Orlando, FL.............407-657-0595
Freeman, Nancy J
3600 Sprucedale Dr, Annandale, VA 703-750-0025
Fukuda, Fujie
101B Sea Oats Dr, Juno Beach, FL.............407-626-2164
Gaijin Studios/Karl Story
5581 Peachtree, Chamblee, GA 404-986-0453
Galileo/Mike Wittenstein
680 14th St, Atlanta, GA 404-425-4536
Geniac, Ruth
9555 Trulock Ct, Orlando, FL.............407-677-4801
● **GLASGOW & ASSOC, DALE (P 67)**
4493 Andy Ct, Woodbridge, VA 703-590-1702
Graphic Zone, The/David Brown
10 Office Park Cir #100, Birmingham, AL...........205-870-5300
Graphics at the Speed of Light/Robert Vann
PO Box 952344, Lake Mary, FL 407-330-5230
Hall, Susan
7500 NW First Ct #110, Plantation, FL.............305-923-5111
Hamilton Jr, Robert
5900 Riverdale Rd #7, College Park, GA...........404-994-9067
Harlan, Steve
10510 Reeds Landing Cir, Burke, VA 703-250-2410
Hathaway, Andrew J
104 Childers St, Clarksville, TN.............615-648-0340
Hattersley, Lissa
4530 B Ave G, Austin, TX.............512-458-2697
Herring, David
6246 Woodlake Dr, Atlanta, GA.............404-945-8652
High, Philip
3480 Greenlawn Dr, Lexington, KY 606-272-3060
Hillier, Karen
2301 Bristol, Bryan, TX 409-822-5528
Holmes, Michael
409A N Tyler, Dallas, TX 214-942-4061
Humphrey Jr, John J
2046 Albert Circle, Wilmington, NC 919-452-7062
Imagen/Al Schmidt
2631 Commerce, Dallas, TX 214-748-7288
In You Wendo Design/Wendy Meyer
5636 Souchak Dr, W Palm Beach, FL.............407-686-4847
Jackson Design/Beth Middleworth
220 25th Ave N #205, Nashville, TN 615-327-2387

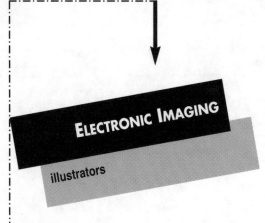

ELECTRONIC IMAGING

illustrators

Kinard, Lou
311 E 8th St #B, Charlotte, NC 704-347-8873
Leicht, Christina
12342 Hunter's Chase Dr #1313, Austin, TX........512-219-0156
Lewczak, Scott
1600 E Jefferson Ct, Sterling, VA 703-435-5982
Lowry Graphics, David
PO Box 121861, Nashville, TN 615-298-5841
Luckett, Julie L
810 Bellevue Rd #280, Nashville, TN 615-646-9581
MacMedia Systems of Orlando/
Jon Anthony Blumhagen
1022 S Lee Ave, Orlando, FL 407-425-9160
Maikoetter, Mitch
13 Lone Star Trail, Wimberley, TX 512-847-1518
Maile, Richard W
3232 Valley View St, Powder Springs, GA 404-439-6747
McLain, W Clay
9211 E Lake Highlands Rd, Dallas, TX 214-324-0168
Media Management Systems Inc/
Wynne Ragland
One Meca Way #600, Norcross , GA 404-564-5606
Mediascape/Laura Smith
PO Drawer 3948, West Palm Beach, FL 407-687-0099
Morgan Communicating Art Ltd/Craig Morgan
966 Highland View NE, Atlanta, GA 404-874-0743
Morris, Don
848 Greenwood Ave NE, Atlanta, GA 404-875-1363
Morrow, Beret
14500 Marsh Ln #283, Addison, TX 214-233-0972
Morrow, Michael
5508 Dorset Shoals Rd, Douglasville, GA 404-949-2745
Nelson, Pamela
7700 Bailey Cove Rd, Huntsville, AL 205-881-3623
Nuthouse Studios/Jeff Seldin
11365 SW 123rd Terrace, Miami, FL 305-254-6535
Of Mice & Graphics Inc/Henry Patton
1570-D Dekalb Ave, Atlanta, GA 404-377-6504
Paul Aho Artworks/Paul Aho
1610 S Dixie Hghwy, West Palm Beach, FL 407-833-9117
Performing Artists Gallery/Philip Rosenberg
209 SW 2nd St, Ft Lauderdale, FL 305-728-8887
Phillips, Chet
6527 Del Norte Lane, Dallas, TX 214-987-2234
Pixel Media/Brad Gerleman
6441 Oriole, Dallas, TX 214-350-2485
Posey, David
815 Devon Pl, Alexandria, VA 703-836-8162
Q-Burn Records/Torin Hill
429 Moreland Ave NE, Atlanta, GA 404-222-8410
Rauchman & Assocs/Robert Rauchman
5210 SW 60th Pl, Miami, FL 305-663-9432
Richards, Robin M Nance & Courtland
PO Box 59734, Birmingham, AL 205-871-8923
Roanoke College/Elizabeth K Heil
221 College Lane, Salem, VA 703-375-2361
Romeo, Richard
1066 NW 96th Ave, Ft Lauderdale, FL 305-472-0072
Rose, Lee
4250 T C Jester Blvd, Houston, TX 713-686-4799
Rupp Art & Design, Katherine
8511 Cheltenham Cir, Louisville, KY 502-425-9266
Rusciano, Patty
3522 Ridge Rd, Durham, NC 919-489-7661
S A & A Inc/Suzanne Anderson & Assocs
17 Executive Park #100, Atlanta, GA 404-636-0134

Sakai, Kazuya
1804 Roxton, Richardson, TX 214-234-0502
SANA/Edgehill Studio
17 Executive Park Dr NE #100, Atlanta, GA 404-636-0134
Sanson, Wells, Rogers/Jeff Sanson
1212 N Post Oak St , Houston, TX 713-688-0637
Savage Art & Design, David/
Mr David Savage
PO Box 1422, Boca Raton, FL 407-394-4644
Schuster, Robert
45 Porter Ln, Ft Thomas, KY 513-369-1713
Scriber, Cheryl
4201 Durham Cir, Stone Mountain, GA 404-271-0719
Sense Interactive/Dave Crossley
1412 W Alabama, Houston, TX 713-523-5757
Sherwin, Cynthia A
2515 NE Expressway #L-10, Atlanta, GA 404-248-1453
Short, Robbie
2903 Bentwood Dr, Marietta, GA 404-565-7811
Simerman, Tony
Castle Harbor Way, Centreville, VA 703-802-4950
Stargardt, Fred
8014 Timberlane Dr, Tampa, FL 813-884-4725
Stat Cat/Bill Jenkins
2400 McKinney Ave, Dallas, TX 214-720-0606
● **STRUTHERS, DOUG (P 52,53)**
6453 Southpoint Dr, Dallas, TX 214-931-0838
Studio One/Carol F Hines
1102 Buckinghgam Ave, Norfolk, VA 804-423-1028
Thomas Marketing Communications, Steve
409 East Blvd, Charlotte, NC 704-332-4624
Thompson, Jim
Rte 2/Box 419, Statesville, NC 704-876-4492
Thonen, Rod
8621 Fort Hunt Rd, Alexandria, VA 703-781-0412
Tillander, Michelle
3530 Bapaume Ave, Norfolk, VA 804-857-7269
Toney, Allen
3064 Wallace Cir, Huntington, WV 304-523-7744
Truly Computer Graphics
417 Wakefield Dr, League City, TX 800-829-4990
Turner, Dave
800 Clearview Dr, Nashville, TN 615-297-5377
Uni-Graphix/Monica Harrion
2715 Kingsroad Ave, Jackson, MS 601-355-4008
Vincent, Wayne
957 N Livingston St, Arlington, VA 703-532-8551
Walker, Kevin
15 McMakin Dr, Greenville, SC 803-298-4419
Walter, Melissa Jo
1231 31st St South, Birmingham, AL 205-703-0418
Weithers, Arlington
PO Box 585, Tuskegee, AL 205-727-3514
Wink, David
1391 Willivee Dr, Decatur, GA 404-325-4895
Z-AX-IS/Steven T Shepard
3997 Cocoa Plum Cir, Coconut Creek, FL 305-970-9662
Zimmerman, Robert
16 Gertrude Pl, Asheville, NC 704-252-9689
● **ZUBER-MALLISON, CAROL (P 66)**
2340 Edwin St, Ft Worth, TX 214-906-4162

M I D W E S T

17th St Studios/Gary Olsen
PO Box 855, Dubuque, IA 319-589-5017
Ahrens Photo, Bob
400 N State, Chicago, IL 312-243-5550
Allen, Dave
3188 Eagle's Way Dr, Lafayette, IN 317-474-9704
Apartment 3D/Bob Staake
726 S Ballas Rd, St Louis, MO 314-961-2303
Apartment 3D
726 S Ballas Rd, St Louis, MO 314-961-2303
Asmus, David
3055 Marshall Ave, Cincinnati, OH 513-861-6122
Atomic Art/Tom Miller
6564 Tulip Lane, Middletown, OH 513-779-9818
BDG Production Studio/Julie Mikos
213 W Institute Place #702, Chicago , IL 312-440-0240
Becker, Scott
1528 N Elk Grove Ave, Chicago, IL 312-862-8373
● **BELDING, PAM (P 77)**
1601 Brightwood Dr, Minneapolis, MN 612-476-1338
Berendsen & Associates Inc
2233 Kemper Ln, Cincinnati, OH 513-861-1400

Big Nasty Redhead Inc/Karen Nigida
812 W Van Buren, Chicago, IL 312-721-5137
Birkey, Randal
635 S Home, Oak Park, IL 708-386-5150
Boileau, Lowell
45 Colorado, Highland Park, MI 313-865-3084
Bradfield, Rod
1201 S Center, Terre Haute, IN 812-234-3854
Brewer, Benita
2445-209 S Fairfield, Ft Wayne, IN 219-456-1756
Britton, Arlen
300 Ames, Northfield, MN 507-645-4813
Bush, Diana J
1604 W Farragut Ave, Chicago, IL 312-878-6438
c/o Rita Marie & Friends/William Rieser
405 N Wabash Ave #2709, Chicago, IL 312-222-0337
Carsello Design/Margaret Carsello
117 S Morgan #204, Chicago, IL 312-733-5709
Communigrafix
5550 W Armstrong Ave, Chicago, IL 312-774-3012
Craig, John
RT 2/Box 2224/Tower Rd, Soldiers Grove, WI....608-872-2371
Cully, Mike
401 E Ontario #1109, Chicago, IL 312-440-9208
Cyberplex/Pat Maun
201 SE Main St #215, Minneapolis, MN 612-649-4641
Dammer, Mike
350 W Ontario #601, Chicago, IL 312-943-4995
Design Moves/Laurie Medeiros
405 N Wabash Ave #1312, Chicago, IL 312-661-0999
Digital Ink/David Garon
409 W Maryland St, Duluth, MN 218-724-3020
Diversified Graphics/Greg Phillips
5601 W 74th St, Indianapolis, IN 317-291-6200
Dodson, Liz
1920 S First St #2002, Minneapolis, MN 612-333-8150
Dufour/Anthony Rammer
532 S 8th St, Sheboygan, WI 414-457-9191
Electric Art Company/Ralph Greenhow
3305 W Harrison St, Chicago, IL 312-444-1168
Fearless Eye/Brad Matthieson
310 Delaware St #210, Kansas City, MO 816-221-1047
Frank Miller Illus & Photo/Frank Miller
6016 Blue Circle Dr, Minnetonka, MN 612-935-8888
Goldstein, Edith
1516 Wilmar Rd, Cleveland Heights, OH 216-381-7811
Goodfellow, Stephen
146 Farrand Park, Detroit, MI 313-883-4827
Goodman, John
4116 N Ciarendon #2, Chicago, IL 312-296-0893
GraFX Creative Imaging/Frank Defino Jr
2917 N Latoria Ln, Franklin Park, IL 800-633-7887
Graphic Alchemy/Andrew L McClary
3708 Richelieu Rd, Indianapolis, IN....................317-897-2745
GTP Design Studios/Galen T Pauling
PO Box 3150, Southfield, MI 313-533-7674
● **H-GUN/DIGITAL DIV/**Robert Bial **(P 43)**
2024 S Wabash St 7th Fl, Chicago, IL 312-808-0134
Hall, Lane
1610 Esplanade Ave, Davenport, IA 319-323-6804
Hand to Mouse Arts/Nancy Wirgig McClure
3700 E 34th St, Minneapolis, MN........................612-724-1172
Hanley, Katherine
1831 E 61st St, Indianapolis, IN 317-251-7989
Harman, Richard
207 S Cottonwood, Republic, MO......................417-732-2914
Harris, Martin
9721 W 16th St, St Louis Park, MN 612-546-4573
Hayes, Kevin
4808 S Union Ave, Chicago, IL 312-227-6351
Head Spin Studio/Mike Miller
250 Broadmeadows, Columbus, OH 614-785-0578
Heeter, Carrie
510 Cowley, E Lansing, MI 517-353-5497
Heifner Communications/Greg Heifner
4451 I-70 Drive NW, Columbia, MO 314-445-5855
Hellman Associates/Kathy Forslund
400 First Ave N #218, Minneapolis, MN 612-375-9598
Henriquez Studio/Scott Henriquez
2223 W Roscoe St, Chicago, IL 312-883-9747
Hesselberg, Brenda
554 Cherbourg Ct S, Buffalo Grove, IL 708-634-2388
Howard-Statesman, Deborah
PO Box 178 197, Toledo, OH 419-335-3340
Image Productions
115 W Church, Libertyville, IL 708-680-7100

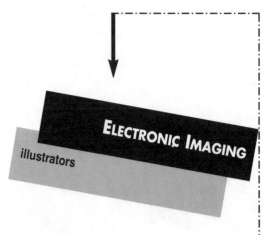

Imageland/Jerry Capozzoli
333 N Michigan Ave, Chicago, IL..........312-984-1003
JD Media Consultants/Dan Brennan
1309 E Dawes Ave, Wheaton, IL...........708-668-0559
Jet Litho/David Graunke
1500 Center Circle, Downer's Grove, IL............708-932-6370
Johnson Design Inc, Dean
604 Ft Wayne Ave, Indianapolis, IN................317-634-8020
Johnson Illustration, Paul
11412 N Port Washington Rd #202-13,
Mequon, WI......................................414-241-4484
Johnson, Diane
73A Hawthorne Rd, Barrington, IL 312-382-4634
K Squared Inc/David Kogan
1242 W Washington Blvd, Chicago, IL........312-421-7345
Kadison-Shapiro, Sari
1432 Elmwood Ave, Evanston, IL...........708-866-8080
Katz, Gary S
7517 W 154th Terr, Overland Park, KS...........913-897-9444
Kirk, Bev
5815 Sovereign Dr, Cincinnati, OH..................513-530-5353
Kitzerow, Scott
3505 N Pine Grove, Chicago, IL 312-935-9234
Kliger, David
2836 N Francisco Ave #2-A, Chicago, IL............312-292-0399
Kohut, Paul
3803 W National Ave #6, Milwaukee, WI 414-649-8806
Kowalski, Stephen
5521 N Glenwood Ave, Chicago, IL...........312-561-8899
KV Graphics/Bob Cavey
710 Canterbury Cir, Chanhassen, MN...........612-949-2902
LaLiberte, Richard
211 E Delaware Pl, Chicago, IL...........312-944-1825
Lange Design, Jim
203 N Wabash #1312, Chicago, IL...........312-606-9313
Legs Akimbo/Charlie Athanas
915 Dempster, Evanston, IL...................708-332-2136
Lim, Deborah
505 N Lake Shore Dr #5606, Chicago, IL...........312-527-3271
Lindbloom, Bruce
7370 Walnut Ct, Eden Prairie, MN...........612-937-9627
Lochray, Tom
3225 Oakland Ave, Minneapolis, MN...........612-823-7630
Logical Art/Jeffrey Weinstein
18447 Adrian, Southfield, MI...................313-559-4392
LunaGrafix/Gala Muench
Route 1/Box 17 28601 S Gem Rd, Harrison, ID..208-689-3877
MacIntosh, Guy
714 Enright Ave, Cincinnati, OH...........513-244-7160
Master's Touch Inc, The
PO Box 474, Elkhart, IN......................219-295-1492
Matthews, Scott
7528 Ethel Ave, St Louis, MO...........314-647-9899
MFS Inc/Patrick J Wendland
2193 Willow Grove Ave, Kettering, OH...........513-299-0503
Midwest Litho Ctr/3D Imaging/Derek Frederickson
125 E Oakton, Des Plaines, IL...........708-296-2000
● **MILLET, CECILE (P 73)**
1280 7th Ave NW, Hutchinson, MN...........612-587-0897
Minneapolis College of Art & Design/Thomas A DeBiasso
Media Arts/2501 Stevens Ave S,
Minneapolis, MN..................................612-874-3638
Minnix, Gary
201 S Cuyler, Oak Park, IL...................708-386-4484
Mitchell, Mindy
2715 Yemans St, Hamtramck, MI 313-874-3023
Mockensturm, Steve
2660 Laybourn Ave, Toledo, OH...........419-474-0484
Moetus, Olaf
3115 Lake Ave #E, Wilmette, IL...........708-251-4237
Murphy, Charles
4146 Pillsbury Ave, Minneapolis, MN...........612-827-8166
Musgrave, Steve
202 S State St #1324, Chicago, IL...........312-939-4717
Nease, Rick
7760 King's Run, Sylvania , OH...........419-882-7805
Niffeneger, Bill
1801 W Larchmont Loft Four, Chicago, IL...........312-528-9302
Nova Media/Thomas Rundquist
206 1/2 S Michigan/PO Box 414,
Big Rapids, MI..................................616-796-7539
Nyberg, Tim
3307 Victoria St N, Saint Paul, MN...........612-490-7559
Obata Design
1610 Menard, St Louis, MO 314-241-1710

Ohio State Univ/Advertising Computing Center for Art & Design/Charles Csuri
1224 Kinnear Rd, Columbus, OH...........614-292-3416
Onli, Turtel
5121 S Ellis, Chicago, IL 312-684-2280
Paul Hertz Media Arts/Paul Hertz
2215 W Fletcher St, Chicago, IL...........312-975-9153
● **PELIKAN PICTURES**/Petra Pepellashi **(P 88,89)**
20500 Civic Center Dr #2800, Southfield, MI.....313-356-2470
Photonics Graphics/Alan J Brown
700 W Pete Rose Wy Lobby B-360,
Cincinnati, OH..................................513-723-4440
Photos Ink/Medical & Legal Visuals
1819 Minnehaha, Minneapolis, MN 612-371-0140
Preslicka, Greg
7730 Grinnell Way, Lakeville , MN...........612-432-2166
Prestige Production & Graphic/John Basso Jr
148 E Aurora Rd (Rte 82), Northfield, OH...........216-467-8400
Printers Inc/Paul Begner
13050 W Custer Ave, Butler, WI...........414-781-1887
Prism Studios/Tom Farmer
2412 Nicollet Ave, Minneapolis, MN 800-659-2001
Product Illustrations Inc/Barbara Zidlicky
233 E Ontario #900, Chicago, IL...........312-943-7311
Pulse Imaging/Tim Clark
3323 N Seminary #3 South, Chicago, IL...........312-525-0170
Puza, Greg
PO Box 1522, Milwaukee, WI...........414-671-1960
Quicksilver Assoc Inc/Kevin Johnson
18 W Ontario St, Chicago, IL 312-943-7622
Rafferty Communications/Jim Rafferty
1518 139th Ln NW, Andover, MN...........612-755-8488
RamPage Electronic Imaging/Tom Patterson
2233 Kemper Ln, Cincinnati, OH...........513-281-8095
RCF Graphics/Craig Fansler
51 Pomeroy Rd, Athens, OH 614-594-5603
RE Miller Communications/Roger E Miller
3520 W 21st St, Minneapolis, MN...........612-925-0781
RedGrafix/Dralene "Red" Hughes
19750 W Observatory Rd, New Berlin, WI........414-542-5547
Ressler, Susan
603 S 10th St, Lafayette, IN...........317-494-3811
Riskind, Jay
505 N Lake Shore Dr #3505, Chicago, IL...........312-644-0638
Rogala, Miroslaw
1524 S Peoria St, Chicago, IL...........312-243-2952
Sacks, Ron/Ron Sacks
1189 Rosebank Dr, Worthington, OH 614-846-1921
Sahulka, Lawrence
5228 11th Ave S, Minneapolis, MN...........612-823-5490
Sandbox Digital Playground/Pam Weston
203 N Wabash #1602, Chicago, IL...........312-372-1170
Sapulich, Joe
8454 W 161st Pl, Tinley Park, IL...........708-532-8766
Sauer & Associates/Christian Sauer
2844 Arsenal, St Louis, MO...........314-664-4646
Schneider, William
15 Morris Ave, Athens, OH 614-594-3205
School of the Art Inst/Claudia Cumbie-Jones
112 S Michigan/Data Bank, Chicago, IL...........312-345-3550
Scientific Illustrators/George Morris
2416 E Washington, Urbana, IL 217-356-0273
Shaff, Tom
1862 Selby Ave, St Paul, MN 612-645-3822
Shaw, Ned
2770 N Smith Pike, Bloomington, IN 812-333-2181
Sherman, John
734 Park Ave, South Bend, IN...........219-287-7369
Siemer, Patrick
1809 W Division St, Chicago, IL...........312-862-4244
Silva, Raul
3507 N Racine #G, Chicago, IL...........312-549-0361
Smyth, Richard F
1235 Glenview Rd, Glenview, IL 708-998-8345
Sowash, Randy
550 Sunset Blvd, Mansfield, OH...........419-756-7139
Strauss Design/Ron Strauss
2469 University Ave W, St Paul, MN...........612-644-7244
Studio One
7300 Metro Blvd Ste 400, Edina, MN...........612-831-6313
Studio One Inc/David Kuettel
7300 Metro Blvd, Edina, MN...........612-831-6313
Synergy Art & Tech/Michael Simmons
230 Tenth Ave SE #213, Minneapolis, MN........612-371-9181

Tate, Clarke
301 Woodford St, Gridley, IL 309-747-3388
Taylor Corporation
8601 Urbandale Ave, Des Moines, IA.........515-276-0992
Taylor, Joseph
2117 Ewing Ave, Evanston, IL...........708-328-2454
Thomas/Bradley Design/Brad Neal
411 Center St, Gridley, IL...........309-747-3266
Three & Associates
2245 Gilbert Ave #200, Cincinnati, OH.........513-281-1600
University of Notre Dame/John Sherman
Dep of Art, Hist & Des, Notre Dame, IN 219-631-5000
Vera, Clara Claudia
1446 W Polk St 2nd Fl, Chicago, IL...........312-227-9372
Warman, Brian
4922 Duebber Dr, Cincinnati, OH...........513-922-6326
Weber, John
3637 Ridgewood Dr, Hilliard, OH...........614-777-0631
Willens + Michigan Corp/Bob Levison
1959 E Jefferson Ave, Detroit, MI...........313-567-8900
Willow Graphics/Lind Babcock
219 Tinkler St, Lafayette, IN 317-742-2203
Willson Creative Group/William Willson
355 N Ashland, Chicago, IL 312-738-3555
Wilson, Lin
1236 W Carmen Ave, Chicago, IL...........312-275-7172
Witte, Mary Stieglitz
3637 Stone Creek Way, Boise, ID 208-383-9493
Wright, Ted
3286 Ivanhoe, St Louis, MO...........314-781-7377
Youngblood, Michael S
RR1 Box 265/Kings Indian Farm, Carbondale, IL.618-457-6497
Zada, Nida
5622 Delmar #510, St Louis, MO 314-454-0818
Zakari, Chantal
2105 N Oakley, Chicago, IL 312-252-2432
Zale Studios/David Zale
3529 Highway Ave, Highland, IN 219-838-4254
Zoot, Ira
1456 N Dayton, Chicago, IL...........312-280-0048

W E S T

Acosta, Ralph
710 E Sunrise Blvd, Long Beach , CA...........310-427-5775
Advanced Concepts/Rich Lovato
4864 Valley Hi Dr, Sacramento, CA...........916-429-2655
Aguilar, Rob
171 Coventry Dr, Campbell, CA 408-378-6402
Ahhhhh Graphics/Andrew Hinshaw
19000 MacArthur Blvd #620, Irvine, CA...........714-476-6366
Allen Design, Mark
129 20th St, Manhattan Beach , CA...........310-396-6471
Allison, Linda
PO Box 2646, San Anselmo, CA...........415-485-0630
Anderson, Darrel
1420 Territory Trail, Colorado Springs, CO.........719-535-0407
Anderson, Sara
3131 Western Ave #516, Seattle, WA...........206-285-1520
Anderson, Terry
5902 W 85th Pl, Los Angeles, CA...........310-645-8469
Andraleria
1817 Shasta Ranch Rd, Mt Shasta, CA 916-926-4681
Andrew Faulkner Design/Andrew Faulkner
3020 Bridgeway #107, Sausalito, CA...........415-332-3521
Anigraf/x/Robert Stein III
517 W Tenth St, Medford, OR 503-857-0614

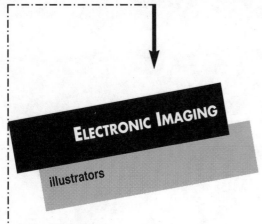

Antigravity Workshop/Dan Lutz
456 Lincoln Blvd, Santa Monica, CA 310-393-9747
Arion, Katherine
1162 N Orange Grove #6, W Hollywood, CA ...213-654-6252
Art for Medicine/Iris Nichols
1509 45th Ave SW, Seattle, WA 206-932-3398
Artlab/Tony DeYoung
1603 Howard St, San Francisco, CA 415-554-0248
Artmarx/Scott Annis
PO Box 22582, Denver, CO 303-758-7905
ASA Darkrooms
905 N Cole Ave #2, Los Angeles, CA................213-463-7513
Ashford, Janet
327 Glenmont Dr, Solana Beach, CA..................619-481-7065
Ashton, Larry
1629 Myrtle Ave, San Diego, CA......................619-543-9658
Atelier Graphics/Dana Trousil
396 Imperial Wy #310, Daly City, CA 415-755-8568
Avery Illustration, Design & Adv/Tony Avery
19132 Magnolia #C17, Huntington Beach, CA...714-962-8862
Avila, Winona
1739 Comstock Lane, San Jose , CA 408-266-7216
Axelrod, Dale
3415 22nd St, San Francisco, CA......................415-824-1549
B Graphic/Richard & Beth Cordero
1442 E Lincoln #180, Orange, CA 714-998-8371
Baby Blues/Rick Kirkman
2432 W Peoria Ave #1191, Phoenix, AZ 602-997-6004
Bahry, Sharon
1880 N El Camino Real #88, San Clemente, CA.714-492-8839
● **BAKER, DON (P 55)**
2814 72nd Ave NW, Seattle, WA......................206-784-1136
● **BAKER, KOLEA (P 54,55)**
2814 72nd Ave NW, Seattle, WA......................206-784-1136
Barnard, Doug
28805 S Lakeshore Dr, Agoura Hills, CA 818-991-9328
Barta, Les
755- Burgundy Lane, Incline Village, NV 702-831-0430
Bartalos, Michael
4222 18th St, San Francisco, CA......................415-863-4569
Bartlett, Michael
24A Varda Landing, Sausalito, CA 415-331-5127
Bates, Betty
1060 Malone Rd, San Jose, CA 408-266-1972
Bates, Karla M
410 Bellevue #309, Oakland, CA......................510-452-0811
Beach, Lou
1114 S Citrus Ave, Los Angeles, CA..................213-934-7335
Beckerman, Carol
4350 Clark Ave, Long Beach, CA......................310-420-2603
Bee Show!, Johnee
3183-G Airway Ave, Costa Mesa , CA 714-708-2270
Beggs Design
619 Maybell Ave, Palo Alto, CA 415-857-9539
Bender, Jon
305 Brookwood Ave, San Jose , CA 408-993-1930
Berlin, Jeff
238A Summit Dr, Corte Madera, CA 415-979-8488
Best Digital/Tony Lobue
1895 Park Ave #203, San Jose, CA 408-261-7410
Biedny, David
PO Box 151498, San Rafael, CA........................415-721-0638
Big Deal Graphics/David Deal
1651 Monte Vista Dr, Vista, CA 619-758-2655
● **BIG PIXEL, THE**/Douglas Buchman **(P 70)**
3188-K Airway Ave, Costa Mesa, CA................714-433-7400

Bio Design/Steve Lustig
17151 Newhope St #205, Fountain Valley, CA..714-434-2810
Black Point Group/Gary Priester
340 Townsend St #410, San Francisco, CA........415-331-4531
Blavatt, Kathleen
743 Sunset Cliffs Blvd, San Diego, CA 619-222-0057
Bliss, Anna Campbell
27 University St, Salt Lake City, UT....................801-364-5835
Bodily, Michael
27151 El Moro, Mission Viejo, CA 714-457-1228
Bollinger, Rebeca
1379 Vallejo St #4, San Francisco, CA...............415-749-1882
Bonauro, Tom
601 Minnesota St #216, San Francisco, CA........415-648-5233
Bookmakers Ltd/David Bolinsky
PO Box 1086, Taos, NM................................505-776-5435
Borruso, John
1259 Guerrero St, San Francisco, CA 415-647-1972
Borruso, John
1259 Guerrero St, San Francisco, CA 415-647-1972
Bradshaw, Stephen
811 York St, Oakland, CA 510-839-6875
Bremmer, Mark
7155 W Walden Place, Littleton, CO.................303-932-8759
● **BRICE, JEFF (P 54)**
2814 72nd Ave NW, Seattle, WA......................206-784-1136
Brocke Graphic Design, Robert
425 30th St #25, Newport Beach, CA...............714-673-4281
Brody, Bill
PO Box 82533, Fairbanks, AK..........................907-479-4139
Brownwood, Bruce
15402 Saranac Dr, Whittier, CA.......................310-947-5770
Bueno, Luz
548 Cragmont, Berkeley, CA 510-524-2163
Buhler, Ray Varn
Blue Mountain Rd, Wilseyville, CA....................209-293-4169
Cal Poly State Univ/Eric R Johnson
Art Department, San Luis Obispo, CA................805-544-5121
Capstone Studios
5371 Wilshire Blvd #200, Los Angeles, CA 213-936-1156
Carden, Vince
2308 E Glenoaks Blvd, Glendale, CA.................818-956-0807
Carrier, Alan
1521 Bidwell Dr, Chico, CA.............................916-894-5911
Carter, Mary
Box 421443, San Francisco, CA.......................415-647-5660
Carver Design, Stephen
PO Box 9266, Santa Rosa, CA 707-578-7302
Catalog Design & Production Inc
1485 Bay Shore Blvd #25, San Francisco, CA.....415-468-5500
Chan, Ron
110 Sutter St #706, San Francisco, CA..............415-441-4384
Chandler, Roger
8100 Paseo del Ocaso, La Jolla, CA..................619-551-1135
Chang, Dr Rodney
2119 N King #206, Honolulu, HI......................808-845-6216
Chang, George
4 Commodore Dr, Emeryville, CA.....................510-654-7744
Chase Design, Margo
2255 Bancroft Ave, Los Angeles, CA 213-668-1055
Cherry, Jim
902 E Palm Ln, Phoenix, AZ............................602-340-0715
Chui, John
1500 Laurel, Richmond, CA.............................510-233-2333
Cloud Art & Design, Gregory
2116 Arlington Ave #236, Los Angeles, CA........213-484-9479
Cohen, Hagit
865 14th St #3, San Francisco, CA....................415-552-9677
Cohn, Larry
367 N Clark St, Orange, CA.............................714-997-4552
Colby, Tracy
PO Box 10651, Portland, OR............................503-223-7537
Collen, Mark
151 11th Ave #C21, Seattle, WA 206-682-3583
Communications Design/Jeff Bane
8950 Cal Ctr Dr, Sacramento, CA......................916-362-0400
● **CONRAD & ASSOCIATES, JON/**
Jon Conrad **(P 68)**
221 W Maple Ave, Monrovia, CA......................818-301-9662
Cope, Doug
316 S Del Sol Ln, Irvine, CA............................714-540-9195
Creative Ape/Jonathan Krop
16795 Mission Way, Sonoma, CA......................707-935-7434
Crutchfield, William
2011 S Mesa St, San Pedro, CA........................310-548-4132

Cudlitz, Stuart
1745 Pacific Ave, San Francisco, CA..................415-775-5044
Curtin Design, Paul
1865 Clay St #1, San Francisco, CA 415-885-0546
Dahlquist, Roland
13049 N 18th Dr, Phoenix, AZ.........................602-993-9895
Dakini Designs/Denise Nugent
PO Box 13066, Burton, WA 206-463-5412
Daley, Joann
3101 Peninsula Rd #121, Oxnard, CA................805-985-1608
● **DAMAN STUDIO, TODD (P 25)**
2019 Third Ave #200, Seattle, WA....................206-448-9318
Dangel, Corey
915 Queen Anne N, Seattle , WA 206-286-6423
Davis, Jack
832 Hymettus Ave, Encinitas, CA.....................619-944-7753
Davis, Robert
72 Belcher St, San Francisco, CA......................415-621-0865
Davis, Sally
702 W Halladay, Seattle, WA 206-283-3800
Dawson, Hank
3519 170 Pl NE, Bellevue, WA.........................206-882-3303
Dayal, Antar
1596 Wright St, Santa Rosa, CA.......................707-544-8103
Deen, Georganne
3834 Aloha St, Los Angeles, CA.......................213-665-2700
Dell'Aquila, Mei Ying
2820 Cozumel Cir, Santa Clara, CA...................408-246-8875
Denham, Karl
20568 Ventura Blvd #315, Woodland Hills, CA.818-347-1676
Desert Rat Design/Steve Cooper
Box 125, Springdale, UT 801-772-3327
Dietz, Mike
23962 Dovekie Circle, Laguna Niguel, CA 714-448-0652
Digit Eyes/Rik Henderson
11231 Otsego, N Hollywood, CA.......................818-769-3021
● **DIGITAL ART**/Tim Alt **(P 40,41)**
3166 E Palmdale Blvd #120, Palmdale, CA........805-265-8092
Digital Design Simulations/Dan Biggs
PO Box 565, Bonita, CA 619-421-2107
Doerner Graphics/Dan Doerner
835 Diamond St #A, San Francisco, CA 415-826-6590
Doherty, James
411 S 11th St #4, San Jose, CA........................408-297-5744
Dougherty, Suzanne
13534 Myren Dr, Saratoga, CA........................408-867-1646
Dowlen Artworks/James Dowlen
2129 Grahn Dr, Santa Rosa, CA 707-579-1535
Dream Merchant Graphics/David Donovan
437 Engel Ave, Henderson, NV.........................702-564-3598
Dumptruck Studios/Richard Haffar
63 S 800 East, Salt Lake City, UT 801-328-8309
Dynamic Perspectives/Linda Deschambault
2066 Donald Dr, Moraga, CA...........................510-631-9017
E-Conspiracy/Brian Walls
518 44th St Apt B, Oakland, CA........................510-601-5611
Edge Graphics/Andrea Jennison
565 Parfet St, Lakewood, CO 303-969-9146
Electric Easel/Georgina Curry
10229 N Scottsdale Rd #C, Scottsdale, AZ........602-443-8786
Electric Image Inc/Shawn Steiner
117 E Colorado Blvd #300, Pasadena, CA...........818-577-1627
Electronic Images/Nick Fain
300 Broadway #32, San Francisco, CA...............415-398-3434
Elledge, Leslie
4219 Montgomery St, Oakland, CA...................510-654-4802
Elson, Matt
11901 Sunset Blvd #207, Los Angeles, CA 310-471-4511
Endo, Stan
1522 Rosalia Rd, Los Angeles, CA 213-662-3838
Euphonics/Patrice M Warrender
2685 Burnside Rd, Sebastopol, CA....................800-892-3325
EXP Graphics/Bruce Bennett
432 N Canal St #12, S San Francisco, CA............415-583-8236
Fanning, Rich
174 South O St, Lincoln, CA.............................916-645-7077
Faragher, Patsy
1414 Morro St, San Luis Obispo, CA 805-545-7750
Feder, Eudice
13122 Rangoon St, Arleta, CA..........................818-764-8190
Fenster, Diane
140 Berendos Ave, Pacifica, CA........................415-355-5007
Ferster, Gary
3363 Melendy Dr, San Carlos, CA.....................415-598-0115
Finfirst/David McDougall
5842 6th Ave NW, Seattle, WA.........................206-783-9234

Fink, Mike
4434 Matilija Ave, Sherman Oaks , CA..............818-789-5232

Fisher, Reed
2866 Via Bellota, San Clemente, CA.............714-498-0634

Flax, Carol
2356 Cotner Ave, Los Angeles, CA 310-478-2963

Flom, Eric
829 Lincoln Ave, Alameda , CA 510-769-9391

Foster Digital Imaging/Jeff Foster
530 E Lambert Rd, Brea , CA.............................714-671-0880

Fox, Mark
239 Marin St, San Rafael, CA.........................415-258-9663

From Art to Design Inc/Ron McPherson
1336 9th St, Manhattan Beach, CA.................310-372-7777

Froman, Loralie
4040 18th St, San Francisco, CA......................415-647-5697

Future Perfect/Richard Duardo
1201 E 5th St, Los Avgeles, CA.......................213-620-6234

Gage, Hal
2008 E Northern Lights Blvd, Anchorage, AK 907-272-4356

Gasowski, Igor
1220 Colusa Ave, Berkeley, CA.......................510-524-3777

Gaviota Graphics/Jim Biebl
094 Arlian Ln, Carbondale, CO........................303-963-3309

George, Cathy
PO Box 620727, Woodside, CA........................415-851-2704

Gersch, Wolfgang
255 Stuyvesant Dr, San Anselmo , CA...............415-258-8210

Global One Design & Innovation/Drew Cronk
PO Box 2218, Healdsburg, CA..........................707-431-7664

Goehring, Steven
25456 Bull Run, Alpine, OR...............................503-424-5443

Goldberg, Ken
1006 W Edgeware Rd, Los Avgeles, CA.............213-740-9080

Golden, Helen
460 El Capitan Pl, Palo Alto, CA 415-494-3461

Goldstein, Howard
7031 Aldea Ave, Van Nuys, CA......................818-987-2837

Good, Zane
1900 Iris Way, Escondido, CA...........................619-738-9299

Goyer, Mireille H
7600 W Manchester Ave #1313,
Playa Del Ray, CA..310-827-8791

GP Color Inc/John Portaro
201 S Oxford Ave, Los Angeles, CA.................213-386-7901

Grahame, Donald
1645 Folsom St #3, San Francisco, CA...............415-626-7116

Graphic Design Group/Richard Browski
453 N Genesee Ave, Los Angeles, CA................213-653-2454

Graphics West/Stephen West
21801 Linda Dr, Torrance, CA...........................310-540-3190

Gravity Design/Marcus Badgely
221 San Carlos Ave, Sausalito, CA 415-331-0191

Grey Matter Design/Roxana Villa
16771 Addison St, Encino, CA..........................818-906-3355

● **GROSSMAN, MYRON (P 75)**
12 S Fair Oaks Ave, Pasadena, CA...................818-795-6992

Grossman, Rhoda
216 4th St, Sausalito, CA.................................415-331-0328

Gruel, George
2875 Zapata Ct, Simi Valley, CA 805-529-2727

Guerin, Francois
7545 Charmant Dr #1322, San Diego, CA.........619-457-1546

Hada, Gail
23290 Clearpool, Harbor City, CA 310-539-5114

Haleen, Brentano
PO Box 148, Tesuque, NM 505-986-1799

Hall, Jeffrie
174 W Hollyglen Ln, San Dimas, CA 909-599-3802

Halley Design/Lynda Halley
PO Box 3685, Chatsworth, CA..........................818-407-1642

Hamilton, Bruce & Susan
Rte 1 Box 5C, Glorieta, NM 505-757-6603

Hammer, Bonnie
2105 Broadview Terrace, Los Angeles, CA..........213-851-3681

Hauser, Karl X
1094 Revere Unit A41, San Francisco, CA 415-822-2523

Hebert, Jean-Pierre
4647 Via Huerto, Santa Barbara, CA.................805-964-4699

Henderling, Lisa
370 Glenn St, Ashland, OR 503-488-8036

Hicks, Robin
153 Caselli Ave, San Francisco, CA...................415-252-0277

Higgs, Simon
PO Box 3083, Van Nuys, CA............................818-989-5638

Hillis, Craig F
550 Thomas St, Woodland, CA........................916-668-5848

Hobbs, Pamela
4222 18th St, San Francisco, CA......................415-255-8080

Hoerr, Fred
500 Molino St #315, Los Angeles, CA...............213-680-4188

Hoffman, Patricia
369 Montezuma #316, Santa Fe, NM...............505-983-3165

Holcomb, Mike
3846 Peppertree Dr, Eugene, OR......................503-342-1693

● **HOWE, PHILIP (P 32,33)**
540 First Ave S, Seattle, WA...........................206-682-3453

Huggins, Rucker & Cleo
196 Castro St, Mt View, CA.............................415-960-1951

Hunt, Robert
107 Crescent Rd, San Anselmo, CA...................415-459-6882

Hunter, Nadine
PO Box 307, Ross, CA 415-456-7711

Hutton, Lisa
4606 Castelar St, San Diego, CA.......................619-224-6327

Ikeda, Tomoya
5400 Broadway Terrace #207, Oakland, CA......510-655-9029

Illusion Factory, The/Brian Weiner
23875 Ventura Blvd #104, Calabasas, CA 818-223-8400

Imagine That/Debbie Coleman
2918 S Market, Redding, CA............................916-245-5800

Inner Visions Group/William Boddy
609 N 10th St, Sacramento, CA 916-443-5001

InnerStellar Productions/Francis Hobbs
904 Anita Ave, Big Bear City, CA......................909-585-8495

Iselin, Josephine
28 Lloyd St, San Francisco, CA 415-864-1269

Ivanoff, Deborah
204 Greenfield Dr #F, El Cajon, CA...................619-390-3573

Jane Lily Design/Mason Lyte
610 Anacapa St, Santa Barbara, CA.................805-683-4884

Jasin, Mark
11936 W Jefferson Blvd #C, Culver City, CA 310-390-8663

Joffe, Barbara
7271 S Jersey Court, Englewood, CO................303-843-0346

Johnson, Jay
2260 N Beachwood Dr, Hollywood, CA.............213-464-2606

Jonason, Dave
Camino San Cristobal #4, Galisteo, NM...........505-820-6220

Juliana, Peter
PO Box 985 45812 Old Corral Rd,
Coarsegold, CA...209-683-2726

Kaino, Glenn
17830 S Noran Circle, Cerritos, CA...................310-921-0580

Karas, G Brian
4126 North 34th St, Phoenix, AZ 602-956-5666

Keller, Steve
11936 W Jefferson Blvd #C, Culver City, CA 310-390-8663

KHL Consulting/Kathy Hecht-Lopez
3871 Hatton St, San Diego, CA.........................619-576-4140

Killian, Ted
7660 Cathedral Oaks #7, Goleta, CA.................805-685-4827

King, Stephen
930 Hermes Ave, Encinitas, CA........................619-944-8914

King-Judge, Cynthia
PO Box 4644 , Montebello, CA 213-721-3826

Kirk, Roberta
Highway 101 N/Box 873, Yachats, OR 503-547-4250

Klitsner Industrial Design/Antonio Angulo
636 Fourth St, San Francisco, CA 415-957-1529

Knox, David
2424 N Rose, Mesa, AZ 602-827-9339

Koehnline, James
PO Box 85777, Seattle, WA.............................206-633-2608

LaBerge Graphic Design/Mary Lou LaBerge
PO Box 726, Vancouver, WA 206-573-8283

Landman, Mark
365 Maple Ave, Cotati, CA..............................707-792-1326

Lawhead, Elizabeth
7934 SE 36th, Portland, OR 503-242-0034

● **LAX SYNTAX DESIGN**/Lance Jackson **(P 61)**
1790 5th St, Berkeley, CA 510-849-4313

Lazerus
PO Box 13249, Oakland, CA............................510-339-6263

Lehman, Cassandra
568 N 1st St #1, San Jose, CA..........................408-998-1314

Lens Design, Jenny/Jenny Lens
8111 Remmet Ave #15, Canoga Park, CA.........818-716-1567

Levin, Lon
1317 Avenida De Cortez, Palisades, CA.............818-972-4970

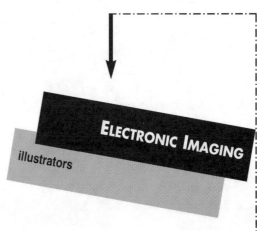

ELECTRONIC IMAGING

illustrators

Levy, Brian
808 Coleman #5, Menlo Park, CA 415-474-7551

Liao Inc, Sharmen
314 N Mission Dr, San Gabriel, CA....................818-458-7699

Linker, James Alan
1235 S Roosevelt St, Tempe, AZ.......................602-921-1516

Lipman, Michael
55 Loring Ave, Mill Valley, CA..........................415-383-4248

Lizarraga, Sergio
1424 Gamma Pl, Anaheim, CA.........................714-778-5692

Lockett, Carolyn L
265 First Ave #3, Salt Lake City, UT..................801-359-7914

Lockwood, Scott
2109 Stoner Ave, Los Angeles, CA....................310-312-9923

Love, Nan
PO Box 5004, Santa Rosa, CA..........................707-527-5683

LSI Graphic Evidence
200 Corporate Pt #300, Culver City, CA.............310-568-1831

Luckwitz, Matthew
1675 Logan St, Denver, CO..............................303-839-8442

Ludtke, Jim
10 Ford St, San Francisco, CA 415-863-6187

Lynch Graphic Design/David Lynch
8800 Venice Blvd #216, Los Angeles, CA..........310-287-0440

Lynx Digital Design/Alejandro Rubalcava
16 Technology #115, Irvine, CA........................714-727-3126

Lyons, Steven
136 Scenic Rd, Fairfax, CA 415-459-7560

MacDonald, Greg
4220 NW 1st, Seattle, WA.............................206-789-0388

MacKenzie Graphic Design/Vic MacKenzie
1913A Ruhland Ave, Redondo Beach, CA...........310-374-7911

MacNicol, Gregory
732 Chestnut St, Santa Cruz, CA 408-459-0880

● **MAD WORKS**/Mitchell Anthony **(P 74)**
960 Maddux Dr, Palo Alto, CA 415-494-3240

Magiera, Rob
8400 Kingscove Dr, Salt Lake City, UT 801-943-3650

Magika/Max Almy
3454 Standish Dr, Encino, CA 818-789-8540

Mar Design, William
220 Montgomery #608, San Francisco, CA........415-989-3935

Markley, Andy
720 Sunrise Ave #24, Roseville, CA...................916-773-6442

Martin Brinkerhoff Associates/Randall Lubert
17767 Mitchell, Irvine, CA...............................714-660-9396

Mason Studio, John
PO Box 3973, Carmel, CA...............................408-625-3868

Mattioli, Angela
10655 Rochester Ave, Los Angeles, CA 310-475-9883

MAX/Max Seabaugh
246 First St #310, San Francisco, CA.................415-543-7775

Maximum Impact Design/Judi Oyama
303 Potrero St #7, Santa Cruz, CA 408-425-1810

● **MC2**/Heidi Merscher **(P 48)**
65 Lwr Arroya Hondo Rd, Arroyo Hondo, NM....505-776-1333

McCord, Jeff
1932 First Ave #819, Seattle, WA.....................206-443-1965

McCormick, Peter
13726 Aleppo Dr, Sun City West, AZ 602-584-8403

McCourt, Tim
257 Hampton Dr, Venice, CA............................310-450-6523

McManus, Robert
3333 W Wethersfield Rd, Phoenix, AZ 602-993-8659

Media Design/Steve McKinstry
3350 Airport Dr Ste #300, Bellingham, WA........206-293-4035

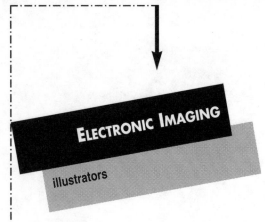

ELECTRONIC IMAGING

illustrators

Media Mark/Mark James Tippin
680 Alberta Ave #D, Sunnyvale, CA.................408-730-9919
Messex, Mike
2014 Vista del Mar, Los Angeles, CA 213-466-2818
Metafor Imaging/Judy Linden
39 E Walnut St, Pasadena, CA...........................818-584-4034
Meyerfeld, Alan
839 Banneker Dr, San Diego, CA.......................619-462-5007
Midnight Oil Studios/Ethan Hutcheson
100 Bush St, San Francisco, CA 415-834-0384
Miller, Jack Paul
370 W Cedar Ave, Burbank, CA 818-841-4668
Mitchell Photo, Josh
8033 Sunset Blvd #490, Los Angeles, CA 213-227-1000
Monroy, Bert
1052 Curtis St, Albany, CA 510-524-9412
Moojedi, Kamran
900 W Sierra Madre #122, Azusa, CA.............818-969-5508
Morse, Bill
173 18th Ave, San Francisco, CA.......................415-221-6711
Multimedia Professionals
5835 Avenida Encinas #127, Carlsbad, CA 619-929-8788
Myman, Barbara T
PO Box 81, Point Reyes Station, CA...................415-663-8059
Nairn Gaphic Design, Doreen
PO Box 55756, Sherman Oaks, CA....................818-501-5065
Nasty Productions/Glenn Grillo
23337 Califa St, Woodland Hills, CA.................310-285-8509
Nexus Design & Marketing/Craig Calsbeek
1316 3rd St Promenade #109,
Santa Monica, CA...310-394-6037
No Steroids Design/Rob Schultz
1409 N Alta Vista Blvd #105, Los Angeles, CA...213-850-8209
O'Shea, Kevin
20414 32nd Pl S, Seattle, WA 206-870-0772
Ogdemli/Feldman Design/Daniel Feldman
11911 Magnolia Blvd #39, N Hollywood, CA....818-760-1759
Okada, Corinne
4120A 22nd St, San Francisco, CA....................415-826-6648
On the Wave Visual Communications/
John Ulliman
2339 Ward St #8, Berkeley, CA 510-649-8514
Orange Coast Comm College/
Electronic Media/Joan Salinger
2701 Fairview Rd/Box 5005, Costa Mesa, CA...714-432-5691
Osiecki, Lori
123 W 2nd St, Mesa, AZ.................................602-962-5233
Osiow, Andrew
2735 River Plaza Dr #131, Sacramento, CA.......916-922-1384
Otus, Erol
509 Bonnie, El Cerrito, CA 510-528-2053
● **PACE, JULIE (P 79)**
678 Wellsley Dr, Lake Arrowhead, CA...............909-337-0731
Paintpot Productions/Tony Readhead
3777 May St, Mar Vista, CA.............................310-398-5612
Palermo, David
3379 Calle Santiago, Carlsbad, CA...................619-944-9907
Pandemonium Design/Juliette Beckstrand
342 W 200nd St #220 , Salt Lake City, UT 801-531-9407
Parsons Design, Glenn
8522 National Blvd #108, Culver City, CA.........310-559-6571
Partners by Design/Marty Safir
11240 Magnolia Blvd Ste 205,
N Hollywood, CA...818-509-0555
Paschal, Richard
1296 N Chester Ave, Pasadena, CA 818-398-1799

Paternoster, Nance
546 Wisconsin St, San Francisco, CA.................415-641-1922
Patton Brothers Design/Tom Gulatta, Court Patton
3768 Miles Court, Spring Valley, CA 619-463-4562
Paul, Edie
859 Hollywood #136, Burbank, CA...................818-505-1874
Payne, Lawrence
13265 SW Aragon St, Beaverton, OR................503-644-7158
Peji, Bennett
1110-B Torrey Pines Rd, La Jolla, CA.................619-456-8071
Pellerin, Dana
6655 N Fresno St #107, Fresno, CA 209-439-3197
Perry, Rebecca
15532 Antioch St #510, Pacific Palisades, CA ...310-459-0071
Peters Design, David
2141 Walnut Ave, Venice, CA 310-390-3528
Peterson, Marty
160 Copperdle Ln/Coal Crk Cnyn, Golden, CO..303-642-0453
Phase 2 Digital Arts/Barry Chall
2111 147th Ave, Hayward, CA.......................510-276-1633
Philip, Nick
615 Cole St, San Francisco, CA 415-221-7289
Phototime/Mark Toal
138 Stanford Shopping Ctr, Palo Alto, CA.........415-326-7687
Pina, C David
409 S Beachwood Dr, Burbank, CA...................818-972-9239
Pitts, Tom
1045 17th St #A, San Francisco, CA.................415-861-7053
Pixel Propogator/Jeffrey Schier
3928 Shafter, Oakland, CA.............................510-653-5825
Podevin, Jean-Francois
5812 Newlin Ave, Whittier, CA 310-693-3601
Pravda, Kit Monroe
2148 Sand Hill Rd, Menlo Park, CA 415-854-1050
Price, Joan
1469 Canyon Rd, Sante Fe, NM.......................505-986-3823
Psychic Dog Illustration/Linda Modaff
8210 Creighton St, Los Angeles, CA.................310-641-6916
Puckett Design, David
16 Prairie Falcon, Aliso Viejo, CA.....................714-837-4417
Rahner, Andrea
22 S 40 Pier/Waldo Pnt Harbor, Sausalito, CA ...415-331-0249
Rainbow Technology/Earl Jarred
3900 E Camelback #406, Phoenix, AZ 602-956-5936
Rearick, Kevin
293 Goldenwood Cir, Simi Valley, CA...............805-584-9259
Rico, Antonio
4138 Geary Blvd #1, San Francisco, CA............415-668-9514
Roxburgh, Ed
4316 W Point Loma Blvd - Unit , San Diego, CA..619-225-1438
Rozasy, Frank
2228 3rd St #12, Santa Monica, CA.................310-399-1891
Ryane, Nathen
San Diego, CA..619-260-2472
Sachs, Jim
28971 Banoff Dr Box 1182,
Lake Arrowhead, CA.......................................909-337-5838
Sammel, Chelsea
25 Esquiline Rd, Carmel Valley, CA...................408-659-1813
Scan Magazine/Stewart McSherry
4782 Panorama, San Diego, CA.......................619-295-8829
Schminke, Karin
236 Beverly, Laguna Beach, CA 714-497-1921
Scopinich, Robert
21026 Pacific Coast Hwy, Malibu, CA...............310-456-7569
Scripps Research Institute
10666 N Torrey Pines Rd, La Jolla, CA 619-455-9100
Seelig, Derek
201 S Francisca #10, Redondo Beach, CA 310-318-9587
Seigel/Inocencio/Matthew Seigel
33 Vandewater St #302, San Francisco, CA.......415-433-5817
Sexton Design/Rob Sexton
16812 Red Hill Ave, Irvine, CA 714-474-7525
Shultz, David
1118 E Platte Ave, Colorado Springs, CO...........719-473-1641
Sky Tree/Kit Croucher
50 Kipling Dr #3, Mill Valley, CA.......................415-383-7157
Slavin, Daniel
150 S Harwood St, Orange, CA.......................714-744-0118
Slide Factory/Carlos Lara
300 Broadway #14, San Francisco, CA.............415-957-1369
Sloan, Rick
9432 Appalachian Dr, Sacramento, CA.............916-364-5844
Smith, CJ
881 Church St, Woodburn, OR 503-981-0095

● **SMITH, MARTY (P 65)**
2402 Michelson Dr #220, Irvine, CA 714-962-0461
Smith, Steve
400 West South Boulder, Louisville, CO..............303-666-7907
Smool, Carl
1528 Valentine Place S, Seattle, WA.................206-328-7920
Sorensen, Marcos
3740 25th St #305, San Francisco, CA.............415-282-5796
Sorensen, Vibeke
2322-D La Costa Ave, Carlsbad, CA 619-943-0170
Stehney, Regina
25591 Via Inez, San Juan Capistrano, CA.........714-496-1652
Steinkamp, Jennifer
1299 Inverness Dr, Pasadena , CA 818-578-1011
Stevens, Alex
2816 Burkshire Ave, Los Angeles, CA...............310-473-6939
Stone, April
1178 Gardenside Lane, Cupertino, CA...............408-725-1293
Stratton, Mary M
7708 Etiwanda, Reseda, CA 818-757-1921
Stubbs, Diane N
3355 Spring Mountain Rd #19, Las Vegas, NV ...702-871-2711
Studer, Gordon
1576 62nd St, Emeryville, CA 510-655-4256
Studio M D
1512 Alaskan Way, Seattle, WA......................206-682-6221
Studio MD/Glenn Mitsui
1512 Alaskan Way, Seattle, WA......................206-682-6221
Sutton, Jeremy
245 Everett Ave, Palo Alto, CA.........................415-325-3493
Swaine, Mike
6735 N 10th Pl, Phoenix, AZ...........................602-264-5400
Syd Mead Inc/Roger Servick
1716 N Gardner St, Los Angeles, CA.................213-850-5225
Tanenbaum Inc, Robert
5505 Corbin Ave, Tarzana, CA 818-345-6741
● **THE BLANK COMPANY**/Jerry Blank
(P 59)
1048 Lincoln Ave, San Jose, CA.......................408-289-9095
Thornock, Christopher
1677 E Beverly Dr, Pasadena, CA 818-791-4714
Threinen, Cher
475 San Gorgonio St, San Diego, CA.................619-226-6050
Tishman, Jill Rosean
PO Box 1592, Sante Fe, NM.............................505-986-9987
Torinus, Sigi
185 Collingwood, San Francisco, CA.................415-558-8346
● **TRACER DESIGN**/Chad Little (P 42)
4206 N Central Ave, Phoenix, AZ.....................602-265-9030
Tracy, Donna
2011 Vista Cerro Gordo St, Los Angeles, CA 213-666-4087
Trici Venola & Co/Trici Venola
410 California Ave #3, Santa Monica, CA..........310-395-5475
TSA Design Group/Jeff Turner
4505 1/2 N Sepulveda Blvd, Sherman Oaks, CA818-501-5554
Turner & DeVries
701 N Kalaheo Ave, Kailua, HI.........................808-261-2179
Tuveson, Christine
1119 Hi-Point St, Los Angeles, CA.....................213-936-5851
Tyler, Wayne R
3035 E Middleton Way, Salt Lake City, UT 801-272-9320
Valesco, Frances
135 Jersey St, San Francisco, CA.......................415-647-5607
Valle, Robin
513-1/2 N Spaulding Ave, Los Angeles, CA 213-653-1238
Vanderbos, Joseph
211 40 St Way #2, Oakland, CA.....................510-653-3751
Vandervoort, Gene
3201 S Ramona Dr, Santa Ana, CA.................714-549-3194
Veridian/Ronnie Sampson
268 Ninth Ave, San Francisco, CA.....................415-979-4980
Vesna, Victoria
Univ California/Art Studio, Santa Barbara, CA....805-893-2852
VideoGraphicArts/Anne Farrell
131 Huddleson, Santa Fe, NM..........................505-983-5126
Vietor, Noel
1250 Prospect St, La Jolla, CA 619-551-0520
Voris, Rebecca J
7031 Serenity Cir, Anchorage, AK....................907-243-5234
Walker, Todd
2890 N Orlando Ave, Tucson, AZ 602-327-1569
Weisman, David
332 Bayview St, San Rafael, CA.......................415-455-9628

West Design, Jeffery
736 N 17th St, San Jose, CA............................408-971-0504
Whitaker, Corinne
1074 Pine Oak Lane, Pasadena, CA 818-795-0103
Wilson, Nancy Patton
1305 Boulevard Way #115,
Walnut Creek, CA.......................................510-934-1014
Wokuluk, Jon
1301 S Westgate Ave, Los Angeles, CA.............310-473-5623
Wolf, Dixon
2008 Kiva rd, Santa Fe, NM 505-982-0155
Woo, Don
8 Avocet Dr #205, Redwood Shores, CA............415-593-4803
Works, The/Joel Shively
820 E Ocean Blvd, Long Beach, CA 310-436-0343
Worthington, Nancy
PO Box 2558, Sebastopol, CA.........................707-823-3581
Wright, Michael Ragsdale
2021 S Alameda #10, Los Angeles, CA 213-748-4022
Wrinkle, James
1141 Calle Pensamiento,
Thousand Oaks, CA.....................................805-495-5732
Yazzolino, Brad
6451 SE Morrison Ct, Portland, OR..................503-238-3776
Young, Emily
2173 NE Multnomah, Portland, OR 503-281-3923
Zilberts & Assocs, Ed
5690 DTC Blvd #190, Englewood, CO 303-220-5040

CANADA

Apocalyptik Sparkz Design/Kenny K Lim
2020 Victoria Dr, Vancouver, BC......................604-251-3076
Coates, Peter
41 Wroxetere, Toronto Canada, ON 416-466-2324
Cummins, Karla
3110-850 McGill Rd, Kamloops, BC...................604-374-1982
Faith Inc/Paul Sych
1179A King St W #112, Toronto, ON.................416-539-9977
Ghiglione, Kevin
F1 9; 96 Spading Ave, Toronto, ON..................416-365-9965
Hatt, Shelley
85 Evelyn St, Ottawa, ON..............................613-237-3304
Herman, Michael P
1627A 13th Ave SW, Calgary, AB.....................403-228-3495
Infographie Canada/Joseph LeFevre
3728 Parc Ave #100, Montreal, QB..................514-527-8638
Lasry, Ronen
35 George St W, Ottawa, ON..........................613-230-7672
Mably, Greg
52 Saguenay Ave, Toronto Canada, ON............416-784-3576
Mani, Anand
1250 Comox St #406, Vancouver, BC................604-669-9507
Martin, Sean
512 Richmond St E, Toronto, ON......................416-367-2446
McDonnell, Patrick
3420 Westmore, Montreal, QB........................514-483-5489
Media Foundation/Geoff Coates
1243 W 7th Ave, Vancouver Canada, BC...........604-736-9401
Niemann, Andrew
1290 Astoria St, Victoria Canada, BC 604-383-9367
Overdrive/James Wilson
65 Liberty St, Toronto Canada, ON 416-537-2803
Rainbow Productions/Steve Arscott
7895 Tranmere Dr #4, Mississauga, ON 416-678-6596
Shoffner, Terry
11 Irwin Ave, Toronto, ON 416-967-6717
Showmakers Production/Bill Frymire
6 E Third Ave, Vancouver, BC..........................604-875-9880
Soft Image Inc/Elizabeth Jones
3510 Blvd St Laurent #214, Montreal, QB 514-845-1636
Struve-Dencher, Goesta
#303 750 E 7th Ave, Voncouver, BC 604-872-3439
Tivadar, Bote
512 Richmond St E, Toronto, ON......................416-367-2446
Tuttle, Jean
c/o 51 Camden St, Toronto Canada, ON 212-967-7699
Warnell, Ted
2511 15 A St SW, Calgary, AB 403-244-7395
Zgodzinski, Rose
512 Richmond St E, Toronto, ON......................416-367-2446

INTERNATIONAL

Martins, Marcos
Rua Nascimento Silva 107 #201,
Rio de Janeiro Brazil, ...021-521-0534

Designers

NEW YORK CITY

4-Front Video Design Inc/Jack M Beebe
1500 Broadway #509, New York, NY 212-944-7055
AFCG Inc/Floyd Gillis
305 E 46th St, New York, NY 212-688-3283
Aliman, Elie
134 Spring St, New York, NY 212-925-9621
Allied Graphic Arts/Carol Delia
1515 Broadway, New York, NY.........................212-642-9500
Alper Studio, A J
224 Elizabeth St #19, New York, NY 212-935-0039
Anagram
133-48 115th St, S Ozone Park, NY..................718-848-6176
Andersen, Rolf
45 W 84th St, New York, NY...........................212-787-3305
Antler & Baldwin Design Group
7 E 47th St, New York, NY..............................212-751-2031
Applied Graphics/David Rickerd
50 W 23rd St, New York, NY...........................212-627-4111
Applied Graphics Technologies/Brad Dorin
50 W 23rd St, New York, NY...........................212-627-4111
Art Resources & Technologies/Uri Dothan
37 W 20th St #1209-10, New York, NY 212-255-5399
Art-Pro Graphics/Sam Newman
47 E 19th St, New York, NY 212-473-4100
Associated Images/Brian Lee
545 5th Ave 3rd Fl, New York, NY....................212-687-2844
Associated Images/Tito Saubidet
545 Fifth Ave, New York, NY 212-687-2844
Atomique Film International/Alyce Wittenstein
110-20 71st Rd #PH5, Forest Hills, NY 718-520-0354
Axion, Pierce/Lior Azoulai
690 Ocean Parkway, Brooklyn, NY....................718-435-6339
Bakst, Edward
160 W 96th St, New York, NY.........................212-666-2579
Baldino Design, Patt
305 Madison Ave #956, New York, NY 212-986-5987
Batsry, Irit
PO Box 1561/Cooper Station, New York, NY.....212-260-3071
BBP Graphic Design Inc/Stephen Thompson
309 W 136th St, New York, NY........................212-926-0489
BCD Ink Ltd
108 E 16th St, New York, NY 212-420-1222
Beeton, Maija
9-01 44th Dr #2C, Long Island City, NY.............718-786-6396
Belman, Vickie
210 E 181st St #4A, Bronx, NY........................718-367-2688
Bessen Tully & Lee
220 E 23rd St 12th Fl, New York, NY.................212-213-1911
Billian, Cathy
456 Broome St, New York, NY.........................212-431-4716
Black & White Dog Studio/Julia Gran & Michael Dowdy
3240 Henry Hudson Pkwy #6H, Bronx, NY.........718-601-8820
Black Inc, Roger
PO Box 860/Radio City Station, New York, NY...212-649-2000
Black Lamb Studios/Bekka Lindstrom
31-33 Walker St 2nd Fl, New York, NY 212-334-1209
Blake Design/Susan Blake
750 Eighth Ave #503, New York, NY 212-575-4705
Blauweiss, Stephen
32-15 41st St, Long Island City, NY 718-204-8335
Bottiglieri, Dessolina
126 E 12th St #6C, New York, NY....................212-473-4592
Boyd, Cathy
547 Henry St, Brooklyn, NY............................718-875-7367
Bradford, Peter
928 Broadway #709, New York, NY 212-982-2090
Bramble Design/Bill Bramble
327 W 11th St, New York, NY 212-929-6289
Brianstorm Unlimited, Mark
180 Varick St 13th Fl, New York, NY..................212-647-9278
Broadway Video Graphics/Peter Rudoy
1619 Broadway 10th Fl, New York, NY..............212-265-7600
Cabimat/David B Kelley
88 W Broadway 2nd Fl, New York, NY..............212-385-8191
Cacioppo Production Design/Tony Cacioppo
42 E 23rd St 5th Fl , New York, NY...................212-777-1828

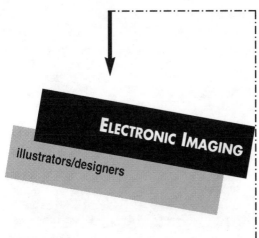

ELECTRONIC IMAGING
illustrators/designers

Calderhead & Phin/Richard Calderhead
821 Broadway 11th Fl, New York, NY................212-673-6200
Calfo/Aron Inc
156 Fifth Ave #500, New York, NY....................212-627-4054
Cantwell, James
235 Lincoln Pl #1D, Brooklyn, NY 718-789-5026
Center for Advanced Whimsy, The/
Rodney Alan Greenblatt
61 Crosby St, New York, NY 212-219-0342
Cerulli, Andrew
82-31 63rd Ave, Middle Village, NY..................718-803-2321
Chattum Design Group/Shirley Chetter
10 E 23rd #200, New York, NY........................212-673-4514
Chollick, Phyllis
8507 Homelawn St, Jamaica, NY......................718-658-0670
Class Line Graphics/Karen Miles
368 Broadway #209, New York, NY..................212-571-3678
Cohen Design Inc, Hayes
133 Cedar Rd, E Northport, NY.........................516-368-2031
Cohen, Adam
96 Greenwich Ave #3, New York, NY...............212-691-4074
Coleman Lipuma Segal & Morrill
305 E 46th St 11th Fl, New York, NY 212-421-9030
Concept & Design/Robert Mentken
51 E 97th St, New York, NY 212-534-5101
Concrete Couple Productions/John Fekner
31-14 80th St, Jackson Heights, NY...................718-651-3388
Cooke, Charles
416 W 23rd St #2D, New York, NY...................212-989-7026
Corrente, Linda
45 E 62nd St, New York, NY 212-486-3015
Creative Freelancers
25 W 45th St #703, New York, NY...................212-398-9540
Cynosure Films/Robert Luttrell
131 W 21st St, New York, NY..........................212-645-8216
David Curry Design/David Curry
338 E 53rd St #2C, New York, NY....................212-935-8130
DDB Needham Worldwide/Erik Hanson
437 Madison Ave, New York, NY......................212-415-2000
Deck Design, Barry
611 Broadway, New York, NY..........................212-777-6627
Decotech/Edwin Boria
148 E 89th St, New York, NY...........................212-860-7627
Demark Keller & Gardner, Inc/
Jeffrey Gardner
141 W 28th St 9th Fl, New York, NY 212-714-1731
Design Heads/Ray Sysko
270 Park Ave S #4E, New York, NY..................212-598-9710
Design Plus
156 Fifth Ave #712, New York, NY....................212-645-2686
Design Provisions/Haggai Shamir
190 Grand St, New York, NY...........................212-431-7129
Design Space/Kathleen Schenck
476 Broadway 7th Fl, New York, NY.................212-925-9696
Designed to Print & Assoc
130 W 25th St, New York, NY.........................212-924-2090
di Liberto, Lisa
413 3rd St #4, Brooklyn, NY............................718-768-3212
Di Re, John
265 Riverside Dr, New York, NY.......................212-749-6294
Digital Photo Works/Marc Silverman
62 W 70th St, New York, NY 212-595-5464
Dixon Design/Ted Dixon
594 Broadway #902, New York, NY 212-226-5686
Dogmatic, Irene
508 E 12th St #26, New York, NY....................212-505-7154

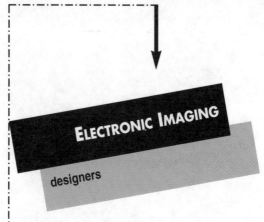

ELECTRONIC IMAGING

designers

Doublespace/David Sterling
170 Fifth Ave 2nd Fl, New York, NY 212-366-1919
Drenttel Doyle Partners/William Drenttel
1123 Broadway, New York, NY......................212-463-8787
Dressler, Marjory
4 E 2nd St, New York, NY 212-254-9758
E3 Inc/Kathleen Bordelon
16 W 22nd St, New York, NY........................212-727-7099
Edwards Design Inc, Sean Michael/Thomas S Duane
28 W 25th St 5th Fl, New York, NY 212-924-5700
Ehrentreu, Devora
488 Crown St, Brooklyn, NY.........................718-604-1176
Eisenberg, Sheryl
465 Washington St, New York, NY 212-966-4910
Electronic Publishing Ctr/Irving S Berman
110 Leroy St 2nd Fl, New York, NY................212-727-2655
Emerson/Wajdowicz Studios Inc
1123 Broadway, New York, NY......................212-807-8144
Ferri-Grant, Carson
255 W 90th St PH, New York, NY...................212-362-8567
Fleishman, Carole
16 Abingdon Sq #1B, New York, NY 212-645-9071
Frankfurt Gips Balkind Technology
244 E 58th St, New York, NY.......................212-421-5888
Friday Saturday Sunday Inc
210 E 15th St, New York, NY.......................212-353-2060
Friedman Consortium, Harold
404 E 5th Ave #14A, New York, NY 212-688-6434
Gadsden-Grant Graphic/Ingrid Williamson
676 E 57th St, Brooklyn, NY.......................718-251-0577
Gartel, Laurence
270-16 B Grand Central Pkwy, Floral Park, NY....718-229-8540
Gentile Studio
333 E 46th St, New York, NY.......................212-986-7743
Gershoni, Gil
332 Bedford Ave, Brooklyn, NY....................718-599-5671
Glazer & Kalayjian Inc
301 E 45th St, New York, NY.......................212-687-3099
Glitch Graphics/Jessica Perry
185 E 85th St #26M, New York, NY 212-722-0535
Goode , Michael
160-15 Powells Cove Blvd, Whitestone, NY 718-767-3906
Goodman/Orlick Design/Judi Orlick
150 5th Ave #407, New York, NY 212-620-9142
Gorewitz, Shalom
310 W 85th St #7C, New York, NY.................212-724-2075
Graham, Thomas
408 77th St #D4, Brooklyn, NY 718-680-2975
Graphic Chart & Map Co
236 W 26th St #801, New York, NY 212-463-0190
Graphic Expression, The
330 E 59th St, New York, NY.......................212-759-7788
Graphics by Nostradamus
250 W 57th St #1128A, New York, NY.............212-581-1362
Graphics for Industry/Mark Pahmer
8 W 30th St 7th Fl, New York, NY 212-889-6202
Greland Graphics, Gerald
70 Hudson Ave, Brooklyn, NY......................718-624-5841
Grid, Steve Chang
18 W 21st St 6th Fl, New York , NY................212-255-1806
Hahn, Alexander
PO Box 20164 /Tompkins Sq Sta, New York, NY 212-420-1368
Hamann, Brad
330 Westminster Rd, Brooklyn, NY 718-287-6086
Hamm, Julian
58-14 Seabury St, Elmhurst, NY...................718-899-5199

Handler Group Inc
22 W 23rd St 3rd Fl, New York, NY 212-645-3900
Harris Design, George
301 Cathedral Pkwy #2N, New York, NY..........212-864-8872
Hartley, Jill
361 Union St #3, Brooklyn, NY.....................718-625-6494
Hays, Sorrel
697 West End Ave, New York, NY 212-663-6164
Healy, Anne
306 W 109th St #4E, New York, NY...............212-864-0746
Heavy Meta/Barbara Glauber
596 Broadway #1212, New York, NY..............212-966-2493
Heun, Christine
337 W 20th St #2M, New York, NY 212-645-5536
High Priority Consulting/Bonnie Kane
45 First Ave #5-O, New York, NY...................212-228-6000
Himmelstein, Shelley
216 E 29th St, New York, NY.......................212-679-5614
Honkanen, William
PO Box 20402/Tompkins Sq Stn,
New York, NY 212-982-2435
Hoon, Samir
235 E 40th St, New York, NY.......................212-818-1921
Horizon Images/Nona Abiathar McCarley
534 9th Ave #D2, New York, NY 212-529-5452
Horn, Jenny
211 E 18th St #3-O, New York, NY................212-777-3015
Hot-tech Multimedia Inc/Lawrence Kaplan
46 Mercer St, New York, NY........................212-925-3010
Hudson, Ross
3655 34th St, Long Island City, NY................718-482-8500
ID Studios/Isauro De la Rosa
2 Jane St #4-A, New York, NY......................212-243-5175
Idesign/Robert M Stewart
165 E 35th St #11A, New York, NY................212-683-0244
Image Axis Inc/Kevin O'Neill
38 W 21st St, New York, NY........................212-989-5000
Ink Tank/B O'Connell
2 W 47th St 14th Fl, New York, NY 212-869-1630
Inkwell Inc
5 W 30th St, New York, NY 212-279-2066
Interface Arts/Robert Lyons
Liberty Studios/238 E 26th St, New York, NY....212-532-1865
Isley Design, Alexander
361 Broadway #111, New York, NY 212-941-7945
J & M Studio/Consulting/Jacqueline Skiles
236 W 27th St, New York, NY......................212-675-7932
Jensen Communications Group Inc/Bill Jensen
145 Sixth Ave #PH, New York, NY.................212-645-3115
Josephs, Carolyn
1045 E 59th St, Brooklyn, NY......................718-531-8375
Julie Research Laboratories/Loebe Julie
508 W 26th St, New York, NY......................212-633-6625
Juzwik, Amy Stahl
43 E 60th St, New York, NY........................212-759-5475
K Landman Design/Kathy Landman
156 Fifth Ave 3rd Fl, New York, NY 212-924-4254
Kass Communications
529 W 42nd St, New York, NY......................212-868-3133
Kauftheil/Rothchild/Henry Kauftheil
220 W 19th St #1200, New York, NY 212-633-0222
Kellum, Ron
151 First Ave #PH1, New York, NY.................212-979-2661
Keltner, Stephen
109 Sterling Pl, Brooklyn, NY.......................719-951-5325
Kendrick, Dennis
99 Bank St #3G, New York, NY....................212-924-3085
Kensinger Studio/Edward Kensinger
210 Elizabeth St, New York, NY....................212-343-9500
Ketchoyian Design, Suzanne
395 Broadway, New York, NY.......................212-924-2771
Klein, Nikoai
109 N 9th St #3R, Brooklyn, NY...................718-384-3193
Klineman, Peggy
310 W 47th St #4B, New York, NY 212-757-3460
Klinkowstein, Tom
101 Thompson St #34, New York, NY.............212-925-8213
Kode/William Kochi
20 W 20th St #308, New York, NY 212-366-0800
Kollberg-Johnson Assoc Inc
7 W 18th St 3rd Fl, New York, NY.................212-366-4320
Lance Studios/David Wasserman
151 W 46th St, New York, NY......................212-382-0290
Langenstein, Michael
56 Thomas St 2nd Fl, New York, NY 212-964-9637

Lazar, Jeff
155 E 23rd St #406, New York, NY................212-533-0010
Levine, Les
20 E 20th St, New York, NY........................212-673-8873
Lord, Rosalind
175 W 12th St #16N, New York, NY 212-807-7959
Lovejoy, Margot
166-04 81st Ave, Queens, NY......................718-969-3199
Lovitt, Anita
308 E 78th St, New York, NY.......................212-628-8171
M & Co
225 Lafayette St #904, New York, NY.............212-348-2408
M Plus M Inc/Takaaki Matsumoto
17 Cornelia St, New York, NY.......................212-807-0248
M2 Design
225 E 43rd St , New York, NY......................212-599-1616
Madridejos, Fernando
130 W 67th St #1B, New York, NY 212-724-7339
Mary Perillo Inc/Mary Perillo
125 Cedar St #8S, New York, NY 212-608-3943
Matador Design/Karl Peiler
36 Horatio St #4B, New York, NY 212-255-3879
● **MATSURI CORPORATION**/Terry Terui
(P 72)
150 W 56th St #4611, New York, NY 212-582-6001
Max Grafix/Jaime Gutierez
87-43 Kingston Pl, Jamaica Estates, NY.............718-658-5265
Maya Technology/Mark Gilliland
156 Chambers St, New York, NY 212-267-1195
McDaniel, Jerry
155 E 38th St, New York, NY.......................212-697-6170
McKenzie, Crystal
30 E 20th St #502, New York, NY 212-598-4567
McLeod, Lisa
628 E 9th St #5AB, New York, NY 212-529-5452
McNicholas, Florence
1419 Eighth Ave #1, Brooklyn, NY 718-965-0203
Merchandising Workshop/Roland Millman
550 W 43rd St, New York, NY......................212-239-4646
Metamorphosis Conputer Concept/Sonya Shannon
c/o Sch of Vis Arts 209 E 23rd, New York, NY ...212-874-4930
MetroGrafik/John Cavanagh
330 Seventh Ave, New York, NY 212-695-7778
Meyerowitz + Co, Michael
175 Fifth Avenue #609, New York, NY.............212-353-1561
Mezzina/Brown Inc/Alistair J Gillett
401 Park Ave S, New York, NY......................212-251-7718
Midnight Oil Studios/Kathryn Klein
156 Fifth Ave #623, New York, NY 212-366-9117
Mitchell, Diane
175 W 73rd St, New York, NY......................212-877-7624
Montana Graphics/Martha Towler
34-30 78 St #1J, Jackson Heights, NY 718-651-1549
Morris/Stylism, Dean
307 E 6th St #4B, New York, NY 212-420-0673
Mulhern, Mike
125 Cedar St #9N, New York, NY 212-732-2692
Mullican, Matt
182 Grand St, New York, NY.......................212-431-3629
Mulligan, Donald
418 Central Park W #81, New York, NY.............212-666-6079
Myrvik Productions, Ron
34 E 29th St, New York, NY........................212-685-0726
Mythic Graphics/Bonnie Robin Cohen
230 Riverside Drive #8H, New York, NY...........212-866-6059
Nappi Inc, Maureen
229 W 78th St #84, New York, NY 212-877-3168
Nessim & Assocs, Barbara
63 Greene St, New York, NY........................212-677-8888
Neumann, William
119 W 23rd St #206, New York, NY 212-691-7405
Nexvisions/Guy Fery
150 Fifth Ave #202, New York, NY.................212-255-4646
Ng, Michael
58-35 155th St, Flushing, NY.......................718-461-8264
Nichols, Mary Ann
80 Eighth Ave #900, New York, NY 212-727-9818
Noneman & Noneman Design
230 E 18th St, New York, NY.......................212-473-4090
Notovitz Design Inc
47 E 19th St 4th Fl, New York, NY.................212-677-9700
NovaWorks/Bill Keogh
630 Third Ave 9th Fl, New York, NY................212-557-9199
Novus Visual Communications Inc
18 W 27th St, New York, NY........................212-689-2424

Number Seventeen/Emily Oberman
33 Greenwich Ave #3G, New York, NY 212-243-2353
O Design Group Inc
27 W 24th St 10th Fl, New York, NY 212-727-1140
O&J Design/Andrzej Olejniczak
9 W 29th St 5th Fl, New York, NY 212-779-9654
Oko & Mano Inc/Alex Arce
10 E 2nd St, New York, NY 212-387-9209
Olsen Muscara Design/Dag Olsen
60 Madison #1010, New York, NY 212-684-4580
Ong, Philip
24 Monroe Pl #8C, Brooklyn, NY 718-596-0752
Our Designs Inc/Randy Tibbott
27 W 20th St #400, New York, NY 212-647-1164
Packard, Wells
235 W 22nd St #1A, New York, NY 212-691-5016
Pantuso, Mike
350 E 89th St, New York, NY 212-534-3511
Parham Santana Inc/John Parham
7 W 18th St 7th Fl, New York, NY 212-645-7501
Peled, Einat
62-28 Cromwell Crescent, Rego Park, NY 718-275-6549
Pendergrast, James
151 E 20th St #4E, New York, NY 212-473-3606
Pentagram Design
212 Fifth Ave 17th Fl, New York, NY 212-683-7000
Platinum Design
14 W 23rd St 2nd Fl, New York, NY 212-366-4000
Porter, Jill
240-29 51st Ave, Douglaston, NY 718-279-2077
Preda, Dan
598 Ninth Ave #2A, New York, NY 212-265-2145
Principato, Salvatore
220 Sullivan St #4H, New York, NY 212-477-8161
● **PUNIN, NIKOLAI (P 60)**
161 W 16th St #18E, New York, NY 212-727-7237
Quon Design Office, Mike
568 Broadway #703, New York, NY 212-226-6024
Rae, William
662 Warren St, Brooklyn, NY 718-398-4423
Rick Barry Desktop Design Studio/Rick Barry
1631 W 12th St, Brooklyn, NY 718-232-2484
Rifai, Christian
225 Central Park W #1407A, New York, NY......212-799-9375
Rio Station Graphics/Silvio Da Silva
47 Murray St 3rd Fl, New York, NY 212-964-9418
RJG Design/Robert J George
366 Sterling Place, Brooklyn, NY.....................718-783-8514
Robinson, Lenor
201 E 69th St #6E, New York, NY 212-734-0944
Rodriguez, Claudio
304 Mulberry St #4B, New York, NY 212-941-0573
Rogers Seidman Design/Gene Seidman
20 W 20th St #703, New York, NY 212-741-4687
Romanello, Kim Marie
251 W 19th St #4D, New York, NY 212-255-8276
● **ROMERO DESIGN, JAVIER**/Javier Romero
(P 57)
24 E 23rd St 3rd Fl, New York, NY 212-420-0656
Romney, Michael
201 E 77th St PH C, New York, NY..................212-288-0618
Rosebush Visions Corp/Judson Rosebush
154 W 57th St #826/Carnegie, New York, NY ..212-398-6600
Rosenworld/Laurie Rosenwald
54 W 21st St #705, New York, NY 212-675-6707
Ross Culbert & Lavery
15 W 20th St 9th Fl, New York, NY 212-206-0044
Rowland, Lauren
176 Ludlow St #1H, New York, NY 212-254-2731
Rubin, Susan
4700 Broadway #4A, New York, NY 212-569-4860
Rubyan, Robert
270 Park Ave S #7C, New York, NY 212-460-9217
Saiki & Assoc
154 W 18th St, New York , NY 212-255-0466
Salsgiver Coveney Assoc I/Karen Salsgiver
41 Union Square West #707, New York, NY 212-206-7870
Sanders, Phil
563 Van Duzer St, Staten Island, NY...............718-720-0388
Santoro Design, Joe
503 W 27th St, New York, NY 212-563-2506
Sauers, Valerie
295 St. Johns Pl #6C, Brooklyn, NY.................718-857-5863
Sawyer Studios Inc/Arnie Sawyer
115 W 27th St 8th Fl, New York, NY 212-645-4455

Schaffer, Stephanie
175 W 87th St, New York, NY 212-580-1570
Schell, Paul
1608 E 51st St, Brooklyn, NY.......................718-951-8976
Schwartz, Roberta
22 Cheever Pl, Brooklyn, NY........................718-624-4408
Seawright, James
155 Wooster St, New York, NY 212-674-6868
Seibold, J Otto
38 W 21st St #1101, New York, NY 212-366-4949
Show & Tell
39 W 38th St, New York, NY 212-840-2912
Shtern, Adele
11-21 47th Rd #3L, L I City, NY 718-706-6363
Simone, Luisa
222 E 27th St #18, New York, NY 212-679-9117
Siren Design Inc/Joan Kristensen
1204 3rd Ave #120, New York, NY 212-580-3238
Smart Design Inc/David Stowell
7 W 18th St 8th Fl, New York, NY 212-807-8150
Smith Design, Edward
1133 Broadway #1519, New York, NY 212-255-1717
Solomon Video Productions, Rob
331 W 57th St #242, New York, NY 212-541-9507
Spano Design/Michael Spano
35 Regina Ln, Staten Island, NY 718-966-8979
Sterling Group/Kurt Leunis
800 Third Ave 27th Fl, New York, NY 212-371-1919
Stewart Design Group
215 Park Ave S #1303, New York, NY 212-979-2248
Stillman Design Assocs
17 E 89th St #1-B, New York, NY 212-410-3225
Studio Francesca/Janet Giampietro
405 E 14th St #5F, New York, NY 212-677-8502
Studio Q/Carolyn Quan
285 W Broadway #620, New York, NY 212-343-1108
Swann Design/Swann Smith
340 W 19th St #10, New York, NY 212-807-8261
Terra Incognita Prods/Dan Weissman
147 W 15th St #106, New York, NY 212-929-8277
Thorpe, Peter
254 Park Ave S #6D, New York, NY 212-477-0131
Tomzak, Paul
34-33 87th St, Jackson Heights, NY...............718-429-6382
Tower Graphics/Catherine Tower
23 West 73rd St #714, New York, NY 212-874-6229
Trollbeck, Jakob
7 Cornelius St, New York, NY 212-924-6394
Twelve Point Rule Ltd/Bryan Thatcher
36 W 20th St 8th Fl, New York, NY 212-929-7644
Von Ulrich, Mark
One Union Sq W #903, New York, NY 212-989-9325
Wallace/Church Assocs/Craig Swanson
330 E 48th St, New York, NY 212-755-2903
Wang Studio Inc, Tony
18 W 21st St 6th Fl, New York, NY 212-229-1900
Waters Design/Colleen Syron
3 W 18th St 8th Fl, New York, NY 212-807-0717
Waters Design Assoc Inc
3 W 18th St 8th Fl, New York, NY 212-807-0717
WBMG Inc
207 E 32nd St, New York, NY 212-689-7122
Weisser, Carl
163 Joralemon St #1500, Brooklyn, NY...........718-834-0952
Weissman, Walter
463 West St #B332, New York, NY 212-989-9694
Wijtvliet, Ine
440 E 56th St, New York, NY 212-319-4444
● **WILEY, PAUL (P 56)**
410 W 24th St #12i, New York, NY 212-627-8071
Williamson Janoff Aerographics/Dean Janoff
514 W 24th St 3rd Fl, New York, NY 212-807-0816
Windsor Digital/Maggie Everts
8 W 38th St, New York, NY 212-944-9090
Wu, Brian
210 W 10th St #6F, New York, NY 212-691-0352
Yourke, Oliver
525-A Sixth Ave, Brooklyn, NY 718-965-0609
Zeisse, Brook
120 W 75th St, New York, NY 212-249-3227
Zlowodzka, Joanna
144 W 10th St #18, New York, NY 212-620-7981
Louey/Rubino Design Group/Gene Nicholais
215 Park Ave S #702, New York, NY.............212-777-4220

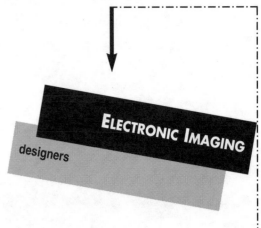

ELECTRONIC IMAGING

designers

E A S T

Erin Morrissey
176 Portsmouth Ave, Stratham, NH 603-778-3924
AB Graphics/Arnold Bombay
69 Farrington Rd, Matawan, NJ 908-566-5101
Abbey Photographers
416 E Central Blvd, Palisades Park, NJ...............201-947-1221
AdViz Advanced Visualizations/Allen H Cosgrove
709 Rt 206 N #333, Bellemead, NJ 908-281-0421
Alco Computer Graphic/Alex Guben
302 Carlton Terr, Teaneck, NJ 201-836-1294
Alcorn, Kathryn
434 S Main St, Hightstown, NJ 609-448-6698
Archambault Group, The/Donna Archambault
One Scenic Drive, Portland, CT 203-342-1023
Artattack Studios/Richard Kaufman
2025 Wallace St, Philadelphia, PA 215-232-8722
Artlandish Graphic Design/Bevi Chagnon
7417 Holly Ave, Takoma Park, MD...................301-585-8805
Ascroft, Robert
287 Orchard Creek Ln, Rochester, NY 716-227-1976
Auras Design/Robert Sugar
1746 Kalorama Rd NW, Washington, DC 202-745-0088
BAIRDesign
38 Sunset Ave, W Bridgewater, MA 508-580-9903
Baxley, Diane
19 Keeler Pl, Richfield, CT...............................203-743-2750
Bellamy, Mike
70 Willow Ave, Hackensack, NJ 201-487-2342
Berry, Rick
93 Warren St, Arlington, MA...........................617-648-6375
Blake + Barancik Design
135 S 18th St #403A, Philadelphia, PA..............215-977-9540
Blakeslee Group
916 N Charles St, Baltimore, MD 410-727-8800
Bober , Marlene
449 Pleasant St, Framingham, MA 508-872-4165
Bodkin Design Group
25 Sylvan Rd S, Westport, CT 203-221-0404
Bolognese & Assocs Inc, Leon
135 Connecticut, Freeport, NY 516-379-3405
Book 1/Gary Folven
110 High St, Hingham, MA 617-749-1692
Boris, Daniel
1835 K St NW #500, Washington, DC 202-833-4333
Brady Design Consultants, John
3 Gateway Ctr #17, Pittsburgh, PA 412-288-9300
Brand Design
814 N Harrison St, Wilmington, DE...................302-888-1648
Breth, Jill Marie
434 Bayview Ave, Cedarhurst, NY 516-569-6685
Brownstone Graphics/Kathryn Sikule
303 Hudson Ave, Albany, NY 518-434-8707
Burdick, Gary
9 Parker Hill, Brookfield, CT 203-775-2894
Buszka, Kimberly
96 W Central St, Natick, MA...........................508-655-6807
Cantrell Design, David
85 Spencer Ave, Lancaster, PA 717-396-1134
Carrara Design/Richard Carrara
218 Lorraine Ave, Upper Montclair, NJ 201-783-2872
Centre Grafik/Arthur Bromley
900 W Valley Rd #802, Wayne, PA 215-688-2949
Chernishov, Anatoly & Irene
4 Willow Bank Ct, Mahwah, NJ 201-327-2377

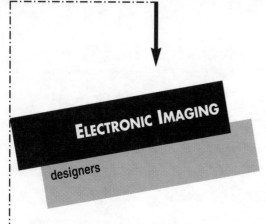

ELECTRONIC IMAGING

designers

Christopher Designs/Christopher Magalos
3308 Church Rd, Cherry Hill, NJ 609-667-7433
Citron Assocs, Lisabeth
110 Bi County Blvd #124, Farmingdale, NY 516-420-4200
Cliggett Design Group/Jack Cliggett
3406 Baring St, Philadelphia, PA 215-222-8511
Coffman, Claudia
82 Manhattan Ave #4, Jersey City, NJ 201-963-7030
College of New Rochelle/Steve Bradley
SAS Dept of Art, New Rochelle, NY 914-235-0516
Collins & Collins/James Collins
PO Box 10736, State College, PA 814-234-2916
Computer Graphics Design/Paul Roseneck
Box 1717 , Schenectady, NY 518-381-6570
Conant, Chrissy
504 University Ave #1, Ithaca, NY 607-273-2150
Connacher, Nat
65 High Ridge Rd, Stamford, CT 203-323-1330
Creative Video Design & Prod/Marie Leuchte
480 Neponset St, Canton, MA 617-821-1110
Crocker Inc/Bruce Crocker
17 Station St, Brookline, MA 617-738-7884
Crowder Design, N David
165 Edgeway, Arnold, MD 410-974-0027
Cutler Productions, Howard M
10 Harstrom Pl, Rowayton, CT 203-857-4332
D'Art Studio Inc
PO Box 299, N Scituate, MA 617-545-7313
Dadabase Design/Gary Zamchick
56 Hillside Ave, Tenafly, NJ 201-568-3727
Daly, Perla Linda
30 Gregory Ave, Mt Kisco, NY 914-241-8249
Dan, Noah
10222 Parkwood Dr, Kensington, MD.............301-564-1566
David Cundy Inc/David Cundy
142 Danbury Rd, Wilton, CT 203-761-1412
Davidson, Peter
9R Conant Ave, Acton, MA 508-635-9780
Davis, Mills
2704 Ontario Rd NW, Washington, DC 202-667-6400
De Cerchio, Joe
62 Marlborough Ave, Marlton, NJ 609-596-0598
Delago, Ken
706 Steamboat Rd #35, Greenwich, CT 203-661-6547
DePersico Inc/Paul DePersico
878 Sussex, Broomall, PA 215-328-3155
DeRiggi, Danamarie
622 Inman Ave #3, Colonia, NJ................908-382-1519
Design at Work/Steve Hoskins
220 Stony Run Ln #3D, Baltimore, MD 410-235-2168
Design Comp Inc
5 Colony St, Meriden, CT 203-235-6696
Design Imaging
1730 M St NW #505, Washington, DC 202-296-3797
design M design W/James Westhall
22 Second St, Troy, NY 618-273-5994
Design Point/Glen Walter
8 Webb St, Middleton, MA 508-777-2300
Design Team/Nancy Ness
PO Box 1322, Huntington, NY 516-427-4024
DiGiacomo, Carole
HCI #26, Mt Tremper, NY 914-688-7938
Digital Valley Design/Francis Shepherd
PO Box 204, Chadds Ford, PA 610-347-6799
DMC & Co/Donna Chernin
169 Central St, Acton, MA 508-266-1000

Double Click Design/Susan Mason
9 Marble Terrace, Hastings-on-Hudson, NY.........914-478-4508
Dreamlight Inc/Michael Scaramozzino
50 Clifford St, Providence, RI.....................401-861-8002
Dunn & Rice Inc/John Dunn
16 N Goodman St, Rochester, NY 716-473-2880
Dysart Creative/Patty Dysart
2 Wolcott Park, Medford, MA 617-391-3516
E Fitz Art/Ellen Fitzpatrick Pinkman
22 Elm Ave, Hackensack, NJ 201-342-4034
Edgerton, Brian
10 Old Rte 28/PO Box 364, Whitehouse, NJ......908-534-9400
El-Darwish, Mahmoud
9112 Kittery Ln, Bethesda, MD..................301-469-6383
Falcone & Assoc
13 Watchung Ave, Chatham, NJ 201-635-2900
Fastforward Communications Inc/
Jaime Martovamo
401 Columbus Ave, Valhalla, NY 914-741-0555
Feinen, Jeff
4702 Sawmill Rd, Clarence, NY................716-759-8406
Flink Design Inc, HGans
11 Martine Ave, White Plains, NY.............914-328-0888
Flood, David Williams
43 Booraem Ave, Jersey City, NJ 201-420-7475
Flying Turtle Productions/Laurie Holmes
721 Crestbrook Ave, Cherry Hill, NJ...........609-751-7213
Ford Myers & Co/Ford Myers
326 W Lancaster Ave, Ardmore, PA 215-649-0100
Fullmoon Creations/Lisa Gingras
81 S Main St, Doylestown, PA 215-345-1233
Galeano, Margaret
35 Burwell Ave, Milford, CT....................203-876-1544
Gallagher, Matthew
1 Old Manor Rd, Holmdel, NJ..................908-888-3953
Gambino, Donald
12 Carolyn Way, Purdys, NY 914-277-1837
Gates, Jeff
2000 Hermitage Ave, Silver Spring, MD 301-949-0436
Gavin, Bill
268 Orchard St, Millis, MA 508-376-5727
Glessner, Marc
24 Evergreen Rd, Somerset, NJ 908-249-5038
Goodman, Alan
1350 Main St/Bk of Boston Bldg, Springfield, MA 413-736-1616
Gorelick Design, Alan
26 Cromwell Dr/PO Box 1476, Morristown, NJ ..201-898-1991
Graboff, Paul
702 Erie Ave, Takoma Park, MD 301-587-0535
Graffito Inc/Tim Thompson
601 N Eutaw St #704, Baltimore, MD 410-837-0070
Graphic Consortium/Dominic Gazzara
924 Cherry St, Philadelphia, PA 215-923-3200
Graphic Images/Michele Tokach
735 Grant St, Hazelton, PA 717-455-2144
Graphics Unlimited/Walter Schwarz
41 Bog Meadow Rd PO Box 848, Sharon, CT.....203-364-0415
Graphix Group Inc/Misty Pilz
324 Elm St #203B, Monroe, CT 203-261-7665
Green Grphc Dsgn & Adv, Mel
145 Richdale Rd, Needham, MA 617-449-6777
Green, Howie
6 St James Ave, Boston, MA....................617-542-6063
Greenboam & Casey/Robert Greenboam
335 Greenwich Ave, Greenwich, CT 203-629-2331
Greenfield Belser Ltd/Burkey Belser
1818 N St NW #110, Washington, DC 202-775-0333
Group C/Brad Collins
109 Livingston St, New Haven, CT 203-865-0006
Hanson Assocs Inc
133 Grape St, Philadelphia, PA 215-487-7051
Harnett, Gena
740 Madison Ave #7B, Albany, NY.............518-432-0959
Harris, Ellen
45 Marion St #20, Brookline, MA 617-739-1867
Hathaway Point Design/Kate Wolinsky
Hathaway Point Rd RD 3, St Albans, VT........802-527-7572
Haywood & Sullivan/Michael Sullivan
955 Massachusetts Ave, Cambridge, MA 617-864-7744
Hearn, Walter
22 Spring St, Pauling, NY........................914-855-9655
Heller, Donald
7 George St, Danbury, CT 203-730-1818
Hellerman, William
Box 850, Philmont, NY 518-672-4775

Hermine Design Group
3 Lockwood Ave, Old Greenwich, CT 203-698-1732
Hierro Studio Inc/Claudia & Gregory Hierro
162 Tamboer Dr, N Haledon, NJ................201-427-3647
Higgins, Michael Francis
138 E Butler Ave, New Britain, PA 215-340-1205
Holmes Agency/Greg Holmes
9153 Brookville Rd, Silver Spring, MD 301-589-1251
Hothouse Designs Inc/Brian Sheridan
481 Main St #100, New Rochelle, NY 914-636-0075
Hundertmark, Karen
4129 Grandview Dr, Gibsonia, PA 412-443-8527
● **ICON GRAPHICS**/Keith Meehan/Steve Bon Durant **(P 76)**
34 Elton St, Rochester, NY 716-271-7020
Iconcepts/Jeff Cummings
118 Magazine St, Cambridge, MA 617-661-7289
ICONS/Glenn Johnson
76 Elm St #313, Boston, MA 617-522-0165
Immaculate Concepts/Charles Linskey
12 Norfolk Rd, Arlington, MA 617-646-9631
Information/Context/Joseph Coates
35 Partridge Lane, Madison, CT 203-421-5517
Informed Solutions Inc/David Josephson
420 Cedar Ln, Teaneck, NJ 201-836-7368
Inverse Media/Chris Thomas
PO Box 1072, Southport, CT 203-255-9620
J Gibson & Co/Carol Prince
1320 19th St NW #600, Washington, DC 202-835-0177
Jager DiPaola Kemp Design/Michael Shea
308 Pine St, Burlington, VT 802-864-5884
James, Rick
1668 E Bishopwood Blvd, Harleysville, PA.......215-361-2644
Kaczmarkiewicz, Maryanne
3323 Beaver Dr, Yorktown Hgts, NY...........914-962-7563
Kaprow, Alyce
26 Hope St, Newton, MA 617-969-0288
Kelly Michener Advertising/Jeff Wright
416 W Marion St, Lancaster, PA 717-393-9776
Kelly, Sloan W
43 Wyoming Ave, Malden, MA 617-324-9877
King, Cindy
182 Dartmouth St, Highland Park, NJ.........908-819-9382
Klopp, Karyl
5209 8th Ave/Cnstitution Qtrs,
Charlestown, MA 617-242-7463
Kluhspies, Christine
PO Box 67 , Cresskill, NJ..........................201-568-4495
Kubota & Bender
184 Laurel Ridge, S Salem, NY 914-533-6391
Laird Illustration & Design/Thomas L Laird
706 Scott St, Philipsburg, PA 814-342-2935
Lapham/Miller Assoc
34 Essex St, Andover, MA 508-475-8570
Lavin, Arnie
23 Glenlawn Ave, Seacliff, NY 516-676-1228
Lebbad, James A
24 Independence Way, Titusville, NJ...........212-645-5260
Lehner & Whyte Design/Illustration
8-10 S Fullerton Ave, Montclair, NJ............201-746-1335
Leveile Stawarz & Holl/RJ Holl
35 Old Chicopee St, Chicopee, MA 413-594-8188
Lowry, Rose
119 Little Michigan Rd , Jaffrey, NH...............603-532-8433
Lukas, Chris
PO Box 951, Nyack, NY 914-358-8137
Lyons, Lisa
9 Deane St, Gardiner, ME 207-582-1602
M Design/Mark Maddalena
125 A Ridge Rd, New City, NY..................914-638-0482
● **MACNEILL AND MACINTOSH/**
Scott A MacNeill **(P 69)**
74 York St, Lambertville, NJ......................609-397-4631
Marcolina Design Inc/Dan Marcolina
1100 E Hector St, Conshohocken, PA 215-940-0680
Marseille, Anne
122 Edward St, New Haven, CT 203-865-8304
Maryland Inst Coll Art & Design/Lew Fifield
1300 W Mt Royal/Visual Comms,
Baltimore, MD 410-225-2239
Matthew, Matthew K
5801A Chinquapin Parkway, Baltimore, MD 410-433-0035
Mauro, George
9 Fairfield Ave, Little Falls, NJ..................201-890-0880
McCoy, Dan
Box 573/1079 Main St, Housatonic, MA 413-274-6211

McGroarty Advertising/Kevin McGroarty
945 Wyoming Ave, West Pittston, PA 717-655-5646
Medvec, Emily
151 Kentucky Ave SE, Washington, DC 202-546-1220
Melanson Assoc, Donya
437 Main St, Charlestown, MA 617-241-7300
Mendoza Design
3104 Grindon Ave, Baltimore, MD 410-426-3671
Merrill, John
123 Albany Shaker Rd, Albany, NY 518-447-5660
Meyer, Bonnie
259 Collignon Way #2A, River Vale, NJ.............201-666-5763
Midnight Oil Studios
51 Melcher St, Boston, MA 617-350-7970
Minelli Design/Mark & Peter Minelli
381 Congress St 5th Fl, Boston, MA 617-426-5343
Minerbi, Joanne Sheri
11102 Old Coach Rd, Potomac, MD 202-625-6982
Mitnick Design Inc, Joel/Joel Mitnick
20 S Van Brunt St, Englewood, NJ 201-567-6801
Mobius Artists Group/Nancy Adams
354 Congress St, Boston, MA 617-542-7416
Monderer Design Inc, Stewart
10 Thatcher St #112, Boston, MA 617-720-5555
Moser/Media, Michael
2000 P St NW #500, Washington, DC 202-293-1780
Motionart Studios/Pell Osborn
27 Common St, Boston, MA 617-242-1222
Movidea Inc/Nancy Ballantone
502 Commercial St, Boston, MA 617-367-1187
Muskie, Stephen O
23 Lookout Hill, Peterborough, NH 603-924-6541
Naudin, Suzanne
94 Corey Rd, Boston, MA................................617-731-4482
Neptune, Frantz
3140 Sheffield Pl, Holland, PA 215-860-0851
New Edge Technologies/Nick Michael
PO Box 3016, Peterborough, NH 603-547-2263
New Media Group/Michael Endres
10018 Tenbrook Dr, Silver Spring, MD...............301-681-6100
New Overbrook Press/Charles Altschul
356 Riverbank Rd, Stamford, CT 203-329-7251
Nordstrom, Jennifer
45 Church St #D15, Montclair, NJ....................201-509-2399
Novich, Bruce
PO Box 501, Ardsley, NY................................914-693-3693
Nu Vox Animation & Design/Michael Tiedeman
264 Independence Dr, Chestnut Hill, MA............617-469-2891
NY Inst of Technology/Peter Voci
School of Arch & Fine Arts, Old Westbury, NY.....516-876-2752
Optisonics Productions/Jim Brown
186 8th St, Cresskill, NJ.................................201-871-4192
Ostro Design/Michael Ostro
147 Fern St, Hartford, CT 203-231-9698
Otitis Media Productions/Mary Scott
30 Thorne Ln, Port Jefferson, NY 516-928-9645
Our Studio/Nancy Jo Funaro
23 Katherine Dr, Hamden, CT 203-248-8194
Page & Page Slidemarket/Allen Lee Page
703 Maple Hill Dr, Woodbridge, NJ 908-750-4171
Palmer, Laura Leigh
3924 Denfield Ct, Kensington, MD 301-946-5026
Pasinski Assocs, Irene
6026 Penn Cir S, Pittsburgh, PA 412-661-9000
Peirce, George E
133 Ramapo Ave, Pompton Lakes, NJ 201-831-8418
Pentleton, Carol
685 Chestnut Hill Rd, Chepachet, RI 401-568-0275
Peskor, Sharon
11 Smith St, Bloomfield, NJ.............................201-748-1433
Petronio, Frank
113 Commonwealth, Rochester, NY...................716-461-5583
plus design inc/Anita Meyer
10 Thatcher St #109, Boston, MA 617-367-9587
Polniaszek, John
1604 Northcrest Rd, Silver Spring, MD 410-997-9500
Post Press, The/Martha Carothers
16 Thompson Ln, Newark, DE 302-731-7577
Post, Marilyn
123 W Ville Ave, Danbury, CT 203-797-0353
Prendergast, Michael
12 Merrill St, Newburyport, MA........................508-465-8598
Prism Printing & Design, Inc./Bill Metcalfe
10E Community Pl, Warren, NJ..........................908-755-1111
Radius Group/Dennis Kunkler
19 Old Chicopee St, Chicopee, MA 413-594-2681

Renaissance Communications
7835 Eastern Ave, Silver Spring, MD 301-587-1505
Reuter & Associates/Bill Reuter
653 Washington Blvd, Baltimore, MD 410-385-1213
Riley, George
PO Box 1025, Yarmouth, ME 207-846-5787
River Design Inc/Elaine Dominguez
722 W Nyack Rd, W Nyack, NY 914-735-8457
Roman Assoc, Helen
177 Newtown Tpke, Weston, CT 203-222-1608
Routch Design/Monique Saner
1000 Conestoga Rd #C-266, Rosemont, PA 610-527-6250
Ruby Shoes Studio/Susan Tyrrell
124 Watertown St #1E, Watertown, MA..............617-923-9965
Rutgers Dept of Visual Arts/Philip Orenstein
Berrue Circle/Livingston Camp, Piscataway, NJ....908-445-9078
Santo, Vincent
2 Skibo Ln, Mamaroneck, NY 914-698-4667
Scabrini Design Inc, Janet
90 Maywood Rd, Norwalk, CT 203-853-6676
Scuderi, John
87 Terrace Ave, Elmont, NY 516-488-2959
Selbert Design, Clifford
2067 Massachusetts Ave 3rd Fl, Cambridge, MA.617-497-6605
Serrano Company/Paul Serrano
733 15th St NW #430, Washington, DC 202-628-2105
Silverleaf Design & Illustration/Mara Levin
360 Woodland St, Holliston, MA........................508-429-8194
Silverman, Gail G
49 New England Dr, Ramsey, NJ.......................201-825-2129
Simmons, Jessica
34 Masonic St, Rockland, ME 207-594-5113
Skylight Graphics/Sandra Gola
397 Hudson St, Hackensack, NJ 201-440-3909
Sloan, Richard C
242 Thunder Rd, Holbrook, NY 516-472-3898
Smalley, David
190 Old Colchester Rd, Quaker Hill, CT.............203-444-1150
Smith-Evers, Nancy
147 Franklin St, Stoneham, MA 617-438-5716
Sparkman & Assoc, Don
1120 Connecticut Ave #270, Washington, DC 202-785-2414
St John, Bob
PO Box 1043 , York Beach, ME 508-762-2417
Staada, Glenn
490 Schooley Mt Rd, Hackettstown, NJ..............908-852-4949
Stafford, Rod
1491 Dewey Ave, Rochester, NY.......................716-647-6200
Stellarvisions/Stella Gassaway
4041 Ridge Ave/Bldg 17 #105, Philadelphia, PA.215-848-9272
Stemrich, J David
1334 W Hamilton St, Allentown, PA 215-776-0825
Stuart, Preston
8 Apple Tree Ln, Darien, CT 203-655-7688
Studio 46/Rita Squier
Routes 20/22 Box 619, New Lebanon, NY.........518-794-8747
Supon Design Group/Supon Phornirunlit
1000 Connecticut Ave NW #415,
Washington, DC 202-822-6540
Symington, Gary
145 Newbury St, Portland, ME 207-774-4977
Tedesco, Thomas
24 Elizabeth St, Port Jervis, NY 914-856-8889
● **TEICH, DAVID (P 71)**
41 Tamara Dr/PO Box 246, Roosevelt, NJ 609-448-5036
Tepper, Marlene
35 Parkview Ave, Bronxville, NY 914-779-9080
Thin Air/Laureen Trotto
25 Seir Hill Rd, Norwalk, CT.............................203-849-8104
Turbitt, Kevin
96 Wyndham Ave, Providence, RI 401-331-0031
University of Maryland/Sarah Geitz
Fine Arts Bldg/5401 Wilkins, Baltimore, MD 410-455-9644
Valla, Victor R
19 Prospect St, Falls Village, CT 203-824-5014
Van Campen, Tim
39 Knox St, Thomaston, ME 207-354-8979
Van Dyke Columbia Printing/Leonard Drabkin
370 State St, North Haven, CT...........................203-288-7957
Viele Inc, Maureen
Terrace Hill #102 112 Prospect, Ithaca, NY.........607-272-4172
Villanova, Joseph
2810 Atlantic Ave, Stottville, NY 518-828-0141
Visual Graphic Communications
177 Newtown Tpke, Weston, CT 203-222-1608

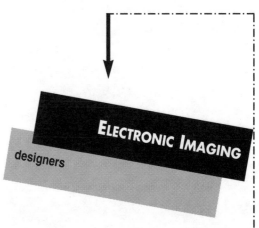

Wasserman, Cary
6 Porter Rd, Cambridge, MA............................617-492-5621
WCVB TV Design/Michael Tiedemann
5 TV Place, Needham, MA 617-449-0400
Westinghouse CSS/John Wolowiec
11 Stanwix St #936 , Pittsburgh, PA..................412-642-3246
Whitney-Edwards Design
3 N Harrison/PO Box 2425, Easton, MD............410-822-8335
Wickham & Assoc Inc
1700 K St NW #1202, Washington, DC 202-296-4860
Wiggin Design Inc/Gail Wiggin
23 Old Kings Hwy S, Darien, CT.......................203-655-1920
Wilson, Barry
654-1/2 Morton Pl NE, Washington, DC 703-276-5386
Wilson, Mark
18 River Rd, W Cornwall, CT............................203-672-6360
Wolf Design, Stephen
904 Garden St, Hoboken, NJ 201-659-2422
Wolf, Anita
43 Irving St, Cambridge , MA...........................617-576-2067
Wolf-Hubbard, Marcie
1507 Ballard St, Silver Spring, MD 301-585-5815
WYD Design Inc/Lynn Swigart
61 Wilton Rd, Westport, CT..............................203-227-2627
York Graphic Services/Kurt Weber
3600 W Market St, York, PA.............................717-792-3551
Young Assoc, Robert
78 N Union St, Rochester, NY 716-546-1973

S O U T H

Aartvark Studios/Chris Charles
220 W Broughton St, Savannah, GA..................912-233-8275
Abramson, Elaine
PO Box 330008, Ft Worth, TX 817-292-1855
Akins Design, Charles
2276 Virginia Pl NE Apt C, Atlanta , GA 404-231-1312
Aktulun Design/Kenan Aktulun
111 Southwood Rd, Austin, TX 512-447-6522
Alexander, Martha
Box 130144, Houston, TX 713-529-0472
AppleTree Technologies Inc/Larry Hall
3020 Mercer Univ Dr, Atlanta, GA.....................404-457-4500
Art Department, The
1779 Independence Blvd, Sarasota, FL...............813-355-7266
Art House Graphic Design/Alicia Ankele
PO Box 8002, Coral Springs, FL.........................305-753-4499
Art Source Design
PO Box 2193, Grapevine, TX 817-481-2212
Banchero, J
6506 Telegraph Rd, Alexandria, VA 703-719-5654
Barnum Design/David Barnum
6363 NW 6th Way #475, Ft Lauderdale, FL 305-491-8863
Becker, Lisa
981 Lakewood Court, Ft Lauderdale, FL..............305-389-8109
Bill & Virgies Publications/Bill Cowie
PO Box 11372, Ft Lauderdale, FL 305-568-1944
Bornstein, Stephen
2381 NE 193rd St, N Miami Beach, FL 305-933-2300
Carrera Design, Joann
4622 SW 89th Ct, Miami, FL............................305-553-7642
Center For Disease Control/Nancy Gathany
1600 Clifton Rd Bldg 6 #G27, Atlanta, GA........404-639-3841
Chiavenna Design/Barbara Clavenna
6000 Stone Lane, Birmingham, AL.....................205-991-8909
Cook Design, Robert
PO Box 66627, Houston, TX.............................713-227-8350

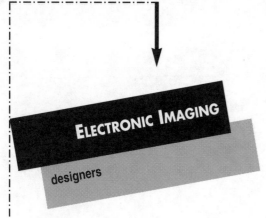

Cowen Design/Melinda Cowen
   3841 NE 2nd Ave, Miami, FL 305-573-9838
Crane, Gary
   1511 W Little Creek Rd, Norfolk, VA...............804-523-7520
Cuevas, George
   4640 NW Seventh St, Miami, FL......................305-672-4142
Customizer, The/Shirley Dellerson
   2402 Embassy Dr, W Palm Beach, FL........407-689-4747
DeRose, Andrea Legg
   1942 Shiver Dr, Alexandria, VA 703-768-3193
Dixon, David
   8 Gentry Carson Dr, Gray, TN.......................615-283-0484
● **DOROTHEA TAYLOR-PALMER**/
T P Design **(P 62)**
   490 Stone Mtn-Lithonia Rd #124,
   Stone Mountain, GA 404-413-8276
Downtown Design/Larry Goode
   2414 A South Lamar, Austin, TX 512-707-1192
Drebelbis, Marsha
   8150 Brookriver Dr #208-S, Dallas, TX 214-951-0266
Duffy Designs/Kerri Duffy
   5711 E 97th Pl, Tulsa, OK 918-299-7624
E M 2 Design/Mark Silvers
   530 Means St NW #402, Atlanta, GA............404-221-1741
Eagle Graphics/Greg Fawcett
   2133 Polo Gardens Dr #207, W Palm Beach, FL.407-791-0882
Eastlake/Michael Pond
   200 26th NW #Z106, Atlanta, GA.................404-875-5236
Encompassed Graphics/Patrick McKeon
   101 South Jennings Suite 5304, Ft Worth , TX 817-336-1118
● **FEARLESS DESIGNS**/Jeff Tull **(P 58)**
   622 E Main St #206, Louisville, KY 502-584-1333
Floyd Design/Mike Cusick
   1465 Northside Dr NW #110, Atlanta, GA........404-351-4518
Forma/David Chapin
   108 E Hargett St, Raleigh, NC 919-832-1244
Foster, Kim A
   1801 SW 11th St, Miami, FL 305-642-1801
Fuller Dyal & Stamper Inc/Herman Ellis Dyal
   1711 S Congress Ave #300, Austin, TX 512-447-7733
Garrison, Gary
   4921 Jefferson Hwy, New Orleans, LA...............504-734-0916
Geniac, Ruth
   9555 Trulock Ct, Orlando, FL........................407-677-4801
Grafiks Comm/Debi Wittig
   PO Box 162522, Altamonte Springs, FL 407-682-3088
Graphic Actualization/Danny Garrett
   3219 Barton View Dr, Austin, TX 512-892-6810
Graphic Zone, The/David Brown
   10 Office Park Cir #100, Birmingham, AL...........205-870-5300
Graphx/Eric Diehl
   79 Squire Dr, W Palm Beach, FL....................407-790-5583
Gravlee, Bob
   1323 Hampton Ridge Rd, Norcross, GA 404-717-9699
Group 1 Design/Carleton J Giles
   PO Box 50274, Columbia, SC 803-799-4322
Hall, Susan
   7500 NW First Ct #110, Plantation, FL............305-923-5111
Harlan, Steve
   10510 Reeds Landing Cir, Burke, VA 703-250-2410
Hastings, Pattie Belle
   1925 College Ave NE, Atlanta, GA.................404-373-5220
Hatcher, Mary Josie
   104 Childers St, Clarksville, TN 615-648-0340
HDM Design/Herman Maldonado
   4261 Duke St #A6, Alexandria, VA.................703-823-0945

Herman, Ben
   701 Pennsylvania, Ft Worth, TX......................817-332-7679
Herring, David
   6246 Woodlake Dr, Atlanta, GA.....................404-945-8652
Icon Associates/Andy King
   PO Box 147050, Gainesville, FL 904-371-8142
Image Assocs/Alan E Sypher
   1400 NW 101 Terrace, Plantation, FL................305-370-2159
Image Assocs Inc/John Muruca
   4909 Windy Hill Dr, Raleigh, NC.....................919-876-6400
Image Center/Lisa Phillips
   1011 Second St SW, Roanoke, VA 703-343-8243
Image Resources Inc
   PO Box 616688, Orlando, FL..........................407-843-4200
Images/Julius Friedman
   1835 Hampden Ct, Louisville, KY 502-459-0804
Imagic/Joe Huggins
   1570 Northside Dr NW # 240, Atlanta, GA.......404-355-0755
In You Wendo Design/Wendy Meyer
   5636 Souchak Dr, W Palm Beach, FL................407-686-4847
InkSpot Designs/Michael D Colanero
   2701 SE 7th St, Pompano Bch, FL..................305-946-2693
Inman, Rebecca
   2751 SW 71st Terrace #804, Davie, FL.............305-452-0181
Interface Comms/Mark Harrison
   3399 Peachtree Rd NE #200, Atlanta, GA.........404-237-4345
International Imaging Inc/Red Maxwell
   1531-J Westbrook Plaza Dr, WInston-Salem, NC..919-760-0770
Jackson Design/Beth Middleworth
   220 25th Ave N #205, Nashville, TN................615-327-2387
James Design Studio, Inc/Jim Spangler
   7520 NW 5th St #203, Plantation, FL...............305-587-2842
Jariya, Peat
   13164 Memorial Dr #222, Houston, TX.............713-523-5175
Joel Bennett Design/Joel Bennet
   2606-514 Phoenix Dr, Greensboro, NC.............910-547-0009
Johnson Galleries, Stephen
   1610 S Dixie Hwy, W Palm Bch, FL.................407-659-1431
Kantor, Jessica
   1703 Maple Shade Ln, Richmond, VA................804-353-3113
Kirwan Graphic Design/Kathryn Kirwan
   8720 Bridlewood Way , Seminole, FL................813-397-8987
Krumme, Leah
   1252 E Madison Ave, Stuart, FL.....................407-287-7119
Lawler, Pat
   1427 13th St, Huntsville, TX...........................409-294-1319
Lewczak, Scott
   1600 E Jefferson Ct, Sterling, VA 703-435-5982
Lowery, Ron
   409 Spears Ave, Chattanooga, TN..................615-344-3701
Lowry Graphics, David
   PO Box 121861, Nashville, TN 615-298-5841
Max Place/Peter Sugarman
   Rte 4 Box 54C, Louisa, VA 703-894-0511
McKinley, Liz
   6278 N Federal Hwy #381, Ft Lauderdale, FL.....305-568-9334
Meggs, Philip
   10211 Windbluff Dr, Richmond, VA..................804-740-2729
Mitchell, Bono
   2118 N Oakland St, Arlington, VA 703-276-0612
Monte, Steven W
   1124 Kensington Dr, DeSoto, TX 214-230-4324
Morales Design, Frank
   12770 Coit Rd #905, Dallas, TX.....................214-233-0667
Morrow, Michael
   5508 Dorset Shoals Rd, Douglasville, GA...........404-949-2745
Myers, Jeff
   5250 Gulfton/Bldg 4/Studio A, Houston, TX 713-661-9532
O'Connor, Buster
   4320 W University Ave, Gainesville, FL...............904-338-7519
Olson, Carl
   3325 Laura Way, Winston, GA.........................404-949-1532
Orr Communications Inc/Janine Orr
   6201 Leesburg Pike #408, Falls Church, VA 703-532-6696
Parker, Nick
   Intergraph Corp MS CR2-901 , Huntsville, AL......205-730-6208
PC Graphics Inc/George Riddick
   401 E Main St, Richmond, VA 804-780-0660
Pearlstone & Assoc
   PO Box 6528, San Antonio, TX........................210-826-1897
Peterson & Co/Brian Peterson
   2200 N Lamar #310, Dallas, TX......................214-954-0522
Phelps, Greg
   1245 North Expressway, Brownsville, TX............210-541-4909

Pietrodangelo Prod Group Inc/
Danny Pietrodangelo
   216 E Oakland Ave, Tallahassee, FL 904-681-2392
Pixel Media/Brad Gerleman
   6441 Oriole, Dallas, TX 214-350-2485
Posey, David
   815 Devon Pl, Alexandria, VA..........................703-836-8162
Random Access Media Works/Gregory Roberts
   4001-2 Confederate Pt Rd , Jacksonville, FL.......904-778-8568
Rattan Design, Joseph
   4445 Travis St #104, Dallas, TX.....................214-520-3180
Rauchman & Assocs/Robert Rauchman
   5210 SW 60th Pl, Miami, FL 305-663-9432
RBL Publications/Beki Levantini
   PO Box 2566, Boca Raton, FL........................407-482-2024
Renda Gaphic Design, Molly
   331 W Main St #607, Durham, NC..................919-682-5559
Richards Group, The/Richards Brock Miller Mitchell & Assoc
   7007 Twin Hill #200, Dallas, TX.....................214-987-4800
Richards, Robin M Nance & Courtland
   PO Box 59734, Birmingham, AL 205-871-8923
Romeo, Richard
   1066 NW 96th Ave, Ft Lauderdale, FL..............305-472-0072
Rose, Lee
   4250 T C Jester Blvd, Houston, TX..................713-686-4799
Rutkovsky, Paul
   227 Westridge Dr, Tallahassee, FL 904-575-3339
Ryan Design, Thomas
   PO Box 24449, Nashville, TN..........................37202
S A & A Inc/Suzanne Anderson & Assocs
   17 Executive Park #100, Atlanta, GA...............404-636-0134
Sakai, Kazuya
   1804 Roxton, Richardson, TX.........................214-234-0502
SANA/Edgehill Studio
   17 Executive Park Dr NE #100, Atlanta, GA 404-636-0134
Savage Art & Design, David/Mr David Savage
   PO Box 1422, Boca Raton, FL.........................407-394-4644
Schulwolf, Frank
   524 Hardee Rd, Coral Gables, FL....................305-665-2129
Seiwell, Yvonne
   5650 Camino Del Sol #307, Boca Raton, FL 407-393-7157
Sense Interactive/Dave Crossley
   1412 W Alabama, Houston, TX 713-523-5757
Short, Robbie
   2903 Bentwood Dr, Marietta, GA.....................404-565-7811
Smith Design, Laura/Laura Smith Risi
   237 1/2 9th St, W Palm Beach, FL...................407-833-6037
Specialty Media/Timothy Pedersen
   3301 Ponce de Leon Blvd #300, Coral Gables, FL.305-446-3388
Streetworks Studio
   13908 Marble Stone Dr, Clifton, VA 703-631-1650
Text & Graphics/Bianca Thomas
   3200 NE 36th St #609, Ft Lauderdale, FL 305-565-9526
Thomas Mrktng Comms, Steve
   409 East Blvd, Charlotte, NC 704-332-4624
Tolan, Stephanie
   14232 N Dallas Parkway #1404, Dallas, TX.......214-392-2774
Trinity Photo Graphic/Timothy Cracchiolo
   8625 Tourmaline Blvd, Boynton Beach, FL 407-736-3916
Turpin Design Assocs
   1762 Century Blvd #A, Atlanta, GA..................404-320-6963
TW Design/D J Teeslink
   3490 Piedmont Rd NE #1200, Atlanta, GA 404-237-3958
Ultra Design Group/Brian Smith
   46 Haywood St #336, Asheville, NC 704-254-7985
Uni-Graphix/Monica Harrion
   2715 Kingsroad Ave, Jackson, MS 601-355-4008
Unisys Corporation
   5550-A Peachtree Pkwy, Norcross, GA 404-368-6000
Urban Taylor & Assocs/Terry Stone
   116 Alhambra Cir, Coral Gables, FL..................305-444-1804
Visual Communications/Frank L Wright Jr
   3 Commercial Pl, Norfolk, VA...........................804-629-2301
Wages Design/Robert Wages
   1201 W Peachtree NE #3630, Atlanta, GA........404-876-0874
Walter, Melissa Jo
   1231 31st St South, Birmingham, AL.................205-703-0418
Weber Designs Inc, Elaine
   1024 N Palmway, Lake Worth, FL 407-585-7455
Weithers, Arlington
   PO Box 585, Tuskegee, AL.............................205-727-3514
Whitehead, Dennis R
   1410 N Nelson St, Arlington, VA......................703-524-6814
Wondermedia/Aaron Bowles
   203 Elden St #401-D, Herndon, VA 703-318-7889

Worth, Dennis
6885 NW 75th Pl, Parkland, FL 407-994-2660
Z-AX-IS/Steven T Shepard
3997 Cocoa Plum Cir, Coconut Creek, FL............305-970-9662

MIDWEST

Abrams Design, Kym
213 W Institute Pl #698, Chicago, IL 312-654-1005
Admark Inc/Gary Piland
3630 SW Burlingame Rd, Topeka, KS...........913-267-4712
Ahrens Photo, Bob
400 N State, Chicago, IL 312-243-5550
Alessi Design, Anthony
15737 Thomas Lane, Oak Forest, IL 708-535-6542
Alternative Design/Arvell Jones
14615 Evergreen, Detroit, MI 313-532-8322
Amber Productions Inc
929 Harrison Ave #205, Columbus, OH.............614-299-7192
Anderson Design Company/Charles S Anderson
30 N First St, Minneapolis, MN........................612-339-5181
Antenna/James Sholly
6167 B Carvel Ave, Indianapolis, IN.................317-254-1432
Apartment 3D/Bob Staake
726 S Ballas Rd, St Louis, MO 314-961-2303
Baker, Brenna, Madison/Tom Baker
230 10th Ave S #211, Minneapolis, MN...........612-342-4480
Bangert, Colette & Charles
721 Tennessee St, Lawrence, KS 913-842-2085
Barfuss Creative Services
1331 Lake Dr SE, Grand Rapids, MI...................616-459-8888
Barth, Henrietta
2113 Warwick Lane, Shaumburg, IL.................708-582-1063
BGN Unltd/Bryan Gerard Nelson
2420 31st Ave S, Minneapolis, MN 612-724-3430
Biner Design/Shawn & Russ Biner
2208 Crystal Way, Crystal Lake, IL 815-477-1650
Bradford-Cout & Jansen/Alan Jansen
9933 Lawler Ave, Skokie, IL.............................708-673-4777
Bridge Communication Group/Chris Brogdon
233 E Ontario St #901, Chicago, IL 312-787-0245
Britton, Arlen
300 Ames, Northfield, MN 507-645-4813
Carsello Design/Margaret Carsello
117 S Morgan #204, Chicago, IL 312-733-5709
Ceisel, Beth
5705 N Lincoln Ave, Chicago, IL.......................312-275-3590
Chmelewski, Kathleen
2006 Boudreau Dr, Urbana, IL 217-328-4303
Color Image/Jim Trotter
12342 Conway Rd, St Louis, MO.......................314-878-0777
Colorpointe Design
21700 Northwestern Hwy #565, Southfield, MI ..313-559-2880
ComCorp/George Kubricht
542 S Dearborn, Chicago, IL 312-939-6424
Commbine/P Scott Makela
3711 Glendale Terrace, Minneapolis, MN 612-922-2271
Communica/Jeff Kimble
31 N Erie St, Toledo, OH 419-244-7766
Condon Norman Design/Phylane Norman
11 W Main #300, Carpentersville, IL.................708-426-5981
Corbin Design/Jeffrey Corbin
109 E Front St #304, Traverse City, MI 616-947-1236
Creative Resource Center Inc/Michael Lundeby
6321 Bury Dr #10, Eden Prairie, MN 612-937-6000
Creative Text + Page/John Townsend
2231 Wilmette, Kalamazoo, MI 616-383-1442
Culver & Assocs
533 N 86th St, Omaha, NE 402-393-5435
Cyberplex/Pat Maun
201 SE Main St #215, Minneapolis, MN...........612-649-4641
De Goede & Others Inc/Jan De Goede
3826 N Marshfield Ave, Chicago, IL 312-525-6500
DeBrey Design/Robert J DeBrey
6014 Blue Circle #D, Minneapolis, MN 612-935-2292
Design Center
15119 Minnetonka Blvd, Minnetonka, MN 612-933-9766
Design Co/Patricia Hayes Kaufman
201 Western Ave N, St Paul, MN 612-221-1030
Design II
811 W John St, Champaign, IL..........................217-356-6388
Design Moves/Laurie Medeiros
405 N Wabash Ave #1312, Chicago, IL............312-661-0999
Design Office of George Wong/George Wong
935 W Chestnut #500, Chicago, IL 312-733-2391
Designwerks/Bert Stoutz
3540 W 66th St, Chicago, IL.............................312-776-7370

Desktop Design/Scott Rendleman
1565 Colorado Ct, Aurora, IL 708-896-7140
Digital Ink/David Garon
409 W Maryland St, Duluth, MN 218-724-3020
Double D Associates Inc/Chris Anderson
12579 W Custer Ave, Butler , WI......................414-783-4800
DouPonce, Kirk
801 Coit NES, Grand Rapids, MI 616-456-1118
Duffy Design Group, The/Nancy Kullas
901 Marquette Ave S #3000, Minneapolis, MN..612-339-3247
Dufour/Anthony Rammer
532 S 8th St, Sheboygan, WI 414-457-9191
Dynamic Graphics Inc/Peter Force
6000 N Forest Park Dr, Peoria, IL......................309-688-8800
Eclipse/Michael Spaw
8117 Rosewood, Prairie Village, KS 913-864-2515
Eldred, Dale
4334 McBee, Kansas City, MO..........................816-931-3859
Exographic
23 Lincoln Way, Valparaiso, IN.........................219-462-5810
Falk Design Group, Robert
4425 W Pine, St Louis, MO...............................314-531-1410
Fontastik Inc/Cliff Ping
911 Main St #1717, Kansas City, MO................816-474-4366
Forcade & Associates/Mark Levine
1213 Mulford St, Evanston, IL 708-328-1318
Forcade & Associates/Tim Forcade
1440 Lawrence Ave, Lawrence, KS....................913-843-1605
Gardner Design/Nancy Gardner
100 N 6th St #901-A, Minneapolis, MN.............612-332-2270
Glass Design, Michael
213 W Institute Pl #612, Chicago, IL 312-787-5977
GTP Design Studios/Galen T Pauling
PO Box 3150, Southfield, MI.............................313-533-7674
Gumucio Creative Comm/Louis Gumucio
869 Happfield Dr, Arlington Hghts, IL.................708-670-8835
● **H-GUN/DIGITAL DIV**/Robert Bial **(P 43)**
2024 S Wabash St 7th Fl, Chicago, IL 312-808-0134
Hafeman Design Group
935 W Chestnut #203, Chicago, IL 312-829-6829
Hanke Computer Services/Jeff Hanke
10211 S Cedar Lake Rd #209,
Minnetonka, MN 612-545-8545
Harman, Richard
207 S Cottonwood, Republic, MO.....................417-732-2914
Hedstrom/Blessing Inc/Rebecca McManus
5500 Wayzata #650, Minneapolis, MN 612-591-6200
Hellman Associates/Kathy Forslund
400 First Ave N #218, Minneapolis, MN............612-375-9598
Hoekstra Graphics, Grant
18 Nottingham Dr, Lincolnshire, IL....................708-948-7378
House of Graphics/Jim Hadlich
1991 Stowater Ave, St Paul, MN.......................612-738-1143
Howard-Statesman, Deborah
PO Box 178 197, Toledo, OH 419-335-3340
Hughes Design/Edward Hughes
2122 Orrington Ave, Evanston, IL 708-869-2330
Iconos/Gary Brandenberg
118 E 26th St #201, Minneapolis, MN...............612-879-0504
Identity Center
1340 Remington Rd #5, Schaumburg, IL............708-THE-BEST
Ill Inst of Technology/Patrick Whitney
Inst of Design/10 W 35th, Chicago, IL...............312-567-3250
Image Productions
115 W Church, Libertyville, IL 708-680-7100
Imageland/Jerry Capozzoli
333 N Michigan Ave, Chicago, IL 312-984-1003
Ingram Design/Debbie Ingram
12688 Oneida Woods Trail, Grand Ledge, MI 517-627-8705
Jackson Studios
1101 Southeastern Ave, Indianapolis, IN 317-639-5124
Jarvis & Assoc, Nathan Y
13307 Park/Hills Dr, Grand View, MO..............816-765-0617
JD Media Consultants/Dan Brennan
1309 E Dawes Ave, Wheaton, IL.......................708-668-0559
Jilling, Helmut
2735 Eighth St, Cuyahoga Falls, OH..................216-928-1330
Johnson Design Inc, Dean
604 Ft Wayne Ave, Indianapolis, IN...................317-634-8020
Juenger, Richard
1324 S 9th St, St Louis, MO 314-231-4069
K Squared Inc/David Kogan
1242 W Washington Blvd, Chicago, IL...............312-421-7345
Kanar, Jodi
471 White Oak, Riverwoods, IL 708-729-6255

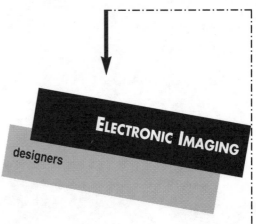

ELECTRONIC IMAGING
designers

Kirk, Bev
5815 Sovereign Dr, Cincinnati, OH 513-530-5353
Kitzerow, Scott
3505 N Pine Grove, Chicago, IL.........................312-935-9234
Krogstad Design/Richard Krogstad
3735 Fairhomes Rd, Deephaven, MN.................612-476-1825
Krueger Wright Design
3744 Bryant Ave S, Minneapolis, MN 612-827-7570
KV Graphics/Bob Cavey
710 Canterbury Cir, Chanhassen, MN.................612-949-2902
LaLiberte, Richard
211 E Delaware Pl, Chicago, IL..........................312-944-1825
LaMotte, Les
3002 Keating Ct, Burnsville, MN 612-894-1879
Lange Design, Jim
203 N Wabash #1312, Chicago, IL....................312-606-9313
Larsen Design Office/Nancy Whittlesey
7101 York Ave S #120, Minneapolis, MN 612-835-2271
Laughing Dog Creative Inc/Frank Grubich
900 N Franklin #600, Chicago, IL.......................312-951-8399
LightSource Images Inc/Lois Stanfield
PO Box 4001, Hopkins, MN...............................612-797-0770
Liska & Assoc
676 N St Claire #1550, Chicago, IL 312-943-4600
LK Design Source/Linnea Koch
2 S 946 Shagbark Rd, Batavia, IL.......................708-879-9390
Lutheran Laymens League/John Lautermilch
2185 Hampton Ave, St Louis, MO 314-965-9000
Lyon Design, Catt
305 W McMillan St, Cincinnati, OH....................513-241-2448
M Graphics/Mary Brandenburgh
150 Emerald St SE, Minneapolis, MN.................612-378-3007
MacDonald, Al
1221 Jarvis Ave, Elk Grove, IL...........................708-437-8850
MAF Graphic Design/Mary Ann Forys
62 Harris Ave, Clarendon Hills, IL.......................708-252-7214
Marketing Comms By Design/Robin Huffine-Schaffer
4606 E State Blvd #A, Ft Wayne, IN 219-482-2815
● **MILLET, CECILE (P 73)**
1280 7th Ave NW, Hutchinson, MN...................612-587-0897
Minneapolis Coll of Art & Design/Russ Mrcvek
2501 Stevens Ave S/Comp Grphcs,
Minneapolis, MN...612-874-3700
Mitchell, Mindy
2715 Yemans St, Hamtramck, MI 313-874-3023
Moetus, Olaf
3115 Lake Ave #E, Wilmette, IL.........................708-251-4237
Mofoto Graphics/Bob Moore
3651-A Shenandoah, St Louis, MO 314-231-1430
Muller & Co, John/Scott Chapman
4739 Belleview, Kansas City, MO.......................816-531-1992
Murray Lienhart Rysner & Co
58 W Huron Ave, Chicago, IL............................312-943-5995
Murrie, Lienhart, Rysner & Assoc/Jim Lienhart
50 W Huron St, Chicago, IL...............................312-943-5995
Muskovitz, Rosalyn
3731 Highgate, Muskegon, MI 616-798-3225
Natl Ctr for Supercomputing Applications/Donna Cox
405 N Mathews Ave, Urbana , IL 217-244-2005
Nehman Kodner Design/Gary Kodner
1507 McCausland, St Louis, MO 314-644-0114
Nelson, Jeffrey
107 Westmoreland Ave, Rockford, IL 815-963-9191
Nesnadny & Schwartz Inc/Joyce Nesnadny
10803 Magnolia Dr, Cleveland, OH 800-866-7721

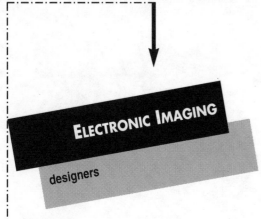

ELECTRONIC IMAGING

designers

Nicholas Assocs Design Consultants/Nicholas Sinadinos
213 W Institute Pl #704, Chicago, IL 312-951-1185
Nielsen, Ron
1313 W Randolph #326, Chicago, IL 312-226-2661
Nielson Design Group/Tim Nielson
114 E Front St #205, Traverse City, MI 616-946-0925
Obata Design
1610 Menard, St Louis, MO 314-241-1710
**Ohio State Univ/Advertising Computing
Center for Art & Design**/Charles Csuri
1224 Kinnear Rd, Columbus, OH 614-292-3416
Oldach, Mark
3525 North Oakley, Chicago, IL 312-477-6477
Optimum Group/Randy Zimmerman
9745 Mangham Dr, Cincinnati, OH 513-563-2700
Pagliaro Design, Joseph
434 Hidden Valley Ln, Cincinnati, OH 513-761-7707
● **PELIKAN PICTURES**/Petra Pepellashi
(P 88,89)
20500 Civic Center Dr #2800, Southfield, MI 313-356-2470
Perich & Partners Ltd
117 N First, Ann Arbor, MI 313-769-2215
Peterson Associates/Marshall Bohlin
23 N Lincoln St, Hinsdale, IL 708-920-1092
Phlange Design/Dean LaGrow
W 65 N 425 Westlawn Ave, Cedarburg, WI 414-377-9424
Pinzke, Nancy
1935 N Kenmore, Chicago, IL 312-528-2277
Pressley Jacobs Design Inc
101 N Wacker Dr #100, Chicago, IL 312-263-7485
Prestige Production & Graphic/John Basso Jr
148 E Aurora Rd (Rte 82), Northfield, OH 216-467-8400
Printers Inc/Paul Begner
13050 W Custer Ave, Butler, WI 414-781-1887
Prism Studios/Tom Farmer
2412 Nicollet Ave, Minneapolis, MN 800-659-2001
Qually & Co Inc
2238 Central St #3, Evanston, IL 708-864-6316
Ralston Design/Jill Ralston
2236 W Logan Blvd, Chicago, IL 312-772-2626
Ramba Design
1776 Columbus Rd, Cleveland, OH................216-621-1776
RCF Graphics/Craig Fansler
51 Pomeroy Rd, Athens, OH 614-594-5603
RedGrafix/Dralene "Red" Hughes
19750 W Observatory Rd, New Berlin, WI 414-542-5547
Redmond Design, Patrick
PO Box 75430, St Paul, MN........................612-646-4254
Reed Sendecke/Kay Krebsbach
240 W Gilman St, Madison, WI 608-256-5000
Rickabaugh Graphics/Eric Rickabaugh
384 W Johnstown Rd, Gahanna, OH 614-337-2229
Roscoe, Kurt
450 Sycamore Lane #108, Aurora , OH 216-562-2107
Ross-Ehlert
225 W Illinois, Chicago, IL 312-644-0244
Samata Associates/Pat Samata
101 S First St, Dundee, IL 708-428-8600
Sandbox Digital Playground/Pam Weston
203 N Wabash #1602, Chicago, IL 312-372-1170
Sargent Design, Ann
432 Ridgewood Ave, Minneapolis, MN 612-870-9995
Segura Inc/Carlos Segura
361 W Chestnut, Chicago, IL 312-649-5688
Sense of Design/Lynda Dahlheimer
5800 Baker Rd, Minnetonka, MN 612-935-8827

Shoulder-High Eye Productions/
Ron Shook
4057 N Damen, Chicago, IL 312-883-1693
Sirko Design, R/Robert Sirko
621 Fox Point Dr, Chesterton, IN 219-926-8759
Skjei Design Co, M/Michael Skjei
2222 Girard Ave S #9, Minneapolis, MN..........612-374-3528
Spangler Design Team/Mark Spangler
4500 Park Glen Rd #200, St Louis Park, MN......612-927-5425
Spectrum Image Group/Byron Sletton
10835 Midwest Ind Blvd, St Louis, MO 314-423-8111
Spiece, Jim
1811 Woodhaven #7, Ft Wayne, IN 219-747-3916
Stanard Inc, Michael
1000 Main St, Evanston, IL 708-869-9820
Steward Studio/Glenn Steward
225 W Ohio St 3rd Fl, Chicago, IL 312-645-0737
Strauss Design/Ron Strauss
2469 University Ave W, St Paul, MN 612-644-7244
Stribiak & Assoc/John Stribiak
506 N Clark St #2N, Chicago, IL 312-644-1285
Studio One
7300 Metro Blvd Ste 400, Edina, MN 612-831-6313
Studio One Inc/David Kuettel
7300 Metro Blvd, Edina, MN........................612-831-6313
Sugiyama Design/Kazlinn Sugiyama
5924 N Washtenaw Ave, Chicago, IL 312-271-9026
Synthesis Concepts/Liane Sebastian
360 N Michigan Ave, Chicago, IL 312-609-1111
Systems/Will Bannister
849 W Lill Ave #K, Chicago, IL 312-327-2143
Tassian, George Org
702 Gwynne Bldg, Cincinnati, OH 513-721-5566
Thiel Visual Design/John Thiel
325 E Chicago St, Milwaukee, WI 414-271-0775
THIRST/Rick Valicenti
855 W Blackhawk, Chicago, IL 312-951-5251
Thomas Group, The/Tony Thomas
4421 Granada Blvd #415, Cleveland, OH 216-663-0469
Thomas, Paul
1100 Owendale #1, Troy, MI 313-528-0009
Three & Associates
2245 Gilbert Ave #200, Cincinnati, OH 513-281-1600
Tiberi Graphics/Mindy Tiberi
3725 W Morse Ave, Lincolnwood, IL 708-933-9424
Tillis, Harvey S
1050 W Kinzie, Chicago, IL 312-733-7336
Towers Perrin
200 W Madison St #3300, Chicago, IL 312-609-9842
Triad/Artbytes/Greg Dodge
7654 W Bancroft, Toledo, OH 419-841-2272
Truckenbrod, Joan
14 Cari Ct, DeKalb, IL 815-756-2447
Tulloss Design/Thea Tulloss
114 N Third St 2nd Fl, Minneapolis, MN 612-341-9333
VSA Partners Inc/Dana Arnett
542 S Dearborn #202, Chicago, IL 312-427-6413
Walkoe Design, Don
2228 N Dayton St, Chicago, IL 312-477-1360
Walton & Assocs/Kevin Walton
14345 23rd N, Plymouth, MN 612-476-6683
Warman, Brian
4922 Duebber Dr, Cincinnati, OH 513-922-6326
Weiss Associates, Jack
1103 Mulford St, Evanston, IL 708-866-7480
Westgate Graphic Design Inc/Timothy True
1111 Westgate Ave, Oak Park, IL 708-848-8323
Willens + Michigan Corp/Bob Levison
1959 E Jefferson Ave, Detroit, MI 313-567-8900
Willman Design/Brian Hennigan
308 E 8th St, Cincinnati, OH 513-241-7403
Willow Graphics/Lind Babcock
219 Tinkler St, Lafayette, IN 317-742-2203
Willson Creative Group/William Willson Crtv Group
355 N Ashland, Chicago, IL 312-738-3555
Windy City Communications/Wayne Stuetzer
350 W Hubbard St #450, Chicago, IL 312-464-0390
Zada, Nida
5622 Delmar #510, St Louis, MO 314-454-0818
Zakari, Chantal
2105 N Oakley, Chicago, IL 312-252-2432
Zender & Assocs Inc
2311 Park Ave, Cincinnati, OH 513-961-1790
Zoot, Ira
1456 N Dayton, Chicago, IL 312-280-0048

1185 Design/Peggy Burke
259 University Ave, Palo Alto, CA 415-325-4804
A2Z Creative/Bob Colwell
1700 E Elliot Rd #6, Tempe, AZ 602-345-6706
Abaci Gallery of Computer Art/Daria Barclay
312 NW Tenth Ave, Portland, OR....................503-228-8642
Aerodrome Picture Inc/Jenny Bright
1041 N Mansfield Ave #201, Hollywood , CA....213-957-1116
Ahhhhh Graphics/Andrew Hinshaw
19000 MacArthur Blvd #620, Irvine, CA............714-476-6366
Alben & Faris/Lauralee Alben
317 Arroyo Seco, Santa Cruz, CA 408-426-5526
Aldus Corp
411 First Ave S/Customer Svc, Seattle , WA........206-622-5500
Alvarez Group
3171 Cadet Ct, Los Angeles, CA....................213-876-3491
Anderson, Lisa
15750 Crestwick Dr, La Mirada, CA................310-943-7851
April Greiman Inc/Lorna Turner
620 Moulton Ave #211, Los Angeles, CA 213-227-1222
Art for Medicine/Iris Nichols
1509 45th Ave SW, Seattle, WA....................206-932-3398
Artmarx/Scott Annis
PO Box 22582, Denver, CO 303-758-7905
Arts in Action/Dena M Paponis
1010 N Kings Rd #311, Los Angeles, CA............213-656-7376
ASA Darkrooms
905 N Cole Ave #2, Los Angeles, CA..............213-463-7513
Asal, Ahamad
PO Box 36585, Los Angeles, CA 213-936-9563
Ashford, Janet
327 Glenmont Dr, Solana Beach, CA..............619-481-7065
Astavision Post Prod/Scott Kinnimon
910 N Citrus Ave, Hollywood, CA..................213-465-3333
Atelier Graphics/Dana Trousil
396 Imperial Wy #310, Daly City, CA 415-755-8568
B.N. Architect/Ben Newcomb
6016 Hillegass Ave, Oakland, CA 510-601-1020
Bagel Design Studio, Richard
2042 Cummings Dr, Santa Rosa, CA 707-575-4110
Bailey Inc, Robert
0121 SW Bancroft St, Portland, OR................503-228-1381
Baker Design Assocs/Yee-Ping Cho
1450 20th St, Santa Monica, CA 310-453-6613
Beckerman, Carol
4350 Clark Ave, Long Beach, CA..................310-420-2603
Bee, Paula
380 16th Pl #A, Costa Mesa, CA 714-261-2765
Beggs Design
619 Maybell Ave, Palo Alto, CA 415-857-9539
Berlin, Jeff
238A Summit Dr, Corte Madera, CA 415-979-8488
Bielenberg Design, John
333 Bryant St #130, San Francisco , CA 415-495-3371
Black Point Group/Gary Priester
340 Townsend St #410, San Francisco, CA........415-331-4531
Blair Photography, Richard
2207 Fourth St, Berkeley, CA......................510-548-8350
Blavatt, Kathleen
743 Sunset Cliffs Blvd, San Diego, CA 619-222-0057
Blik Design, Tyler
3607 Fifth Ave #C, San Diego, CA 619-497-0033
Boelts Brothers Design/Eric & Jackson Boelts
14 E 2nd St, Tucson, AZ 602-792-1026
Bonauro, Tom
601 Minnesota St #216, San Francisco, CA........415-648-5233
Borges, Ruth
17104 Casimir Ave, Torrance, CA 310-532-9258
Boss Film Studios/Bob Mazza
13335 Maxella Ave, Marina Del Rey, CA 310-823-0433
Bright & Associates Inc
901 Abbot Kinney Blvd, Venice, CA 310-450-2488
Brocke Graphic Design, Robert
425 30th St #25, Newport Beach, CA 714-673-4281
Brogren/Kelly & Assoc
234 Columbine St #320, Denver, CO................303-399-3851
Brooks Communication Design/Brooks Cole
4000 Brideway #405, Sausalito, CA................415-331-5855
Brown, Gloria
Box 473061, Aurora, CO 800-372-9465
Buhler, Ray Varn
Blue Mountain Rd, Wilseyville, CA 209-293-4169
Burns & Assocs Inc
2700 Sutter St, San Francisco, CA................415-567-4404

Buz Design Group/Seri McClendon
8952 Ellis Ave, Los Angeles, CA..........310-202-0140
Cahan & Assocs/Bill Cahan
818 Brannan St #300, San Francisco, CA..........415-621-0915
Cal Poly State Univ/Eric R Johnson
Art Department, San Luis Obispo, CA 805-544-5121
Call, Ed
757 Armada Terr, San Diego, CA..........619-225-8569
Calvin Lew Design/Calvin Lew
3112 Pinole Valley Rd, Pinole, CA..........510-758-7815
Capstone Studios
5371 Wilshire Blvd #200, Los Angeles, CA 213-936-1156
Carden, Vince
2308 E Glenoaks Blvd, Glendale, CA..........818-956-0807
Carpe DM Design/David S Maltz
PO Box 1807, Boulder, CO 303-770-5285
Carson Design, David
128 1/2 Tenth St, Del Mar, CA 619-481-0609
Carter, Mary
PO Box 421443, San Francisco, CA 415-647-5660
Carver Design, Stephen
PO Box 9266, Santa Rosa, CA..........707-578-7302
Catalog Design & Production Inc
1485 Bay Shore Blvd #25, San Francisco, CA.....415-468-5500
Chase Design, Margo
2255 Bancroft Ave, Los Angeles, CA..........213-668-1055
Chui, John
1500 Laurel, Richmond, CA..........510-233-2333
Cies/Sexton Photo Lab/Ed Cies
1247 Santa Fe Dr, Denver, CO..........303-534-4000
Clark Design, James
200 W Mercer #102, Seattle, WA 206-283-3000
Cloud Art & Design, Gregory
2116 Arlington Ave #236, Los Angeles, CA........213-484-9479
Communications Design/Jeff Bane
8950 Cal Ctr Dr, Sacramento, CA..........916-362-0400
Complete Post Inc/Neal Rydall
6087 Sunset Blvd, Hollywood, CA..........213-467-1244
Coy, Los Angeles/John Coy
9520 Jefferson Blvd, Culver City, CA 310-837-0173
Creative Computing/Roslyn Wilkins
10734 Jefferson Blvd #276, Culver City, CA 310-839-2591
Cronan, Michael Patrick
1 Zoe St, San Francisco, CA 415-543-6745
Crutchfield, William
2011 S Mesa St, San Pedro, CA 310-548-4132
Culbertson Design, Jo
222 Milwaukee #402, Denver, CO 303-355-8818
Curry Design Inc/Steve Curry
1501 Main St Mezzanine, Venice, CA 310-399-4626
Curtin Design, Paul
1865 Clay St #1, San Francisco, CA..........415-885-0546
CWA Inc/Calvin Woo
4015 Ibis St, San Diego, CA..........619-299-0431
Dakini Designs/Denise Nugent
PO Box 13066, Burton, WA 206-463-5412
David Slavin Design/David Slavin
400 S Beverly Dr #305, Beverly Hills, CA 310-277-7036
Davis & Assoc, Bruce
7535 Ruffner Ave, Van Nuys, CA 818-988-0655
Davis, Jack
832 Hymettus Ave, Encinitas, CA..........619-944-7753
Design Communications/Robert Coleman
650 Linda Ave, La Habra, CA..........310-690-2104
Design Graphics/Shelly Adler
2647 S Magnolia, Los Angeles, CA..........213-749-7347
Design Matters/Carol Porter
840 Somerset Ct, San Carlos, CA..........415-508-0882
Design Source/Doug Hines
1121 Broadway #G-1, Boulder, CO..........303-447-8604
Designers Desktop/Miriam Davis
2869 Ridgeway Dr, National City, CA..........619-267-2580
DESIGNS +/Gaye L Graves
561 Homer Ave, Palo Alto, CA..........415-321-5046
Designs for Business/Joseph Ammirato
415 West Foothill Blvd #115, Claremont, CA 714-626-4842
Devereaux, Carole
15025 NW Rock Creek Rd, Portland, OR 503-621-9770
Deverich, Michael
3976 Albright Ave, Los Angeles, CA..........310-398-5854
Diehl, Michael
3251 Primera Ave, Los Angeles, CA..........213-851-3111
Digital Design Simulations/Dan Biggs
PO Box 565, Bonita, CA..........619-421-2107
Digital Zone/John Pierre D'Zahr
PO Box 5562, Bellevue, WA..........206-623-3456

Dream Merchant Graphics/David Donovan
437 Engel Ave, Henderson, NV 702-564-3598
Duin, Ed
712 100th Ave SE, Bellevue, WA 206-462-0500
DuPuis Design/Steven DuPuis
29800 Agoura Rd #103, Agoura Hills, CA..........818-706-2864
DuPuis Press/John Malmquist
29800 Agoura Rd #150, Agoura Hills, CA..........818-865-0400
Dyna Pac
7926 Convoy Ct, San Diego, CA..........619-560-0117
Electric Paint & Design/Jon Paul Davidson
PO Box 8822, Incline Village, NV 702-831-3490
Emm, David
P O Box 1089, Sierra Madre, CA..........818-355-1504
Engle & Murphy
236 E Third St #210, Long Beach, CA..........310-983-7270
Evenson Design Group
4445 Overland Ave, Culver City, CA..........310-204-1995
Fink, Mike
4434 Matilija Ave, Sherman Oaks , CA..........818-789-5232
Fisher & Day Design/Brian Day
646 Fillmore St, San Francisco, CA..........415-621-4934
Flaherty, Michael
6151 Rancho Mission Rd #302, San Diego, CA..619-282-4616
Fox, John
211 E Carrillo #205, Santa Barbara, CA..........805-564-8499
Fox, Mark
239 Marin St, San Rafael, CA..........415-258-9663
Frazier, Craig
600 Townsend #412W, San Francisco, CA........415-863-9613
From Art to Design Inc/Ron McPherson
1336 9th St, Manhattan Beach, CA 310-372-7777
Full Circle Production/Marigold Fine
1009 Third St, Santa Cruz, CA 408-459-8300
G-Nine Productions/Bill Groshelle
2030 Scott St, San Francisco, CA..........415-922-0531
Gable Design Group Inc/Tony Gable
1809 Seventh Ave #310, Seattle, WA 206-628-3744
Gage, Hal
2008 E Northern Lights Blvd, Anchorage, AK 907-272-4356
Gauldin/Farrington Design
4860 San Fernando Rd #201, Glendale, CA 818-244-7704
Gee Design, Earl
501 Second St #700, San Francisco, CA..........415-543-1192
Gensurowsky, Yvonne
312 N Sparks Ave, Burbank, CA..........818-953-9440
Gentry, David
1145 Wisconsin St, San Francisco, CA..........415-824-2920
George & Company/George McGinness
4044 Pacific St, Highland, CA 909-862-6168
Gerrard Associates/Cathy & Lewis Gerrard
5452 Evanwood Ave, Agoura , CA..........818-706-3959
Gillian/Craig Assoc
165 Eighth St #301, San Francisco, CA..........415-558-8988
Goldstein, Howard
7031 Aldea Ave, Van Nuys, CA 818-987-2837
Goss, John C
1022 Keniston, Los Angeles, CA 213-938-9526
Gould, Tom & Stephanie
2966 Dove St, San Diego, CA..........619-298-8605
Grafica/Carol Girardi
7053 Owensmouth Ave, Canoga Park, CA 818-712-0071
Graphic Data/Carl Gerle
PO Box 99991, San Diego, CA..........619-274-4511
Greene Productions, Jim
PO Box 2150, Del Mar, CA..........619-454-4133
Gregg, Mutsumi
249 Alpine St #53, Pasadena, CA 818-449-8909
Greiman Inc/April Greiman
620 Moulton Ave #211, Los Angeles, CA 213-227-1222
Grey Matter Design/Roxana Villa
16771 Addison St, Encino, CA..........818-906-3355
Griggs, Peter
1832 Conifer Way, Redding , CA..........916-243-5886
Gunnar Swanson Design Office
739 Indiana Ave, Venice, CA..........310-399-5191
Hada, Gail
23290 Clearpool, Harbor City, CA..........310-539-5114
Hagner, Dirk
27931 Paseo Nicole, San Juan Capistrano, CA...714-493-3596
Hall Kelley Organization/Hall Kelley
929 E Duane Ave, Sunnyvale, CA 408-720-0431
Halley Design/Lynda Halley
PO Box 3685, Chatsworth, CA 818-407-1642
Harrington & Assocs
11480 Burbank Blvd, N Hollywood, CA 818-508-7322

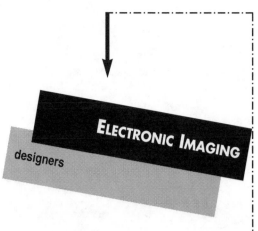

ELECTRONIC IMAGING

designers

Have A Happy/Charlotte Davis
20219 Edinburgh Dr, Saratoga, CA 408-748-4477
Haveman, Josepha
47 Del Mar Ave, Berkeley, CA 510-848-3776
Henderson, C William
24764 Soquel Rd, Los Gatos, CA 408-353-1673
Herigstad, Dale
3112 Ledgewood Dr, Los Angeles, CA..........213-463-1339
Hershman, Lynn
1935 Filbert, San Francisco, CA..........415-567-6180
Hillis, Craig F
550 Thomas St, Woodland, CA..........916-668-5848
Homer & Associates Inc/Peter Conn
1420 N Beachwood Dr, Hollywood, CA..........213-462-4710
Hornall Anderson Design Works
1008 Western Ave 6th Fl, Seattle, WA 206-467-5800
House Graphics/John Rickey
3511 Camino del Rio S #200, San Diego, CA 619-563-5752
How Studio/Howard Lim
1002 14th St #8, Santa Monica, CA..........310-394-3005
Huggins, Rucker & Cleo
196 Castro St, Mt View, CA..........415-960-1951
Hui, Sandy
268 E Coral View St, Monterey Park, CA 213-888-1801
Hume, Kelly
14645 Sunrise Dr, Bainbridge, WA 206-780-9000
Ikon Communications Inc/Martin Gueulette
3760 S Robertson #202, Culver City, CA..........310-204-5711
Illusion Factory, The/Brian Weiner
23875 Ventura Blvd #104, Calabasas, CA 818-223-8400
Image Advantage/Phil Gordon
351 Caspian Way #15, Imperial Beach, CA 619-424-5277
Image Maker, The/Anthony Helmstetter
13430 N Scottsdale Rd, Phoenix, AZ..........602-948-7643
Image World San Jose/
Knowledge Industry Publctns Inc
San Jose, CA..........914-328-9157
Imageland/Mits Kataoka
10250 Santa Monica Blvd, Los Angeles, CA 310-788-5444
Imagics Design Group/Carrie Toder
207 Escalona Dr, Santa Cruz, CA 408-426-1531
Independent Graphics/Polly Kessen
914 Bright Star St, Thousand Oaks, CA 805-492-1837
InnerStellar Productions/Francis Hobbs
904 Anita Ave, Big Bear City, CA 909-585-8495
International Corporate Video Inc/
James Fasso
1020 Serpentine Ln #114, Pleasanton, CA..........510-426-8230
InterVision/Will Doolittle
401 E Tenth Ave #160, Eugene, OR 503-343-2278
Ivanoff, Deborah
204 Greenfield Dr #F, El Cajon, CA..........619-390-3573
Iwasaki, Karen
3400 Barham Blvd, Los Angeles, CA..........213-850-5700
Iylem: Artists Using Sci & Technology
PO Box 749, Orinda, CA..........510-482-2483
Jane Lily Design/Mason Lyte
610 Anacapa St, Santa Barbara, CA..........805-683-4884
Jasin, Mark
11936 W Jefferson Blvd #C, Culver City, CA 310-390-8663
JFM Graphic Design/Wesley Gates
1135 Kimberly Rd, Twin Falls, ID 208-734-0322
Jones Design, Brent A
328 Hayes St, San Francisco, CA..........415-626-8337
Keedy
574 S Ogden Dr, Los Angeles, CA 213-939-6355

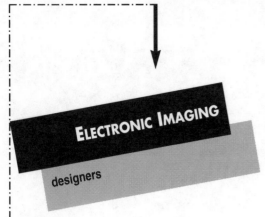

ELECTRONIC IMAGING

designers

Kent, Eleanor
544 Hill St, San Francisco, CA...................415-647-8503
Keswickhamilton/Kimberlee Keswick
3519 W 6th St, Los Avheles, CA...................213-380-3933
KHL Consulting/Kathy Hecht-Lopez
3871 Hatton St, San Diego, CA...................619-576-4140
King-Judge, Cynthia
PO Box 4644 , Montebello, CA...................213-721-3826
Klein Design/Richard Klein
6290 Sunset Blvd #505, Hollywood, CA...................213-466-0633
Knox, David
2424 N Rose, Mesa, AZ...................602-827-9339
Kraft, Heather
1042 Guerrero St, San Francisco, CA...................415-648-8892
Kuntz, Diane
817 Euclid St, Santa Monica, CA...................310-451-3601
LaBerge Graphic Design/Mary Lou LaBerge
PO Box 726, Vancouver, WA...................206-573-8283
Lancaster Design
1810 14th St, Santa Monica, CA...................310-450-2999
LaTona, Kevin
159 Western Ave W #454, Seattle, WA...................206-285-5779
● **LAX SYNTAX DESIGN**/Lance Jackson **(P 61)**
1790 5th St, Berkeley, CA...................510-849-4313
Leeds Design, Richard
21 McLellan Ave, San Mateo, CA...................415-341-0193
Lens Design, Jenny/Jenny Lens
8111 Remmet Ave #15, Canoga Park, CA...................818-716-1567
Leonhardt Group
1218 Third Ave #620, Seattle, WA...................206-624-0551
Levin Design, Lisa
124 Locust Ave, Mill Valley, CA...................415-389-9813
Levin, Lon
1317 Avenida De Cortez, Palisades, CA...................818-972-4970
Lightform Design/Bill Lae
1771 Bel Air Rd, Bel Air, CA...................310-476-8617
Lockwood, Scott
2109 Stoner Ave, Los Angeles, CA...................310-312-9923
Look Twice/Scott Kim
4016 Farm Hill Blvd #103, Redwood City, CA...................415-329-9081
Looking/John Clark
660 S Ave 21 #5, Los Angeles, CA...................213-226-1086
Louey/Robino Design Group/Robert Louey
2525 Main St, Santa Monica, CA...................310-396-7724
Love, Nan
PO Box 5004, Santa Rosa, CA...................707-527-5683
LSI Graphic Evidence
200 Corporate Pt #300, Culver City, CA...................310-568-1831
Luminarts/Helena Powell
5875 Doyle St #15, Emeryville, CA...................510-547-3191
Lusk, Danielle
22101 Mulholland Dr, Woodland Hills, CA...................818-884-7891
Lynch Graphic Design/David Lynch
8800 Venice Blvd #216, Los Angeles, CA...................310-287-0440
M A D/Liz Rico
329 Bryant St #3E, San Francisco, CA...................415-495-7968
MacNicol, Gregory
732 Chestnut St, Santa Cruz, CA...................408-459-0880
Magika/Max Almy
3454 Standish Dr, Encino, CA...................818-789-8540
Magnum Design/Ed Boyle
2762 Octavia St Top Fl, San Francisco, CA...................415-922-1728
Majlessi, Heda
1616 Summit Ave #202, Seattle, WA...................206-323-2694
Mar Design, William
220 Montgomery #608, San Francisco, CA...................415-989-3935

Mass Productions/Howard Petrick
1510 Guerrero St, San Francisco, CA...................415-648-3789
Mattingly Design, George
820 Miramar Ave, Berkeley, CA...................510-525-2098
Mattioli, Angela
10655 Rochester Ave, Los Angeles, CA...................310-475-9883
Maximum Impact Design/Judi Oyama
303 Potrero St #7, Santa Cruz, CA...................408-425-1810
McCarroll Advertising & Design/Bob McCarroll
1438 W Broadway #B-210, Tempe, AZ...................602-894-0607
McGuire, Steve
60 Alhambra Hills Dr, Martinez, CA...................510-228-6096
Media Culture/Frank Wiley
215 Fair Oaks, San Francisco, CA...................415-824-3993
Media Mark/Mark James Tippin
680 Alberta Ave #D, Sunnyvale, CA...................408-730-9919
Mednick Group, The/Scott Mednick
7972 Mulholland Dr, Los Angeles, CA...................213-656-6501
Mercer, Celia
2507 Beachwood Dr #3, Los Angeles, CA...................213-461-9604
Meyerfeld, Alan
839 Banneker Dr, San Diego, CA...................619-462-5007
Midnight Oil Studios/Ethan Hutcheson
100 Bush St, San Francisco, CA...................415-834-0384
Mock Design Assoc/Mark Mock
1738 Wynkoop #303, Denver, CO...................303-292-0801
Mok Designs, Clement
600 Townsend St #PH, San Francisco, CA...................415-703-9900
Montano, Daniel
1616 17th St #369, Denver, CO...................303-628-5440
Montgomery, George
3351 Oak Glen Dr, Los Angeles, CA...................213-850-1644
Montoya-Padilla Communications/Elaine Montoya & Rebecca Padilla
2313 Kimo NE #2D, Albuquerque, NM...................505-884-3366
Moran Studios/John Moran
711 W 17th St #J-2, Costa Mesa, CA...................714-722-0992
Morla Design
463 Bryant St, San Francisco, CA...................415-543-6548
Mortensen Design/Gordon Mortensen
416 Bush St, Mountain View, CA...................415-988-0946
Multimedia Professionals
5835 Avenida Encinas #127, Carlsbad, CA...................619-929-8788
NBC Magic
3000 W Alameda Ave, Burbank, CA...................818-840-4863
Neumeier Design Team
915 Waverly St, Palo Alto, CA...................415-323-7225
Nexus Design & Marketing/Craig Calsbeek
1316 3rd St Promenade #109, Santa Monica, CA...................310-394-6037
No Steroids Design/Rob Schultz
1409 N Alta Vista Blvd #105, Los Angeles, CA...................213-850-8209
● **O'VERY/COVEY/**
Mikel Covey & Traci O'Very **(P 10)**
1577 Sherman Ave, Salt Lake City, UT...................801-582-8505
Odam Design, John
2163 Cordero Rd, Del Mar, CA...................619-259-8230
Ogdemli/Feldman Design/Daniel Feldman
11911 Magnolia Blvd #39, N Hollywood, CA...................818-760-1759
Omnivore/Val Cohen
27 Grenada Ave, Long Beach, CA...................310-987-5244
One on One Design/Peter Kaye
6137 Carpenter Ave, North Hollywood, CA...................818-760-0448
Otus, Erol
509 Bonnie, El Cerrito, CA...................510-528-2053
Pacific Data Images Inc
111 Karlstadt Dr, Sunnyvale, CA...................408-745-6755
Pacific Motion/Nick Felton
60 Park Terrace, Mill Valley, CA...................415-383-5554
Paglietti & Terres Design Co/Mark Paglietti
80 Gilman Ave #1, Campbell, CA...................408-379-8004
Palermo, David
3379 Calle Santiago, Carlsbad, CA...................619-944-9907
Parsons Design, Glenn
8522 National Blvd #108, Culver City, CA...................310-559-6571
Peji, Bennett
1110-B Torrey Pines Rd, La Jolla, CA...................619-456-8071
Pennington, Juliana
546 Shotwell St, San Francisco, CA...................415-285-4195
Peter Green Design Studios/Chris Savage
4219 W Burbank Blvd, Burbank, CA...................818-953-2210
Phase 2 Digital Arts/Barry Chall
2111 147th Ave, Hayward, CA...................510-276-1633
Phideaux Communications/Mark Berry
202 South St, Sausalito, CA...................415-331-8015

Pittman Hensley/Donna Pittman
2114 Hillhurst Ave, Los Angeles, CA...................213-665-5168
Pollman Marketing Arts Inc
2060 Broadway #210, Boulder, CO...................303-440-4827
Priestly, Joanna
1801 NW Upshur, Portland, OR...................503-274-2158
Primo Angeli Inc
590 Folsom St, San Francisco, CA...................415-974-6100
Print Expression/Phil Tekunoff
2939 E Broadway, Tucson, AZ...................602-327-0077
PRINTZ/Charles Wyke-Smith
340 Townsend St, San Francisco, CA...................415-543-5673
Proforma
510 3rd St #520, San Francisco, CA...................415-243-4350
Purington, Camille & Mark
3539 E Easter, Littleton, CO...................303-843-0877
Reagan, Russell
956 Moreno, Palo Alto, CA...................415-856-9593
Rearick, Kevin
293 Goldenwood Cir, Simi Valley, CA...................805-584-9259
Reineck & Reineck
1425 Cole St, San Francisco, CA...................415-566-3614
Reiser, Beverly
6979 Exeter Dr, Oakland, CA...................510-482-2483
Relf, Geoff Graphics Group
3511 Camino del Rio S, San Diego, CA...................619-280-0922
ReVerb/Lorraine Wild
5514 Wilshire Blvd #900, Los Angeles, CA...................213-936-7305
Richardson or Richardson/Valerie Richardson
1301 E Bethany Home Rd, Phoenix, AZ...................602-266-1301
Ridgley Curry & Assoc Inc/Ridgley Curry
87 E Green St #309, Pasadena, CA...................818-564-1215
Romero Gaphics, Artie
1157 N Circle Dr, Colorado Springs, CO...................719-637-7012
Rose Design, David
14647 Cole Dr, San Jose, CA...................408-377-2770
Runyan Hinsche Assocs
PO Box 12260, Marina Del Rey, CA...................310-823-0975
Rupert Design, Paul
708 Montgomery St, San Francisco, CA...................415-391-2966
Russell Communications Group/Rob Foss
6076 Bristol Parkway #202, Culver City, CA...................615-254-5374
Ruth Kedar Designs/Ruth Kedar
433 College Ave, Palo Alto, CA...................415-326-3706
S M Sheldrake Graphics/Susan Sheldrake
744 Faxon Ave, San Francisco, CA...................415-334-3004
Sackett Design/Mark Sackett
864 Folsom St, San Francisco, CA...................415-543-1590
Scan-Graphics Inc/Jerry Seehof
PO Box 3179, Huntington Beach, CA...................714-840-5380
Schiada-Smith, Laurie
2402 Michelson Dr #220, Irvine, CA...................714-851-6981
Schubert, Christa
2924 Ladoga Ave, San Francisco, CA...................310-421-0124
Seaman Design, Robin
1309 Highland Ave, Glendale, CA...................818-240-2674
Sebastian, Benno
8800 Kester Ave #139, Van Nuys, CA...................818-892-9321
Sexton Design/Rob Sexton
16812 Red Hill Ave, Irvine, CA...................714-474-7525
Shawver Associates/Mark Shawver
1249 Quarry Lane, Pleasanton, CA...................510-484-4052
Shields, Charles
415 E Olive Way, Fresno, CA...................209-497-8060
Shultz, David
1118 E Platte Ave, Colorado Springs, CO...................719-473-1641
Silberstein, Sammy
13825 Cumpston St, Van Nuys, CA...................818-787-5866
Sincyr Creative Services/Roger Cyr
1315 S Sycamore, Los Angeles, CA...................213-935-6597
Sky Tree/Kit Croucher
50 Kipling Dr #3, Mill Valley, CA...................415-383-7157
Slavin, Daniel
150 S Harwood St, Orange, CA...................714-744-0118
Sloan, Rick
9432 Appalachian Dr, Sacramento, CA...................916-364-5844
Smetts Design, Bonnie
1798 5th St, Berkeley, CA...................510-644-1313
Smith, David
650 Butte St, Sausalito, CA...................415-332-9577
Smool, Carl
1528 Valentine Place S, Seattle, WA...................206-328-7920
Spectris/Sergio Villarreal
252 Durian St, Vista, CA...................619-471-6855
Spiegelman & Assoc/Marjorie Spiegelman
1735-C Union, San Francisco, CA...................415-731-9381

stat™media/Gary W Birch
7077 E Shorecrest Dr, Anaheim Hills, CA............714-280-0038
Stieglitz, Carol
Santa Monica, CA......................................310-395-5978
Stratton, Mary M
7708 Etiwanda, Reseda, CA 818-757-1921
Studio M D
1512 Alaskan Way, Seattle, WA 206-682-6221
Swaim, Howard
605 S Adams St, Glendale, CA 818-500-0337
Takessian Creative, Adam
111 W Pennsylvania Ave #4E, San Diego, CA....619-260-0695
Takigawa Design/Jerry Takigawa
591 Lighthouse Ave #15, Pacific Grove, CA........408-372-7486
Taylor Design, Robert W/Clyde Mason
2930 Center Green Court #200, Boulder, CO......303-443-1975
● **THE BLANK COMPANY**/Jerry Blank **(P 59)**
1048 Lincoln Ave, San Jose, CA......................408-289-9095
The San Diego Convention &
Visitors Bureau/Jim Hance
401 B St #1400, San Diego, CA 619-557-2852
Threinen, Cher
475 San Gorgonio St, San Diego, CA 619-226-6050
Total Vision Inc/Paul Rother
3015 Main St, Santa Monica, CA 310-450-1315
Tracy, Donna
2011 Vista Cerro Gordo St, Los Angeles, CA 213-666-4087
Tribotti Designs
22907 Bluebird Dr, Calabasas, CA 818-591-7720
TSA Design Group/Jeff Turner
4505 1/2 N Sepulveda Blvd, Sherman Oaks, CA818-501-5554
Turner & DeVries
701 N Kalaheo Ave, Kailua, HI........................808-261-2179
Tuveson, Christine
1119 Hi-Point St, Los Angeles, CA 213-936-5851
Univ of Calif/San Diego/Kim MacConnel
Vis Arts 0327/9500 Gilman Dr, La Jolla, CA......619-534-5784
Vanderbyl Design/Michael Vanderbyl
539 Bryant St 4th Fl, San Francisco, CA............415-543-8447
VanderLans, Rudy
48 Shattuck Sq #175, Berkeley, CA 510-845-9021
Vasulka, Steina & Woody
Rte 6 Box 100, Sante Fe, NM 505-471-7181
Vaughn Wedeen Creative
407 Rio Grande NW, Albuquerque, NM 505-243-4000
Vigon, Jay
11833 Brookdale Ln, Studio City, CA 213-650-0505
Viper Optics/JoAnn Gillerman
950 61st St, Oakland, CA 510-654-2880
Visionary Art Resources/Roy Montibon
3972 Barranca Pkwy #J438, Irvine, CA............714-241-0604
Volan Design Associates/Mike Krell
1800 38th St, Boulder, CO..............................303-449-3838
Wait Design, Wes
6512 SW Barnes Rd, Portland, OR....................503-297-2181
Walcott-Ayers Group, The
1230 Preservation Pk Way, Oakland, CA 510-444-5204
Weinberg, Jan
215-1/2 S Detroit St, Los Angeles, CA 213-857-1103
Weller Inst for the Cure/Don Weller
PO Box 726, Park City, UT..............................801-649-9859
West Design Studio, Harlan
PO Box 7213, Thousand Oaks, CA 805-493-4049
West Design, Suzanne
555 Bryant St #282, Palo Alto, CA....................415-324-8068
Wilcher Design/Jim Wilcher
18210 Redmond Way, Redmond, WA 206-882-2300
Williams & Ziller Design
330 Fell St, San Francisco, CA 415-621-0330
Wokuluk, Jon
1301 S Westgate Ave, Los Angeles, CA............310-473-5623
Woods + Woods/Alison Woods
414 Jackson St #304, San Francisco, CA............415-399-1984
Words Worth/Cheryl Jencks
1280 Sixth St, Monterey, CA 408-375-0288
Zinn, Elizabeth
1241 Honokahua St, Honolulu, HI....................808-395-9732

CANADA

2 Dimensions Inc/Derek Armstrong
260 Sorauren Ave, Toronto, ON......................416-539-0766
4 Designerly Types/Sean Gaherty
592 Oriole Parkway #3, Toronto, ON................416-481-8511
Bang Co, The/Bill Douglas
176 1/2 Queen St W, Toronto, ON 416-408-1972

Blue Jacket Studios/Greg Frazier
Toronto, ON..416-538-3797
Cabana Seguin Design Inc/Colombe Boudreau
1420 Sherbrooke St W #600, Montreal, QB......514-285-1311
Constructive Communications/Tracy Watts
3405 American Dr Unit 5, Mississauga, ON 416-678-2500
Dubreuil, Michel
307 Ste Catherine Ouest #310, Montreal, QB.....514-843-6791
Faith Inc/Paul Sych
1179A King St W #112, Toronto, ON 416-539-9977
Fine Art/Andrew Wysotski
1224 Cedar St, Oshawa, ON 416-721-9488
Mason & Associates Inc, Dave
406-1040 Hamilton St, Vancouver, BC..............604-684-6060
Overdrive/James Wilson
65 Liberty St, Toronto Canada, ON 416-537-2803
Paterson, Nancy
475 W Mall #1513, Etobicoke, ON 416-621-3290
Riordon Design Group Inc
1001 Queen St W, Mississauga, ON 416-271-0399
RoboShop Design/Simon Tuckett
25 Ontario #201, Toronto, Ontario, CN 416-360-0160
SRG Design/Steven Gilmore
1185 Haro St #2, Vancouver, BC 604-681-4090
Tarzan Communication Graphique/Steve Spazuk
20 Marie Anne W, Montreal, QB 514-843-3911
Taylor & Browning Design Assoc/Paul Browning
10 Price St, Toronto, ON 416-927-7094
Western Front/Hank Bull
303 E 8th Ave, Vancouver, BC 604-876-9343

Representatives

NEW YORK CITY

American Artists
353 W 53rd St #1W, New York, NY..................212-682-2462
Ario Inc/Yoshiji Nishimoto
185 E 3rd St #1A, New York, NY 212-979-7378
● **ARTCO (P 50-53)**
232 Madison Ave #402, New York, NY 212-889-8777
Bernhard Studio, Ivy
270 Lafayette #401, New York, NY 212-925-1111
● **BERNSTEIN & ANDRIULLI (P 64)**
60 E 42nd St #822, New York, NY 212-682-1490
Bruml, Kathy
New York, NY..212-874-5659
Buck & Barney Kane, Sid
566 Seventh Ave #603, New York, NY..............212-221-8090
Charles, Bill
212 E 25th St #4, New York, NY......................212-213-6810
Chislovsky Design Inc, Carol
853 Broadway #1201, New York, NY 212-677-9100
Collignon, Daniele
200 W 15th St, New York, NY 212-243-4209
Creative Freelancers
25 W 45th St #703, New York, NY 212-398-9540
Dedell Inc, Jacqueline
58 W 15th St 6th Fl, New York, NY 212-741-2539
Gomberg, Susan
41 Union Sq W, New York, NY........................212-206-0066
Gordon Associates, Barbara
165 E 32nd St, New York, NY 212-686-3514
Greenberg Associates, R
350 W 39th St, New York, NY 212-239-6767
Grien, Anita
155 E 38th St, New York, NY 212-697-6170
H K Portfolio
666 Greenwich St #747, New York, NY 212-675-5719
Heyl, Fran
230 Park Ave #2525, New York, NY 800-327-0333
Holmberg, Irmeli
280 Madison Ave #1010, New York, NY 212-545-9155
Johnson, Bud & Evelyne
201 E 28th St, New York, NY 212-532-0928
Josell Communications/Jessica Josell
185 West End Ave #22C, New York, NY............212-877-5560
Lindgren & Smith
41 Union Sq W #1228, New York, NY 212-929-5590
Mendola Ltd
420 Lexington Ave #PH, New York, NY 212-986-5680
Mintz, Les
111 Wooster St #PH C, New York, NY 212-925-0491
Morgan Associates, Vicki
194 Third Ave, New York, NY 212-475-0440
Newborn Group/Joan Sigman
336 E 54th St, New York, NY 212-832-7980

Pinkstaff, Marsha
25 W 81st St 15th Fl, New York, NY..................212-799-1500
Pushpin Assoc
215 Park Ave S #1300, New York, NY..............212-674-8080
● **R/GA PRINT**/Jimm Burris **(P 44,45)**
350 W 39th St, New York, NY 212-239-6767
Rapp Inc, Gerald & Cullen
108 E 35th St #1, New York, NY......................212-889-3337
● **RENARD REPRESENTS (P 63)**
501 Fifth Ave #1407, New York, NY 212-490-2450
S I International
43 E 19th St, New York, NY 212-254-4996
Saunders, Michele
84 Riverside Dr #5, New York, NY 212-496-0268
State of the Arts Graphic Svc/Marlene Cohen
540 E 20th St, New York, NY 212-777-2599
Stockland Martel
5 Union Sq W 6th Fl, New York, NY 212-727-1400
Weiss Group, The/Betsy Harkavy
1991 Broadway #8C, New York, NY 212-799-8220

EAST

● **ARTCO (P 50-53)**
227 Godfrey Rd, Weston, CT 203-222-8777
Birenbaum, Molly
7 Williamsburg Dr, Cheshire, CT 203-272-9253
Caton, Chip
15 Warrenton Ave, Hartford, CT 203-523-4562
Elliott/Oreman
265 Westminster Rd, Rochester, NY 716-244-6956
Lynch, Alan
11 Kings Ridge Rd, Long Valley, NJ 908-813-8718
Putcher, Terry
1214 Locust St, Philadelphia, PA 215-569-8890
Sonneville, Dane
PO Box 155, Passaic, NJ................................201-472-1225
Stemrich, J David
1334 W Hamilton St, Allentown, PA 215-776-0825
Visual Graphic Communications
177 Newtown Tpke, Weston, CT......................203-222-1608
Wolfe Ltd, Deborah
731 N 24th St, Philadelphia, PA 215-232-6666

SOUTH

Aldridge Reps Inc
758 Brookridge Dr NE, Atlanta, GA 404-872-7980
Alexander/Pollard Inc
848 Greenwood Ave NE, Atlanta, GA................404-875-1363
Baker, Michael
2548 Walsh Ct, Ft Worth, TX 817-924-4356
Bender, Brenda
4170 S Arbor Cir, Marietta, GA......................404-924-4793
Brooke & Company/Brooke Davis
4323 Bluffview Blvd, Dallas, TX 214-352-9192
Church, Diane
2325 DeSoto Dr, Ft Lauderdale, FL..................305-764-3953
Godfrey, Ally
4902 Worth St #2A, Dallas, TX........................214-827-2559
Jett & Associates, Clare
1408 S Sixth St, Louisville, KY 502-561-0737
Mills & Co, L Jane
600 N Bishop, Dallas, TX 214-946-6569
Photocom Inc/Melanie Spiegel
3005 Maple Ave #104, Dallas, TX....................214-720-2272
Simpson, Elizabeth
1222 Manufacturing, Dallas, TX......................214-761-0001

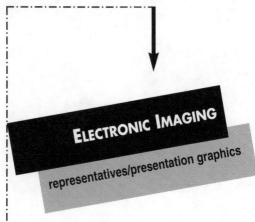

ELECTRONIC IMAGING

representatives/presentation graphics

● **TONAL VALUES INC (P 60)**
111 NE 42nd St, Miami Beach, FL 305-576-0142

MIDWEST

Artisan Professional Freelance/Bejan Douraghy
575 W Madison St #1110, Chicago, IL.............312-902-2669
Bartels Associates, Ceci
3286 Ivanhoe, St Louis, MO 314-781-7377
Gelman, Candice
1330 Deerfield Ct, Highland Park, IL..............708-831-3038
Judy Neis/The Neis Group
11440 Oak Drive, Shelbyville, MI 616-672-5756
Kastaris & Assoc, Harriet
3301-A S Jefferson Ave, St Louis, MO 314-773-2600
Lux & Associates Inc, Frank
20 W Hubbard #3E , Chicago, IL 312-222-1361
Mind Meld Reps/Michaeline Siera
605 W Madison Ave #4213, Chicago, IL...........312-902-4935
Munro/Goodman/Steve Munroe
405 N Wabash #3112, Chicago, IL.................312-321-1336
Potts & Assocs, Carolyn
4 E Ohio #11, Chicago, IL 312-944-1130
Zarley, Tim
2420 81st Circle, Urbandale, IA 515-270-1987

WEST

Another Girl Rep/Barbara Hauser
Box 421443, San Francisco, CA 415-647-5660
● **BAKER, KOLEA (P 54,55)**
2814 72nd Ave NW, Seattle, WA 206-784-1136
Bookmakers Ltd
PO Box 1100 #148, Taos, NM 505-776-5435
Bookmakers Ltd/Gayle McNeil
PO Box 1086, Taos, NM 505-776-5435
Butler Creative Resources/Sherry Butler
5525 E Thomas Rd #F11, Phoenix, AZ 602-941-5215
Creative Resource, The/Sylvia Franks
12056 Summit Cir, Beverly Hills, CA................310-276-5282
Guenzi Agents Inc, Carol
1863 S Pearl St, Denver, CO 303-733-0128
Held, Cynthia
6516 W 6th St, Los Angeles, CA 213-655-2979
Holland & Co, Mary
6638 N 13th St, Phoenix, AZ 602-263-8990
Lilie, Jim
110 Sutter St #706, San Francisco, CA.............415-441-4384
Marie & Friends, Rita
Los Angeles, CA 213-934-3395
Martha Productions Inc/Martha Productions Inc
11936 W Jefferson Blvd #C, Culver City, CA 310-390-8663
● **NEWMAN & ASSOCIATES, CAROLE/**
Sandra Canniford **(P 91)**
1119 Colorado Ave #110, Santa Monica, CA....310-394-5031
Pate & Assocs, Randy
PO Box 2160, Moorpark, CA 805-529-8111
Pepper, Missy
35 Stillman #206, San Francisco, CA 415-543-6881
Richard Salzman Represents/Fauzia Osmond
716 Sanchez St, San Francisco, CA 310-285-8267
Salzman, Richard W
716 Sanchez St, San Francisco, CA 415-285-8267
Sanders Agency, Liz
30166 Chapala Ct, Laguna Niguel, CA.............714-495-3664
Search West/Doug Yackel
100 Pine St #2500, San Francisco, CA.............415-788-1770

Swanstock/Mary V Swanson
Box 2350, Tucson, AZ..........................602-622-7133
Wiley, David
870 Market St #1053, San Francisco, CA..........415-989-2023

CANADA

Kane, Dennis
135 Rose Ave A#1105, Toronto Canada, ON.....416-447-0180
Reactor Worldwide Inc
c/o 51 Camden St, Toronto Canada, ON 212-967-7699
Three in a Box
512 Richmond St E, Toronto Canada, ON..........416-367-2446
Viner, Kevin
66 Broadway Ave #1713, Toronto, ON.............416-485-9365

Presentation Graphics

NEW YORK CITY

Artichoke Arts/Glenn Cho
333 E 34th St, New York, NY 212-532-2188
Associated Images/Brian Lee
545 5th Ave 3rd Fl, New York, NY 212-687-2844
Boyd, Cathy
547 Henry St, Brooklyn, NY......................718-875-7367
Brilliant Image Inc/Jerry Kahn, David Lapidus
7 Pennsylvania Plz 11th Fl, New York, NY..........212-736-9661
Broad Street Productions/Tom Adler
10 W 19th St 11th Fl, New York, NY 212-924-4700
Carol Chiani Digital Workshop/Carol Chiani
20 E 46th St #402, New York, NY 212-867-4371
Chartmakers
33 W 60th St 2nd Fl, New York, NY 212-247-7200
● **COPYTONE**/Mark Bevilacque **(P 103)**
115 W 45th St 10th Fl, New York, NY 212-575-0235
Critical Image/Michael Berger
4 W 37th St 2nd Fl, New York, NY 212-594-2586
Desktop Publishing/Melody Reed
353 W 57th St #2006, New York, NY 212-246-6354
Donath Communications, Ellen
400 E 89th St #12-N, New York, NY 212-289-4134
Fleisher, Audrey
430 W 24th St, New York, NY 212-463-3722
Genigraphics Inc/William Miller
444 Park Ave S 3rd Fl, New York, NY 212-684-4364
Hornbacher, Sara
40 Harrison St #9B, New York, NY 212-964-9582
Interactive Computer Entertainment/Michael Spano
35 Regina Ln, Staten Island, NY 718-966-0084
K Landman Design/Kathy Landman
156 Fifth Ave 3rd Fl, New York, NY 212-924-4254
Mariah Productions Ltd/Ken Book
545 8th Ave 3rd Fl, New York, NY 212-947-0090
Next Wave Productions/Juan Moreiras
134 Tenth Ave/Grnd Lev, New York, NY 212-989-2727
Novack, Dev
2075 Palisades Ave, Riverdale, NY 718-884-2819
Pyramid Recording
12 E 32nd St 3rd Fl, New York, NY 212-686-8687
Visual Communications Group/Leslie Bradley
777 Third Ave 6th Fl, New York, NY 212-546-1990

EAST

Action Graphics/Tyrone Thompson
4725 Dorsey Hall Dr #A-402, Ellicott City, MD 410-992-0749
Applied Optical Media Corp/Paul Dellavigne
1450 Boot Rd, Bldg 400, West Chester, PA 215-429-3701
Associated Media Productions
Sherman Plaza Rrl Rte Box 1778, Brewer, ME.....207-989-5125
Autographix/Peter Vorgone
21 North Ave, Burlington, MA....................617-272-9000
Blakeslee Group
916 N Charles St, Baltimore, MD 410-727-8800
BOGH AV Productions
110 Jefferson Blvd, Warwick, RI 401-737-1911
Book 1/Gary Folven
110 High St, Hingham, MA 617-749-1692
Brownstone Graphics/Kathryn Sikule
303 Hudson Ave, Albany, NY....................518-434-8707
Business & Professional Software Inc/Judy Farrell
139 Main St, Cambridge, MA 800-342-5277
Channel 3 Video/Jeffrey Page
PO Box 8781, Warwick, RI 401-421-1616
CIP/Provideo Inc
209 Butler St, York, PA 717-848-3655

Colorado Design Associates/Meredith Dittmar
2226 39th Pl NW, Washington, DC 202-337-6725
Commonwealth Creative Group/Bob Fields
27 Strathmore Rd, Natick, MA 508-651-7556
Computer Graphics World/Paul McPherson
10 Tara Blvd 5th Fl, Nashua, NH..................603-891-0123
Creative Productions Inc/Gus Nichols
200 Main St, Orange, NJ 201-676-4422
CSR Productions Inc
One Bala Ave #G10, Bala Cynwyd, PA 215-668-6353
Electric Image Center/Steve Laifer
77 Rte 5, Edgewater, NJ 201-224-6676
Electro/Grafiks/Paul Wilson
1238 Callow Hill St #405, Philadelphia, PA 215-923-6440
Fisher Yates Communication/Tim Potter
6 Phoenix St, Canandaigua, NY..................716-394-7880
George Washington University/Prof S Molina
Art & Visual Communication/2300 I St NW,
Washington, DC 202-994-7455
Graphic Accent/Bruce Alexander
446 Main St PO Box 243, Wilmington, MA 508-658-7602
Graphic Consortium/Dominic Gazzara
924 Cherry St, Philadelphia, PA 215-923-3200
Greystoke
43 Havenwood Ln, Grand Island, NY 716-773-1709
Images for Medicine Inc
142-A Jericho Tpke, Mineola, NY 516-741-5165
Imageworks
3679 Concord Rd PO Box 3325, York, PA 717-757-5710
Impact Studios/Lisa Smart
1084 N Delaware Ave, Philadelphia, PA............215-426-3988
Independent Media Corp
2655 Philmont Ave #203, Huntingdon Valley, PA .215-938-7710
Kent, David H
3140 Broadway Ave, Pittsburgh, PA 412-882-1337
Laird Illustration & Design/Thomas L Laird
706 Scott St, Philipsburg, PA......................814-342-2935
Light & Power Productions/Bill Moffitt
26 N Broadway, Schenectady, NY 518-381-6788
Lyons Inc
715 Orange St, Wilmington, DE...................302-654-6146
MAI2 / The Graphics Room/Tony Rosser
11 Spruce Hill Rd, Armonk, NY 914-273-9294
Main Point Productions/Will Stanton
Box 76/RD#2/Lobachsville Vllg, Oley, PA 215-987-9320
Media Vision Productions/Janet May
240 New Britain Ave, Hartford, CT.................203-249-2424
Media Works/Jay M Arancio
185 Hickory Corner Rd, Milford, NJ 908-996-7855
Miller Mauro Group Inc/Joe Mauro
2005 Concord Pike #200, Wilmington, DE 302-426-6565
Mirage Visuals/Curt Brooks
206 Monroe st, Monroeville, PA 412-372-4181
Neptune, Frantz
3140 Sheffield Pl, Holland, PA 215-860-0851
Page & Page Slidemarket/Allen Lee Page
703 Maple Hill Dr, Woodbridge, NJ 908-750-4171
Polaroid Resource Ctr
784 Memorial Dr, Cambridge, MA................800-225-1618
Posch, Michael
716 Grand St, Hoboken, NJ 201-792-7003
Preston Productions
160 Locke Dr, Marlborough , MA 508-624-4100
Q1 Productions
15 Campus Dr, Somerset, NJ 908-563-2233
Robert L Biel Associates
87 Greenwich Ave, Greenwich, CT 203-622-6630
Serrano Company/Paul Serrano
733 15th St NW #430, Washington, DC 202-628-2105
Sound Image Inc/Jim Goodell
42 Richards St, Worcester, MA...................508-756-0673
Speaker Support Group/Jeff McPhie
19 Wall St, Princeton, NJ........................609-921-3400
Specter Video
4148 Windsor St, Pittsburgh, PA..................412-521-1444
Spectratone
230 Ferris Ave, White Plains, NY 914-946-3336
Stuart, Preston
8 Apple Tree Ln , Darien, CT 203-655-7688
Technical Photographers/Kevin Porter
8 Westwind Rd, Andover, MA 508-470-3129
Thin Air/Laureen Trotto
25 Seir Hill Rd, Norwalk, CT 203-849-8104
Triad Media
PO Box 778, Frederick, MD 301-663-1471

Videos Multimedia Inc/Bill Mutsler
   South River Rd/Rte 130 Bldg C, Cranbury, NJ 609-395-1120
Visual Services Inc/Daniel Gallo
   25 Van Zant St, E Norwalk, CT 203-852-1010
Wuilleumier Inc, Will
   607 Boylston St, Boston, MA......................617-266-0103
Wurster, George
   128 Berwick, Elizabeth, NJ.........................908-352-2134

S O U T H

Absolute Inc/Sharda Fahepuria
   PO Box 1210, Duluth, GA...........................404-416-0717
Ad Graphics Inc/Richard Thompson
   6601 Lyons Rd #C11, Coconut Creek, FL..........305-421-4669
Advance Concepts Inc/Jim Brennsteiner
   8229 Boone Blvd #102, Vienna, VA................703-448-0445
American Computer Imaging/Bill Glassgold
   351 Hiatt Dr, Palm Beach Gardens, FL.............407-624-6820
Art in Media Productions/Rick Sturnberg
   7212 McNeil Dr #206, Austin, TX.................512-250-5535
Bachmann Photography/Bill Bachmann
   PO Crawer 568248, Orlando, FL....................407-333-9988
Baldridge Studios Inc/Donna Baldridge
   347 N Main St, Memphis, TN........................901-522-1185
Blair Inc/Tom Eckhardt
   5751-A General Washington Dr, Alexandria, VA .703-642-2000
Color Place, The/Jim Moss
   1330 Conant St, Dallas, TX.........................214-631-7174
Colordynamics/Ed Leach
   150 E Bethany Rd, Allen, TX.......................214-390-6500
Commercial Projection Service
   5053 Ocean Blvd #53, Sarasota, FL................813-349-2831
Compu/Fox Productions
   135 SE Second Ave, Miami, FL 305-381-8221
Corporate Graphics Inc/Pat Murphy
   3111 Monroe Rd, Charlotte, NC....................704-335-0534
Desktop Graphic Solutions/Jeff Adams
   5565 Rockpoint Dr, Clifton, VA....................703-803-7373
French & Prtnrs, Paul
   503 Gabbettville Rd, La Grange, GA................706-882-5581
Garman Audio/Video/Steve Garman
   7701 N Broadway #A-6, Oklahoma City, OK 405-842-3230
Harmon's Audio-Visual Services
   14330 S Tamiami Trl, Ft Myers, FL................813-482-7220
High, Philip
   3480 Greenlawn Dr, Lexington, KY 606-272-3060
Horizon Entertainment Inc/Gary Seline
   7102 Grand, Houston, TX..........................713-747-6433
Image Assocs Inc/John Muruca
   4909 Windy Hill Dr, Raleigh, NC..................919-876-6400
Island Video Productions/Tony Dabney
   4211 Ave T 1/2 , Galveston, TX....................409-762-2252
Kaetron Software Corp/Rick Pharr
   25211 Grogan's Mill Rd #260,
   The Woodlands, TX................................713-298-1500
Lentz & Associates
   1000-104 Brighthurst Dr, Raleigh, NC 919-828-6761
Magnum Communications Inc
   1333 Maryland, Irving, TX 214-554-0533
Marketing Network, The/Mark Murtha
   PO Box 15320, Winston-Salem, NC................919-766-9914
Media Events Concepts Inc
   10139 Metropolitan Dr, Austin, TX................512-832-1142
Miller Multi Media/Bruce Miller
   615 Sycamore St, Decatur, GA 404-378-9029
Mind's Eye Graphics/Paul Lambert
   2246 E Dabney Rd, Richmond, VA 804-353-7958
Neo Geo/Diana Georgiou
   130 E Marks St, Orlando, FL.......................407-423-9524
On Video Inc/Jeff Schumm
   2435 Southwell, Dallas, TX.........................214-406-9292
PO Box 5126/James L Webster
   4220 Amnicola Hwy, Chattanooga, TN 615-622-1193
Postmasters
   50 Vantage Way #100, Nashville, TN...............615-256-7678
Random Access Media Works/Gregory Roberts
   4001-2 Confederate Pt Rd, Jacksonville, FL.........904-778-8568
Roberts Renditions/Lillian Roberts
   PO Box 6, Vanceboro, NC 919-244-2692
Slide Shoppe/James Bradford
   1412 Texas St, Ft Worth, TX........................817-870-1197
Slide Step/Fred Lines
   22 Seventh St NE, Atlanta, GA 404-873-5353
Trans Media Creative Assoc/David Thornton
   4720 Hewes Ave #104, Gulfport, MS 601-864-9051

Tri-Comm Productions/Carol Bartholomew
   11 Palmetto Pkwy, Hilton Head Island, SC 803-681-5000
Trimble Production Studios/Grady Trimble
   612 Walnut, Little Rock, AR.......................501-666-8742
Video Promotions Intl/Chris Bentley
   PO Box 351985, Palm Coast, FL 800-486-1874
Visual Solutions/Michael Fell
   1821 Michael Faraday Dr #202, Reston, VA 703-318-8122
Willi Communications/Caroline Willi
   11229 SW 88th #D211, Miami, FL..................305-595-0059
Words & Pictures/Diane Maldin
   4372 Spring Valley Rd, Dallas, TX 214-702-9119

M I D W E S T

Admark Inc/Gary Piland
   3630 SW Burlingame Rd, Topeka, KS...............913-267-4712
Advent AudioVisual Design
   5629 Fraley Ct, Columbus, OH......................614-538-1622
Artform Communications
   325 W Huron, Chicago, IL 312-664-9402
Audio Visual Systems
   955 W Washington Blvd, Chicago, IL 312-733-3370
AVS Inc/Rob Ramseier
   2109 Ward Ave, La Crosse, WI.....................608-787-1010
Baker, Brenna, Madison/Tom Baker
   230 10th Ave S #211, Minneapolis, MN............612-342-4480
Bauman Communications Inc/Chuck Joffe
   602 Main St #300, Cincinnati, OH..................513-621-6806
Benchmark Display Graphics/Gary Zuckerman
   113 N May St, Chicago, IL.........................312-733-0070
Bergdorf Productions
   150 Springside Dr, Akron, OH 216-668-2009
Blue Rose Studio/Anthony & Laurel Kashinn
   Stonecroft 2206, Grafton, WI 414-377-1669
BT&D Audio/Visual Inc
   2356 Hassell #A, Hoffman Estates, IL...............708-882-0369
Cameron & Co/Jim Cameron
   2233 University Ave #150, St Paul, MN 612-645-4002
Carlson Marketing Communications/Fred Carlson
   Route 4/Box 384-B, Mora, MN 612-679-4105
Carsello Design/Margaret Carsello
   117 S Morgan #204, Chicago, IL 312-733-5709
Cinecraft Productions Inc/Neil McCormick
   2515 Franklin Blvd, Cleveland, OH 216-781-2300
Clicks Inc/Susan Vince
   3363 Commercial Ave, Northbrook, IL..............708-291-1020
Combined Services/David Patsy
   9 W 14th St, Minneapolis, MN.....................612-871-5503
Communitronics Corp/Rita Anderson
   1907 S Kings Hwy, St. Louis, MO 314-771-7160
Creative Concepts/Doug Crane
   94-D Westpark Rd, Centerville, OH.................513-436-2020
Creative Visuals Inc/Lois Gower
   731 Harding St NE, Minneapolis, MN...............612-378-1621
● **CUSTOM COLOR CORP**/Guy Clark
   (P 102)
   300 W 19th Terrace, Kansas City, MO 800-821-5623
Dawber & Co
   329 Enterprise Ct, Bloomfield Hills, MI 313-253-0700
Digicliques/Gerry Moore
   4320 High View Pl, Minnetonka, MN 612-377-8202
Double D Associates Inc/Chris Anderson
   12579 W Custer Ave, Butler , WI..................414-783-4800
Emerald City Animation/Post Prod/Robert Southerly
   2660 Horizon Dr, Grand Rapids, MI................616-949-9283
Grace & Wild Studios/Steven Wild
   23689 Industrial Pk Dr, Farmington Hills, MI 313-471-6010
Henneman-Hopp Design/Linda Henneman
   119 N 4th St #301, Minneapolis, MN...............612-338-3226
Imageworks/John Biemann
   7850B W Appleton Ave, Milwaukee, WI 414-536-9393
IVL Post/Bob Rohde
   Target Center/600 First Ave N, Minneapolis, MN 612-673-1250
Johnson Productions, Ric/Ric Johnson
   820 35th St #C, Des Moines, IA 515-277-4308
LaMotte, Les
   3002 Keating Ct, Burnsville, MN...................612-894-1879
Lighthouse Productions Inc
   1900 Hicks St, Rolling Meadows, IL................708-506-1414
Little & Co/Mary Mouss
   1010 S 7th St #550, Minneapolis, MN 612-375-0077
Main Frame Inc/Peggy Rooks
   82 Ionia NW #150, Grand Rapids, MI.............616-451-9882
Microvision
   11126 O St, Omaha, NE..........................402-592-4350

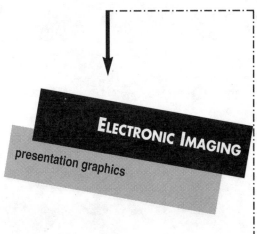

Monotype Typography/Bill Davis
   150 S Wacker Dr #2630, Chicago, IL 312-855-1440
Phar-Mor Productions/Diane Olenik
   20 Federal Plaza W, Youngstown, OH 216-746-6641
Presentation Services
   2131 Easthill Dr, Muskegon, MI 616-780-4147
Quicksilver Assoc Inc/Tim Wais
   18 W Ontario St, Chicago, IL 312-943-7622
Radmar Inc
   1263-B Rand Rd, Des Plaines, IL....................708-298-7980
Rafferty Communications/Jim Rafferty
   1518 139th Ln NW, Andover, MN..................612-755-8488
Rehabilitation Inst of Chicago/Craig Perryman
   345 E Superior, Chicago, IL........................312-908-6183
SOS/Dave Barbau
   1214 N LaSalle St, Chicago , IL 312-649-9504
Soundlight Productions/Terry Luke
   1915 Webster, Birmingham, MI 313-642-3502
Studio Stitchers/Aaron Helfman
   346 N Justine Ave, Chicago, IL 312-455-1932
Universal Images/Etta Menlo
   26011 Evergreen #200, Southfield, MI.............313-357-4160
Video Post & Graphics Inc/Claude Kennedy
   14 Cambridge Ave, Dayton, OH 513-276-3113
Williams Marketing Services
   PO Box 40139, Ft Wayne, IN......................219-432-6962
Worthwhile Films
   333 W Mifflin, Madison, WI 608-251-8855

W E S T

Aldus Corp
   411 First Ave S/Customer Svc, Seattle , WA 206-622-5500
Amicus Communications
   1012 N Sycamore Ave, Los Angeles, CA 213-874-3073
Andrews, Jackie
   CA 415-546-5666
Anning & Associates
   1134 Tangerine Way, Sunnyvale, CA 408-738-1411
Artform Communications/Carol Koby
   1700 Montgomery St #420, San Francisco, CA ..415-781-1515
Artype Publications/Steve McNair
   12049 W Jefferson Blvd, Culver City, CA...........310-391-0403
Carden, Vince
   2308 E Glenoaks Blvd, Glendale, CA 818-956-0807
Chartmasters
   201 Filbert, San Francisco, CA.....................415-421-6591
Chess Productions
   1906 Morse St, Santa Clara, CA 408-292-8191
Crucible/Brian Raney
   1717 N Seabright Ave #1, Santa Cruz, CA 408-423-4600
Curved Space/Kai Krause
   115 S Topanga Canyon Blvd #115, Topanga, CA310-455-2082
DLH Studios/Dennis Hocking
   2900 Adams #B-29, Riverside, CA 909-687-6654
Dolphin Multi Media/Cynthia Kondratieff
   2440 Embarcadero Way, Palo Alto, CA............415-962-8310
Dyma Engineering Inc/Barry Samuels
   PO Box 1535, Los Lunas, NM 505-865-6700
Electronic Sweat Shop
   990 N Hill St #200, Los Angeles, CA 213-225-1918
Ewert's Photo & Audio Visual
   2090 Duane Ave, Santa Clara, CA 408-727-3686
Farace Photography & Comm/Joe Farace
   14 Inverness Dr E #B-104, Englewood, CO 303-799-6606
Film Bank
   425 S Victory Blvd, Burbank, CA..................818-841-9176

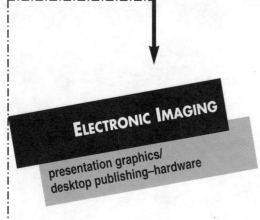

ELECTRONIC IMAGING

presentation graphics/
desktop publishing–hardware

Flora & Co/Brian Flora
PO Box 8263, Albuquerque , NM 505-255-9988
Graphic Media Inc/Doug New
411 SW Second, Portland, OR 503-223-2262
Great Projections Inc /Maureen King
1955 W Grant Rd #150, Tucson, AZ 602-620-6677
Halcyon Software/Don Hsi
1590 La Pradera, Campbell, CA 408-378-9898
House Graphics/John Rickey
3511 Camino del Rio S #200, San Diego, CA 619-563-5752
IBIS Graphics/Ralph Bentley
3443 SE Grant Ct, Portland, OR 503-231-1781
Image Architects
101 First St #337, Los Altos, CA 415-968-1141
Image World San Jose/Knowledge Industry Publctns Inc
San Jose, CA..914-328-9157
Independent Graphics/Polly Kessen
914 Bright Star St, Thousand Oaks, CA 805-492-1837
Independent Media Producers
1141-A Old County Rd, Belmont, CA..................415-595-1217
Keller, Steve
11936 W Jefferson Blvd #C, Culver City, CA 310-390-8663
Lasergraphics Inc
20 Ada, Irvine, CA..714-727-2651
LateNite/Ken Jenkins
4630 E Elwood St #1, Phoenix, AZ 602-966-0465
Lumeni Productions/Tony Valdez
1362 Flower, Glendale, CA..............................213-462-2110
Mast/Keystone Inc
4673 Aircenter Cir, Reno, NV..........................702-827-8110
Masters Presentation Graphics/Julie Bradley
2405 Lakeview Ave, Los Angeles, CA................213-660-7010
Mind Links/Nicholas MacConnell
1135 Stratford, Del Mar, CA............................619-481-7535
Mytilene Enterprises/Tarey Dunn
555 Clayton St #22, San Francisco, CA..............415-431-2246
Phosphor Inc/Michael A Eddy
5951 Canterbury Dr Unit 7, Culver City, CA........310-645-0122
Presenting Solutions
245 Calle Pueblo, San Clemente, CA 714-492-8200
Progressive Image/Bob Rowan
3296 Scotia Rd, Newport, WA..........................509-447-4982
S.L.A.D.E. Corporation/Stephanie Slade
PO Box 10176, Beverly Hills, CA......................213-466-7171
Software Assist US/Richard Wilson
76 Perine Pl, San Francisco, CA 415-921-3417
TEM Associates Inc
2000 Powell St #1500, Emeryville, CA..............510-655-9485
TSA Design Group/Jeff Turner
4505 1/2 N Sepulveda Blvd, Sherman Oaks, CA818-501-5554
Vanguard Video Productions/Christopher Robinson
PO Box 250, Ashland, OR................................503-488-0201
Veridian/Ronnie Sampson
268 Ninth Ave, San Francisco, CA 415-979-4980
WordPerfect Corp/Sales Dept
1555 N Technology Way, Orem, UT...................801-225-5000
XAOS/Helene Plotkin
600 Townsend St #271E, San Francisco, CA.......415-558-9831

CANADA

Northwest Communications
2339 Columbia St #100, Vancouver, BC 604-873-9330
Omni Media Productions Ltd/Peter Murray
235 Martindale Rd #6, St Catharines, ON 416-684-9455

Desktop Publishing-Hardware

NEW YORK CITY

Comstock Inc/Sales Dept
30 Irving Pl, New York, NY..............................212-889-9700
JPM Enterprises/Jessie Miller
PO Box 937, New York, NY 212-724-6467
Willow Peripherals/Elucian Royall
190 Willow Ave, Bronx, NY 718-402-9500

EAST

Atex Commercial & Color/Bill Younker & Keith Rendell
15 Crosby Dr , Bedford, MA.............................617-275-2323
AVPE Systems/Deborah Katz
2 Robbins Rd, Westford, MA............................508-692-2020
Baka Inc/Peter Krakow
200 Pleasant Grove Rd, Ithaca, NY 607-257-2070
Brother International Corp/Bill Hall
200 Cottontail Ln/S Vantage, Somerset, NJ 908-356-8880
Canon USA Inc/Don Carter
One Canon Plaza, Lake Success, NY 516-488-6700
CELCO (Constantine Engineering Labs Co)
70 Constantine Dr, Mahwah, NJ 201-327-1123
Color Age
900 Technology Pk Dr Bldg 8, Billerica, MA 508-667-8585
Dai/Nippon Screen Mfg Co Ltd
175 Rte 46 W, Fairfield, NJ 201-882-1922
Data Translation/Patrick Rafter
100 Locke Dr, Marlborough , MA......................508-481-3700
Electrim/Gaylord Olson
PO Box 2074, Princeton, NJ 609-683-5546
Facit Inc/Cathy Thomas
400 Commercial St, Manchester, NH..................603-647-2700
GCC Technologies Inc/Matt Allard
209 Burlington Rd, Bedford, MA 800-422-7777
High Technology Solutions/Marty Gottschalk
Box 3426, Poughkeepsie, NY 914-473-5700
Hitachi Denshi America LTD/Pam McDermott
150 Crossways Park Dr, Woodbury, NY..............516-921-7200
● **IKEGAMI**/Joe C Guzman **(P 108)**
37 Brook Ave, Maywood, NJ............................201-368-9171
Intercon Assocs/Soto Flouris
95 Allens Creek Rd Bldg 2 #200, Rochester, NY..716-244-1250
Kodak
343 State St, Rochester, NY.............................800-445-6325
Linotype Hell/Joel Friedman
425 Oser Ave, Hauppauge, NY.........................516-434-2000
Lundy Computer Graphics/Shirley Murray
One Robert Ln, Glen Head, NY 516-671-9000
Microlytics Inc/Walt Paycga
2 Tobey Village Office Pk, Pittsford, NY 716-248-9150
Mitsubishi Intl Corp/Mike Kurahashi
701 Westchester Ave #101W,
White Plains, NY 914-997-4999
Needham Graphics Inc/Jim Friend
633 Highland Ave, Needham, MA 617-449-8404
Nikon Inc/EID/Joe Carfora
1300 Walt Whitman Rd, Melville, NY.................516-547-4200
Number Nine Computer Corp/Mktg Director
18 Hartwell Ave, Lexington, MA 617-492-0999
Optimax/Wong Wilford
373 E Rte 46W, Fairfield, NJ 201-882-0028
PCPC/Robert Leeds
34 Jerome Ave #214, Bloomfield, CT.................203-243-5320
Pixelworks Inc/Rattan Dhar
7 Park Ave, Hudson, NH 603-880-1322
Polaroid Resource Ctr
784 Memorial Dr, Cambridge, MA....................800-225-1618
Scion Corporation/Bonnie Polesky
152 W Patrick St, Frederick, MD.......................301-695-7870
Scitex America Corp/Paul Thiel
8 Oak Park Dr, Bedford, MA............................617-275-5150
Shima Seiki USA Inc/Helen Estakharian
22 Abeel Rd, Cranbury, NJ..............................609-655-4788
Varitype Inc/A Tegra Co
11 Mt Pleasant Ave, E Hanover, NJ...................800-631-8134
Xerox Imaging Systems/Chris Cahill
9 Centennial Dr, Peabody, MA.........................800-248-6550
Xerox Imaging Systems Inc/Janet Knudsen
9 Centennial Dr, Peabody, MA.........................800-248-6550

SOUTH

Barco Inc/Earlene Bentley
1000 Cobb Pl Blvd, Kennesaw, GA...................404-590-7900

Birmy Graphics Corp
255 East Dr #H, Melbourne, FL 407-768-6766
Image Technology/Vance Gayle
5605 Roanne Way, Greensboro, NC 919-299-2452
Intergraph Corporation/Marla Robinson
Internal Mail #LR 23 B4, Huntsville, AL...............205-730-2000
MacProducts
608 W 22nd St, Austin, TX...............................800-622-8721
Memory Technology Texas
3007 N Lamar, Austin, TX 512-451-2600
NISCA Inc/Julia Maxey
1919 Old Denton Rd #104, Carrollton, TX..........214-242-9696
Perceptive Solutions Inc/Lonnie Sciambi
2700 Flora St, Dallas , TX................................214-954-1774
QMS/Ann Strople
One Magnum Pass, Mobile, AL 205-633-7223
Quadram/Lisa Rizk
One Quad Way, Norcross, GA..........................404-923-6666
Third Wave Computing
1219 W Sixth St, Austin, TX.............................800-284-0486
Truvel Corporation/Mary Jefferson
520 Herndon Pkwy, Herndon, VA.....................703-742-9500

MIDWEST

Alliance Peripheral Systems
6131 Deramus, Kansas City, MO 800-235-2752
ASI Image Studios/Krys Ciaciarulo
10 Second St NE #214, Minneapolis, MN..........612-379-7117
Data Service Company/Robert Chudek
3110 Evelyn St, St Paul, MN............................612-636-9469
Image Systems Corp/Diana Scheff
11595 K-Tel Dr, Hopkins, MN 612-935-1171
LaserMaster Corp/Bruce Butler
7156 Shady Oaks Rd, Eden Prairie, MN.............612-944-9330
Management Graphics Inc/Victor Hallberg
1401 E 79th St, Minneapolis, MN 612-854-1220
Mirror Technologies/John Norberg
5198 W 76th St, Edina, MN.............................800-643-2680
Monotype Inc/Steve Farkas
2100 Golf Rd, Rolling Meadows, IL 708-427-8800
Printware Inc/Tom Pick
1270 Egan Industrial Rd, St Paul, MN 612-456-1400
Purup Pre-Press America/Sharon Kruskopf
1340 Mendota Hts Rd, St Paul, MN...................612-686-5600
Ran-Ger Technologies Inc/Gerald Farnstrom
313 E Second St, Hastings, MN 612-437-2233
Screaming Color Inc/Jill Schwartz
125 N Prospect, Itasca, IL 708-250-9500
Xitron Inc/Wendy Treichel
1428 E Ellsworth Rd, Ann Arbor, MI 313-971-8530

WEST

AccuData Inc/Barton Pugh
PO Box 65667, Salt Lake City, UT 801-485-7400
Acer America Corp/Olin King
2641 Orchard Pkwy, San Jose, CA....................408-432-6200
Advanced Micro Devices Inc/Chris Henry
MS 1028/PO Box 3453, Sunnyvale, CA...............408-732-2400
Advanced Technologies Intl
355 Sinclair-Frontage Rd, Milpitas, CA...............408-942-1780
American Research Corp/Tern Wang
1101 Monterey Pass Rd, Monterey Park, CA...214-264-6531
ASP Computer Products Inc/Robert Spivack
160 San Gabriel Dr, Sunnyvale, CA 408-746-2965
Autologic Inc/Ruta Medina
1050 Rancho Conejo Blvd, Thousand Oaks, CA..805-498-9611
Avery Dennison Corp/Denise Marsh
20955 Pathfinder Rd, Diamond Bar, CA.............909-869-7711
Aztek/Phil Lippincott
15 Marconi, Irvine, CA 714-770-8406
Bridgette Inc
2616 Willow Bend Pl/Sales, El Cajon, CA..........800-257-1666
CalComp Inc/Heidi Schweizer
2411 W La Palma Ave, Anaheim, CA.................714-821-2000
Calera Recognition Systems Inc/
Deborah Herath
475 Potrero Ave, Sunnyvale, CA 408-720-8300
Chinon America Inc/Scott Elrich
615 Hawaii Ave, Torrance, CA 310-533-0274
Colorprep Inc/Raymond Neoh
320 Hatch Dr, Foster City, CA..........................415-358-8898
Computer Peripherals Inc/Natalie Hernandez
667 Rancho Conejo Blvd, Newbury Park, CA......805-499-5751
Cornerstone Technology/Sally Blodgett
1990 Concourse Dr, San Jose , CA 408-435-8900

CSS Laboratories/Pam Rocke
1641 McGaw Ave, Irvine, CA 714-852-8161
Deico Electronics/Sales Office
2800 Bayview Dr, Fremont, CA 510-651-7800
Extended Systems/Kristin Tolle
PO Box 4937, Boise, ID 208-322-7575
FWB Inc/David Lamont Booth
240 Polk #215, San Francisco, CA 415-474-8055
Info Systems/David Ujita
9740 Irvine Blvd, Irvine, CA 714-583-3000
Kensington Microware/Mary Shank
2855 Campus Dr, San Mateo, CA 415-572-2700
KeyTronic/Linda Franklin
PO Box 14687, Spokane, WA 509-928-8000
Koala Acquisitions/Earl Liebich
PO Box 1924, Morgan Hill, CA 408-776-8181
Kurta Corp/Viva La Mantia
3007 E Chambers, Phoenix, AZ 602-276-5533
Lasergraphics Inc
20 Ada, Irvine, CA 714-727-2651
Mainstay/Lance Merker
591-A Constitution Ave, Camarillo, CA 805-484-9400
Marstek Inc/David Hsieh
15225 Alton Pkwy, Irvine, CA 714-453-0110
Mass Microsystems
1507 Centre Pointe, Milpitas, CA 408-946-9207
MicroNet Technology Inc/Exec Committee
80 Technology, Irvine, CA............................714-453-6000
Microtek Lab Inc/Gina Scalese
3715 Doolittle Dr, Redondo Beach, CA...........310-297-5000
Migraph Inc
32700 Pacific Hwy S #14, Federal Way, WA 206-838-4677
Mind Links/Nicholas MacConnell
1135 Stratford, Del Mar, CA 619-481-7535
Mouse Systems Corp/Dave Todd
47505 Seabridge Dr, Fremont, CA 510-656-1117
Novell/Toby Correy
70 Garden Ct, Monterey, CA 408-649-3896
Oce Graphics
1221 Innsbruck Dr/Marketing, Sunnyvale, CA.....415-966-9400
Orange Micro Inc/Greg Petersen
1400 N Lakeview Ave, Anaheim, CA 714-779-2772
Pacific Data Products/Cindy Walton
9125 Rehco Rd, San Diego, CA 619-552-0880
PCPI (Personal Computer Products Inc)/Lori Castro
10865 Rancho Bernardo Rd, San Diego, CA.......619-485-8411
Peripheral Land Inc/Frank Jaramillo
47421 Bayside Pkwy, Fremont, CA.................510-657-2211
Presentation Technologies/Jim Creede
779 Palomar Ave, Sunnyvale, CA...................408-730-3700
Printronix/Sue Seamon
17500 Cartwright Rd, Irvine, CA....................714-863-1900
Quintar Co/Craig Douglas
370 Amapola Ave #106, Torrance, CA 310-320-5700
Qume/Val Cureton
3475-A N First St, San Jose, CA 408-473-1500
Radius Inc/Carol Johnson
1710 Fortune Dr, San Jose, CA 408-434-1010
Scantronix/Rick Camoirano
14311 Cerise Ave #108, Hawthorne, CA..........310-644-8585
Sigma Designs Inc/Jess Herrera
47900 Bayside, Fremont, CA........................510-770-0100
Skill Set Graphix/Manny Pavia
5757 Wilshire Blvd M101, Los Angeles, CA 213-937-5757
Strategic Mapping Inc/Steve Poizner
3135 Kifer Rd, Santa Clara, CA.....................408-985-7400
Three G Graphics Inc/Glenn Giaimo
104 Second Ave S #104, Edmonds, WA 206-774-3518
Thunderware Inc/Brian Smith
21 Orinda Way, Orinda, CA 510-254-6581
TPS Electronics
2495 Old Middlefield Way, Mountain View, CA..415-856-6833
UMAX Technologies Inc/Aileen Yang
3353 Gateway Blvd, Freemont, CA.................510-651-8883
● **WACOM TECHNOLOGY CORP/**
Jeff Nichols **(P 112)**
501 SE Columbia Shores #300, Vancouver, WA..206-750-8882
Winners Circle Systems
2618 Telegraph Ave, Berkeley, CA 510-845-4814
Wyse Technology
3471 N First St, San Jose, CA........................408-473-1200
XRS Corp/Anne Ego
4030 Spencer St #101, Torrance , CA 310-214-1900

CANADA

Corel Systems/Paul Bodnoff
1600 Carling Ave, Ottawa, ON 613-728-8200

Desktop Publishing-Software

NEW YORK CITY

CG Graphic Arts Supply Inc/Stephen Lambros
481 Washington St, New York, NY 212-925-5332
Compugraphia
190 Avenue B, New York, NY.......................212-614-0283
Comstock Inc/Sales Dept
30 Irving Pl, New York, NY...........................212-889-9700
JPM Enterprises/Jessie Miller
PO Box 937, New York, NY...........................212-724-6467
MetroCreative Graphics Inc/Chris Foley
33 W 34th St, New York, NY.........................212-947-5100
US Lynx Inc/Paul Williams
853 Broadway, New York, NY.......................212-673-3210
Voyager/Jane Wheeler
578 Broadway #406, New York, NY 212-431-5199

EAST

● **AGFA**/Div of Miles Inc **(P 109)**
200 Ballardvale St, Wilmington, MA...............508-658-5600
Alacrity Systems Inc/Martha Cerch
43 Newburg Rd, Hackettstown, NJ 908-813-2400
Application Techniques Inc/Ken Perrin
10 Lomar Park Dr #2, Pepperell, MA...............508-433-5201
Archetype Inc/Cheryl Jarosh
100 Fifth Ave, Waltham, MA........................617-890-7544
Atex Commercial & Color/Bill Younker & Keith Rendell
15 Crosby Dr , Bedford, MA.........................617-275-2323
Baka Inc/Peter Krakow
200 Pleasant Grove Rd, Ithaca, NY 607-257-2070
Bestinfo Inc/Harry Vane
100 Matson Ford Rd Bldg 5, Radnor, PA...........215-293-7100
Bidco Manufacturing Corp/Harvey Bidner
8 Commercial St, Hicksville, NY 516-433-0740
Cabarga, Leslie
258 W Tulpehocken St, Philadelphia, PA............215-438-9954
Caddylak Systems Inc/David Wilson
131 Heartland Blvd, Brentwood, NY................516-254-2000
Canon USA Inc/Don Carter
One Canon Plaza, Lake Success, NY 516-488-6700
Cartesia/Barbara Fordyce
PO Box 757, Lambertville, NJ 800-334-4291
CELCO (Constantine Engineering Labs Co)
70 Constantine Dr, Mahwah, NJ.....................201-327-1123
Coddbarrett Associates/Mary Codd
105 Gano St, Providence, RI..........................401-273-9898
Color Age
900 Technology Pk Dr Bldg 8, Billerica, MA.......508-667-8585
Compumation/Jim Betlyon
100 N Patterson St, State College, PA 814-238-2120
Computer Assocs Intl Inc/Marc Sokol
One Computer Assocs Plz, Islandia, NY.............516-342-6224
CPS Technologies Inc/John Attas
PO Box 648, Rockaway, NJ............................201-586-9330
Crown & Shield Software Inc/Mary Lynn Davis
29 Crafts St #200, Newton, MA 617-965-3383
CyberChrome Inc/Art Satterthwaite
25 Business Park Dr E, Branford, CT..................203-488-9594
Dataware Technologies Inc
222 Third St #3300, Cambridge, MA.................617-621-0820
Durbin Associates/Harold Durbin
3711 Southwood Dr, Easton, PA......................215-252-6331
ECRM/Rudy Bundy
554 Clark Rd, Tewksbury, MA.........................508-851-0207
Electrim/Gaylord Olson
PO Box 2074, Princeton, NJ 609-683-5546
Firstdesk Systems Inc/William Hulbig
7 Industrial Park Rd, Medway, MA...................800-522-2286
Font World/Mark Seldowitz
2021 Scottsville Rd, Rochester , NY 716-235-6861
FreeSoft Co
105 McKinley Rd, Beaver Falls, PA 412-846-2700
Funk Software/Joe Ryan
222 Third St, Cambridge, MA.........................617-497-6339
Graphsoft/Diane Reynolds
10270 Old Columbia Rd #100, Columbia , MD..410-290-5114
Hampstead Computer Graphics/Dan Healy
PO Box 469, E Hampstead, NH........................603-329-5076
High Technology Solutions/Marty Gottschalk
Box 3426, Poughkeepsie, NY 914-473-5700

ELECTRONIC IMAGING

desktop publishing–hardware/
desktop publishing–software

Hitachi Denshi America LTD/
Pam McDermott
150 Crossways Park Dr, Woodbury, NY..............516-921-7200
Howtek Inc/Jean Vosler
21 Park Ave, Hudson, NH 603-882-5200
Hyphen Inc/Donna Jones
181 Ballardvale St, Wilmington, MA 508-988-0880
Imaging Automation Inc/Bruce Monk
7 Henry Clay, Merrimack, NH.........................603-598-3400
Inner Media Inc/Nancy Rosenberg
60 Plain Rd, Hollis, NH 603-465-3216
InterCAP Graphics Systems/Ronn Wheiler
116 Defense Hwy 4th Fl, Annapolis, MD............410-224-2926
Intercon Assocs/Soto Flouris
95 Allens Creek Rd Bldg 2 #200, Rochester, NY..716-244-1250
Interleaf Inc/Larry Bohn
9 Hillside Ave/Prospect Pl, Waltham, MA............617-290-4990
Intex Solutions Inc/Jim Wilner
35 Highland Cir, Needham, MA.......................617-449-6222
Intl Media Services/Stuart Allen
718 Sherman Ave, Plainfield, NJ......................908-756-4060
IOC/Westbrook Technologies/Larry Bastanza
PO Box 910, Westbrook, CT............................203-399-7111
Linotype Hell/Joel Friedman
425 Oser Ave, Hauppauge, NY........................516-434-2000
Media Cybernetics Inc/Cindy Batz
8484 Georgia Ave #200, Silver Spring, MD.......301-495-3305
Micro Dynamics Ltd/Beth Farrell
8555 16th St #701, Silver Spring, MD 301-589-6300
Microlytics Inc/Walt Paycga
2 Tobey Village Office Pk, Pittsford, NY 716-248-9150
Microseeds Software
2-A Dorset Ln, Williston, VT 802-879-3365
Miles 33 International/Marino DeCecchis
101 Merritt 7, Norwalk, CT 203-846-9933
Mitsubishi Intl Corp/Mike Kurahashi
701 Westchester Ave #101 W, White Plains, NY.914-997-4999
Needham Graphics Inc/Jim Friend
633 Highland Ave, Needham, MA....................617-449-8404
Number Nine Computer Corp/Mktg Director
18 Hartwell Ave, Lexington, MA......................617-492-0999
Optical Access Intl/Mike Maloney
500 W Cummings Park, Woburn, MA...............617-937-3910
Optronics/Intergraph Div/Colin Murphy
7 Stuart Rd, Chelmsford, MA..........................508-256-4511
PaperDirect Inc/Alan Kipuse
205 Chubb Ave, Lyndhurst, NJ 800-A-PAPERS
PCPC/Robert Leeds
34 Jerome Ave #214, Bloomfield, CT.................203-243-5320
Penta Software Inc/David Lewis
107 Lakefront Dr, Hunt Valley, MD 410-771-TYPE
Phoenix Technologies Ltd
846 University Ave, Norwood, MA....................617-551-4000
Pipeline Associates Inc/Michael Tangreti
2740 Rte 10W, Morris Plains, NJ 201-267-3840
Pixelink Corp/Bill Ingraham
8 Kane Industrial Dr, Hudson, MA....................508-562-4803
Precision Type/Marketing Dept
47 Mall Dr, Comack, NY 516-543-3636
Prism Enterprises Inc/Ashley Brooks
360 Domer Ave #600, Laurel, MD 301-604-6611
Quality Software Inc/Susan Adams
60 Lewis St, Newton, MA..............................617-965-2231
Regional Typographers Inc/Alan Coopersmith
131 Henry St, Freeport, NY 516-378-4422

ELECTRONIC IMAGING

desktop publishing–software

Scitex America Corp/Paul Thiel
8 Oak Park Dr, Bedford, MA.............................617-275-5150
SH Pierce & Co/Steve Hollinger
One Kendall Sq/Bldg 600 #323, Cambridge, MA ...617-395-8350
Shima Seiki USA Inc/Helen Estakharian
22 Abeel Rd, Cranbury, NJ...............................609-655-4788
Software Complement/Karen Cohen
8 Pennsylvania Ave, Matamoras, PA 717-491-2492
Software Consulting Services/Dan Zito
3162 Bath Pike, Nazareth , PA 215-837-8484
● **SPECULAR INTERNATIONAL/**
Dave Trescot **(P 114,115)**
479 West St, Amherst, MA 413-253-3100
Spinnaker Software/Andrea Snader
201 Broadway 6th Fl, Cambridge, MA...............617-494-1200
Teach Services/Timothy Hullquist
182 Donivan Rd, Brushton, NY...........................518-358-2125
Technology Group/Jeff Ramsay
36 S Charles St, Baltimore, MD...........................410576-2040
Umansky, Steven
84 Shore Rd, Port Washington, NY.....................516-621-5543
Varitype Inc/A Tegra Co
11 Mt Pleasant Ave, E Hanover, NJ.....................800-631-8134
Video/Visuals Inc/Bob Lewis
61 Chapel St, Newton, MA..................................617-527-7800
Xerox Imaging Systems/Chris Cahill
9 Centennial Dr, Peabody, MA............................800-248-6550

SOUTH

AlSoft Inc
22557 Aldine Westfield #122, Spring, TX....713-353-4090
Altsys Corp/Tom Milks
269 W Renner Pkwy, Richardson, TX 214-680-2060
Church, Diane
2325 DeSoto Dr, Ft Lauderdale, FL.....................305-764-3953
Compatible Systems Engineering Inc/Larry Spevac
7630 Little River Tpke #216, Annandale, VA 703-941-0917
Computer Support Corp/Gail Campbell
15926 Midway Rd, Dallas, TX............................214-661-8960
CoOperative Printing Solutions Inc/Craig Nettleship
5950 Live Oaks Pkwy #175, Norcross, GA.........404-840-0810
Deneba Software/Doug Levy
7400 SW 87th Ave, Miami, FL 305-596-5644
EDCO Services Inc/Ed Cohen
4107 Gunn Hwy, Tampa, FL...............................813-962-7800
Image Technology/Vance Gayle
5605 Roanne Way, Greensboro, NC 919-299-2452
InSight Systems Inc/John Evans
10017 Coach Rd, Vienna, VA.............................703-938-0250
Intergraph Corporation/Marla Robinson
Internal Mail #LR 23 B4, Huntsville, AL 205-730-2000
Jariya, Peat
13164 Memorial Dr #222, Houston, TX...............713-523-5175
Kinetic Presentations Inc/Raymond Schumann
240 Distillery Commons, Louisville, KY 502-583-1679
Medina Software Inc
PO Box 521917, Longwood, FL..........................407-2601676
Micrographx/Grant Wickes
1303 Arapaho, Richardson, TX 2149946336
Microvitec Inc/Diana Davis
4854 Old National Hwy #110, Atlanta, GA 404-767-0706
National Digital Corp/Frank Roache
7700 Leesburg Pike #125, Falls Church , VA 703-356-5600
NISCA Inc/Julia Maxey
1919 Old Denton Rd #104, Carrollton, TX 214-242-9696

Olduvai Corp/Michael Moore
9200 S Dadeland Blvd #725, Miami, FL.............305-670-1112
QMS/Ann Strople
One Magnum Pass, Mobile, AL 205-633-7223
Quadram/Lisa Rizk
One Quad Way, Norcross, GA............................404-923-6666
Softsync Inc/Rod Campbell
SW 37th Ave #765, Coral Gables, FL 305-444-0080
Solution Technology Inc/Jim Hill
1101 S Rogers Cir #14, Boca Raton, FL.............407-241-3210
Taylored Graphics/Max Taylor
21332 Aaron Ct, Lutz, FL...................................813-948-7808
Techware Inc/Matt Landan
PO Box 151085, Altamonte Springs, FL 800-347-3224
Tempra Software/Derek Hodges
402 S Kentucky Ave #210, Lakeland, FL 813-682-1128
Third Wave Computing
1219 W Sixth St, Austin, TX...............................800-284-0486
Truvel Corporation/Mary Jefferson
520 Herndon Pkwy, Herndon, VA 703-742-9500
VGC Corporation/Tom Broncato
5701 NW 94th Ave, Tamarack, FL.....................305-722-3000
Videotex Systems Inc/Bob Gillman
8499 Greenville Ave #205, Dallas, TX.................214-343-4500
VideoTutor Inc/Genevieve Baker
903 Texas Trail, Austin, TX................................512-288-6674
Weaver Graphics
5165 S Hwy A1A, Melbourne Beach, FL.............407-728-4000
ZSoft/Technical Support
450 Franklin Rd #100, Marietta, GA..................404-428-0008

MIDWEST

Ahrens Photo, Bob
400 N State, Chicago, IL...................................312-243-5550
Amgraf Inc/Marjie Garner
1501 Oak St, Kansas City, MO..........................816-474-4797
ArborText/Merredith Faye
1000 Victor's Way #400, Ann Arbor, MI.............313-996-3566
ASDG Inc/Jolaine Benson
925 Stewart St, Madison, WI.............................608-273-6585
Baudville Inc/Karen Westover
5380 52nd St SE, Grand Rapids, MI 800-728-0888
Blackhawk Data Corp
7234 W North Ave #411, Elmwood Park, IL......708-453-9590
Coda Music Technology/Peggy Wagner
6210 Bury Dr, Eden Prairie, MN 800-843-2066
Computer Presentations Inc/Marijo Viox
1117 Cypress St, Cincinnati, OH 513-281-3222
Concept Publishing Systems/Annette Budde
809 Park Ave, Beaver Dam, WI.........................414-887-3731
D'Pix Inc/Mark Eno
929 Harrison Ave #205, Columbus, OH 614-294-6062
Data Service Company/Robert Chudek
3110 Evelyn St, St Paul, MN.............................612-636-9469
Datalogics Inc/Mike Masirka
441 W Huron, Chicago , IL 312-266-4444
DesignSoft/Marty Dell
PO Box 1130, Wheaton, IL...............................800-426-0265
Desktop Plus/Mary Hietala
20 E 107th St Cir, Bloomington, MN 612-888-4307
Dynamic Graphics Inc/Roger Van Zanbergen
6000 N Forest Park Dr, Peoria, IL 800-255-8800
Dynamic Graphics Inc/Peter Force
6000 N Forest Park Dr, Peoria, IL 309-688-8800
Foresight Resources Corp/Carol Johnston
10725 Ambassador Dr, Kansas City, MO 816-891-1040
Graphic Enterprises/Don Frank
439 Market Ave N, Canton, OH 216-452-2033
Graphic Services/Steve Kelly Sr
1520 Second Ave, Des Moines, IA 515-244-0112
Image Processing Software/Peggy Bostwick
PO Box 5016, Madison, WI...............................608-233-5033
ImageTech Inc/Chris Oesterling
29444 Northwestern Hwy #L-500,
Southfield, MI 313-353-7900
Laser Masters Inc/Lee Newsom
9955 W 69th St, Eden Prairie, MN 612-944-9264
LaserMaster Corp/Bruce Butler
7156 Shady Oaks Rd, Eden Prairie, MN.............612-944-9330
Lattice Inc/Liz Dunham
3010 Woodcreek Dr #A, Downers Grove, IL 708-769-4060
Le Baugh Software Corp/Robert Gaddi
PO Box 7787, Omaha, NE 402-731-4304
Mactemps/Pat DeJager
821 Marquette Ave #1010, Minneapolis, MN ...800-MAC-TEMP

Microsystems Engineering
2500 Highland Ave #350, Lombard, IL...............708-261-0111
Microware Systems Corp/Mary Jo Marturello
1900 NW 114th St, Des Moines, IA 515-224-1929
Microware Systems Corp
1900 NW 114th St, Des Moines, IA 515-224-1929
Monotype Inc/Steve Farkas
2100 Golf Rd, Rolling Meadows, IL....................708-427-8800
Monotype Typography/Bill Davis
150 S Wacker Dr #2630, Chicago, IL................312-855-1440
Multi-Ad Services/Marc Radosivic
1720 W Detweiler Dr, Peoria, IL 309-692-1530
Multi-Ad Services Inc/Robert Jennings
7668 Golden Triangle Dr, Eden Prairie, MN.........612-944-7933
Purup Pre-Press America/Sharon Kruskopf
1340 Mendota Hts Rd, St Paul, MN...................612-686-5600
Right Brain Inc/John Nelson
4263 Brigadoon Dr, Shoreview, MN 612-229-6299
Screaming Color Inc/Jill Schwartz
125 N Prospect, Itasca, IL 708-250-9500
SoftCraft Inc/Jim Marshall
16 N Carroll St #220, Madison, WI....................608-257-3300
Square One Desktop Publishing/Pati Anderson
2208 W 21st St, Minneapolis, MN 612-377-1152
SYSTAT Inc
1800 Sherman Ave #801, Evanston, IL...............708-864-5670
Timeworks Inc/Curtiz Gangi
625 Academy Dr, Northbrook, IL........................708-559-1300
Wace Resource Center/Rocco Matera
200 E Ohio St, Chicago , IL 312-915-0695
Wayzata Technology/Shelley Knudson
2515 E Highway 2, Grand Rapids, MN 218-326-0597
Wheeler Arts/Paula Wheeler
66 Lake Park, Champaign, IL.............................217-359-6816
Xitron Inc/Wendy Treichel
1428 E Ellsworth Rd, Ann Arbor, MI....................313-971-8530

WEST

AccuData Inc/Barton Pugh
PO Box 65667, Salt Lake City, UT 801-485-7400
Acius Inc/Suzanne Whitney-Smedt
20883 Stevens Creek Blvd, Cupertino, CA.........408-252-4444
Adobe Systems Inc/Alice Peterson
1585 Charleston Rd/PO Box 7900,
Mountain View, CA...415-961-4400
Advanced Micro Devices Inc/Chris Henry
MS 1028/PO Box 3453, Sunnyvale, CA............408-732-2400
Aladdin Systems Inc/David Schargel
165 Westridge Dr, Watsonville, CA 408-761-6200
Aldus Corp
411 First Ave S/Customer Svc, Seattle , WA........206-622-5500
American Research Corp/Tern Wang
1101 Monterey Pass Rd, Monterey Park, CA 213-264-6531
Aristosoft
7040 Koll Ctr Pkwy #160, Pleasanton, CA..........510-426-5355
Artbeats/Phil Bates
2611 S Myrtle Rd, Myrtle Creek, OR 503-863-4429
Auto-trol Technology Corp/Guri Stark
12500 N Washington, Denver, CO......................303-452-4919
Avalanche Development Co/Bill Zoellick
947 Walnut St, Boulder, CO 303-449-5032
Avery Dennison Corp/Denise Marsh
20955 Pathfinder Rd, Diamond Bar, CA.............909-869-7711
Aztek/Phil Lippincott
15 Marconi, Irvine, CA 714-770-8406
BCA/Desktop Designs/Carl Ballay
PO Box 2191, Walnut Creek, CA.......................510-946-1716
Bioscan Inc/Dennis Flanagan
190 W Dayton #103, Edmonds, WA 206-775-8000
Blue Sky Research/Dave Heinen
534 W Third Ave, Portland, OR...........................503-222-9571
Blueberry Software/Kevin Dwan
260 Petaluma Ave, Sebastopol, CA 707-829-2499
Borland International
100 Borland Way, Scotts Valley, CA 408-431-1000
Bridgestone Multimedia/Bill Lamphear
1979 Palomar Oaks Way, Carlsbad, CA 619-431-9888
Caere Corporation/Larry Lunetta
100 Cooper Ct, Los Gatos, CA..........................408-395-7000
CalComp Inc/Heidi Schweizer
2411 W La Palma Ave, Anaheim, CA.................714-821-2000
Calera Recognition Systems Inc/Deborah Herath
475 Potrero Ave, Sunnyvale, CA 408-720-8300
Chipsoft Inc/Customer Svc
6330 Nancy Ridge Dr #103, San Diego, CA.......619-453-4446

Claris Corp
5201 Patrick Henry Dr, Santa Clara, CA 800-3CLARIS
Colorprep Inc/Raymond Neoh
320 Hatch Dr, Foster City, CA 415-358-8898
Columbia Software Inc/Frank Mahdavi
18908 MuirKirk Dr, Northridge, CA 818-363-2574
CPI Inc/Dave Hansberry
1953 Landings Dr, Mountain View, CA 800-345-3540
Creative Media Services/Linda Harris
2936 Domingo Ave, Berkeley, CA 510-843-3408
Curved Space/Kai Krause
115 S Topanga Canyon Blvd #115, Topanga, CA310-455-2082
Dayna Communications Inc/Steven Clegg
849 W Levoy Dr, Salt Lake City, UT 801-269-7200
Delrina Technology Inc
6830 Via Del Oro #240, San Jose, CA 800-268-6082
DeltaPoint Inc/Mike Malloy
2 Harris Ct #B1, Monterey, CA 408-648-4000
DeScribe Inc/Marketing
4234 N Freeway Blvd #500, Sacramento, CA ...916-646-1111
Design Science Inc
4028 E Broadway/Marketing, Long Beach, CA...310-433-0685
Destiny Technology Corp/David Larrimore
3255-1 Scott Blvd #201, Santa Clara, CA 408-562-1000
● **DIGITAL STOCK PROFESSIONAL/**
Charles Smith (P 99)
400 S Sierra Ave #100, Solana Beach, CA......619-794-4040
Digital Technology Intl/Marc Thompson
500 W 1200 South, Orem, UT 801-226-2984
Dream Maker Software/Dan Sutphin
925 W Kenyon Ave #16, Englewood, CO 800-876-5665
Dubl-Click/Robert Spector
22521 Styles St, Woodland Hills, CA 818-888-2068
Dynaware USA Inc/Sho Hayami
950 Tower Ln #1150, Foster City, CA 415-349-5700
Etgin Co/Gary Duarte
336 Greenbrae Dr, Sparks, NV 702-355-7588
Extended Systems/Kristin Tolle
PO Box 4937, Boise, ID 208-322-7575
Finot Group, The/Dan Finkelstein
1504 Franklin St #310, Oakland, CA 800-748-6480
Font Company, The
7850 E Evans Rd #111, Scottsdale, AZ 800-442-3668
Fractal Design Corp/Andrea Godkin
PO Box 2380, Aptos, CA 408-688-8800
Fujitsu Computer Product of America/
Steve Butterfield
2904 Orchard Pkwy, San Jose, CA....................408-894-3512
Gibson Research Corp/Gene Senegal
35 Journey, Aliso Viejo, CA 714-362-8800
Halcyon Software/Don Hsi
1590 La Pradera, Campbell, CA 408-378-9898
Innovative Data Design/Jane Schwarz
1820-L Arnold Industrial Way, Concord , CA.......510-680-6818
Island Graphics Corp/David Neumann
4000 Civic Center Dr, San Rafael, CA 800-255-4499
Jones Design, Brent A
328 Hayes St, San Francisco, CA 415-626-8337
Kensington Microware/Mary Shank
2855 Campus Dr, San Mateo, CA......................415-572-2700
KeyTronic/Linda Franklin
PO Box 14687, Spokane, WA 509-928-8000
Koala Acquisitions/Earl Liebich
PO Box 1924, Morgan Hill, CA 408-776-8181
LaserGo Inc/Truc Nguyen
9369 Carroll Park Dr #A, San Diego, CA............619-450-4600
LaserTools Corp/Sales Line
1250 45th St #100, Emeryville, CA 800-767-8004
Mainstay/Lance Merker
591-A Constitution Ave, Camarillo, CA 805-484-9400
Marstek Inc/David Hsieh
15225 Alton Pkwy, Irvine, CA..........................714-453-0110
Mass Microsystems
1507 Centre Pointe, Milpitas, CA......................408-946-9207
Metro ImageBase Inc/Frank Whitehead
18623 Ventura Blvd #210, Tarzana, CA.............818-881-1997
Migraph Inc
32700 Pacific Hwy S #14, Federal Way, WA 206-838-4677
Mind Links/Nicholas MacConnell
1135 Stratford, Del Mar, CA............................619-481-7535
Mouse Systems Corp/Dave Todd
47505 Seabridge Dr, Fremont, CA....................510-656-1117
On the Go Software/Robert Crumpler
4225 Executive Sq #1200, La Jolla, CA............619-558-4114

Optisys
8620 N 22nd Ave #109, Phoenix, AZ 800-327-1271
Orange Micro Inc/Greg Petersen
1400 N Lakeview Ave, Anaheim, CA 714-779-2772
OWL Software/John Campbell
7633 Fulton Ave, North Hollywood, CA..............818-765-5311
Pacific Micro/Tom Anderson
201 San Antonio Cir #C250, Mtn View, CA.......415-948-6200
Page Studio Graphics/Roger Vershon
3175 N Price Rd #1050, Chandler, AZ 602-839-2763
PCPI (Personal Computer Products Inc)/
Lori Castro
10865 Rancho Bernardo Rd, San Diego, CA.......619-485-8411
Personal TEX Inc
12 Madrona St, Mill Valley, CA........................415-388-8853
PIXAR/Pam Kerwin (P 116)
1001 W Cutting Blvd, Richmond, CA 510-236-4000
Postcraft Intl Inc/Paul Sibek
27811 Ave Hopkins #6, Valencia, CA................805-257-1797
Pre-Press Technologies/Marketing
2443 Impala Dr, Carlsbad, CA 619-931-2695
Presentation Technologies/Jim Creede
779 Palomar Ave, Sunnyvale, CA 408-730-3700
Promark Ltd/William Riley
6207 Pan American NE, Albuquerque, NM 505-345-7701
Quark Inc/Peter Warren
1800 Grant, Denver, CO.................................303-894-8888
Quintar Co/Craig Douglas
370 Amapola Ave #106, Torrance, CA 310-320-5700
Ray Dream Inc/Eric Hautemont
1804 N Shoreline Blvd #240, Mt View, CA........415-960-0765
RIX Softworks Inc/Paul Harker
17811 Sky Park Cir #B, Irvine , CA....................714-476-8266
Salient Software Inc/Hattie Johnson
10201 Torie Ave, Cupertino, CA 408-253-9600
Scantronix/Rick Camoirano
14311 Cerise Ave #108, Hawthorne, CA............310-644-8585
ScenicSoft Inc/Jerry Nakal
11400 Airport Rd, Everett, WA 206-776-7760
SeeColor Corp/Ronald LaForge
PO Box 3148, Federal Way, WA 206-946-1948
Sigma Designs Inc/Jess Herrera
47900 Bayside, Fremont, CA............................510-770-0100
Skill Set Graphix/Manny Pavia
5757 Wilshire Blvd # M101, Los Angeles, CA 213-937-5757
SKW Computers Inc/Larry Wang
49 W El Camino Real, Mountain View, CA 415-390-9888
Step 2 Software/Loren Brown
1501 Fourth Ave #2270, Seattle, WA 800-867-8372
● **STRATA STUDIO**/Reed Terry
(P 110,111)
2 W St George Blvd #2100, St George, UT 801-628-5218
Strategic Mapping Inc/Steve Poizner
3135 Kifer Rd, Santa Clara, CA........................408-985-7400
Studio Advertising Art/Rick Barker
PO Box 43912, Las Vegas, NV 702-641-7041
Sumeria/Jerry Borrell
329 Bryant St, San Francisco, CA 415-904-0800
Symantec Corp/Pamela Barnett
10201 Torre Ave, Cupertino, CA......................408-253-9600
Symmetry Software Corp/Brad Nelson
8603 E Royal Palm Rd #110, Scottsdale , AZ......602-998-9106
T/Maker Co/John Tompane
1390 Villa St, Mountain View, CA 415-962-0195
Team Xerox
4674 Cardin St, San Diego, CA........................619-571-8885
Three G Graphics Inc/Glenn Giaimo
104 Second Ave S #104, Edmonds, WA 206-774-3518
Three-D Graphics/Dan Weingart
860 Via de la Paz, Pacific Palisades , CA 310-459-7949
Three-D Visions/David Ulmer
2780 Skypark Dr, Torrance, CA 310-325-1339
Thunderware Inc/Brian Smith
21 Orinda Way, Orinda, CA............................510-254-6581
TriMetrix Inc/Cheryl Mauer
444 NE Ravenna Blvd #210, Seattle, WA 800-548-5653
UniDisc Inc/Chris Andrews
36 Seascape Village, Aptos, CA 408-684-2191
Visitel/Allen Smithson
Box 2019, Cottonwood , AZ............................602-634-5100
Wasatch Computer Technology/Rod Tate
123 E 200 South, Salt Lake City, UT 801-575-8043
Winners Circle Systems
2618 Telegraph Ave, Berkeley, CA....................510-845-4814

WordPerfect Corp/Sales Dept
1555 N Technology Way, Orem, UT..................801-225-5000
WordStar International Inc
PO Box 6113, Novato, CA..............................800-227-5609
Wyse Technology
3471 N First St, San Jose, CA..........................408-473-1200
XAOS Tools/Frances Doaz
600 Townsend # 270E, San Francisco, CA 415-487-7027
Zedcor/Peter Gariepy
4500 E Speedway #22, Tucson, AZ..................800-482-4567
Zenographics
4 Executives Cir #200, Irvine , CA 714-851-6352

C A N A D A

Avanti Computer Systems/Richard Wallin
2788 Bathurst St #305, Toronto, ON 416-785-0424
Corel Systems/Paul Bodnoff
1600 Carling Ave, Ottawa, ON 613-728-8200
Delrina Technology Inc
895 Don Mills, 500-2 Park Ctr , Toronto, ON 416-441-3676
Eicon Technology Corp/Maks Wulkan
2196 32nd Ave, Lachine, QB 514-631-2592
Exoterica Corp/Wayne Lucky
383 Parkdale Ave #406, Ottawa, ON.................613-722-1700
Image Club Graphics/Brad Zumwalt
1902 11th St SE #5, Calgary, AB 403-262-8008
International Digital Fonts
1431 Sixth St NW, Calgary, AB........................403-284-2288
Mortice Kern Systems Inc/Ruth Songhurst
35 King St N, Waterloo, ON 519-884-2251
SoftQuad Inc/Steve Downie
56 Aberfoyle Cres, Toronto, ON........................416-239-4801
Ultimate Technographics Inc/Joanne David
800 Renee Levesque Blvd #2660, Montreal, QB ...514-954-9050

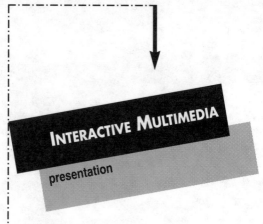

INTERACTIVE MULTIMEDIA

presentation

Presentation

NEW YORK CITY

Admaster
104 E 25th St, New York, NY..........................212-673-8100
Applied Imagination Inc/Thomas Tafuto
826 Broadway 6th Fl, New York, NY..............212-529-9696
Artichoke Arts/Glenn Cho
333 E 34th St, New York, NY..........................212-532-2188
Bedrock Design/Janet Tingey
223 W 21st St #4G, New York, NY...................212-929-1664
Blauweiss, Stephen
32-15 41st St, Long Island City, NY..............718-204-8335
Bramble Design/Bill Bramble
327 W 11th St, New York, NY..........................212-929-6289
Burnett Group/Johanna Skilling
39 E 20th St 7th Fl, New York, NY.................212-254-3344
C & C Visual/Glen Botkin
1500 Broadway #400, New York, NY...............212-869-4900
Carol Chiani Digital Workshop/Carol Chiani
20 E 46th St #402, New York, NY...................212-867-4371
Center for Advanced Whimsy, The /
Rodney Alan Greenblatt
61 Crosby St, New York, NY..........................212-219-0342
Communications Plus/Geoffrey Fraize
102 Madison Ave, New York, NY....................212-686-9570
Coppola, Richard
50-16 213th St, Bayside, NY..........................718-428-3636
Creative Ways Inc/Stuart Sternbach
305 E 46th St, New York, NY..........................212-935-0145
Cunningham, Peter
53 Gansevoort St, New York, NY....................212-633-1077
Curious Pictures/Tina Contis
23 Watts St, New York, NY..............................212-966-1020
Darino Films/Edward Darino
222 Park Ave S, New York, NY.......................212-228-4024
Data Motion Arts/Tony Caio
165 W 46th St #611, New York, NY.................212-768-7411
de Kerangal, Pierre
130 W 3rd St #4R, New York, NY...................212-387-9204
Edson, Jennifer
347 W 57th St #27A, New York, NY...............212-333-5578
Edwin Schlossberg Inc/Heather Liston
641 Sixth Ave 5th Fl, New York, NY..............212-989-3993
Epure, Serban
60-11 Broadway #5L, Woodside, NY.............718-335-7685
Equitable Production Group
787 Seventh Ave, New York, NY....................212-554-1389
Fat Baby Productions/Jessica Perry
185 E 85th St #26M, New York, NY...............212-722-0535
Ferri-Grant, Carson
255 W 90th St PH, New York, NY...................212-362-8567
Fusion Communications/Craig A Mengel
510 E 74th St, New York, NY..........................212-879-8787
Graphic Art Resource Assoc
257 W 10th St #5E, New York, NY.................212-929-0017
Harris Design, George
301 Cathedral Pkwy #2N, New York, NY.......212-864-8872
Hersh, Seth
301 E 47th St #4-O, New York, NY................212-826-3818
Hoberman, Perry
167 N 9th St #10, Brooklyn, NY.....................718-388-8241
Holter, Catherine
721 Broadway #606, New York, NY...............212-998-1605
Idesign/Robert M Stewart
165 E 35th St #11A, New York, NY...............212-683-0244

Image Group/Adriane Truex
305 E 46th St, New York, NY..........................212-752-3010
K Landman Design/Kathy Landman
156 Fifth Ave 3rd Fl, New York, NY...............212-924-4254
Kayser, Alex
211 W Broadway, New York, NY....................212-431-8518
Levy, Jaime
Box 448 Prince St Station, New York, NY.........212-777-4801
Mariah Productions Ltd/Ken Book
545 8th Ave 3rd Fl, New York, NY.................212-947-0090
Mattlin, Brian
254 W 25th , New York, NY............................212-627-4963
Micro Interactive/Steve Baum
1123 Broadway #1200, New York, NY.............212-366-1391
Morawa, Amy Lynne
375 South End Ave #4K, New York, NY..........212-938-5732
Morgan Guaranty Trust Company/J Terri Ferrari
23 Wall St, New York, NY...............................212-483-1308
MTI/Robert Marmiroli
885 Second Ave Level C, New York, NY..........212-355-0510
Neographic, Inc/Gregg Trueman
18 E 16th St, New York, NY............................212-633-6649
Ogilvy Mather Direct/Interactive
Marketing Div/Martin Nisenholtz
309 W 49th St 5th Fl, New York, NY..............212-237-6010
Premier Post/Keith Shapiro
630 Ninth Ave 14th Fl, New York, NY............212-757-1711
Raabe, Dan
80 Montague St, Brooklyn, NY.......................718-260-9666
Renaissance Art Center/Andrew Edwards
29 W 38th St 3rd Fl, New York, NY...............212-575-6100
Romulus Productions Inc/Peter Crown
100 Greene St, New York, NY.........................212-226-6226
Schneemann, Carolee
114 W 29th St, New York, NY........................212-695-6615
Solomon Video Productions, Rob
331 W 57th St #242, New York, NY...............212-541-9507
Spano Design/Michael Spano
35 Regina Ln, Staten Island, NY...................718-966-8979
Tom Nicholson Associates/Tom Nicholson
295 Lafayette St 8th Fl, New York, NY..........212-274-0470
Tower Graphics/Catherine Tower
23 West 73rd St #714, New York, NY...........212-874-6229
Tumble Interactive Media/Cal Vornberger
910 West End Ave #3D, New York, NY..........212-316-0200
Two Twelve Assocs/David Peters
596 Broadway #1212, New York, NY.............212-925-6885
Video Works/Jim Riche
24 W 40th St, New York, NY..........................212-869-2500
Videograf/Michael Frenchman
144 W 27th St, New York, NY........................212-206-8477
Wallin, Michael
26-04 25 Rd, Astoria, NY..............................718-278-7196
Wu Multimedia, Brian
229 E 21st St #25, New York, NY.................212-674=6480
Zang Design, Ulla
36 Plaza Street #7B, Brooklyn, NY...............718-399-0454
Zink Communications/Marianne Callaghan
245 W 19th St, New York, NY........................212-929-2949

EAST

AB Communications/Adam Brooks
14 Dana St, Somerville, MA............................617-330-2275
Action Graphics/Tyrone Thompson
4725 Dorsey Hall Dr #A-402, Ellicott City, MD....410-992-0749
Active 8/Ken Valley
601 N Eutaw St #704, Baltimore, MD..............410-962-0272
AdViz Advanced Visualizations/Allen H Cosgrove
709 Rt 206 N #333, Belle Mead, NJ...............908-281-0421
Altered Image/Frank Thomas Ward
7 Deer Park Dr # D, S Brunswick, NJ............908-274-2220
Amigo Business Computer/Bill Teller
192 Laurel Rd, E Northport, NY......................516-757-7334
Applied Imagination Inc/Thomas Tafuto
RR 1 Box 1765, Moretown, VT.......................802-496-3520
Applied Optical Media Corp/Paul Dellavigne
1450 Boot Rd, Bldg 400, West Chester, PA........215-429-3701
Arkay Technologies/Ray Khorram
5 Tsienneto Rd #2, Derry, NH........................603-425-2149
Arkwright/Leanne Butcher
538 Main St, Fiskeville, RI..............................401-821-1000
AV Design/Joe Wall
1823 Silas Deane Hwy, Rocky Hill, CT...........203-529-2581
Baker MediaSource/Wai Wong
2001 Joshua Rd, Lafayette, PA.......................215-832-0340

Becker, Peter
Hawley Road, N Salem, NY............................212-249-4493
Business Presentation Services/Mary Clupper
40 Cameron Ave, Somerville, MA...................617-666-1161
Capital Presentations/Wolfgang Esh
10 Post Office Rd #2N, Silver Spring, MD..........301-588-9540
Ciné-Med Inc/Kevin McGovern
127 Main St N/PO Box 1007, Woodbury, CT....203-263-0006
Coffman, Claudia
82 Manhattan Ave #4, Jersey City, NJ............201-963-7030
Colony Multimedia/Timothy Tolman
150 Chestnut St, Providence, RI.....................401-274-0933
Computer Creations/James Houghtaling
14 Balsam Cir, Whitesboro, NY......................315-768-8471
Computer Graphics Resources/
One Dock St, Stamford, CT............................203-327-3635
Corporate Communications/Monte Davis
223 Vose Ave, S Orange, NJ..........................201-378-3327
Cramer Productions/Tom Martin
355 Wood Rd, Braintree, MA..........................617-849-3350
DCP Communications Grp Ltd/Alison Connors
301 Wall St, Princeton, NJ.............................609-921-3700
DePersico Inc/Paul DePersico
878 Sussex, Broomall, PA...............................215-328-3155
DI Group/Sharon Lynch
651 Beacon St, Boston, MA............................617-267-6400
Digital Video Arts/George Breen
715 Twining Rd #107, Dresher, PA................215-576-7920
Documentation Development Inc/Ron Miskie
125 Cambridge Park Dr, Cambridge, MA..........617-864-7300
Domin, Jacqueline
26 Monroe St, Honeoye Falls, NY..................716-624-3318
Eaglevision Inc/Jeff Weil
880 Canal St, Stamford, CT............................203-359-8777
Egeland Wood & Zuber/Elizabeth Wood
176 Freemans Bridge Rd, Scotia, NY..............518-374-3131
Ehrlich Multimedia/Alexander Ehrlich
1 Maynard Dr, Park Ridge, NJ........................201-262-4499
ESPN Inc/Steve Ullman
1 ESPN Plaza, Bristol, CT...............................203-585-2084
Expanded Video Inc/Ted Miles
477 Congress St, Portland, ME.......................207-773-7005
ExperTech Corp/Kathleen Harmeyer
303 Wynell Ct, Timonium, MD.......................410-561-3238
Flying Turtle Productions/Laurie Holmes
721 Crestbrook Ave, Cherry Hill, NJ...............609-751-7213
Forsight Inc/Ronald Gregory
POB 0427/1700 Rockville Pike#150,
Rockville, MD..301-816-4900
Forward Design/John Forward
1115 E Main St Box 61, Rochester, NY...........716-288-0250
Genigraphics/Adam Yorks
129 Portland St 3rd Fl, Boston, MA................617-723-5550
Gil Bellin Productions/Gil Bellin
175 Westminster Dr, Yonkers, NY...................914-968-2892
Girard Video/Jenny Gsell
1331 F St NW #250, Washington, DC.............202-393-6666
Greater Media Cable Adv/Michael Savino
95 Higgins St, Worcester , MA........................508-853-1515
Guymark Studios
3019 Dixwell Ave/PO Box 5037, Hamden, CT...203-248-9323
Hogpenny Productions Inc/Brian Treutlein
121 E 14th St/Ship Bottom, Long Beach
Island, NJ..609-494-6640
Hyperview Systems Corp/Dick Ficke
28 Jacome Way, Middletown, RI....................401-849-8900
Iconcepts/Jeff Cummings
118 Magazine St, Cambridge, MA..................617-661-7289
ICT Productions/Noel Izon
4320 Hamilton St, Hyattsville, MD.................301-864-6333
Idea Works
90 Bridge Dr, Newton, MA.............................617-244-0101
Image Production/Jenny Coffelt
50 Water St, S Norwalk, CT............................203-853-3486
Imergy/Suzanne Mason
12 S Main St, S Norwalk, CT..........................203-853-6200
Instructional Design International Inc/Lin Lougheed
1775 Church St NW, Washington, DC.............202-332-5353
Interactive Design/Jan Diamondstone
1501 N Walnut St, Wilmington, DE.................302-429-0143
Lake Champlain Productions/Steve Murphy
4049 Williston Rd, S Burlington, VT...............802-863-9780
Lehigh Interactive Multimedia/
James W Kauffman
937 High St, Bethlehem, PA...........................215-867-8448

Lewis Cohen & Co/Lewis Cohen
  141 Ruxton Rd, Mt Kisco, NY 914-241-3638
Light & Power Productions/Bill Moffitt
  26 N Broadway, Schenectady, NY 518-381-6788
Magical Media/Peter Jensen
  139 E Spring Ave, Ardmore, PA 215-642-4225
Marc Multimedia Ltd/Yale Marc
  61 Chapel St, Newton, MA 617-527-7800
Media Construction Inc/Michael Gallelli
  4700 Crescent St, Bethesda, MD 301-229-4188
Media Vision/James Lush
  80 Great Hill Rd, Seymour, CT 203-735-0002
Mendoza Design
  3104 Grindon Ave, Baltimore, MD 410-426-3671
● **META 4 MULTIMEDIA**/Paul Lemberg (P 95)
  245 W Norwalk Rd, W Norwalk, CT 203-857-0654
Meyer Software/Linda Nunez
  616 Continental Rd, Hatboro, PA 800-643-2286
Micro Visual Systems Inc/Tom Nickel
  385 Elliot St, Newton, MA 617-964-0640
Midi Inc/Jack Noon
  100 Thanet Cir #105, Princeton, NJ 609-924-4817
Mirage Visuals/Curt Brooks
  206 Monroe st, Monroeville, PA 412-372-4181
Modern Video Productions Inc/Jim Burt
  1650 Market St 3rd Fl, Philadelphia, PA 215-569-4100
MPTV Video Production Unit/Mark Patterson
  2116 Noble St, Pittsburgh, PA 412-271-6788
Multimedia Workshop/Michael D Murie
  Box 44-37, Somerville, MA 617-776-2469
National Education Training Group/Steve Maxwell
  9 Oak Park Dr, Bedford , MA 800-227-1127
New England Technology Group/Steve Gregory
  One Kendall Sq Bldg 200, Cambridge, MA 617-494-1151
NFL Films Video/Hal Lipman
  330 Fellowship Rd, Mt Laurel, NJ 609-778-1600
NUS Training Corp/Philip Taplin
  910 Clopper Rd, Gaithersburg, MD 301-258-2530
Optimedia Systems/George Hoffman
  373 Rte 46 W, Fairfield, NJ 201-227-8822
Paul, Dayan
  779 Mabie St, New Milford, NJ 201-265-3842
Pixel Light Communication/Carmine DeFalco
  271 Rte 46 W # F205, Fairfield, NJ 201-808-1389
Pragma Design/Mark Esterly
  23 Jay Dr, Londonderry, NH 603-437-2010
PREFIT Corporation/Joseph Chernicoff
  PO Box 326, Willow Grove, PA 215-657-3976
PVS (Professional Video Services)/Liz Lokey
  2030 M St NW, Washington, DC 202-775-0894
Rampion Visual Productions/Steven Tringali
  316 Stuart St, Boston, MA 617-574-9601
REF Associates/Dr Robert Fidoten
  118 Greyfriar Dr, Pittsburgh, PA 412-963-8785
Robilotto, Philip
  777 Stone Rd Route 9W, Glenmont, NY 518-767-3196
Scabrini Design Inc, Janet
  90 Maywood Rd, Norwalk, CT 203-853-6676
Schneider, Pat
  Six Lincoln Gardens, Long Branch, NJ 908-222-6840
Schwartz, Ariel
  113 Winchester Ln, Newtown, PA 215-860-0888
Scion Corporation/Bonnie Polesky
  152 W Patrick St, Frederick, MD 301-695-7870
Seroka Group Inc, The/Paul Seroka
  Putnam Green #38-H, Greenwich, CT 203-862-9700
Sonic Images Productions Inc/Jolie Barbiere
  4590 MacArthur Blvd NW, Washington, DC 202-333-1063
sound/IMAGE
  PO Box 1442, Cambridge, MA 617-494-5400
Speaker Support Group/Jeff McPhie
  19 Wall St, Princeton, NJ 609-921-3400
Specter Video
  4148 Windsor St, Pittsburgh, PA 412-521-1444
Spicer Productions Inc/Sharon Jackson
  1708 Whitehead Rd, Baltimore, MD 410-298-1200
Stellarvisions/Stella Gassaway
  4041 Ridge Ave/Bldg 17 #105, Philadelphia, PA215-848-9272
Szabo, Michelle
  1 Little John Ln, Danbury, CT 203-791-8599
Tangent Design/Communications/Otto Timmons
  25 Sylvan Rd S #N, Westport, CT 203-221-1013
Tepper, Marlene
  35 Parkview Ave, Bronxville, NY 914-779-9080

Tiani, Alex
  PO Box 4530, Greenwich, CT 203-661-3891
Transparent Media/Warren Schloss
  2729 N 47th St, Philadelphia, PA 215-477-5868
Trend Multimedia/Tony Rosa
  4 Holly Ave, West Keansburg, NJ 908-787-0786
V Graph Inc/Rob Morris
  PO Box 105, Westtown, PA 215-399-1521
Video/Visuals Inc/Bob Lewis
  61 Chapel St, Newton, MA 617-527-7800
Videos Multimedia Inc/Bill Mutsler
  South River Rd/Rte 130 Bldg C, Cranbury, NJ ..609-395-1120
Visual Impressions/Bill Robbins
  429 State St, Rochester, NY 716-546-1917
Visual Logic/Jack Harris
  724 Yorklyn Rd #150, Hockessin, DE 302-234-5707
Visual Services Inc/Daniel Gallo
  25 Van Zant St, E Norwalk, CT 203-852-1010
Vox-Cam Associates Ltd
  813 Silver Spring Ave, Silver Spring, MD 301-589-5377
Wave Inc/Walter Henritze
  11 California Ave, Framingham, MA 508-795-7100
Wolf Design, Stephen
  904 Garden St, Hoboken, NJ 201-659-2422

SOUTH

(TC)2 Textile/Clothing Technology/Frank H Hughes
  211 Gregson Dr, Cary, NC 919-380-2156
4th Wave Inc/John N Latta
  PO Box 6547, Alexandria, VA 703-360-4800
Aktulun Design/Kenan Aktulun
  111 Southwood Rd, Austin, TX 512-447-6522
ASV Corp/Barbara Fishburn
  2075 Liddell Dr NE, Atlanta, GA 404-876-3445
BERMAC Communications Inc/Phil Chadwick
  4545 Fuller Dr #336, Irving, TX 214-255-9007
Blair Inc/Tom Eckhardt
  5751-A General Washington Dr, Alexandria, VA 703-642-2000
Bransby Productions/Sharon Grilliot
  2124 Metro Cir, Huntsville, AL 205-882-1161
Cheshire Designs/Julia Martinroe
  2492 Fair Knoll Ct NE, Atlanta, GA 404-633-8330
Cinema Concepts/Stewart Harnell
  2030 Powers Ferry Rd #214, Atlanta, GA 404-956-7460
Computerized Video Service/Johnnie Byers
  Rte 2 Box 135, Comanche , OK 405-439-5932
Corporate Media Services/Ralph Preston
  632 W Summit Ave, Charlotte, NC 704-377-1601
CPN Televison Studios/Mike McKown
  14375 Myerlake Cir, Clearwater, FL 813-530-5000
Creative Video/Guy Davidson
  1465 Northside Dr # 110, Atlanta, GA 404-351-1721
Development Communications/Welby Smith
  1800 N Beauregard St, Alexandria, VA 703-845-7370
Editworks/Patrick Furlong
  3399 Peachtree Rd NE #200, Atlanta, GA.......404-237-9977
Ensemble Productions/Barry McConatha
  1036 Cornstalk Ln, Auburn, AL 205-826-3045
Floyd Design/Mike Cusick
  1465 Northside Dr NW #110, Atlanta, GA.......404-351-4518
Galileo/Mike Wittenstein
  680 14 St, Atlanta, GA 404-425-4536
Georgia Inst of Technology/Ray Haleblian
  Multimedia Technology Labratory
  400 10 St, Atlanta, GA 404-894-4195
HP Productions/Heidi Pfisterer
  8229 boone Blvd #390, Vienna, VA 703-356-3099
IBM Media/Mike Griner
  3100 Windy Hill Rd, Marietta, GA 404-835-3426
Image Center/Lisa Phillips
  1011 Second St SW, Roanoke, VA 703-343-8243
Insight Productions/Scott Long
  PO Box 12498, Research Triangle Pk, NC 919-544-5700
Interactive Arts/Rick Ligas
  501 E 4th St #511, Austin, TX 512-469-0502
Interlight International/W R "Jerry" Lindquist
  2365 Centerville Rd, Tallahassee, FL 904-385-3310
ITC/Joan Dasher
  13515 Dulles Technology Dr, Herndon, VA.......703-713-3335
James Design Studio, Inc/Jim Spangler
  7520 NW 5th St #203, Plantation, FL 305-587-2842
JHT Multimedia/Gary Twitchell
  5514 Lake Howell Rd, Winter Park, FL 407-657-2727
Knowles Video Inc/Carl Knowles
  2003 Apalachee Pkwy #204, Tallahassee, FL......904-878-2298

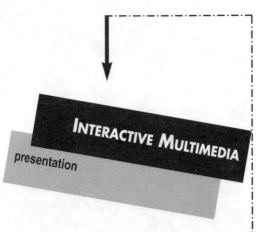
INTERACTIVE MULTIMEDIA
presentation

Layton Intl/Eugene Vasconi
  2013 Wells Branch Pkwy #201, Austin, TX 512-251-0074
Lee Martin Productions/Lee Martin
  11250 Pagemill Rd, Dallas, TX 214-556-1991
Marketing Network, The/Mark Murtha
  PO Box 15320, Winston-Salem, NC.............919-766-9914
Marshall Productions Inc/Bill Hite
  404 BNA Dr #102, Nashville, TN 615-399-8895
Max Place/Peter Sugarman
  Rte 4 Box 54C, Louisa, VA 703-894-0511
Micron/Green/Millard Pate
  1240 NW 21st Ave, Gainesville, FL 904-376-1529
Mindflex/Steve Tanner
  566 Dutch Valley RD, Atlanta, GA 404-892-6232
Moxie Media Inc/Martin Glenday
  5734 Jefferson Hwy/POB 10203,
  New Orleans, LA 504-733-6907
MultiMedia Resources (Level 4 Comm)/
  Doris Seitz-Boothe
  14875 Landmark #102, Dallas, TX.............214-650-1986
Multivision Video & Film/Bob Berkowitz
  7000 SW 59th Pl, S Miami, FL................305-662-6011
Odyssey Communications Group/Rebecca Stevenson
  6309 N O'Connor LB83, Irving, TX 214-432-9070
Park Avenue Teleproductions Inc/Melanie Cox
  3500 Mayland Ct, Richmond, VA..............804-346-3232
Pearlman Productions/Andrew Arten
  2401 W Bellfort, Houston, TX................713-668-3601
PERC/Doug Kalish
  4774 Bogie Rd, Duluth, GA 404-447-1876
Picture Conversion Inc/Go Babaoglu
  5109 Leesburg Pike #212, Falls Church, VA.......703-998-5777
Pixel Graphics/Douglas H Parr
  PO Box 63230, Pipe Creek, TX 210-535-9585
Post Edge Inc/Mike Duncan
  2040 Sherman St, Hollywood , FL 305-920-0800
Presentation Resource/Dee Hardie
  5 W Cary St, Richmond, VA 804-648-7854
Pyramid Studios
  1710 Altamont Ave, Richmond, VA 804-353-0700
Quinn Associates Inc/Margaret Quinn
  110 W Jefferson St, Falls Church, VA 703-237-7222
Random Access Media Works/Gregory Roberts
  4001-2 Confederate Pt Rd , Jacksonville, FL 904-778-8568
Results Video/Larry Emerson
  6515 Escondido #A-1, El Paso, TX 915-581-0184
Richard Kidd Productions Inc/John Kindervag
  5610 Maple Ave, Dallas, TX.................214-638-5433
S Presentation Design B4-1D-47/Greg Spencer
  5400 Legacy Dr, Plano, TX 214-604-7501
Scriber, Cheryl
  4201 Durham Cir, Stone Mountain, GA 404-271-0719
Shooting Stars Post Inc/John Samaha
  3106 W North "A" St, Tampa, FL 813-873-0100
Sirius Image/Larry Goddard
  2062 Weems Rd, Tucker, GA 404-939-2004
Smart Concepts Ltd/Steve Sembritsky
  Tulsa, OK 918-747-6006
Super Computer/Greg Sherman
  366 Lytle St, W Palm Beach, FL 407-833-0558
TM Century Inc
  2002 Academy, Dallas, TX..................800-879-2100
Tri-Comm Productions/Carol Bartholomew
  11 Palmetto Pkwy, Hilton Head Island, SC 803-681-5000
Turner, Pam
  4700 Nine Mile Rd, Richmond, VA.............804-222-1699

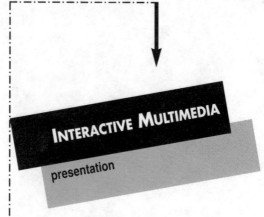

INTERACTIVE MULTIMEDIA
presentation

United States Video/Frank Garber
2070 Chain Bridge Rd #485, Vienna, VA..........703-848-1990
Video Editing Services Inc/Arthur Rouse
215 E High St, Lexington, KY 606-255-9049
Video Promotions Intl/Chris Bentley
PO Box 351985, Palm Coast, FL 800-486-1874
Videofonics Inc/Debbie Valentine
1101 Capital Blvd, Raleigh, NC................919-821-5614
Videotexting/Jaf Fletcher
275 Commercial Blvd, Laudrdale-by-the-Sea, FL....305-771-5999
Visioneering International/Robert Foah
66 12th St NE, Atlanta, GA 404-876-7841
Waterworks Productions/Denise Hodgson
1717 W Sixth St #150, Austin, TX...............512-476-0400
Weithers, Arlington
PO Box 585, Tuskegee, AL 205-727-3514
Z-AX-IS/Steven T Shepard
3997 Cocoa Plum Cir, Coconut Creek, FL 305-970-9662

MIDWEST

Applied Integration/David Feller
Box 200, Chelsea, MI 313-475-3266
Abraham Animation/Richard Abraham
2001 Atwood Ave, Madison, WI.................608-241-3900
Ad-Visual Communications/Gary DiCenzo
1461 Oakwood, Sylvan Lake, MI 313-682-4424
Adpro Inc/Rob Robbins
331 N Main St, Sycamore, IL....................815-895-2000
Advanced Video Communications/William Thermos
750 Pasquinelli Dr #244, Westmont, IL 708-323-7464
AGS&R Communications/IN
1835 S Calhoun, Fort Wayne, IN................219-744-4255
AlexanderDesign Inc/Roy Alexander
2117 W Arthur Ave, Chicago, IL................312-508-9313
**American Cablevision Teleproductions/
WTEN**/Tim Renshaw
3030 Roosevelt Ave, Indianapolis, IN 317-632-2288
Arneson, Peter
1817 Second Ave S, Minneapolis, MN..........612-870-8243
Associates Creative Inc/Elliot Goldberg
24471 West Temik #100, Southfield, MI 313-354-3003
Audio-Visual Associates/Bud Osborne
4760 E 65th St, Indianapolis, IN 317-255-6457
AVS Inc/Rob Ramseier
2109 Ward Ave, La Crosse, WI................608-787-1010
Bazooka Graphics/Div Swell Pictures/
Dave Mueller
233 E Wacker Dr, Chicago, IL 312-649-9000
Beacon Software/John Grozik
P O Box 234422, Milwaukee, WI..............414-355-4460
Bovey & Assocs, Ruth
1443 W Altgeld, Chicago, IL 312-327-2886
Bridge Communication Group/Chris Brogdon
233 E Ontario St #901, Chicago, IL 312-787-0245
Bromley Chapin Designs/Al Chapin
400 Renaissance Ctr #2250, Detroit, MI 313-259-2661
Cameron & Co/Jim Cameron
2233 University Ave #150, St Paul, MN 612-645-4002
Carlson Marketing Communications/Fred Carlson
Route 4/Box 384-B, Mora, MN...............612-679-4105
Cinecraft Productions Inc/Neil McCormick
2515 Franklin Blvd, Cleveland, OH 216-781-2300
Combined Services/David Patsy
9 W 14th St, Minneapolis, MN................612-871-5503
Computer Artworks Inc/Daniel Hayes
700 W Pete Rose Way, Cincinnati, OH 513-421-9000

Connecting Images/Terry Hickey
2469 University Ave N, St Paul, MN 612-644-1977
Creative Images/Steve Hendrickson
1400 Energy Park Dr #23, St Paul, MN 612-644-2157
Cyberplex/Pat Maun
201 SE Main St #215, Minneapolis, MN..........612-649-4641
**Dar Electronic Research &
Development**/Anire Dar
919 Blair Ave, Neenah, WI 414-725-6543
Double D Associates Inc/Chris Anderson
12579 W Custer Ave, Butler , WI 414-783-4800
Dynacom Inc/Jerry Sullivan
1512 N Fremont #101, Chicago, IL 312-951-5510
EDR Media/Niki Dias
23330 Commerce Park Rd, Beachwood, OH.......216-292-7300
Ewert, Pat
430 Oakland Dr, Highland Park, IL 708-433-4577
Filmack Studios/Robert Mack
1327 S Wabash Ave, Chicago, IL 312-427-3395
Forcade & Associates/Mark Levine
1213 Mulford St, Evanston, IL 708-328-1318
Frame One Inc/Paul Bernan
676 St Clair #425, Chicago, IL 312-440-1328
Golan Productions/Ari Golan
507 W North Ave, Chicago, IL 312-274-FIL
Golden Dome Prdctns/WNDU-TV/Chuck Huffman
PO Box 1616, South Bend, IN 219-631-1616
Goldsholl: Film Group/Deborah Goldsholl
420 Frontage Rd, Northfield, IL 708-446-8300
Hartley Metzner Hunick Comms Inc/Colleen S Hartley
3076 S Calhoun Rd, New Berlin, WI 414-784-1010
Iconos/Gary Brandenberg
118 E 26th St #201, Minneapolis, MN............612-879-0504
Industrial Video/Ray Grubic
1601 N Ridge Rd, Lorain, OH 216-233-4000
**Information Technology Design
Assoc Inc**/Daniel Klassen
7156 Shady Oak Rd, Eden Prairie, MN.............612-944-8244
Interactive Personalities/Dan Yaman
708 N First St #241, Minneapolis, MN............612-926-5924
Iowa Teleproduction Center Inc/Brad Morford
4800 Corporate Drive, West Des Moines, IA.......515-225-7800
Jones-Rasikas/Arnold Jones
800 Bond NW, Grand Rapids, MI 616-456-1855
Kasper Studio/Walt Kasper
17903 Huron River Dr, New Boston, MI............313-753-9100
Koplar Communications Ctr/Jim Wright
4935 Lindell Blvd, St Louis, MO 314-454-6320
L-E-O Systems Inc/Jill O'Neil
1505 E David Rd, Dayton, OH.................513-298-1503
LaVoie's Photography/Rex LaVoie
01423 US 127 S, Bryan, OH 419-636-4602
Lee DeForest Communications Ltd/Sherri Shallenberg
300 W Lake St, Elmhurst, IL 708-834-7855
LightSource Images Inc/Lois Stanfield
PO Box 4001, Hopkins, MN...................612-797-0770
Mainstream Communications Inc/Becky Beck
955 James Ave S #235, Bloomington, MN........612-888-9000
Marketing Connection, The/George Klink
93 Piquette St, Detroit, MI 313-873-7744
Media Process Group Inc/Robert Hercules
770 N Halsted St #507, Chicago, IL 312-850-1300
Meeting Media Enterprises Ltd/Phil Weintraub
3100 Dundee Rd #703, Northbrook, IL 708-564-8160
Mills-James Prdctns/Teleproduction Ctr/
Jim Zangmeister
3545 Fishinger Blvd, Columbus, OH 614-777-9933
Motivation Media Inc/John Moxley
1245 Milwaukee Ave, Glenview, IL 708-297-4740
Nygard & Associates/Jim Nygard
400 N First St #120, Minneapolis, MN...........612-371-9228
Odom, Laddie Scott
2135 W Giddings #3W, Chicago, IL 312-728-4942
Optimum Group/Randy Zimmerman
9745 Mangham Dr, Cincinnati, OH 513-563-2700
Paradigm Communication Group/Keith Meyer
250 Production Plz, Cincinnati, OH 513-381-7100
Passin, Jim
1900 W Berwyn Ave, Chicago, IL 312-334-0408
Pegasus Productions/Shane Hieronymus
222 S Meramec Ave #300, St Louis, MO..........314-727-7707
Platinum Productions/Rob Gibbons
8100 N High St, Columbus , OH 614-888-4181
Post Effects/Jacqui Jones
400 W Erie St #101, Chicago, IL 312-944-1690

Post Production Services Inc/Dave Dittgen
602 Main St #900, Cincinnati, OH 513-621-6677
Power Presentations/Jerry McAfee
6564 Robin Hood Dr, Indianapolis, IN 317-784-2005
Prism Studios/Jack Graff
2412 Nicollet Ave, Minneapolis, MN............800-659-2001
Promedia Productions/Steve Keller
2593 Hamline Ave, Roseville, MN...............612-631-3681
Quantum Group/Janice Glassfort
101 W 10th St, Indianapolis, IN 317-639-6001
Quicksilver Assoc Inc/Tim Wais
18 W Ontario St, Chicago, IL 312-943-7622
Rafferty Communications/Jim Rafferty
1518 139th Ln NW, Andover, MN 612-755-8488
Reelworks Animation Studio/Peter Lang
318 Cedar Ave S #300, Minneapolis, MN 612-333-5063
Rehabilitation Inst of Chicago/Craig Perryman
345 E Superior, Chicago, IL 312-908-6183
Reitz Data Communications/Larry Reitz
8225 Farnsworth Rd #A, Waterville, OH 419-878-3334
RJ Bauer Studio/Robert Bauer
503 Beacon Hill #1, Troy, MI..................810-510-6968
Rogala, Miroslaw
1524 S Peoria St, Chicago, IL 312-243-2952
Roscor Corp/Bennett Grossman
1061 Freehanville Dr, Mt Prospect , IL 708-299-8080
Sanders & Co/Denita Clifford
3610 N Meridian, Indianapolis, IN 317-926-2841
Second Sight Productions/Thom Bixbe
950 N Meridian St # 33, Indianapolis, IN 317-237-7888
Sinkler Or Sinkler/Jim Wilieko
1214 N LaSalle, Chicago, IL 312-649-9504
Sirko Design, R/Robert Sirko
621 Fox Point Dr, Chesterton, IN 219-926-8759
Soundlight Productions/Terry Luke
1915 Webster, Birmingham, MI 313-642-3502
Still Life & Kicking/Martha Coleman
6573 City West Pkwy, Minneapolis, MN 612-943-1029
Telemation Productions/John Dussling
100 S Sangamon, Chicago, IL 312-421-4111
Thirteen Fifty-one Prdctns/Tom Kwilosz
1351 W Grand, Chicago, IL 312-421-0400
Typecasters Ltd/Ted Bailey
6167 28th St #18, Grand Rapids, MI 616-940-1041
Universal Images/Etta Menlo
26011 Evergreen #200, Southfield, MI...........313-357-4160
VGI Productions/Video Genesis Inc/Howard Schwartz
4949 Galaxy Pkwy, Cleveland, OH 216-464-3635
VIDCAM Inc/Craig C Smith
10683 S Saginaw #E, Grand Blanc, MI............313-694-0996
Video Nova Inc/Dennis Sennett
534 Parkview, Detroit, MI 313-331-1975
Visual Motivation Inc/Robert Jackson
10901 Valley View Rd, Eden Prairie, MN 612-943-0060
VPL Productions/Vincent Lucarelli
415 Assembly Dr, Bolingbrook, IL 708-739-6439
Zender & Assocs Inc
2311 Park Ave, Cincinnati, OH................513-961-1790

WEST

3D Magic/FTI/Dave Altenau
55 Hawthorne St 10th Fl, San Francisco, CA 415-543-2517
Advanced Imaging/Richard Lowenberg
PO Box 1770, Telluride, CO.................303-728-6960
After Science/Joe Lewis
228 Main St #2, Venice, CA 310-314-9810
Alan C Ross Productions/Alan C Ross
202 Culper Ct, Hermosa Beach, CA 213-374-9895
Alben & Faris/Lauralee Alben
317 Arroyo Seco, Santa Cruz, CA 408-426-5526
Allen Communication/Marty Newey
140 Lakeside Pl II/5225 Wiley,
Salt Lake City, UT 801-537-7800
Amazing Media/Keith Metzger
55 San Anselmo Ave, San Anselmo, CA 415-453-0686
**Annenberg Ctr at Eisenhower/
TV Svcs**/Karen Snyder
39000 Bob Hope Dr, Rancho Mirage, CA 619-773-4554
Arts in Action/Dena M Paponis
1010 N Kings Rd #311, Los Angeles, CA 213-656-7376
Avtex Research Corp/Mark J Bunzel
2105 S Bascom Ave #300, Campbell , CA..........408-371-2800
Bailey Productions/Terry Bailey
POB 2284, Van Nuys, CA...................818-781-9520

Bay Area Video Coalition/Luke Hones
1111 17th St, San Francisco, CA......................415-861-3282
Bechtold Studios/Bob Bechtold
430 S Niagara St, Burbank, CA.................818-562-1751
Bender, Jon
305 Brookwood Ave, San Jose , CA.................408-993-1930
Berlin, Jeff
238A Summit Dr, Corte Madera, CA 415-979-8488
Brilliant Media/Lisa Hoffman
350 Townsend 1st Fl, San Francisco, CA.........415-777-1479
Bruce Hayes Productions/Bruce Hayes
959 Wisconsin St, San Francisco, CA.........415-282-2244
Brunn, K David
12375 Mt Jefferson Terrace #4K,
Lake Oswego, OR......................503-636-0352
Button Interactive/Don Button
624 Parkman Ave, Los Angeles, CA 213-483-6062
C Design/Pat Coleman
915 Oak Ln #4, Menlo Park, CA.................415-321-2094
Calico/Jan Nagle
8843 Shirley Ave, Northridge, CA 818-701-5862
Campos, Mike
PO Box 308, Sausalito, CA 415-892-1573
CD-ROM Strategies/Dr Ash Pahwa
Six Venture #208, Irvine, CA....................714-453-1702
Center for Electronic Art/Harold Hedelman
950 Battery St #3D, San Francisco, CA.............415-956-6500
Christensen Media
3638 Auburn Blvd #F, Sacramento, CA 916-487-4722
Cimarron Productions/Michael Theis
3131 S Vaughn Way #134, Aurora, CO.............303-368-0988
Cimity Arts/Special Visual Effects/Barbara Cimity
800 S Robertson Blvd, Los Angeles, CA.............310-659-4504
Cinema Network (Cinenet)/Jim Jarrard
2235 First St #111, Simi Valley, CA.............805-527-0093
Cloud Art & Design, Gregory
2116 Arlington Ave #236, Los Angeles, CA.......213-484-9479
Communicate/Emilie Caruso
3201 Wilshire Blvd, Santa Monica, CA 310-998-9228
Communication Bridges/Jon Leland
180 Harbor Dr #204A, Sausalito, CA 415-331-3133
Communications Concepts Co/Marijane Lynch
PO Box 3018, San Rafael, CA.................415-456-6495
Communications Video/Arthur Marchetti
PO Box 2125, Walnut Creek, CA 510-930-7788
Computer Graphics Group/Joel Spaunberg
23181 Via Celeste, Trabuco Canyon, CA 714-858-8978
Computer Graphics Group/William Brown
1325 W Imola #121, Napa, CA.................707-257-7975
Computer Graphics Group/Jim Boren
1450 Manhattan Beach Blvd #B,
Manhattan Beach, CA.................310-372-3228
Creative Alliance/David Ginsberg
6523-1/2 Leland Way, Los Angeles, CA 213-463-6647
Creative Media Development/Doug Crane
710 SW Ninth Ave, Portland, OR.................503-223-6794
Cronan, Michael Patrick
1 Zoe St, San Francisco, CA 415-543-6745
Crucible/Brian Raney
1717 N Seabright Ave #1, Santa Cruz, CA.......408-423-4600
Dana White Productions/Dana C White
2623 29th St, Santa Monica, CA.................310-450-9101
DAROX Interactive/Darox Interactive
7825 Fay Ave #250, La Jolla, CA.................619-456-3577
Delta Teleproductions
3333 Glendale Blvd #3, Los Angeles, CA 213-663-8754
Depixion
27125 Wapiti Dr, Evergreen, CO.................303-670-7822
DESIGNS +/Gaye L Graves
561 Homer Ave, Palo Alto, CA 415-321-5046
Dixon & Associates/Michael Dixon
3359 Karen Ave, Long Beach , CA 310-496-3376
Dollar & Associates/John Dollar
450 S 41st St, Boulder, CO 303-494-8626
Double Vision Productions/Betty Vivado
1955 Railroad Dr #A, Sacramento, CA.............916-921-6986
Dourmashkin Productions/Barbara Dourmashkin
3852 Camino de Solana, Sherman Oaks, CA.....818-995-3997
Editel/Los Angeles/Michael Moreale
729 N Highland Ave, Hollywood, CA 213-931-1821
Effective/Visual Imagery/Mia Corrales
151 Kalmus #K1, Costa Mesa, CA....................714-641-5222
Electric Canvas/Richard Hornor
1001 Art Rd, Pilot Hill, CA.................916-933-3990

EMS Communications/John Sandberg
14251 Chambers Rd, Tustin, CA......................714-669-8001
EP Graphics/Burbank/Eddie Pong
2525 N Naomi, Burbank, CA.................818-953-9375
Evon, Susan
55 Valparaiso #1, San Francisco, CA.................415-441-5951
Fearey, Patricia
1300 Noe St, San Francisco, CA 415-821-9253
Form & Function/Jonathan Gibson
1595 17th Ave, San Francisco, CA.................415-664-4010
Full Spectrum Productions/John McCauley
150 E Dana St, Mountain View, CA 415-967-1883
FX Plus Design/Elan Soltes
3025 W Olympic Blvd, Santa Monica, CA 310-315-2175
Gasowski, Igor
1220 Colusa Ave, Berkeley, CA.................510-524-3777
General Parametrics Corp/Michael Callaghan
1250 Ninth St, Berkeley, CA.................510-524-3950
Gentry, David
1145 Wisconsin St, San Francisco, CA.................415-824-2920
George & Company/George McGinness
4044 Pacific St, Highland, CA.................909-862-6168
Glaser Media Group/Russell Glaser
655 Skyway #225, San Carlos, CA.................415-593-6607
Glenn & Associates/Bernice T Glenn
2020 S Robertson Blvd, Los Angeles, CA.............310-837-9063
GR Graphics Audio Visual/Guillermo Rodriguez
1850 N 15th Ave, Phoenix, AZ.................602-252-6525
Grafica Multimedia Inc/Michael McGrath
940 Emmett Ave #11, Belmont, CA.................415-595-5599
GRAFX/Doug Wolfgram
1046 Calle Recodo #A, San Clemente, CA.........714-361-3475
Gragame-Harding Productions/Marabeth S Harding
1645 Folsom St #3, San Francisco, CA.................415-626-7116
Graphic Evidence/Peter Rothenberg
200 Corporate Pt #300, Culver City, CA.............310-568-1831
Graphix Zone/Angela Aber
38 Corporate Pk #100, Irvine, CA.................714-833-3838
Group Video Productions/Jerry Casey
10364 Rockingham Dr, Sacramento, CA.............916-362-3964
Haukom Associates/Richard Haukom
2120 Steiner St, San Francisco, CA.................415-922-0214
Haydock Associates
49 Shelley Dr, Mill Valley, CA.................415-383-6986
Herigstad, Dale
3112 Ledgewood Dr, Los Angeles, CA 213-463-1339
Hershey Multimedia/John Hershey
431 Bryant St, San Francisco, CA.................415-252-5607
Hitachi Sales Corp of America/Eric Kamayatsu
401 W Artesia Blvd, Compton, CA.................310-537-8383
Hollywood Interactive/Robert Sherman
32215 Pacific Coast Hwy, Malibu, CA 310-858-0577
House Film Design/James House
7033 W Sunset Blvd #301, Los Angeles, CA.......213-467-3429
HSC SellMEDIA Creative Svcs/David Coy
1661 Lincoln Blvd #101, Santa Monica, CA.......310-392-8441
HyperMedia Group Inc, The/Jim Edlin
5900 Hollis St #O, Emeryville, CA.................510-601-0900
Image Corp/Dave Hoffman
1617 Wazee St, Denver, CO 303-573-6206
Image Maker, The/Anthony Helmstetter
13430 N Scottsdale Rd, Phoenix, AZ.................602-948-7643
Imagetects/Verdonna Ahrens
PO Box 4, Saratoga, CA 408-252-5487
Immersive Technologies Inc/Martin Perlmutter
2866 McKillop Rd, Oakland, CA 510-261-0128
Impact Communications Group/Brad Vinikow
18627 Brookhurst St #314, Fountain Valley, CA..714-963-6760
Impact Corporate Communications/Fred Gerhart
4601 Telephone Rd #117, Ventura, CA.............805-644-3645
Inland Audio Visual Co/Larry Bergman
27 W Indiana, Spokane, WA 800-584-8916
InnerStellar Productions/Francis Hobbs
904 Anita Ave, Big Bear City, CA 909-585-8495
Intellichoice Inc/Steve Gross
1135 S Saratoga-Sunnyvale Rd, San Jose, CA.....408-554-8711
Interactive Audio/Gary Levenberg
544 Natoma St, San Francisco, CA.................415-431-0778
Interactive Design Inc/John Laney
1309 NE Ravenna Blvd, Seattle, WA.................206-523-7879
Interactive Media Technologies/Allan Ayars
13402 N Scottsdale Rd #B155, Scottsdale, AZ ...602-443-3093
Interactive Muse, The/Shoshanah Dubiner
2701 Larkin St #200, San Francisco, CA...........415-885-6873

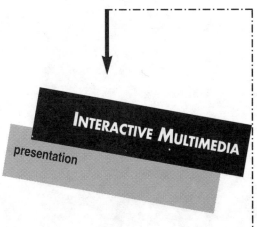

INTERACTIVE MULTIMEDIA

presentation

InterNetwork Inc/Payson R Stevens
411 7th St , Del Mar, CA.................619-755-0439
Intrepid Productions/William Sorensen
7 Mt Lassen Dr #A116, San Rafael, CA 415-491-4050
INVIEW Corp/Richard N Carter
PO Box 208, Del Mar, CA.................619-792-6473
Isis Inc/Jack Walkins
3755 Wild Horse Rd, Reno, NV.................702-688-6288
JAG Broadcast Video/John Gay
PO Box 12442, Reno , NV.................702-688-6277
Jane Sallis & Associates/Jane Sallis
148 NE 57th , Seattle, WA.................206-522-5522
JHL & Assocs/John Luxenberg
26501 Mimosa Ln, Mission Viejo, CA 714-770-5656
Jim Muse Presentation Technology/Jim Muse
10525 Lawson River, Fountain Valley, CA.........714-965-5205
JMTV /John Thill
2808 Roosevelt St, Carlsbad, CA 619-434-3363
Johnson, Jay
2260 N Beachwood Dr, Hollywood, CA.............213-464-2606
Kallisto Productions Inc/Alan Weiler
2173 Francisco Blvd #E, San Rafael, CA.............415-257-4777
Kenwood Group, The/Wayne Leonard
139 Townsend St #505, San Francisco, CA.........415-957-5333
Klang Sound
75 Wood St, San Francisco, CA 415-931-3382
Klitsner Industrial Design/Antonio Angulo
636 Fourth St, San Francisco, CA.................415-957-1529
Laser Pacific Corp /Steve Mitchell
540 Hollywood Way, Burbank, CA.................818-842-0777
Laser Pacific Media Corp/Leon Silverman
809 N Cahuenga Blvd, Los Angeles, CA.............213-462-6266
Laxer, Jack
16952 Dulca Ynez Ln, Pacific Palisades, CA 310-459-1213
Louey/Robino Design Group/Robert Louey
2525 Main St, Santa Monica, CA.................310-396-7724
Main Street Multimedia/David Watkinson
3005 Main St #419, Santa Monica, CA.............310-396-4084
Main Street Video Productions Inc/Mark Tyson
119 N Wahsatch Ave, Colorado Springs, CO.....719-520-9969
Martin Brinkerhoff Associates/Randall Lubert
17767 Mitchell, Irvine, CA 714-660-9396
Maus Haus/Bob Slote
1209 Howard Ave, Burlingame, CA.................415-343-2996
McCamant Productions Animation/David McCamant
1520 Akard Dr, Reno , NV.................702-746-8276
McLean Media/Lois McLean
80 Liberty Shipway #3, Sausalito, CA.................415-332-6385
Media Architects/MM Resources/Stephan A Rogers
1075 NW Murray Rd #230, Portland, OR.........503-297-5010
Media Lab/Jim Turrentine
400 McCaslin Blvd #211, Louisville, CO.............303-499-5411
Mediamation LP/Bernard Bergman
11498 Laurelcrest Dr, Studio City, CA 818-763-6534
Medior/Tracy Strong
2800 Campus Dr #150, San Mateo, CA...........415-525-4000
MetaLanguage/D Josh Rosen
2901 23rd St, San Francisco, CA.................415-821-6138
Midland Production Corp/Yas Takata
435 S Second St, Richmond, CA.................510-233-4236
Miller, Jack Paul
370 W Cedar Ave, Burbank, CA 818-841-4668
Mok Designs, Clement
600 Townsend St #PH, San Francisco, CA 415-703-9900
Mondo Media/John Evershed
340 Townsend St #408, San Francisco, CA 415-243-8671

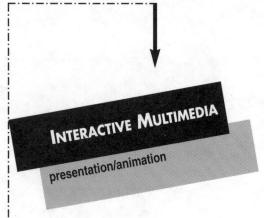

INTERACTIVE MULTIMEDIA

presentation/animation

Moov Design/Ed Coderre
2690 Beachwood Dr, Los Angeles, CA 619-259-6300
Motion City Films/Jerry Witt
1847 Centinela Ave, Santa Monica, CA 310-264-4870
Mr Film/Chris Walker
228 Main St #10, Venice, CA 310-396-0146
Multi Image Productions Inc/Frederic Ashman
8849 Complex Dr, San Diego, CA 619-560-8383
Multimedia Graphic Network/
G & G Designs/John Teschner
225 W Plaza St #400, Solana Beach, CA 619-793-4171
Nolan, Roy
2051 Leese Ln, Novato, CA..............................415-892-3204
Norac Productions /Ashley Bittman
2300 15th St #100, Denver, CO 303-455-8200
North Country Media Group/Doug Bliler
721 Second St, Great Falls, MT......................406-761-7877
Novocom/GRFX Productions/Jack Schaeffer
6314 Santa Monica Blvd, Hollywood, CA...........213-461-3688
Odam Design, John
2163 Cordero Rd, Del Mar, CA 619-259-8230
Ogdemli/Feldman Design/Daniel Feldman
11911 Magnolia Blvd #39, N Hollywood, CA....818-760-1759
On-Q Productions/Vincent Quaranta
618 E Gutierrez St, Santa Barbara, CA..............805-963-1331
Ono Stuff/Michael P Heley
655 Skyway #225, San Carlos, CA 415-593-6722
Pacific Interactive Design Corp/Scott Palamar
1460 Fourth St #200, Santa Monica, CA...........310-458-1898
Pacific Media/Chris Largent
2600 W Olive Ave # 100, Burbank, CA...........818-841-1199
Pacific Rim Video/Roger Proulx
663 Maulhardt Ave, Oxnard, CA......................805-485-9930
Pacific Video Resources/Dana Levy-Wendt
2331 Third St, San Francisco, CA 415-864-5679
Patchwork Productions Inc/Skip Williamson
620 Groton Dr, Burbank, CA 818-955-9875
Pinnacle Post/Vance Martin
2334 Elliot Ave, Seattle, WA 206-443-1000
Planet Productions/Mark Schlichting
2845 Santa Clara, Alameda, CA 510-522-3237
PM Studio/Matthew Leeds
9 Echo Ave, Corte Madera, CA........................415-924-1203
Post Group/Rich Ellis
6335 Homewood Ave, Los Angeles, CA 213-462-2300
Presentation Works/Carol Greenberg
1541 Yardley St, Santa Rosa, CA....................707-545-1400
PRINTZ/Charles Wyke-Smith
340 Townsend St, San Francisco, CA 415-543-5673
Production Co, The/Bob Patrick
601 S Rancho #C-26, Las Vegas, NV 702-263-0933
PSI Inc/Jose Villanueva
18019-A Skypark Cir, Irvine, CA......................714-261-6119
R P & A Inc/Ramin Firoozye
PO Box 77067, San Francisco, CA 415-826-3113
Real Time Media/Will Johnston
1001 Center St #11, Santa Cruz, CA 408-423-3720
Ridgley Curry & Assoc Inc/Ridgley Curry
87 E Green St #309, Pasadena, CA..................818-564-1215
Rock Solid Productions/David Griffin
801 S Main St, Burbank, CA 818-841-8220
Rose Design, David
14647 Cole Dr, San Jose, CA..........................408-377-2770
San Diego Digital Post/Infovideo
9853 Pacific Hts Blvd #B, San Diego, CA...........619-452-9000

San Jose State University/Sam Koplowicz
One Washington Sq, San Jose, CA 408-924-2858
Sea Studios Inc/Mark Shelley
810 Cannery Row, Monterey, CA......................408-649-5152
Seigel/Inocencio/Matthew Seigel
33 Vandewater St #302, San Francisco, CA 415-433-5817
Smetts Design, Bonnie
1798 5th St, Berkeley, CA..............................510-644-1313
Smoke & Mirrors/James A Collins
750 27th St, San Francisco, CA 415-206-1044
Sparks, Joe
181 Downey, San Francisco, CA......................415-731-9112
Spectrum Sight & Sound/Larry Schutte
1800 N Vine St #210, Hollywood, CA...............213-462-0812
stat™media/Gary W Birch
7077 E Shorecrest Dr, Anaheim Hills, CA...........714-280-0038
Stratton, Mary M
7708 Etiwanda, Reseda, CA............................818-757-1921
STS Production/Paul Roden
935 W Bullion St, Murry, UT............................801-263-3959
Studio Productions Inc
650 N Bronson Ave #223, Hollywood, CA.........213-856-8048
Studio, The/Mimi & Larry Elmore
2644 Calmia Ave, Boulder, CO........................303-442-6030
Subjective Technologies Inc/Jack Lampl
1106 Second St #103, Encinitas, CA................619-942-0928
Telemation Denver/Dan Boyd
8745 E Orchard Rd #500,
Greenwood Village, CO..................................303-290-8000
Top Communications/Ferris Top
1201 SW 12th Ave #210, Portland, OR.............503-222-2773
Total Vision Inc/Paul Rother
3015 Main St, Santa Monica, CA 310-450-1315
Triad Media Services/Kirk Storms
3055 Triad Dr, Livermore, CA..........................510-449-0606
TVA/Television Associates Inc/Lauri A Clark
2410 Charleston Rd, Mt View, CA....................415-967-6040
Vegas Valley Prods/KVVU Bdcst Corp/Richard Walsh
25 TV5 Dr, Henderson, NV..............................702-435-5555
Video Data Services of Montana/Matthew Scotten
PO Box 1206, Whitefish, MT 406-862-0875
Video Resources/Brad Hagen
1809 E Dyer Rd #307, Santa Ana, CA..............714-261-7266
Videomation/Razzaq A Lodhia
4633 Old Ironsides Dr #270, Santa Clara, CA....408-988-6100
Viper Optics/JoAnn Gillerman
950 61st St, Oakland, CA...............................510-654-2880
Visual Promotions/Paul Tokarski
1905 N Edgemont St, Los Angeles, CA.............213-661-9560
Volan Design Associates/Mike Krell
1800 38th St, Boulder, CO..............................303-449-3838
Volotta Interactive Video/Thomas Volotta
60 Sir Francis Drake Blvd #300, Larkspur, CA.....415-925-2100
WalkerVision Interarts/Pat Walker
PO Box 22533, San Diego, CA........................619-458-9038
West Coast Projections Inc/David Gibbs
11245 W Bernardo Ct #100, San Diego, CA.....619-674-7334
Westcom Creative Group/Steve Ogle
2295 Coburg Rd #105, Eugene, OR.................503-484-4314
Windstar Studios Inc/Dale Mitchell
525 Communications Cir, Colorado Springs, CO.719-635-0422
Winters Productions/Glenn Winters
1855 Foothill Blvd, La Canada, CA 213-682-1205
Wohlmut Media Services/Thomas Wohlmut
2600 Central Ave #L, Union City, CA................510-487-1073
Woodward Associates/Tom Woodward
5069 Tendilla Ave, Woodland Hills, CA 818-348-5669
Zilberts & Assocs, Ed
5690 DTC Blvd #190, Englewood, CO 303-220-5040

CANADA

CAV Productions Inc/Ricky Magder
189 Church St, Toronto, ON............................416-360-7746
Executive Information Base/Harvey Martens
One Yonge St #1801, Toronto, ON 416-594-1117
Hypermedia Productions/Henry W See
4580 Marquette St, Montreal, QB....................514-525-7810
Image Base Videotex Design Inc/Neil Black
517 Wellington West #209, Toronto, ON 416-593-5473
Imagicians/Bonni Evans
PO Box 1005, Manotick, ON 613-692-4306
Phoenix Interactive Design Inc/Kyle MacDonald
470 Rideout St, London, ON...........................519-679-2913
Pyrate Group, The/John V Kennedy
451 St Sulpice #L, Montreal, QB 514-284-0761

Studio Post & Transfer/Colin Minor
5305 104th St, Edmonton, AB 403-436-4444
Xenon Micromedia Productions Inc/J Ferrari
3483 Capricorn Crescent, Mississauga, ON 416-673-7046

Animation

NEW YORK CITY

4-Front Video Design Inc/Jack M Beebe
1500 Broadway #509, New York, NY..............212-944-7055
AFCG Inc/Floyd Gillis
305 E 46th St, New York, NY..........................212-688-3283
Akavia, Eden
PO Box 545, New York, NY.............................212-533-0633
Amendola, Steve
95 Horatio St, New York, NY 212-989-3246
Barrett, Anne S
300 Eighth Ave #6A, Brooklyn, NY 718-965-3162
Bottiglieri, Dessolina
126 E 12th St #6C, New York, NY 212-473-4592
CBS News/Daniel Di Perro
524 W 57th St, New York, NY 212-975-2834
Compugraphia
190 Avenue B, New York, NY 212-614-0283
Curious Pictures/Tina Contis
23 Watts St, New York, NY..............................212-966-1020
D'Andrea Productions
12 W 37th St, New York, NY 212-947-1211
Da Silva Animation
311 E 85th St, New York, NY 212-535-5760
Darino Films/Edward Darino
222 Park Ave S, New York, NY 212-228-4024
Denbo Multimedia/Robert E Denbo
135 W 29th St, New York, NY..........................212-465-2297
Fat Baby Productions/Jessica Perry
185 E 85th St #26M, New York, NY.................212-722-0535
Ferguson, Wichar
18 Gramercy Park S #706, New York, NY.........212-388-7677
Ferri-Grant, Carson
255 W 90th St PH, New York, NY 212-362-8567
Friedman Consortium, Harold
404 E 5th Ave #14A, New York, NY.................212-688-6434
Haxton, David
139 Spring St, New York, NY...........................212-966-1572
High Priority Consulting/Bonnie Kane
45 First Ave #5-O, New York, NY 212-228-6000
Hoberman, Perry
167 N 9th St #10, Brooklyn, NY......................718-388-8241
Hong, Won-Hua
73 Monroe St, New York, NY 212-267-4947
Horizon Images/Nona Abiathar McCarley
534 9th Ave #D2, New York, NY 212-529-5452
Horn, Jenny
211 E 18th St #3-O, New York, NY 212-777-3015
Hot-tech Multimedia Inc/Lawrence Kaplan
46 Mercer St, New York, NY 212-925-3010
Jiempreecha, Wichar
32-86-34th St #3A, Long Island City, NY 718-721-6956
L Squared Studios/Linda Lauro
20 W 20th St #1000, New York, NY................212-675-4681
Lieberman Productions, Jerry
76 Laight St, New York, NY.............................212-431-3452
Mohre, Terry
426 15th St, Brooklyn, NY..............................718-832-4872
Moreau, Sylvain
5 Tudor City Pl #1018, New York, NY...............212-370-6914
Multimedia Library/Diane Schwartz
37 Washington Sq W, New York, NY 800-362-4978
Necro Enema Amalgamated/Eric Swenson
Box 208 Village Station, New York, NY.............212-998-6955
Portfolio Artists Network/Marcus Nispel
601 W 26th St, New York, NY 212-633-6030
Pratt Institute/Isaac Victor Kerlow
Comp Grphcs Ctr/ARC Bldg, Brooklyn, NY.........718-636-3600
Raabe, Dan
80 Montague St, Brooklyn, NY.........................718-260-9666
Sanders, Phil
563 Van Duzer St, Staten Island, NY................718-720-0388
Schmidt, Mechthild
PO Box 453 Radio City Station, New York, NY ...212-245-5254
● **STRUTHERS, DOUG (P 52,53)**
232 Madison Ave #402, New York, NY............212-889-8777
Tom Nicholson Associates/Tom Nicholson
295 Lafayette St 8th Fl, New York, NY..............212-274-0470

Tozzi, Graig
250 E Houston, New York, NY 212-477-1779
Two Twelve Assocs/David Peters
596 Broadway #1212, New York, NY 212-925-6885
Uman, Michael
1781 Riverside Dr #4C, New York, NY 212-304-0756
Virtual Beauty/Jane Nisselson
57 Leroy St, New York, NY...........................212-420-0021
Wallin, Michael
26-04 25 Rd, Astoria, NY 718-278-7196
Waters Design/Colleen Syron
3 W 18th St 8th Fl, New York, NY 212-807-0717
Zang Design, Ulla
36 Plaza Street #7B, Brooklyn, NY 718-399-0454

EAST

Acme Animation Group Ltd/Evan Hirsch
40 Newport Parkway, Jersey City, NJ 201-420-6440
Artificial Reality/Myron Krueger
PO Box 786, Vernon, CT 203-871-1375
Baker MediaSource/Wai Wong
2001 Joshua Rd , Lafayette, PA 215-832-0340
Business Presentation Services/Mary Clupper
40 Cameron Ave, Somerville, MA 617-666-1161
Chipurnoi, Minda
1085 Warburton #508, Yonkers, NY 914-963-8959
Ciné-Med Inc/Kevin McGovern
127 Main St N/PO Box 1007, Woodbury, CT....203-263-0006
Computer Graphics Resources
One Dock St, Stamford, CT 203-327-3635
Cramer Productions/Tom Martin
355 Wood Rd, Braintree, MA 617-849-3350
DCP Communications Grp Ltd/Alison Connors
301 Wall St, Princeton, NJ 609-921-3700
DePersico Inc/Paul DePersico
878 Sussex, Broomall, PA 215-328-3155
DI Group/Sharon Lynch
651 Beacon St, Boston, MA 617-267-6400
Effective Communication Arts/David Jacobson
149 Dudley Rd/PO Box 250, Wilton, CT 203-761-8787
Expanded Video Inc/Ted Miles
477 Congress St, Portland, ME 207-773-7005
Fastforward Communications Inc/Jaime Martovamo
401 Columbus Ave, Valhalla, NY 914-741-0555
Forsight Inc/Ronald Gregory
POB 0427/1700 Rockvll Pike#150,
Rockville, MD 301-816-4900
Girard Video/Jenny Gsell
1331 F St NW #250, Washington, DC 202-393-6666
Greater Media Cable Adv/Michael Savino
95 Higgins St, Worcester , MA 508-853-1515
Image Production/Jenny Coffelt
50 Water St, S Norwalk, CT 203-853-3486
Kelly, Sloan W
43 Wyoming Ave, Malden, MA 617-324-9877
Lake Champlain Productions/Steve Murphy
4049 Williston Rd, S Burlington, VT 802-863-9780
Lewis Cohen & Co/Lewis Cohen
141 Ruxton Rd, Mt Kisco, NY 914-241-3638
Magical Media/Peter Jensen
139 E Spring Ave, Ardmore, PA 215-642-4225
MAI2 / The Graphics Room/Tony Rosser
11 Spruce Hill Rd, Armonk, NY 914-273-9294
Main Point Productions/Will Stanton
Box 76/RD#2/Lobachsville Vllg, Oley, PA 215-987-9320
Mattingly, Matthew
263 Cherry, Newton, MA..............................617-527-0763
Media Vision/James Lush
80 Great Hill Rd, Seymour, CT 203-735-0002
● **META 4 MULTIMEDIA**/Paul Lemberg **(P 95)**
245 W Norwalk Rd, W Norwalk, CT 203-857-0654
Meyer Software/Linda Nunez
616 Continental Rd, Hatboro, PA 800-643-2286
Neptune, Frantz
3140 Sheffield Pl, Holland, PA 215-860-0851
New Media Group/Michael Endres
10018 Tenbrook Dr, Silver Spring, MD 301-681-6100
NFL Films Video/Hal Lipman
330 Fellowship Rd, Mt Laurel, NJ 609-778-1600
Nu Vox Animation & Design/Michael Tiedeman
264 Independence Dr, Chestnut Hill, MA 617-469-2891
Palmer, Laura Leigh
3924 Denfield Ct, Kensington, MD 301-946-5026
Paul, Dayan
779 Mabie St, New Milford, NJ 201-265-3842

Pixel Light Communication/Carmine DeFalco
271 Rte 46 W # F205, Fairfield, NJ 201-808-1389
Posch, Michael
716 Grand St, Hoboken, NJ 201-792-7003
Pragma Design/Mark Esterly
23 Jay Dr, Londonderry, NH 603-437-2010
PREFIT Corporation/Joseph Chernicoff
PO Box 326, Willow Grove, PA 215-657-3976
PVS (Professional Video Services)/Liz Lokey
2030 M St NW, Washington, DC 202-775-0894
Szabo, Michelle
1 Little John Ln, Danbury, CT 203-791-8599
Tangent Design/Communications/Otto Timmons
25 Sylvan Rd S #N, Westport, CT 203-221-1013
● **TEICH, DAVID (P 71)**
41 Tamara Dr/PO Box 246, Roosevelt, NJ 609-448-5036
Tiani, Alex
PO Box 4530, Greenwich, CT 203-661-3891
Tourtellott, Mark
12 Martyn St, Waltham, MA 617-647-9615
Trend Multimedia/Tony Rosa
4 Holly Ave, West Keansburg, NJ....................908-787-0786
Ultitech Inc/William J Comcowich
Foot of Broad St, Stratford, CT 203-375-7300
University of Pennsylvania/Dr Norman Badler
Computer & Info Science, Philadelphia, PA 215-898-5862
Visual Impressions/Bill Robbins
429 State St, Rochester, NY 716-546-1917

SOUTH

4th Wave Inc/John N Latta
PO Box 6547, Alexandria, VA.........................703-360-4800
ASV Corp/Barbara Fishburn
2075 Liddell Dr NE, Atlanta, GA 404-876-3445
Bransby Productions/Sharon Grilliot
2124 Metro Cir, Huntsville, AL 205-882-1161
Cannata Communications Corp/Jack Cannata
7031 Grand Blvd, Houston, TX 713-748-1684
Cheshire Designs/Julia Martinroe
2492 Fair Knoll Ct NE, Atlanta, GA..................404-633-8330
Cinema Concepts/Stewart Harnell
2030 Powers Ferry Rd #214, Atlanta, GA..........404-956-7460
Computed Animation Technology/Trent DiGiulio
2200 N Lamar #301, Dallas, TX 214-871-5055
Computer Studio, The/George Bowers
4000 Cumberland Pkwy 1400 #A1, Atlanta, GA 404-436-6092
Corporate Media Services/Ralph Preston
632 W Summit Ave, Charlotte, NC 704-377-1601
CPN Televison Studios/Mike McKown
14375 Myerlake Cir, Clearwater, FL..................813-530-5000
Cyberkids/David Franks
3721 Sue Ellen Dr, Raleigh, NC 919-832-4774
Design/Efx/Dave Warner
535 Plasamour Dr NE, Atlanta, GA 404-876-9011
Elite Post of Nashville/George Betts
1025 16th Avenue S # 302, Nashville, TN 615-327-8797
Georgia Inst of Technology/Ray Haleblian
Multimedia Technology Lab/
400 10 St, Atlanta, GA 404-894-4195
GMG International/David Gardy
1950 Roland Clarke Pl, Reston, VA...................703-339-8500
Holmes Animation/George Holmes
8652 N May Ave, Oklahoma City, OK...............405-848-5841
Hot Source Media/Michael Vaughan
1916 Wilson Blvd #304, Arlington, VA 703-527-5992
Image Assocs Inc/John Muruca
4909 Windy Hill Dr, Raleigh, NC 919-876-6400
Insight Productions/Scott Long
PO Box 12498, Research Triangle Pk, NC 919-544-5700
Interactive Arts/Rick Ligas
501 E 4th St #511, Austin, TX.........................512-469-0502
ITC/Joan Dasher
13515 Dulles Technology Dr, Herndon, VA 703-713-3335
J Dyer Inc/Ilene Dyer
PO Box 420589, Atlanta, GA..........................404-250-4422
Lee Martin Productions/Lee Martin
11250 Pagemill Rd, Dallas, TX 214-556-1991
MacMedia Systems of Orlando/Jon Anthony Blumhagen
1022 S Lee Ave, Orlando, FL..........................407-425-9160
Marshall Productions Inc/Bill Hite
404 BNA Dr #102, Nashville, TN 615-399-8895
Media Management Systems Inc/Wynne Ragland
One Meca Way #600, Norcross , GA 404-564-5606
Mindflex/Steve Tanner
566 Dutch Valley RD, Atlanta, GA....................404-892-6232

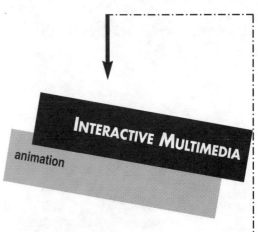

Moxie Media Inc/Martin Glenday
5734 Jefferson Hwy/POB 10203,
New Orleans, LA 504-733-6907
Pixel Graphics/Douglas H Parr
PO Box 63230, Pipe Creek, TX 210-535-9585
Post Edge Inc/Mike Duncan
2040 Sherman St, Hollywood , FL 305-920-0800
Quinn Associates Inc/Margaret Quinn
110 W Jefferson St, Falls Church, VA................703-237-7222
Results Video/Larry Emerson
6515 Escondido #A-1, El Paso, TX 915-581-0184
Richard Kidd Productions Inc/John Kindervag
5610 Maple Ave, Dallas, TX...........................214-638-5433
Shooting Stars Post Inc/John Samaha
3106 W North "A" St, Tampa, FL 813-873-0100
Simerman, Tony
Castle Harbor Way, Centreville, VA...................703-802-4950
Sirius Image/Larry Goddard
2062 Weems Rd, Tucker, GA 404-939-2004
Smart Concepts Ltd/Steve Sembritsky
Tulsa, OK...918-747-6006
● **STRUTHERS, DOUG (P 52,53)**
6453 Southpoint Dr, Dallas, TX 214-931-0838
Summa Logics Corp./Karen Ceretto
8403 Golden Aspen Ct, Springfield, VA...............703-912-7948
Sunbelt Video/Michael Poly
4205-K Stuart Andrew Blvd, Charlotte, NC..........704-527-4152
Super Computer/Greg Sherman
366 Lytle St, W Palm Beach, FL 407-833-0558
Texas Video & Post/Grant Guthrie
8964 Kirby Dr, Houston, TX 713-667-5000
Thonen, Rod
8621 Fort Hunt Rd, Alexandria, VA 703-781-0412
Truly Computer Graphics
417 Wakefield Dr, League City, TX....................800-829-4990
Video Image Productions/Charlie Case
2026 Powers Ferry Rd #100, Atlanta, GA...........404-984-8288

MIDWEST

Abraham Animation/Richard Abraham
2001 Atwood Ave, Madison, WI 608-241-3900
Ad-Visual Communications/Gary DiCenzo
1461 Oakwood, Sylvan Lake, MI 313-682-4424
AlexanderDesign Inc/Roy Alexander
2117 W Arthur Ave, Chicago, IL......................312-508-9313
Analog Video/Dick Madding
192 Walnut Grove Dr, Dayton, OH...................513-299-3495
Associates Creative Inc/Elliot Goldberg
24471 West Temik #100, Southfield, MI 313-354-3003
Ayala Computer Imaging
631 W Stratford Pl, Chicago, IL.......................312-327-1156
Beacon Software/John Grozik
P O Box 234422, Milwaukee, WI 414-355-4460
Bovey & Assocs, Ruth
1443 W Altgeld, Chicago, IL 312-327-2886
Bromley Chapin Designs/Al Chapin
400 Renaissance Ctr #2250, Detroit, MI 313-259-2661
Columbia Coll/Academic Computing/Geof Goldbogen
600 S Michigan Ave, Chicago, IL.....................312-663-1600
Creative Images/Steve Hendrickson
1400 Energy Park Dr #23, St Paul, MN 612-644-2157
Cully, Mike
401 E Ontario #1109, Chicago, IL....................312-440-9208
Dégallier Animation/Mark Degallier
264 Benton Ave, Wayzata, MN 612-476-2388

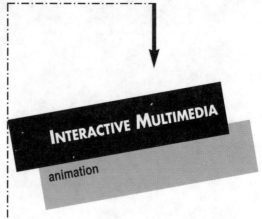

INTERACTIVE MULTIMEDIA

animation

Envision Inc/Jeff Blackwell
   405 N Calhoun, Brookville, WI.....................414-789-8485
Fearless Eye/Brad Matthieson
   310 Delaware St #210, Kansas City, MO.........816-221-1047
Film Craft Video/Sandy Ruby
   37630 Interchange Dr, Farmington Hills, MI.......313-474-3900
Filmack Studios/Robert Mack
   1327 S Wabash Ave, Chicago, IL.................312-427-3395
Frame One Inc/Paul Bernan
   676 St Clair #425, Chicago, IL.................312-440-1328
Golan Productions/Ari Golan
   507 W North Ave, Chicago, IL...................312-274-FIL
Interactive Personalities/Dan Yaman
   708 N First St #241, Minneapolis, MN............612-926-5924
Lightborne Communications/Bob Hanneken
   632 Vine St #817, Cincinnati, OH...............513-721-2272
Lim, Deborah
   505 N Lake Shore Dr #5606, Chicago, IL........312-527-3271
Mainsail Production Services/Ron Beam
   521 Buyers Rd #109, Miamisburg, OH............513-866-7800
Marketing Connection, The/George Klink
   93 Piquette St, Detroit, MI....................313-873-7744
Meeting Media Enterprises Ltd/Phil Weintraub
   3100 Dundee Rd #703, Northbrook, IL............708-564-8160
Mills-James Productions/Teleproduction Center/Jim Zangmeister
   3545 Fishinger Blvd, Columbus, OH..............614-777-9933
Motivation Media Inc/John Moxley
   1245 Milwaukee Ave, Glenview, IL..............708-297-4740
Optimum Group/Randy Zimmerman
   9745 Mangham Dr, Cincinnati, OH...............513-563-2700
Optimus Inc/Cat Gulick
   161 E Grand Ave, Chicago, IL..................312-321-0880
Power Presentations/Jerry McAfee
   6564 Robin Hood Dr, Indianapolis, IN..........317-784-2005
Prism Studios/Jack Graff
   2412 Nicollet Ave, Minneapolis, MN............800-659-2001
Quicksilver Assoc Inc/Tim Wais
   18 W Ontario St, Chicago, IL..................312-943-7622
Rehabilitation Inst of Chicago/Craig Perryman
   345 E Superior, Chicago, IL...................312-908-6183
RJ Bauer Studio/Robert Bauer
   503 Beacon Hill #1, Troy, MI..................810-510-6968
Rogala, Miroslaw
   1524 S Peoria St, Chicago, IL.................312-243-2952
Sacks, Ron/Ron Sacks
   1189 Rosebank Dr, Worthington, OH.............614-846-1921
Siemer, Patrick
   1809 W Division St, Chicago, IL...............312-862-4244
Sinkler Or Sinkler/Jim Wilieko
   1214 N LaSalle, Chicago, IL...................312-649-9504
Still Life & Kicking/Martha Coleman
   6573 City West Pkwy, Minneapolis, MN..........612-943-1029
Synergy Art & Tech/Michael Simmons
   230 Tenth Ave SE #213, Minneapolis, MN........612-371-9181
Triad/Artbytes/Greg Dodge
   7654 W Bancroft, Toledo, OH...................419-841-2272

WEST

3D Magic/FTI/Dave Altenau
   55 Hawthorne St 10th Fl, San Francisco, CA......415-543-2517
After Science/Joe Lewis
   228 Main St #2, Venice, CA....................310-314-9810
Amazing Media/Keith Metzger
   55 San Anselmo Ave, San Anselmo, CA...........415-453-0686

Bailey Productions/Terry Bailey
   POB 2284, Van Nuys, CA........................818-781-9520
Bartlett, Michael
   24A Varda Landing, Sausalito, CA..............415-331-5127
Bender, Jon
   305 Brookwood Ave, San Jose , CA..............408-993-1930
Bravin Learning Technologies/Jonathan Bravin
   PO Box 260678, Encino, CA.....................818-880-6964
Brilliant Media/Lisa Hoffman
   350 Townsend 1st Fl, San Francisco, CA.........415-777-1479
Brookhouse, Winthrop
   473 Jefferies/PO Box 3502, Big Bear Lake, CA...909-866-7020
Brooks Communication Design/Brooks Cole
   4000 Brideway #405, Sausalito, CA.............415-331-5855
Cimity Arts/Special Visual Effects/Barbara Cimity
   800 S Robertson Blvd, Los Angeles, CA..........310-659-4504
CKS Partners/Jill Savini
   Hills Plz/345 Spear St #500, San Francisco, CA .415-905-1647
Communicate/Emilie Caruso
   3201 Wilshire Blvd, Santa Monica, CA..........310-998-9228
Communications Concepts Co/Marijane Lynch
   PO Box 3018, San Rafael, CA...................415-456-6495
Computer Putty/David Poole
   408 West Portola, Los Angeles, CA.............415-948-3319
Creative License–Film & Video/Bonnie MacBird
   131 S Barrington Pl #200, Los Angeles, CA.......310-476-9725
Dana White Productions/Dana C White
   2623 29th St, Santa Monica, CA................310-450-9101
DAROX Interactive/Darox Interactive
   7825 Fay Ave #250, La Jolla, CA...............619-456-3577
David Blum Animation/David Blum
   1660 Puebla Dr, Glendale, CA..................818-567-2279
Depixion
   27125 Wapiti Dr, Evergreen, CO 303-670-7822
Doherty, James
   411 S 11th St #4, San Jose, CA................408-297-5744
Dolphin Multi Media/Cynthia Kondratieff
   2440 Embarcadero Way, Palo Alto, CA...........415-962-8310
Duda Design Inc/Mary T Duda
   16707 Sunset Blvd, Pacific Palisades, CA.......310-459-5531
Electric Paint & Design/Jon Paul Davidson
   PO Box 8822, Incline Village, NV..............702-831-3490
EMA Video Productions/Ed Mellnik
   3210 SW Dosch Rd, Portland, OR................503-241-8663
Fearey, Patricia
   1300 Noe St, San Francisco, CA 415-821-9253
Form & Function/Jonathan Gibson
   1595 17th Ave, San Francisco, CA..............415-664-4010
Foster Digital Imaging/Jeff Foster
   530 E Lambert Rd, Brea , CA...................714-671-0880
Full Spectrum Productions/John McCauley
   150 E Dana St, Mountain View, CA..............415-967-1883
Gaviota Graphics/Jim Biebl
   094 Arlian Ln, Carbondale, CO 303-963-3309
General Parametrics Corp/Michael Callaghan
   1250 Ninth St, Berkeley, CA...................510-524-3950
George & Company/George McGinness
   4044 Pacific St, Highland, CA.................909-862-6168
Gerrard Associates/Cathy & Lewis Gerrard
   5452 Evanwood Ave, Agoura , CA...............818-706-3959
Glaser Media Group/Russell Glaser
   655 Skyway #225, San Carlos, CA..............415-593-6607
Grafica Multimedia Inc/Michael McGrath
   940 Emmett Ave #11, Belmont, CA..............415-595-5599
Graphic Evidence/Peter Rothenberg
   200 Corporate Pt #300, Culver City, CA 310-568-1831
Graphic Media Inc/Doug New
   411 SW Second, Portland, OR..................503-223-2262
Hamilton, Bruce & Susan
   Rte 1 Box 5C, Glorieta, NM....................505-757-6603
Hoffman, Patricia
   369 Montezuma #316, Santa Fe, NM..............505-983-3165
House Film Design/James House
   7033 W Sunset Blvd #301, Los Angeles, CA.......213-467-3429
Human Performance Institute/Paras Kaul
   UMed Ctr 11406 Loma Linda Dr,
   Loma Linnda, CA..............................818-794-4057
HyperMedia Group Inc, The/Jim Edlin
   5900 Hollis St #O, Emeryville, CA.............510-601-0900
IKONIC/Robert May
   188 The Embarcadero 7th fl, San Francisco, CA..415-864-3200
Image Corp/Dave Hoffman
   1617 Wazee St, Denver, CO....................303-573-6206
Image Maker, The/Anthony Helmstetter
   13430 N Scottsdale Rd, Phoenix, AZ............602-948-7643

Image World San Jose/Knowledge Industry Publctns Inc
   San Jose, CA.................................914-328-9157
Impact Communications Group/Brad Vinikow
   18627 Brookhurst St #314, Fountain Valley, CA..714-963-6760
Interactive Design Inc/John Laney
   1309 NE Ravenna Blvd, Seattle, WA.............206-523-7879
Interactive Muse, The/Shoshanah Dubiner
   2701 Larkin St #200, San Francisco, CA.........415-885-6873
Interactive Solutions Inc/William La Commare
   1720 S Amphlett Blvd #219, San Mateo, CA......415-377-0136
Intrepid Productions/William Sorensen
   7 Mt Lassen Dr #A116, San Rafael, CA..........415-491-4050
Invisions Inc/Scott Tuckman
   3641 N 52nd St, Phoenix, AZ..................602-840-1090
Isis Inc/Jack Walkins
   3755 Wild Horse Rd, Reno, NV.................702-688-6288
Ixion Inc/David Hon
   1335 N Northlake Way #102, Seattle, WA 206-547-8801
JRA Interactive /Jane McLaughlin
   3647 Sunset Beach Dr NW, Olympia, WA.........206-866-0533
Kallisto Productions Inc/Alan Weiler
   2173 Francisco Blvd #E, San Rafael, CA.........415-257-4777
Kenwood Group, The/Wayne Leonard
   139 Townsend #505, San Francisco, CA..........415-957-5333
Killingsworth Presentations/Penny Engel
   3834 Long Beach Blvd, Long Beach, CA..........310-595-7796
Klitsner Industrial Design/Antonio Angulo
   636 Fourth St, San Francisco, CA..............415-957-1529
Laguna Productions/Charlie King
   2708 S Highland Dr, Las Vegas, NV.............702-731-5600
Lipman, Michael
   55 Loring Ave, Mill Valley, CA................415-383-4248
Live Marketing/William J Steinmetz
   96 Hawthorne Ave, Los Altos, CA...............415-941-8188
Louey/Robino Design Group/Robert Louey
   2525 Main St, Santa Monica, CA...............310-396-7724
Lumeni Productions/Tony Valdez
   1362 Flower, Glendale, CA....................213-462-2110
Lumigenic Media/Marc Shargell
   133 Beth Dr, Felton, CA......................408-335-4849
Lynx Digital Design/Alejandro Rubalcava
   16 Technology #115, Irvine, CA................714-727-3126
Main Street Multimedia/David Watkinson
   3005 Main St #419, Santa Monica, CA...........310-396-4084
Maus Haus/Bob Slote
   1209 Howard Ave, Burlingame, CA..............415-343-2996
McCamant Productions Animation/David McCamant
   1520 Akard Dr, Reno , NV.....................702-746-8276
Media Lab/Jim Turrentine
   400 McCaslin Blvd #211, Louisville, CO.........303-499-5411
Metropolitan Entertainment/Andy Ullman
   1680 N Vine St #600, Hollywood, CA............213-856-7060
Midland Production Corp/Yas Takata
   435 S Second St, Richmond, CA................510-233-4236
Modern Videofilm Inc/Richard E Greenberg
   4411 W Olive Ave, Burbank, CA................818-840-1700
Mondo Media/John Evershed
   340 Townsend St #408, San Francisco, CA 415-243-8671
Multimedia Graphic Network/G & G Designs/John Teschner
   225 W Plaza St #400, Solana Beach, CA 619-793-4171
Norac Productions /Ashley Bittman
   2300 15th St #100, Denver, CO................303-455-8200
Novocom/GRFX Productions/Jack Schaeffer
   6314 Santa Monica Blvd, Hollywood, CA.........213-461-3688
Nutopia Digital Video/Eric Myers
   300 Valley St #301, Sausalito, CA.............415-331-0714
● **O'VERY/COVEY**/Mikel Covey & Traci O'Very **(P 10)**
   1577 Sherman Ave, Salt Lake City, UT..........801-582-8505
One World Interactive/Charlie Magee
   207 W 5th Ave, Eugene, OR....................503-683-4020
Osiow, Andrew
   2735 River Plaza Dr #131, Sacramento, CA.......916-922-1384
Pacific Video Resources/Dana Levy-Wendt
   2331 Third St, San Francisco, CA 415-864-5679
Patchwork Productions Inc/Skip Williamson
   620 Groton Dr, Burbank, CA...................818-955-9875
Paternoster, Nance
   546 Wisconsin St, San Francisco, CA...........415-641-1922
Paul, Edie
   859 Hollywood #136, Burbank, CA..............818-505-1874
Pinnacle Post/Vance Martin
   2334 Elliot Ave, Seattle, WA.................206-443-1000
Pittman Hensley/Donna Pittman
   2114 Hillhurst Ave, Los Angeles, CA...........213-665-5168

Planet Productions/Mark Schlichting
2845 Santa Clara, Alameda, CA 510-522-3237
Presto Studios/Jack Davis
9888 Carroll Centre Rd #228, San Diego, CA 619-689-4895
Presto Studios/Jack Davis
9888 Carroll Centre Rd #228, San Diego, CA 619-689-4895
PSI Inc/Jose Villanueva
18019-A Skypark Cir, Irvine, CA........................714-261-6119
R P & A Inc/Ramin Firoozye
PO Box 77067, San Francisco, CA 415-826-3113
Real Time Media/Will Johnston
1001 Center St #11, Santa Cruz, CA...............408-423-3720
Ruby-Spears Productions
710 W Victory Blvd #201, Burbank, CA..............818-840-1234
Ryane, Nathen
San Diego, CA 619-260-2472
S E Tice Consulting/Steve Tice
15860 Dartford Way, Sherman Oaks, CA..........818-906-3322
Sharpe, Stuart
437 Mississippi St, San Francisco, CA 415-648-7199
Spectris/Sergio Villarreal
252 Durian St, Vista, CA 619-471-6855
STS Production/Paul Roden
935 W Bullion St, Murry, UT...........................801-263-3959
Studio Productions Inc
650 N Bronson Ave #223, Hollywood, CA.........213-856-8048
StudioGraphics/Richard Bennion
107 South B Sr 3rd Fl, San Mateo, CA 415-344-3855
● **TRACER DESIGN**/Chad Little **(P 42)**
4206 N Central Ave, Phoenix, AZ.....................602-265-9030
Tracy, Donna
2011 Vista Cerro Gordo St, Los Angeles, CA 213-666-4087
Univ of Calif at Santa Barbara/Peter Allen
College of Engineering/Office of Dean #1001,
Santa Barbara, CA 805-893-4803
Video Arts/Kim Salyer
185 Berry St #5400, San Francisco, CA 415-546-0331
Video Data Services of Montana/Matthew Scotten
PO Box 1206, Whitefish, MT 406-862-0875
Video Image Associates/Richard Hollander
5333 McConnell Ave, Los Angeles, CA 310-822-8872
Video Resources/Brad Hagen
1809 E Dyer Rd #307, Santa Ana, CA...............714-261-7266
Videomation/Razzaq A Lodhia
4633 Old Ironsides Dr #270, Santa Clara, CA....408-988-6100
Vision Productions/KSAZ-TV/Bill Lucas
511 W Adams, Phoenix, AZ..............................602-262-5131
Visionary Art Resources/Roy Montibon
3972 Barranca Pkwy #J438, Irvine, CA..............714-241-0604
Visual Promotions/Paul Tokarski
1905 N Edgemont St, Los Angeles, CA.............213-661-9560

CANADA

Ideaction Inc/Luc Courchesne
Univ of Montr/Schl of Ind Dsgn, Montreal, QB.....514-343-7495
Leaping Raster/Kip Hardy
1686 Woodward Dr, Ottawa Canada, ON........613-727-5078
Niemann, Andrew
1290 Astoria St, Victoria Canada, BC 604-383-9367
Pyrate Group, The/John V Kennedy
451 St Sulpice #L, Montreal, QB 514-284-0761
Sanctuary Woods Multimedia/Pierre LeSeach
1006 Government St, Victoria, BC....................604-380-7582
Studio Post & Transfer/Colin Minor
5305 104th St, Edmonton, AB 403-436-4444

Education-Training

NEW YORK CITY

Apple Market Center/NY/Tara Griffin
135 E 57th St, New York, NY 212-339-3700
Applied Imagination Inc/Thomas Tafuto
826 Broadway 6th Fl, New York, NY 212-529-9696
Burnett Group/Johanna Skilling
39 E 20th St 7th Fl, New York, NY 212-254-3344
Edwin Schlossberg Inc/Heather Liston
641 Sixth Ave 5th Fl, New York, NY..................212-989-3993
**Electronic Directions/Publication
Technologies**/Ron Lockhart
220 E 23rd St #503, New York, NY 212-213-6500
Fusion Communications/Craig A Mengel
510 E 74th St, New York, NY 212-879-8787
High-Res Solutions Inc/Jim Casey
23 E 20th St, New York, NY 212-475-6107

Louey/Rubino Design Group/Gene Nicholais
215 Park Ave S #702, New York, NY 212-777-4220
Santoro Design, Joe
503 W 27th St, New York, NY 212-563-2506
Schneemann, Carolee
114 W 29th St, New York, NY.........................212-695-6615
Solomon Video Productions, Rob
331 W 57th St #242, New York, NY.................212-541-9507
Tom Nicholson Associates/Tom Nicholson
295 Lafayette St 8th Fl, New York, NY..............212-274-0470
Videograf/Michael Frenchman
144 W 27th St, New York, NY 212-206-8477
Wallin, Michael
26-04 25 Rd, Astoria, NY................................718-278-7196
Zang Design, Ulla
36 Plaza Street #7B, Brooklyn, NY...................718-399-0454

EAST

Ankeny, Martha Langley
313 West Durham St, Philadelphia, PA 215-247-0442
Applied Imagination Inc/Thomas Tafuto
RR 1 Box 1765, Moretown, VT.........................802-496-3520
Art on Fire/Marian Schiavo
335 38th Rd, Douglaston, NY 718-229-3660
Brownstone Graphics/Kathryn Sikule
303 Hudson Ave, Albany, NY..........................518-434-8707
Documentation Development Inc/Ron Miskie
125 Cambridge Park Dr, Cambridge, MA..........617-864-7300
Effective Communication Arts/David Jacobson
149 Dudley Rd/PO Box 250, Wilton, CT 203-761-8787
Ehrlich Multimedia/Alexander Ehrlich
1 Maynard Dr, Park Ridge, NJ 201-262-4499
ExperTech Corp/Kathleen Harmeyer
303 Wynell Ct, Timonium, MD........................410-561-3238
Ferranti Educational Systems
801 S 18th, Columbia, PA...............................717-684-4398
Flying Turtle Productions/Laurie Holmes
721 Crestbrook Ave, Cherry Hill, NJ.................609-751-7213
George Washington University/Prof S Molina
Art & Visual Communications/2300 I St NW,
Washington, DC...202-994-7455
Henry, Paul
67 The Crossway, Butler, NJ 201-492-0864
Hyperview Systems Corp/Dick Ficke
28 Jacome Way, Middletown, RI......................401-849-8900
Iconcepts/Jeff Cummings
118 Magazine St, Cambridge, MA 617-661-7289
Image Works/Dana Hutchins
537 Congress St #301, Portland, ME 207-774-6399
Imergy/Suzanne Mason
12 S Main St, S Norwalk, CT...........................203-853-6200
Information/Context/Joseph Coates
35 Partridge Lane, Madison, CT 203-421-5517
Infoview Corp/Larry Bell
124 Farmingdale Rd, Wethersfield, CT 203-721-0270
Instructional Design International Inc/Lin Lougheed
1775 Church St NW, Washington, DC...............202-332-5353
Lehigh Interactive Multimedia/James W Kauffman
937 High St, Bethlehem, PA 215-867-8448
Lewis Cohen & Co/Lewis Cohen
141 Ruxton Rd, Mt Kisco, NY..........................914-241-3638
Media Construction Inc/Michael Gallelli
4700 Crescent St, Bethesda, MD.....................301-229-4188
Mendoza Design
3104 Grindon Ave, Baltimore, MD....................410-426-3671
Midi Inc/Jack Noon
100 Thanet Cir #105, Princeton, NJ 609-924-4817
**MIT Ctr for Education Computing
Initiatives**/Evelyn Schlusselberg
1 Amherst St #E40-379, Cambridge, MA...........617-253-0693
MPTV Video Production Unit/Mark Patterson
2116 Noble St, Pittsburgh, PA.........................412-271-6788
Multimedia Workshop/Michael D Murie
Box 44-37, Somerville, MA..............................617-776-2469
National Education Training Group/Steve Maxwell
9 Oak Park Dr, Bedford , MA...........................800-227-1127
NUS Training Corp/Philip Taplin
910 Clopper Rd, Gaithersburg, MD 301-258-2530
Petitto, Andrea
4309 Hosey Rd, Shortsville, NY 716-289-6004
Sonic Images Productions Inc/Jolie Barbiere
4590 MacArthur Blvd NW, Washington, DC 202-333-1063
Syscon Video/Robert F Knospe
523 Fellowship Rd #250, Mt Laurel, NJ 609-234-5510

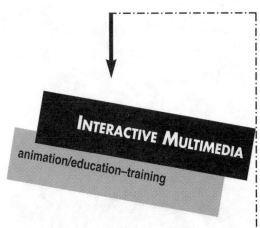

Szabo, Michelle
1 Little John Ln, Danbury, CT 203-791-8599
TEX Interactive Svcs/Dave Kephart
397 Delmar St, Philadelphia, PA 215-483-1126
Thomas Piwowar & Assocs/Thomas Piwowar
1500 Mass Ave NW #34, Washington, DC 202-223-6813
Tourtellott, Mark
12 Martyn St, Waltham, MA............................617-647-9615
Transparent Media/Warren Schloss
2729 N 47th St, Philadelphia, PA 215-477-5868
Ultitech Inc/William J Comcowich
Foot of Broad St, Stratford, CT........................203-375-7300

SOUTH

Bermac Communications Inc/Phil Chadwick
4545 Fuller Dr #336, Irving, TX.........................214-255-9007
Boger, Claire
96 Vicksburg Cove #222, Memphis, TN 407-487-2264
Floyd Design/Mike Cusick
1465 Northside Dr NW #110, Atlanta, GA........404-351-4518
Hamilton Jr, Robert
5900 Riverdale Rd #7, College Park, GA...........404-994-9067
Layton Intl/Eugene Vasconi
2013 Wells Branch Pkwy #201, Austin, TX 512-251-0074
Monte, Steven W
1124 Kensington Dr, DeSoto, TX 214-230-4324
MultiMedia Resources (Level 4 Comm)/
Doris Seitz-Boothe
14875 Landmark #102, Dallas, TX....................214-650-1986
NC Supercomputing Ctr/Chris Landreth
3021 Cornwallis Rd, Research Triangle Pk, NC....919-248-1141
● **NEW MEDIA PRODUCTIONS INC**/
Benjamin Nowak **(P 96)**
4209 Ivy Chase Way NE, Atlanta, GA..............404-257-9220
Super Computer/Greg Sherman
366 Lytle St, W Palm Beach, FL.......................407-833-0558
Trans-Delta Productions/Steve Candelera
3346 Walnut Bend Ln, Houston, TX 800-544-1681
Videotexting/Jaf Fletcher
275 Commercial Blvd, Lauderdale-by-the-Sea, FL....305-771-5999
VisualEdge/Eileen Sosna
9311 W Calusa Club Dr, Miami, FL 305-382-4220

MIDWEST

Computer Artworks Inc/Daniel Hayes
700 W Pete Rose Way, Cincinnati, OH 513-421-9000
Connecting Images/Terry Hickey
2469 University Ave N, St Paul, MN 612-644-1977
Dar Electronic Research & Development/Anire Dar
919 Blair Ave, Neenah, WI 414-725-6543
Dynacom Inc/Jerry Sullivan
1512 N Fremont #101, Chicago, IL 312-951-5510
Hartley Metzner Hunick Comms Inc/Colleen S Hartley
3076 S Calhoun Rd, New Berlin, WI 414-784-1010
LightSource Images Inc/Lois Stanfield
PO Box 4001, Hopkins, MN 612-797-0770
Lim, Deborah
505 N Lake Shore Dr #5606, Chicago, IL...........312-527-3271
Motivation Media Inc/John Moxley
1245 Milwaukee Ave, Glenview, IL 708-297-4740
Onli, Turtel
5121 S Ellis, Chicago, IL 312-684-2280
Ross Roy Communications/Joey Silvian
100 Bloomfield Hills Pkwy, Bloomfield Hills, MI...313-433-6837
Sauer & Associates/Christian Sauer
2844 Arsenal St, St Louis, MO..........................314-664-4646

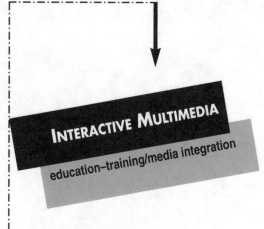

INTERACTIVE MULTIMEDIA
education–training/media integration

Windy City Communications/Wayne Stuetzer
350 W Hubbard St #450, Chicago, IL................312-464-0390

WEST

Alan C Ross Productions/Alan C Ross
202 Culper Ct, Hermosa Beach, CA....................213-374-9895
Allen Communication/Marty Newey
140 Lakeside Pl II/5225 Wiley, Salt Lake City, UT801-537-7800
Amazing Media/Keith Metzger
55 San Anselmo Ave, San Anselmo, CA 415-453-0686
Artlab/Tony DeYoung
1603 Howard St, San Francisco, CA 415-554-0248
Arts in Action/Dena M Paponis
1010 N Kings Rd #311, Los Angeles, CA 213-656-7376
Asymmetrix Corp/Nancy Robertson
110 110th Ave NE #700, Bellevue, WA.............206-637-1500
Avtex Research Corp/Mark J Bunzel
2105 S Bascom Ave #300, Campbell , CA............408-371-2800
Bravin Learning Technologies/Jonathan Bravin
PO Box 260678, Encino, CA............................818-880-6964
Brodsky, Michael
1327 S Carmona Ave, Los Angeles, CA..............310-338-3060
Cohen, Hagit
865 14th St #3, San Francisco, CA 415-552-9677
Communication Bridges/Jon Leland
180 Harbor Dr #204A, Sausalito, CA 415-331-3133
Computer Evolution/Don Roark
2500 Abbot Kinney Blvd #15, Venice, CA 310-821-6184
Cyan Inc/Bonnie McDowall
PO Box 28096, Spokane, WA..........................509-468-0807
Cyber Network Svcs Inc/Eva Way Konigsberg
514 Bryant St, San Francisco, CA.....................415-974-7000
DESIGNS +/Gaye L Graves
561 Homer Ave, Palo Alto, CA 415-321-5046
Dixon & Associates/Michael Dixon
3359 Karen Ave, Long Beach , CA.....................310-496-3376
Duda Design Inc/Mary T Duda
16707 Sunset Blvd, Pacific Palisades, CA............310-459-5531
General Parametrics Corp/Michael Callaghan
1250 Ninth St, Berkeley, CA............................510-524-3950
Gonzales, Mariano
Box 671816, Chugiak, AK 907-688-1250
Goodman Consulting Systems/Philippe Goodman
1615 Hilts Ave #3, Los Angeles, CA 310-470-2998
GRAFX/Doug Wolfgram
1046 Calle Recodo #A, San Clemente, CA 714-361-3475
Graphix Zone/Angela Aber
38 Corporate Pk #100, Irvine, CA.....................714-833-3838
Haukom Associates/Richard Haukom
2120 Steiner St, San Francisco, CA 415-922-0214
Hershey Multimedia/John Hershey
431 Bryant St, San Francisco, CA.....................415-252-5607
Human Interface Technology Lab/Alden Jones
University of Washington, Seattle, WA................206-543-5075
Independent Graphics/Polly Kessen
914 Bright Star St, Thousand Oaks, CA 805-492-1837
Individual Software Inc/Diane Dietzler
5870 Stoneridge Dr #1, Pleasanton, CA.............800-331-3313
Interactive Media Technologies/Allan Ayars
13402 N Scottsdale Rd #B155, Scottsdale, AZ ...602-443-3093
InterNetwork Inc/Payson R Stevens
411 7th St , Del Mar, CA 619-755-0439
Intrepid Productions/William Sorensen
7 Mt Lassen Dr #A116, San Rafael, CA..............415-491-4050
Jane Sallis & Associates/Jane Sallis
148 NE 57th , Seattle, WA 206-522-5522

JHL & Assocs/John Luxenberg
26501 Mimosa Ln, Mission Viejo, CA................714-770-5656
JRA Interactive /Jane McLaughlin
3647 Sunset Beach Dr NW, Olympia, WA.........206-866-0533
Kallisto Productions Inc/Alan Weiler
2173 Francisco Blvd #E, San Rafael, CA............415-257-4777
Kenwood Group, The/Wayne Leonard
139 Townsend St #505, San Francisco, CA 415-957-5333
Lumigenic Media/Marc Shargell
133 Beth Dr, Felton, CA 408-335-4849
Marcus & Assocs, Aaron/Aaron Marcus
1144 65th St #F, Emeryville, CA......................510-601-0994
McLean Media/Lois McLean
80 Liberty Shipway #3, Sausalito, CA...............415-332-6385
Media Lab Inc/Bob Bruce
400 S McCaslin Blvd #211, Louisville, CO..........303-499-5411
Media Learning Systems/Mitch Aiken
1492 W Colorado Blvd, Pasadena, CA...............818-449-0006
Mediamation LP/Bernard Bergman
11498 Laurelcrest Dr, Studio City, CA...............818-763-6534
Medior/Tracy Strong
2800 Campus Dr #150, San Mateo, CA 415-525-4000
Mind Over Macintosh/Bruce Kaplan
600 Corporate Pointe #1170, Culver City, CA 310-216-4000
Moov Design/Ed Coderre
2690 Beachwood Dr, Los Angeles, CA 619-259-6300
Motion City Films/Jerry Witt
1847 Centinela Ave, Santa Monica, CA.............310-264-4870
Multi Image Productions Inc/Frederic Ashman
8849 Complex Dr, San Diego, CA.....................619-560-8383
Nolan, Roy
2051 Leese Ln, Novato, CA.............................415-892-3204
Ono Stuff/Michael P Heley
655 Skyway #225, San Carlos, CA 415-593-6722
Phosphor Inc/Michael A Eddy
5951 Canterbury Dr Unit 7, Culver City, CA........310-645-0122
Pixel Ink, Consultants/Hans Hartman
520 Frderick #13, San Francisco, CA................415-564-0962
Planet Productions/Mark Schlichting
2845 Santa Clara, Alameda, CA 510-522-3237
Presto Studios/Jack Davis
9888 Carroll Centre Rd #228, San Diego, CA ...619-689-4895
Smetts Design, Bonnie
1798 5th St, Berkeley, CA..............................510-644-1313
Studio, The/Mimi & Larry Elmore
2644 Calmia Ave, Boulder, CO 303-442-6030
The Future Image Rep/Alexis J Gerard
1020 Parrott Dr, Burlingame, CA......................415-579-0493
Top Communications/Ferris Top
1201 SW 12th Ave #210, Portland, OR.............503-222-2773
Videomation/Razzaq A Lodhia
4633 Old Ironsides Dr #270, Santa Clara, CA....408-988-6100
Visionary Art Resources/Roy Montibon
3972 Barranca Pkwy #J438, Irvine, CA.............714-241-0604
Weisman, David
332 Bayview St, San Rafael, CA......................415-455-9628
Woodward Associates/Tom Woodward
5069 Tendilla Ave, Woodland Hills, CA 818-348-5669

CANADA

CAV Productions Inc/Ricky Magder
189 Church St, Toronto, ON............................416-360-7746
Hypermedia Productions/Henry W See
4580 Marquette St, Montreal, QB.....................514-525-7810
Image Base Videotex Design Inc/Neil Black
517 Wellington West #209, Toronto, ON 416-593-5473
Imagicians/Bonni Evans
PO Box 1005, Manotick, ON 613-692-4306
Sanctuary Woods Multimedia/Pierre LeSeach
1006 Government St, Victoria, BC.....................604-380-7582

Media Integration

NEW YORK CITY

Applied Imagination Inc/Thomas Tafuto
826 Broadway 6th Fl, New York, NY 212-529-9696
Bedrock Design/Janet Tingey
223 W 21st St #4G, New York, NY 212-929-1664
Cooke, Charles
416 W 23rd St #2D, New York, NY 212-989-7026
Davis, Harold
250 W 104th St #81, New York, NY 212-316-5903
Edwin Schlossberg Inc/Heather Liston
641 Sixth Ave 5th Fl, New York, NY..................212-989-3993

Fusion Communications/Craig A Mengel
510 E 74th St, New York, NY...........................212-879-8787
High-Res Solutions Inc/Jim Casey
23 E 20th St, New York, NY.............................212-475-6107
Idesign/Robert M Stewart
165 E 35th St #11A, New York, NY 212-683-0244
Multimedia Library/Diane Schwartz
37 Washington Sq W, New York, NY 800-362-4978
Sanders, Phil
563 Van Duzer St, Staten Island, NY 718-720-0388
Schneemann, Carolee
114 W 29th St, New York, NY 212-695-6615
Zink Communications/Marianne Callaghan
245 W 19th St, New York, NY 212-929-2949

EAST

AB Communications/Adam Brooks
14 Dana St, Somerville, MA.............................617-330-2275
Active 8/Ken Valley
601 N Eutaw St #704, Baltimore, MD 410-962-0272
Altered Image/Frank Thomas Ward
7 Deer Park Dr # D, S Brunswick, NJ.................908-274-2220
Applied Imagination Inc/Thomas Tafuto
RR 1 Box 1765, Moretown, VT.........................802-496-3520
Chipurnoi, Minda
1085 Warburton #508, Yonkers, NY.................914-963-8959
Colony Multimedia/Timothy Tolman
150 Chestnut St, Providence, RI.......................401-274-0933
Digital Valley Design/Francis Shepherd
PO Box 204, Chadds Ford, PA..........................610-347-6799
Documentation Development Inc/Ron Miskie
125 Cambridge Park Dr, Cambridge, MA............617-864-7300
Ehrlich Multimedia/Alexander Ehrlich
1 Maynard Dr, Park Ridge, NJ..........................201-262-4499
Forward Design/John Forward
1115 E Main St Box 61, Rochester, NY 716-288-0250
Greater Media Cable Adv/Michael Savino
95 Higgins St, Worcester, MA 508-853-1515
Hyperview Systems Corp/Dick Ficke
28 Jacome Way, Middletown, RI.......................401-849-8900
ICT Productions/Noel Izon
4320 Hamilton St, Hyattsville, MD....................301-864-6333
Immaculate Concepts/Charles Linskey
12 Norfolk Rd, Arlington, MA 617-646-9631
Interactive Design/Jan Diamondstone
1501 N Walnut St, Wilmington, DE 302-429-0143
Lehigh Interactive Multimedia/James W Kauffman
937 High St, Bethlehem, PA............................215-867-8448
Media Construction Inc/Michael Gallelli
4700 Crescent St, Bethesda, MD......................301-229-4188
Micro Visual Systems Inc/Tom Nickel
385 Elliot St, Newton, MA................................617-964-0640
Midi Inc/Jack Noon
100 Thanet Cir #105, Princeton, NJ 609-924-4817
Minerbi, Joanne Sheri
11102 Old Coach Rd, Potomac, MD 202-625-6982
MIT Ctr for Education Computing Initiatives/Evelyn Schlusselberg
1 Amherst St #E40-379, Cambridge, MA............617-253-0693
Multimedia Workshop/Michael D Murie
Box 44-37, Somerville, MA..............................617-776-2469
National Education Training Group/Steve Maxwell
9 Oak Park Dr, Bedford , MA............................800-227-1127
New England Technology Group/Steve Gregory
One Kendall Sq Bldg 200, Cambridge, MA 617-494-1151
NFL Films Video/Hal Lipman
330 Fellowship Rd, Mt Laurel, NJ 609-778-1600
Optimedia Systems/George Hoffman
373 Rte 46 W, Fairfield, NJ..............................201-227-8822
plus design inc/Anita Meyer
10 Thatcher St #109, Boston, MA 617-367-9587
Seroka Group Inc, The/Paul Seroka
Putnam Green #38-H, Greenwich, CT 203-862-9700
Sonic Images Productions Inc/Jolie Barbiere
4590 MacArthur Blvd NW, Washington, DC 202-333-1063
Tangent Design/Communications/Otto Timmons
25 Sylvan Rd S #N, Westport, CT 203-221-1013
V Graph Inc/Rob Morris
PO Box 105, Westtown, PA..............................215-399-1521

SOUTH

4th Wave Inc/John N Latta
PO Box 6547, Alexandria, VA...........................703-360-4800
BERMAC Communications Inc/Phil Chadwick
4545 Fuller Dr #336, Irving, TX.........................214-255-9007

Brave Young Artists, The/Rachel Jackson
205 Smyer Terrace, Birmingham, AL.............205-871-4083
Henninger Video Inc/Robert Henninger
2601-A Wilson Blvd, Arlington, VA 703-243-3444
Joel Bennett Design/Joel Bennet
2606-514 Phoenix Dr, Greensboro, NC.............910-547-0009
Layton Intl/Eugene Vasconi
2013 Wells Branch Pkwy #201, Austin, TX.......512-251-0074
● **NEW MEDIA PRODUCTIONS INC**/
Benjamin Nowak **(P 96)**
4209 Ivy Chase Way NE, Atlanta, GA...............404-257-9220
Trans-Delta Productions/Steve Candelera
3346 Walnut Bend Ln, Houston, TX 800-544-1681
United States Video/Frank Garber
2070 Chain Bridge Rd #485, Vienna, VA...........703-848-1990
VisualEdge/Eileen Sosna
9311 W Calusa Club Dr, Miami, FL 305-382-4220

MIDWEST

Applied Integration/David Feller
Box 200, Chelsea, MI 313-475-3266
Ad-Visual Communications/Gary DiCenzo
1461 Oakwood, Sylvan Lake, MI 313-682-4424
AlexanderDesign Inc/Roy Alexander
2117 W Arthur Ave, Chicago, IL.......................312-508-9313
ASC Systems/R Martin
PO Box 566, St Clair Shores, MI 313-882-1133
Associates Creative Inc/Elliot Goldberg
24471 West Temik #100, Southfield, MI 313-354-3003
Atechno-Marketing Inc/Christopher Moon
5170 W 76 St, Edina, MN.................................612-830-1984
Computer Artworks Inc/Daniel Hayes
700 W Pete Rose Way, Cincinnati, OH 513-421-9000
Dynacom Inc/Jerry Sullivan
1512 N Fremont #101, Chicago, IL....................312-951-5510
Hartley Metzner Hunick Comms Inc/Colleen S Hartley
3076 S Calhoun Rd, New Berlin, WI...................414-784-1010
Onli, Turtel
5121 S Ellis, Chicago, IL 312-684-2280
Ross Roy Communications/Joey Silvian
100 Bloomfield Hills Pkwy, Bloomfield Hills, MI 313-433-6837
Triad/Artbytes/Greg Dodge
7654 W Bancroft, Toledo, OH 419-841-2272
Video Nova Inc/Dennis Sennett
534 Parkview, Detroit, MI 313-331-1975
Zender & Assocs Inc
2311 Park Ave, Cincinnati, OH.........................513-961-1790

WEST

Alan C Ross Productions/Alan C Ross
202 Culper Ct, Hermosa Beach, CA 213-374-9895
Allen Communication/Marty Newey
140 Lakeside Pl II/5225 Wiley, Salt Lake City, UT 801-537-7800
Avtex Research Corp/Mark J Bunzel
2105 S Bascom Ave #300, Campbell , CA..........408-371-2800
Bailey Productions/Terry Bailey
POB 2284, Van Nuys, CA 818-781-9520
Bob Olhsson Digital Audio/Bob Olhsson
PO Box 555, Novato, CA..................................415-898-2981
Bravin Learning Technologies/Jonathan Bravin
PO Box 260678, Encino, CA.............................818-880-6964
Brooks Communication Design/Brooks Cole
4000 Brideway #405, Sausalito, CA..................415-331-5855
Button Interactive/Don Button
624 Parkman Ave, Los Angeles, CA 213-483-6062
C Graphics/Cynthia Rudy
2284 S Kingston Ct, Aurora, CO.......................303-337-7974
Cimarron Productions/Michael Theis
3131 S Vaughn Way #134, Aurora, CO...........303-368-0988
Communication Bridges/Jon Leland
180 Harbor Dr #204A, Sausalito, CA 415-331-3133
Creative License–Film & Video/Bonnie MacBird
131 S Barrington Pl #200, Los Angeles, CA 310-476-9725
Cronan, Michael Patrick
1 Zoe St, San Francisco, CA 415-543-6745
Cyan Inc/Bonnie McDowall
PO Box 28096, Spokane, WA...........................509-468-0807
Cyber Network Svcs Inc/Eva Way Konigsberg
514 Bryant St, San Francisco, CA.....................415-974-7000
DAROX Interactive/Darox Interactive
7825 Fay Ave #250, La Jolla, CA 619-456-3577
Dixon & Associates/Michael Dixon
3359 Karen Ave, Long Beach , CA.....................310-496-3376
Duda Design Inc/Mary T Duda
16707 Sunset Blvd, Pacific Palisades, CA...........310-459-5531

Electric Canvas/Richard Hornor
1001 Art Rd, Pilot Hill, CA................................916-933-3990
Evon, Susan
55 Valparaiso #1, San Francisco, CA.................415-441-5951
Fearey, Patricia
1300 Noe St, San Francisco, CA 415-821-9253
George Coates Performance Works/George Coates
110 McAllister St, San Francisco, CA.................415-863-8520
Glaser Media Group/Russell Glaser
655 Skyway #225, San Carlos, CA 415-593-6607
Grafica Multimedia Inc/Michael McGrath
940 Emmett Ave #11, Belmont, CA 415-595-5599
GRAFX/Doug Wolfgram
1046 Calle Recodo #A, San Clemente, CA.........714-361-3475
Hollywood Interactive/Robert Sherman
32215 Pacific Coast Hwy, Malibu, CA...............310-858-0577
HyperMedia Group Inc, The/Jim Edlin
5900 Hollis St #O, Emeryville, CA.....................510-601-0900
IKONIC/Robert May
188 The Embarcadero 7th fl, San Francisco, CA..415-864-3200
Immersive Technologies Inc/Martin Perlmutter
2866 McKillop Rd, Oakland, CA........................510-261-0128
Impact Communications Group/Brad Vinikow
18627 Brookhurst St #314, Fountain Valley, CA..714-963-6760
Interactive Design Inc/John Laney
1309 NE Ravenna Blvd, Seattle, WA.................206-523-7879
Interactive Media Technologies/Allan Ayars
13402 N Scottsdale Rd #B155, Scottsdale, AZ ..602-443-3093
InterNetwork Inc/Payson R Stevens
411 7th St , Del Mar, CA 619-755-0439
Ixion Inc/David Hon
1335 N Northlake Way #102, Seattle, WA 206-547-8801
Jane Sallis & Associates/Jane Sallis
148 NE 57th , Seattle, WA 206-522-5522
JHL & Assocs/John Luxenberg
26501 Mimosa Ln, Mission Viejo, CA................714-770-5656
Laser-Pacific
809 N Cahuenga Blvd, Hollywood, CA..............213-462-6266
Live Marketing/William J Steinmetz
96 Hawthorne Ave, Los Altos, CA.....................415-941-8188
Lumigenic Media/Marc Shargell
133 Beth Dr, Felton, CA 408-335-4849
Magnum Design/Ed Boyle
2762 Octavia St Top Fl, San Francisco, CA.........415-922-1728
McLean Media/Lois McLean
80 Liberty Shipway #3, Sausalito, CA................415-332-6385
Media Architects/MM Resources/Stephan A Rogers
1075 NW Murray Rd #230, Portland, OR 503-297-5010
Media Lab Inc/Bob Bruce
400 S McCaslin Blvd #211, Louisville, CO.........303-499-5411
Media Learning Systems/Mitch Aiken
1492 W Colorado Blvd, Pasadena, CA..............818-449-0006
Mediamation LP/Bernard Bergman
11498 Laurelcrest Dr, Studio City, CA 818-763-6534
Medior/Tracy Strong
2800 Campus Dr #150, San Mateo, CA............415-525-4000
Modern Videofilm Inc/Richard E Greenberg
4411 W Olive Ave, Burbank, CA......................818-840-1700
Moov Design/Ed Coderre
2690 Beachwood Dr, Los Angeles, CA 619-259-6300
Motion City Films/Jerry Witt
1847 Centinela Ave, Santa Monica, CA 310-264-4870
Multi Image Productions Inc/Frederic Ashman
8849 Complex Dr, San Diego, CA 619-560-8383
Multimedia Graphic Network/
G & G Designs/John Teschner
225 W Plaza St #400, Solana Beach, CA 619-793-4171
Nolan, Roy
2051 Leese Ln, Novato, CA..............................415-892-3204
● **O'VERY/COVEY**/Mikel Covey & Traci O'Very **(P 10)**
1577 Sherman Ave, Salt Lake City, UT..............801-582-8505
One World Interactive/Vicki Ayres
215 W Fifth, Eugene, OR.................................503-683-4020
Pacific Interactive Design Corp/Scott Palamar
1460 Fourth St #200, Santa Monica, CA............310-458-1898
Panalog Computer Guides/Dan Nash
4450 Lakeside Dr #200, Burbank, CA...............818-559-4277
Patchwork Productions Inc/Skip Williamson
620 Groton Dr, Burbank, CA............................818-955-9875
Presto Studios/Jack Davis
9888 Carroll Centre Rd #228, San Diego, CA 619-689-4895
R P & A Inc/Ramin Firoozye
PO Box 77067, San Francisco, CA.....................415-826-3113
Ridgley Curry & Assoc Inc/Ridgley Curry
87 E Green St #309, Pasadena, CA...................818-564-1215

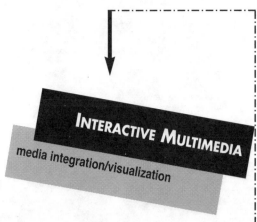

S.L.A.D.E. Corporation/Stephanie Slade
PO Box 10176, Beverly Hills, CA 213-466-7171
Sea Studios Inc/Mark Shelley
810 Cannery Row, Monterey, CA......................408-649-5152
Stanford University/Ed McGuigan
Meyer Library/Media Svcs, Palo Alto, CA 415-725-1176
Subjective Technologies Inc/Jack Lampl
1106 Second St #103, Encinitas, CA 619-942-0928
Telemation Denver/Dan Boyd
8745 E Orchard Rd #500,
Greenwood Village, CO....................................303-290-8000
Vesna, Victoria
Univ California/Art Studio, Santa Barbara, CA....805-893-2852
Wohlmut Media Services/Thomas Wohlmut
2600 Central Ave #L, Union City, CA 510-487-1073

CANADA

CAV Productions Inc/Ricky Magder
189 Church St, Toronto, ON..............................416-360-7746
Hypermedia Productions/Henry W See
4580 Marquette St, Montreal, QB 514-525-7810
Image Base Videotex Design Inc/Neil Black
517 Wellington West #209, Toronto, ON 416-593-5473
Imagicians/Bonni Evans
PO Box 1005, Manotick, ON 613-692-4306
Sanctuary Woods Multimedia/Pierre LeSeach
1006 Government St, Victoria, BC 604-380-7582

Visualization

ADAM Software
1899 Powers Ferry Rd #460, Marietta, GA.........404-980-0888
Animated Design/Andrea Silvestri
5875 Doyle St #15, Emeryville, CA 510-652-6166
Bajuk, Mark
1631 Dennington Ave, Pittsburgh, PA 412-421-4229
Cornell Theory Ctr/Cornell Univ/Wayne Lytle
621 Theory Ctr Bldg, Ithaca, NY 607-254-8793
Digital Image Design Inc/W Bradford Paley
170 Claremount Ave, New York, NY..................212-222-5236
Evon, Susan
55 Valparaiso #1, San Francisco, CA..................415-441-5951
Forcade & Associates/Tim Forcade
1440 Lawrence Ave, Lawrence, KS....................913-843-1605
Frassamito & Assoc, John/Lloyd Walker
1331 Gemini #230, Houston, TX 713-480-9911
G W Hannaway/George Hannaway
839 Pearl St, Boulder, CO................................303-440-9696
Grumman Data Systems/Geoffrey Gardner
1111 Stewart Ave/MSD12-25, Bethpage, NY....516-575-3369
Harris Design, George
301 Cathedral Pkwy #2N, New York, NY...........212-864-8872
Human Performance Institute/Paras Kaul
UMed Ctr 11406 Loma Linda Dr,
Loma Linnda, CA 818-794-4057
IBM TJ Watson Research Ctr/Alan Norton
PO Box 704, Yorktown Hts, NY.........................914-784-7195
Information/Context/Joseph Coates
35 Partridge Lane, Madison, CT 203-421-5517
Intelligenceware
5933 W Century Blvd #900, Los Angeles, CA.....310-216-6177
Ixion Inc/David Hon
1335 N Northlake Way #102, Seattle, WA 206-547-8801
Jet Propulsion Laboratory/Kevin Hussey
4800 Oak Grove Dr/MS168-522, Pasadena, CA 818-354-4016

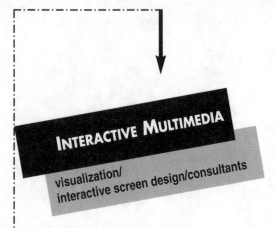

INTERACTIVE MULTIMEDIA

visualization/
interactive screen design/consultants

Joel Bennett Design/Joel Bennet
2606-514 Phoenix Dr, Greensboro, NC.............910-547-0009
Kannofsky, John
5816 1/2 Figaroa Ave, Los Angeles, CA.........213-259-9744
Macko, Nancy
1252 Monte Vista Ave #21, Upland, CA............909-946-3112
Media Lab Inc/Bob Bruce
400 S McCaslin Blvd #211, Louisville, CO........303-499-5411
Minnesota Datametrics Corp/Charles Knox
1000 Ingerson Rd, St. Paul, MN.................612-482-7938
MIT/Heather Matstone
28 Carleton St/Bldg E 32-300, Cambridge, MA..617-253-6966
ModaCAD/Linda Freedman
1954 Cotner Ave, Los Angeles, CA................310-312-6632
NASA Ames Research Ctr/Steve Bryson
MST045-1, Moffett Field, CA 415-604-5000
Novack, Dev
2075 Palisades Ave, Riverdale, NY..............718-884-2819
Octree Corporation/Donald Meagher
7337 Bollinger Rd, Cupertino, CA................408-257-9013
Ono Stuff/Michael P Heley
655 Skyway #225, San Carlos, CA 415-593-6722
School of Computer Science/Andrew Witkin
Carnegie Mellon University, Pittsburgh, PA...412-268-6244
Sound Photosynthesis/Faustin Bray
PO Box 2111, Mill Valley, CA 415-383-6712
Stanford Telecommunications Inc/Jeffrey Freedman
7501 Forbes Blvd #105, Seabrook , MD 301-464-8900
Symbolics
555 Virginia Rd, Concord, MA 508-287-1000
Tara Visual Corporation/Tom Johnson
929 Harrison Avenue, Columbus, OH..............800-458-8731
Univ of Ill at Chicago/Thomas Dr DeFanti
Elec Vis Lab/EECS Dept/MC154, Chicago, IL 312-996-3002
Virtus Corp
117 Edinburgh S #204, Cary, NC 919-467-9700
Visualization Technologies Inc
3355 Richmond Rd #191, Beachwood, OH 216-831-6782
Volotta Interactive Video/Thomas Volotta
60 Sir Francis Drake Blvd #300, Larkspur, CA.....415-925-2100
Wolfram Research Inc/Howard Berg
100 Trade Center Dr, Champaign, IL 217-398-0700

Interactive Screen Design

Alben & Faris/Lauralee Alben
317 Arroyo Seco, Santa Cruz, CA 408-426-5526
Animated Design/Andrea Silvestri
5875 Doyle St #15, Emeryville, CA..............510-652-6166
Axion, Pierce/Lior Azoulai
690 Ocean Parkway, Brooklyn, NY................718-435-6339
Bedrock Design/Janet Tingey
223 W 21st St #4G, New York, NY................212-929-1664
Bender, Jon
305 Brookwood Ave, San Jose , CA 408-993-1930
Brilliant Media/Lisa Hoffman
350 Townsend 1st Fl, San Francisco, CA........415-777-1479
Cambridge Media/David Titus
71 E Wyoming Ave, Melrose, MA 617-665-3053
Chaparral Software/Russell Kohn
9201 W Olympic Blvd #201, Beverly Hills, CA...310-273-4904
Fastcomm Communications Corp/Patty Kingery
45472 Holiday Dr, Sterling, VA 703-318-7750
Fat Baby Productions/Jessica Perry
185 E 85th St #26M, New York, NY................212-722-0535

Forward Design/John Forward
1115 E Main St Box 61, Rochester, NY 716-288-0250
Frassamito & Assoc, John/Lloyd Walker
1331 Gemini #230, Houston, TX 713-480-9911
Gersch, Wolfgang
255 Stuyvesant Dr, San Anselmo , CA 415-258-8210
Gragame-Harding Productions/Marabeth S Harding
1645 Folsom St #3, San Francisco, CA...........415-626-7116
Imaging Technology Inc/Steve Silver
55 Middlesex Tpke, Bedford, MA.................617-275-2700
Imergy/Suzanne Mason
12 S Main St, S Norwalk, CT....................203-853-6200
Immersive Technologies Inc/Martin Perlmutter
2866 McKillop Rd, Oakland, CA..................510-261-0128
Imspace Systems/Gunner Bolz
2665 Arcane Dr #201, San Diego, CA 800-488-5836
Intelligenceware
5933 W Century Blvd #900, Los Angeles, CA.....310-216-6177
Kannofsky, John
5816 1/2 Figaroa Ave, Los Angeles, CA.........213-259-9744
Kinesix Corporation/Morris Covington
9800 Richmond Ave #750, Houston, TX 713-953-8300
Look Twice/Scott Kim
4016 Farm Hill Blvd #103, Redwood City, CA 415-329-9081
Louey/Rubino Design Group/Gene Nicholais
215 Park Ave S #702, New York, NY 212-777-4220
Marcus & Assocs, Aaron/Aaron Marcus
1144 65th St #F, Emeryville, CA................510-601-0994
● **META 4 MULTIMEDIA**/Paul Lemberg **(P 95)**
245 W Norwalk Rd, W Norwalk, CT 203-857-0654
Minerbi, Joanne Sheri
11102 Old Coach Rd, Potomac, MD 202-625-6982
**MIT Ctr for Education Computing
Initiatives**/Evelyn Schlusselberg
1 Amherst St #E40-379, Cambridge, MA...........617-253-0693
On the Wave Visual Communicati/John Ulliman
2339 Ward St #8, Berkeley, CA..................510-649-8514
One World Interactive/Charlie Magee
207 W 5th Ave, Eugene, OR.....................503-683-4020
Packard, Wells
235 W 22nd St #1A, New York, NY 212-691-5016
PCM/George Agnes
1430 Spring Hill Rd #210, McLean, VA...........703-356-4600
Pivotal Graphics/Mac Copas
2153 Otogle Ave #G, San Jose, CA...............408-954-2700
plus design inc/Anita Meyer
10 Thatcher St #109, Boston, MA 617-367-9587
Presto Studios/Jack Davis
9888 Carroll Centre Rd #228, San Diego, CA 619-689-4895
Real Time Media/Will Johnston
1001 Center St #11, Santa Cruz, CA 408-423-3720
SAS Institute/O B Barrett
SAS Campus Dr, Cary, NC.......................919-677-8000
Schwartz, Ariel
113 Winchester Ln, Newtown, PA.................215-860-0888
Seigel/Inocencio/Matthew Seigel
33 Vandewater St #302, San Francisco, CA 415-433-5817
TEX Interactive Svcs/Dave Kephart
397 Delmar St, Philadelphia, PA 215-483-1126
Wavefront Technologies/Tom Sullivan
530 E Montecito St, Santa Barbara, CA..........805-962-8117
Weisman, David
332 Bayview St, San Rafael, CA 415-455-9628
Wolff, Jennifer Snow
429 N Highland Ave NE #2, Atlanta, GA..........404-524-4744
Wolfram Research Inc/Howard Berg
100 Trade Center Dr, Champaign, IL 217-398-0700
Wu Multimedia, Brian
229 E 21st St #25, New York, NY................212-674=6480
Wurman, Richard Saul
180 Narragansett Ave, Newport, RI 401-848-2299

Consultants

NEW YORK CITY

Aerial Image Video Services/John Stapsy
137 W 19th St, New York, NY 212-229-1930
Apple Market Center/NY/Tara Griffin
135 E 57th St, New York, NY...................212-339-3700
Art on Fire/Marian Schiavo
335 38th Rd, Douglaston, NY...................718-229-3660
Burnett Group/Johanna Skilling
39 E 20th St 7th Fl, New York, NY 212-254-3344
Carey, Richard A
735 Kappock St, Riverdale, NY 718-796-9696

CDTD/David Tung
150-21 77th Ave, Kew Garden Hills, NY 718-380-1054
Congleton, Laura
162 W 13th St #64, New York, NY...............212-645-2807
Costanzo, Jim
350 Bleeker St #4P, New York, NY...............212-647-9609
Craig, Andrew B
342 E 22nd St #2C, New York, NY................212-353-1282
Daisley, Dawn
111 W 24th St, New York, NY...................212-255-0569
Denbo Multimedia/Robert E Denbo
135 W 29th St, New York, NY...................212-465-2297
E3 Inc/Kathleen Bordelon
16 W 22nd St, New York, NY....................212-727-7099
Eisenberg, Sheryl
465 Washington St, New York, NY...............212-966-4910
**Electronic Directions/Publication
Technologies**/Ron Lockhart
220 E 23rd St #503, New York, NY...............212-213-6500
Fleisher, Audrey
430 W 24th St, New York, NY...................212-463-3722
Future Light/William Waldman
153 E Hartsdale Ave #2A, Hartsdale, NY 914-723-0592
Garti, Anne Marie
3967 Sedgwick Ave, Bronx, NY..................718-601-9618
Graphic Art Resource Assoc
257 W 10th St #5E, New York, NY 212-929-0017
Hoberman, Perry
167 N 9th St #10, Brooklyn, NY.................718-388-8241
Mascioni, Michael
531 Main St #916, New York, NY 212-838-3226
Mediaware/Stuart Rohrer
175 W 92nd St #5A, New York, NY 212-316-7875
Multimedia Library/Diane Schwartz
37 Washington Sq W, New York, NY 800-362-4978
Newman, Tom
434 5th St, Brooklyn, NY......................718-499-3212
Packard, Wells
235 W 22nd St #1A, New York, NY 212-691-5016
Renaissance Art Center/Andrew Edwards
29 W 38th St 3rd Fl, New York, NY..............212-575-6100
Robbins, Edward
6 Greene St, New York, NY 212-219-0697
Santoro Design, Joe
503 W 27th St, New York, NY 212-563-2506
Schmidt, Mechthild
PO Box 453 Radio City Station, New York, NY ...212-245-5254
Siegel, Jeffrey
71 Sherman Ave, Brooklyn, NY 718-768-6813
Simone, Luisa
222 E 27th St #18, New York, NY...............212-679-9117
Stratton, Robert
237 Eldridge St #6, New York, NY 212-353-3123
Sutton, Eva
239 Ninth Ave #2B, New York, NY 212-242-3599
Traugot-Weldman, Marsha
153 E Hartsdale Ave #2A, Hartsdale, NY 914-723-0592
Two Twelve Assocs/David Peters
596 Broadway #1212, New York, NY 212-925-6885
Walls, David
293 Central Park W #3E, New York, NY...........212-387-4870

E A S T

AB Communications/Adam Brooks
14 Dana St, Somerville, MA....................617-330-2275
Andersen Consulting/Bob DiLullo
100 Campus Dr, Florham Park, NJ 201-301-1289
Ankeny, Martha Langley
313 West Durham St, Philadelphia, PA 215-247-0442
Baker MediaSource/Wai Wong
2001 Joshua Rd , Lafayette, PA 215-832-0340
Boston Media Consultants/David Allen
19 Damon Rd, Scituate, MA.....................617-545-2696
Buck Consultants/Don Zimmerman
500 Plaza Dr, Secaucus, NJ....................201-902-2812
Cambridge Media/David Titus
71 E Wyoming Ave, Melrose, MA.................617-665-3053
Carl M Rodia & Assoc/Carl Rodia
13 Locust St, Trumbull, CT....................203-261-1365
CD Consultants Inc
4404 Keswick Rd, Baltimore, MD................410-243-2755
Center Stage Communications/Doug Fannon
19 Newtown Tpke, Westport, CT.................203-846-8414
Communication Research Inc/Marc Finer
3 Gateway Center, Pittsburgh, PA...............412-765-3535

Connacher, Nat
65 High Ridge Rd, Stamford, CT 203-323-1330
Dietmeier, Homer J
50 Munroe Rd, Lexington, MA 617-862-6505
Digital Valley Design/Francis Shepherd
PO Box 204, Chadds Ford, PA 610-347-6799
Digital Video Arts/George Breen
715 Twining Rd #107, Dresher, PA 215-576-7920
Drucker Associates/David Drucker
22 Lilac Ct, Cambridge, MA 617-876-1505
Fairleigh Dickinson Univ/Madison/Harvey Flaxman
285 Madison Ave, Madison, NJ..............201-593-8500
ImaginThat!/Byron Shafer
1221 Lakeview Terrace , Plainfield, NJ..............908-457-7400
Johnson, James
25 Brahms St, Boston, MA 617-325-7957
Language By Design/Martha Andrews Gan
15 Lansing Dr, Nashua, NH 603-888-3927
Machover, Carl Assoc
152-A Longview Ave, White Plains, NY 914-949-3777
Magical Media/Peter Jensen
139 E Spring Ave, Ardmore, PA 215-642-4225
Marsh, Finnegan
PO Box 18192, Washington, DC 202-234-1160
MINDesign/Mihai Nadin
35 Butts Rock Rd, Little Compton, RI 401-635-1675
Multimedia Research/Dr Barbara Flagg
33 Brown's Ln, Bellport, NY 516-286-8925
Network 90/Sandi Rosenzweig-Cooper
84 Cedar Dr #1100, Great Neck, NY 516-482-7075
Omicron/Cari Kraft
1506 Market St #2300, Philadephia, PA 215-854-0778
OnLine Computer Systems Inc/Lisa Wolin
20251 Century Blvd, Germantown, MD 301-428-3700
Quantum Research Corp/Fabrizio Golino
7315 Wisconsin Ave #631W, Bethesda, MD......301-657-3070
Rampion Visual Productions/Steven Tringali
316 Stuart St, Boston, MA 617-574-9601
Rochester of Inst Techn/Printing/Frank Romano
Box 9887, Rochester, NY 716-475-7023
Schwartz, Ariel
113 Winchester Ln, Newtown, PA 215-860-0888
SLD Communications/Susan Lender Davis
29 Hiram Rd, Framingham, MA 508-877-5269
sound/IMAGE
PO Box 1442, Cambridge, MA..............617-494-5400
Stuart, Preston
8 Apple Tree Ln , Darien, CT 203-655-7688
Szabo, Michelle
1 Little John Ln, Danbury, CT 203-791-8599
Technicad/Philip Gauntt
PO Box 462, Brookside, NJ 201-543-5518
TEX Interactive Svcs/Dave Kephart
397 Delmar St, Philadelphia, PA 215-483-1126
Thomas Piwowar & Assocs/Thomas Piwowar
1500 Mass Ave NW #34, Washington, DC 202-223-6813
Transparent Media/Warren Schloss
2729 N 47th St, Philadelphia, PA 215-477-5868
V Graph Inc/Rob Morris
PO Box 105, Westtown, PA..............215-399-1521
Visual Services Inc/Daniel Gallo
25 Van Zant St, E Norwalk, CT 203-852-1010
Werbickas, Joseph
6 Stafford Way, Marlton, NJ..............609-596-0782

SOUTH

Brave Young Artists, The/Rachel Jackson
205 Smyer Terrace, Birmingham, AL..............205-871-4083
Elographics Inc/Debbie Maxey
105 Randolph Rd, Oak Ridge, TN 615-482-4100
IBM Multimedia/Informations Cntr
4111 Northside Pkwy, Atlanta, GA 800-426-9402
Monte, Steven W
1124 Kensington Dr, DeSoto, TX 214-230-4324
**MultiMedia Resources
(Level 4 Comm)**/Doris Seitz-Boothe
14875 Landmark #102, Dallas, TX..............214-650-1986
Rupp Art & Design, Katherine
8511 Cheltenham Cir, Louisville, KY 502-425-9266
Stuck, Jon D
6138 Edsall Rd #203, Alexandria, VA..............703-354-5158
TMS Inc/Jana Otto
110 W 3rd St, Stillwater, OK 405-377-0880
VisualEdge/Eileen Sosna
9311 W Calusa Club Dr, Miami, FL 305-382-4220

MIDWEST

Applied Integration/David Feller
Box 200, Chelsea, MI 313-475-3266
Arthur Andersen & Co/Shahzad Bashir
33 W Monroe #982, Chicago, IL 312-507-2866
Emerging Technology Consultants Inc/Andrea Epstein
2819 Hamlin Ave N, St Paul, MN 612-639-3973
Iconos/Gary Brandenberg
118 E 26th St #201, Minneapolis, MN..............612-879-0504
Information Arts/Richard Bowers
PO Box 21726, Columbus, OH 614-442-8810
Institute of Design/IIT/John Grimes
10 W 35th St, Chicago, IL 312-808-5300
Interactive Illusions/Michael D Johnston
3846 Abott Ave S, Minneapolis, MN 612-926-0589
Odom, Laddie Scott
2135 W Giddings #3W, Chicago, IL 312-728-4942
Rising Star Graphics/Steve Perry
445 E Illinois St #353, Chicago, IL 312-836-7827
Saztec International/Elvin Smith
6700 Corp Dr #100, Kansas City, MO..............816-483-6900
SF Consulting/Stephen C Fedder
3490 Parkland, W Bloomfield, MI 313-851-1474

WEST

Animated Design/Andrea Silvestri
5875 Doyle St #15, Emeryville, CA..............510-652-6166
Apex Systems Inc/Scott Hamilton
2400 Central Ave #A, Boulder , CO..............303-443-3393
B.N. Architect/Ben Newcomb
6016 Hillegass Ave, Oakland, CA 510-601-1020
Bear River Associates Inc/Anthony Meadow
PO Box 1900, Berkeley, CA 510-644-9400
Bosustow Media Group/Tee Bosustow
7655 Sunset Blvd #114, Hollywood, CA..............213-851-4900
Cannon + Eger
HCRI Box 5164, Keaau, HI 808-966-8565
CD Rom Inc/Roger Dr Hutchison
603 Park Point Dr #110, Golden, CO 303-526-7600
CD Rom Rights/Elissa Hoye
11920 White Water Ln, Malibu, CA 310-457-7156
Chaparral Software/Russell Kohn
9201 W Olympic Blvd #201, Beverly Hills, CA...310-273-4904
Christensen Media
3638 Auburn Blvd #F, Sacramento, CA 916-487-4722
Cimarron Productions/Michael Theis
3131 S Vaughn Way #134, Aurora, CO..............303-368-0988
Communication Design/John Ittelson
328 W Francis Willard Ave, Chico, CA..............916-893-9023
Concentrics Co/Bill McIntosh
320 Galisteo #206, Santa Fe, NM..............505-988-4100
Cordaro, Mary
5803 Rhodes Ave, N Hollywood, CA..............818-766-5882
Czechowicz, Lesley
1140 Holloway Ave, San Francisco, CA 415-334-6303
Dataquest Inc/Bruce Ryon
1290 Ridder Park Dr, San Jose, CA 408-437-8000
Datavision Technologies Corp/Robert Hitchcock
49 Stevenson St #575, San Francisco, CA 415-543-7903
Design Form/PenRose Baldwin
8250 Electric Ave, Stanton, CA..............714-952-3700
Desktop Solutions/Will Newman II
PO Box 1106, Canby, OR..............503-784-0609
Deverich, Michael
3976 Albright Ave, Los Angeles, CA..............310-398-5854
Freeman Associates Inc/Michael Pugh
311 E Carillo St, Santa Barbara, CA..............805-963-3853
Gomes, Lori
2430 Fillmore #602, San Francisco, CA 415-292-6220
Haleen, Brentano
PO Box 148, Tesuque, NM..............505-986-1799
Human Performance Institute/Paras Kaul
UMed Ctr 11406 Loma Linda Dr,
Loma Linnda, CA..............818-794-4057
Interactive Innovations/Brian Blum
2032 Delaware St #4, Berkeley, CA..............510-841-8271
Inview Corp/Richard N Carter
PO Box 208, Del Mar, CA 619-792-6473
Kaliczak & Associates/Janek Kaliczak
3337 Ross, Palo Alto, CA..............415-494-0776
Kolnick Consulting/Cynthia Kolnick
3145 Geary Blvd #230, San Francisco, CA..............415-824-7174
Laser Storage Solutions/Julian Olson
4520 Darcelle Dr, Union City, CA..............510-489-7732

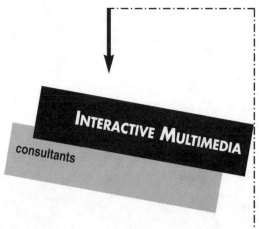

INTERACTIVE MULTIMEDIA
consultants

LSI Logic/Philip Pfeifer
1501 McCarthy Blvd, Milpitas, CA..............408-433-7635
Macko, Nancy
1252 Monte Vista Ave #21, Upland, CA 909-946-3112
Mayer, Judith
2261 Market St #330, San Francisco, CA..............415-648-1008
Media Learning Systems/Mitch Aiken
1492 W Colorado Blvd, Pasadena, CA..............818-449-0006
Media Technics/Patricia McLendon
1701 Via Sombrio, Fremont, CA..............510-656-0720
Mediadesigns/Sue Gradisar
921 Adams NE, Albuquerque, NM..............505-255-5283
Meridian Data Inc/Richard Krueger
5615 Scotts Valley Dr, Scotts Valley, CA..............408-438-3100
Mondo Media/John Evershed
340 Townsend St #408, San Francisco, CA..............415-243-8671
Multi-Media Computing Corp/Nick Arnett
2105 S Bascom Ave #300, Campbell, CA..............408-369-1233
Mundy, Ann
712 Wilshire Blvd #150, Santa Monica, CA..............415-552-2274
One World Interactive/Vicki Ayres
215 W Fifth, Eugene, OR..............503-683-4020
Pacific Interface Inc/Laurin Herr
5703 College Ave #4, Oakland, CA..............510-652-5300
Panalog Computer Guides/Dan Nash
4450 Lakeside Dr #200, Burbank, CA..............818-559-4277
Patton Sight & Sound/Royce Patton
2511 Rodeo Rd, Los Angeles, CA..............213-293-8433
Pixel Ink, Consultants/Hans Hartman
520 Frderick #13, San Francisco, CA..............415-564-0962
PM Studio/Matthew Leeds
9 Echo Ave, Corte Madera, CA..............415-924-1203
Rimmereid, Renell
9333 Sawtooth Ct, San Diego, CA..............619-484-1658
S E Tice Consulting/Steve Tice
15860 Dartford Way, Sherman Oaks, CA..............818-906-3322
San Jose State University/Sam Koplowicz
One Washington Sq, San Jose, CA..............408-924-2858
Shelton Communications/SM Shelton
332 Iowa Ct, Ridgecrest, CA..............619-375-7514
Skilset Grafix
5757 Wilshire Blvd M101, Los Angeles, CA..............213-934-8255
Spatial Data Architects/Henri Poole
220 Sansome St #530, San Francisco, CA..............415-397-6431
Steve Larson Design Associates/Steve Larson
502 Van Ness Ave, Santa Cruz, CA..............408-427-1921
Video Resources/Brad Hagen
1809 E Dyer Rd #307, Santa Ana, CA..............714-261-7266
Video Solutions for Business/Todd O'Neill
530 S Citron St, Anaheim, CA..............714-991-3343
Visions of Naples/Steve Axelrad
293 Ravenna Dr, Long Beach, CA..............310-438-4561
Winkie Interactive/Susan Winkie
4522 Brighton Ave, San Diego, CA..............619-226-8247

CANADA

Campbell Consulting Ltd/Bonni Campbell
669 Shaw St, Toronto, ON..............416-531-4291
DM Infosearch/Brian McIntosh
924 Castlefield Ave, Toronto, ON..............416-785-8113

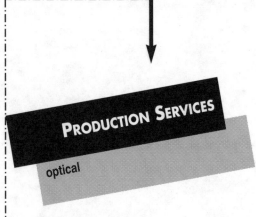

Optical

NEW YORK CITY

Aerial Image Video Services/John Stapsy
137 W 19th St, New York, NY 212-229-1930
Betelgeuse Productions/Jerry Price
44 E 32nd St 12th Fl, New York, NY 212-213-1333
Broad Street Productions/Tom Adler
10 W 19th St 11th Fl, New York, NY 212-924-4700
C & C Visual/Glen Botkin
1500 Broadway #400, New York, NY 212-869-4900
Charlex Inc/Henry Frenzel
2 W 45th St 7th Fl, New York, NY 212-719-4600
Chromavision Corp/Bruce Testa
49 W 27th St, New York, NY 212-686-7366
Communication Technologies/Ellen R Zalk
770 Lexington Ave, New York, NY 212-826-2935
Curious Pictures/Tina Contis
23 Watts St, New York, NY 212-966-1020
Digital House Ltd
330 W 58th St #600, New York, NY 212-333-5950
Editvision/Michel Stylianou
265 Madison Ave 3rd Fl, New York, NY 212-972-6677
Effects House/The Optical House/John Alagna
111 8th Ave #914, New York, NY 212-924-9150
Empire Video
216 E 45th St 11th Fl, New York, NY 212-687-2060
Image Group/Adriane Truex
305 E 46th St, New York, NY 212-752-3010
Laser Edit East Inc/Gary Sharfin
304 E 45th St 2nd Fl, New York, NY 212-983-3255
MTI/Robert Marmiroli
885 Second Ave Level C, New York, NY 212-355-0510
Partners & Agostinelli Productions/Gregory Agostinelli
1123 Broadway #404, New York, NY 212-697-7771
Toy Specialists/Bill Tesar
333 W 52nd St 7th Fl, New York, NY 212-333-2206

EAST

American Holographic Inc
521 Great Rd, Littleton, MA 508-486-9621
Amigo Business Computer/Bill Teller
192 Laurel Rd, E Northport, NY 516-757-7334
Applied Optical Media Corp/Paul Dellavigne
1450 Boot Rd, Bldg 400, West Chester, PA 215-429-3701
Atlantic Video/Chris Cates
650 Massachusetts Ave NW, Washington, DC 202-408-0900
Center City Film & Video/John Gillespie
1503-05 Walnut St, Philadelphia, PA 215-568-4134
Digital United/Mark Magel
27 Observatory Dr, Croton-on-Hudson, NY 914-271-4959
D'Elia-Wittkofski Productions/Curt Powell
One Market St, Pittsburgh, PA 412-391-2900
Heller, Brian
200 Olney St, Providence, RI 401-751-1381
Interface Video Systems/Ken Maruyama
1233 20th St NW, Washington, DC 202-861-0500
KAO Optical Products/Jim Boyer
1857 Colonial Village Ln, Lancaster, PA 717-392-7840
Modern Video Productions Inc/Jim Burt
1650 Market St 3rd Fl, Philadelphia, PA 215-569-4100
NJ Motion Pic & TV Comm
152 Halsey St, PO Box 47023, Newark, NJ 201-648-6279
Oxberry/Div of Cybernetics Products/Steve Hallett
180 Broad St, Carlstadt, NJ 201-935-3000

Pioneer New Media Technologies
600 E Crescent Ave, Upper Saddle River, NJ.......201-327-6400
Telesis Productions Inc/David Rose
277 Alexander St #600, Rochester, NY 716-546-5417
Teletime Video Productions/Warren Manos
37-39 Watermill Ln, Great Neck, NY 516-466-3882
The Window Book Co
61 Howard St, Cambridge, MA 617-661-9515
Visual Services Inc/Daniel Gallo
25 Van Zant St, E Norwalk, CT 203-852-1010
Vox-Cam Associates Ltd
813 Silver Spring Ave, Silver Spring, MD 301-589-5377

SOUTH

Advanced Storage Concepts Inc/Bill Casey
10713 Ranch Rd 620 N #601, Austin, TX 512-335-1077
Ashe/Bowie Productions/Ken Ashe
8531 Fairhaven, San Antonio, TX 210-614-5678
Century III at Universal Studios/Pam Lapp
2000 Universal Studios Plz, Orlando, FL..........407-354-1000
Circle Video Productions/Rich Sublett
631 Mainstream Dr, Nashville, TN 615-244-1717
Computerized Video Service/Johnnie Byers
Rte 2 Box 135, Comanche , OK 405-439-5932
Crawford Communication/Randy Bishop
535 Plasamour Dr NE, Atlanta, GA 404-876-7149
Denon Digital Industries
1380 Monticello Rd, Madison, GA 706-342-3425
F&F Productions Inc/Bob Eisenstaedt
9675 4th St N, St Petersburg, FL 813-576-7676
Greke Film/Video/Animation/Greg Leslie
7405 Kelley, Oklahoma City, OK 800-634-5045
IBM Media/Mike Griner
3100 Windy Hill Rd, Marietta, GA 404-835-3426
Insight Productions/Scott Long
PO Box 12498, Research Triangle Pk, NC 919-544-5700
KXTX-TV/Dennis Brunn
PO Box 190307, Dallas, TX...................214-521-3900
MPL Film & Video/Buddy Morgan
781 S Main St, Memphis, TN..................901-774-4944
Nimbus Info Systems/Bob Hedrick
PO Box 7427, Charlottesville, VA 800-782-0778
Reteaco Inc USA/Len Rubin
7027 Nicki St, Dallas, TX 214-931-9618
TM Century Inc
2002 Academy, Dallas, TX 800-879-2100
Univ of Texas Television/Greg West
7000 Fannin St #1111, Houston, TX..........713-792-5017
Video One Teleproductions/Tom Stout
220 St Michael St, Mobile, AL 205-433-0013
Video Promotions Intl/Chris Bentley
PO Box 351985, Palm Coast, FL 800-486-1874
Visioneering International/Robert Foah
66 12th St NE, Atlanta, GA 404-876-7841

MIDWEST

Adpro Inc/Rob Robbins
331 N Main St, Sycamore, IL 815-895-2000
Advanced Video Communications/William Thermos
750 Pasquinelli Dr #244, Westmont, IL 708-323-7464
Allied Film & Video/Lew Wilson
1322 W Belmont Ave, Chicago, IL 312-348-0373
Beacon Software/John Grozik
P O Box 234422, Milwaukee, WI 414-355-4460
Beatty TeleVisual Productions/Dave Beatty
1287 Wabash Ave, Springfield , IL.............800-777-2043
Busby Productions Inc/Don Flannery
1430 Locust St, Des Moines, IA 515-244-0404
City Animation Co/Emma Justice
57 Park St, Troy, MI 313-589-0600
Cycle Sat/Jake Laate
John K Hanson Dr, Forest City, IA 515-582-6999
DADC–Sony
1800 Fruitridge Ave, Terre Haute, IN..........812-462-8100
Daily Planet Ltd/Fred Berkover
455 Cityfrnt Plz/NBC Twr #2900, Chicago, IL 312-670-3766
Dakota Teleproductions/Curt Friesen
1134 S Sherman Ave, Sioux Falls, SD..........605-331-5630
Dar Electronic Research & Development/Anire Dar
919 Blair Ave, Neenah, WI 414-725-6543
Edge Multimedia/John Sink
804 Burr Oak Dr, Westmont, IL 708-920-1005
Editel Design/Glen Noren
301 E Erie St, Chicago, IL 312-440-2360

Film & Tape Works Inc/Daniel Robichud
211 E Ontario #300, Chicago, IL 312-280-2210
Fredrick Paul Productions Ltd/Jery Jacobs
6547 W North Ave, Oak Park, IL..............708-386-8055
General Television Network/Sheila Minetola
13320 Northend, Oak Park, MI 313-548-2500
Hartley Metzmer Huenink Comm/John Huenink
3076 S Calhoun Rd, New Berlin, WI 414-784-1010
Image Assoc/Ann de Lodder
1625 W Big Beaver Rd, Troy, MI 313-649-2200
Innervision Studios/Bill Faris
11783 Borman Dr, St Louis, MO 314-569-2500
Iowa Teleproduction Center Inc/Brad Morford
4800 Corporate Drive, West Des Moines, IA.......515-225-7800
Ipa/The Editing House/Maggie Magee
1208 W Webster, Chicago, IL 312-871-6033
L-E-O Systems Inc/Jill O'Neil
1505 E David Rd, Dayton, OH...............513-298-1503
Lee DeForest Communications Ltd/Sherri Shallenberg
300 W Lake St, Elmhurst, IL 708-834-7855
Marx Production Ctr/Tom Deming
3100 W Vera Ave, Glendale, WI 414-351-5060
Matrix Video Inc
6975 Washington Ave S, Edina, MN 612-944-9525
Media Group Inc/Gregory Woods
1480 Dublin Rd, Columbus, OH 614-488-0621
Media Group Television/Curt Shaffer
7th Ave & 23rd St, Moline, IL 309-764-6411
Media Loft Inc/Pat Rousseau
333 N Washington Ave #210, Minneapolis, MN ...612-375-1086
Media Process Group Inc/Robert Hercules
770 N Halsted St #507, Chicago, IL 312-850-1300
Motivation Media Inc/Rick Murray
1245 Milwaukee Ave, Glenview, IL 708-297-4740
Northwest Teleproductions/
Chicago/Amy Brierly, Carmen Trombetta
142 E Ontario 4th Fl, Chicago, IL 312-337-6000
Northwest Teleproductions/Edina/Bob Haak
4455 W 77th St, Edina, MN 612-835-4455
Northwestern Mutual Life/Betty Hoff
720 E Wisconsin Ave, Milwaukee, WI.......414-271-1444
Omni Productions/W H Long
655 W Carmel Dr, Carmel, IN 317-844-6664
Prairie Production Group
111 W Goose Alley, Urbana, IL 217-344-4675
Producers Color Service Inc/Susan Flynn
24242 Northwestern Hwy, Southfield , MI..........313-352-5353
Production House/ImageLab
811 St John S/Marketing Dept, Highland Park, IL......708-433-3172
Provideo/Jim Stiener
2302 W Badger Rd, Madison, WI 608-271-1226
Pulse Communications/Fred Monthey
211 N Broadway, Green Bay, WI...........414-436-4777
Quantum Group/Janice Glassfort
101 W 10th St, Indianapolis, IN 317-639-6001
Rainbow Video Productions/Phil Troupe
Route 1/Box 82, Adams, NE 402-788-2556
RDB Productions Inc/Russ Beckner
2100 Heatherwood Ct, Middletown, OH 513-422-9552
Reactor/Mike Saenz
445 W Erie #208, Chicago, IL..............312-573-0800
Rehabilitation Inst of Chicago/Craig Perryman
345 E Superior, Chicago, IL 312-908-6183
Reitz Data Communications/Larry Reitz
8225 Farnsworth Rd #A, Waterville, OH 419-878-3334
Roscor Corp/Bennett Grossman
1061 Freehanville Dr, Mt Prospect , IL.............708-299-8080
RW Video Inc/Gerrit Marshall
4902 Hammersley Rd, Madison, WI.................608-274-4000
Skyview Film & Video/Vivian Craig
541 N Fairbanks, Chicago, IL 312-670-2020
Software Media & CD ROM/Mark Arps
Bldg 223-5N-01/3M Center, St Paul, MN..........612-733-2142
SOS Productions/Mark Braver
753 Harmon Ave, Columbus, OH 614-221-0966
Sound/Video Impressions/Bill Holtane
110 River Rd, Des Plaines, IL 708-297-4360
Stribiak & Assoc/John Stribiak
506 N Clark St #2N, Chicago, IL 312-644-1285
Take 1 Productions Ltd/Teri Murphy
5325 W 74th St, Minneapolis, MN 612-831-7757
Technidisc
2250 Meijer Dr, Troy, MI 313-435-7430
TeleVideo Productions/Mike Sopa
611 S Farwell St, Eau Claire, WI............715-833-9269

TGA Recording Co/Thomas G Alti
295 Urbandale Dr, Benton Harbor, MI..............616-926-7581

Total Video 3/Don Browers
10714 Mockingbird Dr, Omaha, NE..................402-592-3333

Transfer Zone/Roxane B Nusholtz
13251 Northend, Oak Park, MI......................313-548-7580

TV-5 WNEM/John Haupricht
107 N Franklin, Saginaw, MI.........................517-755-8191

USA Teleproductions/Jamie Burns
1440 N Meridian, Indianapolis, IN 317-632-5900

Vanguard Productions Inc/Jerry Beck
7020 Huntley Rd, Columbus, OH 614-436-4610

VIDCAM Inc/Craig C Smith
10683 S Saginaw #E, Grand Blanc, MI...........313-694-0996

Vidcam Productions Inc/Mary Detomaso
7150 Hart St/Unit A-1, Mentor, OH 216-255-5050

Video Genesis/Jim Gerber
4949 Galaxy Pkwy, Cleveland, OH 216-464-3635

Video Impressions/Rick Roesing
1666 N Farnsworth, Aurora, IL.......................708-851-1663

Video Post/M Kappleman
4600 Madison #120, Kansas City, MO.............816-531-1225

Video Wisconsin/Jeff Utschig
18110 W Bluemound Rd, Brookfield, WI...........414-785-1110

W E S T

10th St Video Productions/Jeff Bishop
140 N 16th St/PO Box 2, Boise, ID.................208-336-5222

3D Magic/FTI/Dave Altenau
55 Hawthorne St 10th Fl, San Francisco, CA 415-543-2517

4Media Company/Paul Sehenuk
2813 W Alameda Blvd, Burbank, CA.................818-840-7000

Action Video/Alana Ireland
6616 Lexington Ave, Hollywood, CA..................213-461-3611

American Film Technology/Susan Crampton
11585 Sorrento Valley Rd #104, San Diego, CA....619-259-8112

American Production Services/Ted Hill
2247 15th Ave W, Seattle, WA 206-282-1776

Anderson Video/Frank Bluestein
100 Universal City Pl, Universal City, CA 818-777-7999

Art F/X/Dawn Jones
3575 W Cahuenga Blvd, Los Angeles, CA..........213-876-9469

Big Zig Video/Elliot Porter
329 Bryant St #1B, San Francisco, CA.............415-243-8880

CIS-Hollywood/Godfrey Pye
1144 N Las Palmas, Hollywood, CA.................213-463-8811

Cross, Lloyd
PO Box 672, Gualala, CA.............................707-884-9139

Delta Teleproductions
3333 Glendale Blvd #3, Los Angeles, CA 213-663-8754

Digital Post & Graphics/Terri Williams
1921 Minor Ave, Seattle, WA 206-623-3444

Disc Manufacturing Inc/Rushton Capers
1120 Cosby Way, Anaheim, CA 800-433-DISC

Dream Quest Images/Greg Pappas
2635 Park Center Dr, Simi Valley, CA 805-581-2671

Dub Masters/SP/Caroline Didiego
3923 S McCarran, Reno, NV 702-827-3821

Encore Video/Larry Chernoff
6344 Fountain Ave, Hollywood, CA 213-466-7663

Gannett Productions/Jim Berger
500 Speer Blvd, Denver, CO 303-871-1899

Golden Gaters Productions/Anita Norine
400 Tamal Plaza, Corte Madera , CA 415-924-7500

Gomes, Steve
1901 S Bascom Ave #1190, Campbell, CA........408-559-6369

Gray Matter Advertising/Robert Gray
2434 Monoco St, Channel Islands, CA 818-884-7641

Group Video Productions/Jerry Casey
10364 Rockingham Dr, Sacramento, CA............916-362-3964

Ikonographics/Andrew Millstein
945 N Highland Ave, Hollywood , CA 213-461-0636

ILM (Industrial Light & Magic)/Andrea Merrim
PO Box 2459, San Rafael, CA 415-258-2000

JMTV/John Thill
2808 Roosevelt St, Carlsbad, CA 619-434-3363

KGTV Channel 10/Jack Villarrubia
PO Box 85347, San Diego, CA 619-237-1010

KNTV Production Services/Bill Glenn
645 Park Ave, San Jose, CA..........................408-286-1111

KTVU Retail Services/Rich Hartwit
2 Jack London Sq/PO Box 22222, Oakland, CA.510-874-0228

Laser Pacific Corp /Steve Mitchell
540 Hollywood Way, Burbank, CA...................818-842-0777

Laser Pacific Media Corp/Leon Silverman
809 N Cahuenga Blvd, Los Angeles, CA...........213-462-6266

Main Street Video Productions Inc/Mark Tyson
119 N Wahsatch Ave, Colorado Springs, CO.....719-520-9969

Master Communication/Robert Masters
3429 Kerckhoff Ave, San Pedro, CA................310-832-3303

Master Videoworks/Jeff Killian
3611 S Harbor Blvd #150, Santa Ana, CA........714-241-7724

MDC Teleproductions/
McDonnell Douglas/Milton Moline
4000 Lakewood Blvd, Long Beach, CA 310-496-9040

Oasis Studios
675 Holcomb Ave, Reno, NV.........................702-688-6262

On Tape Productions/Steve Zeifman
724 Battery St, San Francisco, CA..................415-421-5551

One World Interactive/Vicki Ayres
215 W Fifth, Eugene, OR..............................503-683-4020

Pacific Media/Chris Largent
2600 W Olive Ave # 100, Burbank, CA............818-841-1199

Pacific Rim Video/Roger Proulx
663 Maulhardt Ave, Oxnard, CA.....................805-485-9930

Premore Inc
5130 Klump Ave , N Hollywood, CA.................818-506-7714

Rase Productions, Bill
955 Venture Ct, Sacramento, CA....................916-929-9181

San Diego Digital Post/Infovideo
9853 Pacific Hts Blvd #B, San Diego, CA..........619-452-9000

Southwest Television Production
Services/Scott Tuchman
4633 E Van Buren, Phoenix, AZ.....................602-244-2982

Sunset Post Inc/Bob Glassenberg
1813 Victory Blvd, Glendale, CA.....................818-956-7912

Top Communications/Ferris Top
1201 SW 12th Ave #210, Portland, OR...........503-222-2773

TVA/Television Associates Inc/Lauri A Clark
2410 Charleston Rd, Mt View, CA 415-967-6040

Unitel Video/Hollywood/Mark Brown
3330 Cahuenga Blvd W, Los Angeles, CA.........213-878-5800

Varitel Video/Colleen Casey
1 Union St 3rd Fl, San Francisco, CA...............415-495-3328

Video One Inc/Kevin Hamburger
10625 Chandler Blvd, N Hollywood, CA............818-980-0704

Video Pro of Wyoming/Eric Hamm
402 Broadway, Rock Springs, WY 307-382-8776

Vigon, Jay
11833 Brookdale Ln, Studio City, CA................213-650-0505

Visions Plus/Steve Dung
2223 N Main St, Walnut Creek, CA.................510-256-9450

Winters Productions/Glenn Winters
1855 Foothill Blvd, La Canada, CA..................213-682-1205

Yurth Video Production Services/Les Yurth
PO Box 1305, Woodland Hills, CA...................818-999-0080

C A N A D A

Americ Disc Inc/Frank Johansen
2525 Canadien, Drummondville, QB................819-474-2655

Executive Information Base/Harvey Martens
One Yonge St #1801, Toronto, ON 416-594-1117

Oasis Post Production Services/Don McMillan
340 Gerarrd St E, Toronto, ON.......................416-466-5870

Studio Post & Transfer/Colin Minor
5305 104th St, Edmonton, AB 403-436-4444

Video

N E W   Y O R K   C I T Y

AV Workshop Inc/Robin Kazmeroff
333 W 52nd St, New York, NY.......................212-397-5020

AVI Visual Productions/Barney Bloom
915 Broadway 12th Fl, New York, NY 212-505-9155

Barry, Judith
Box 708/Knickerbocker Station, New York, NY...212-254-9220

Betelgeuse Productions/Jerry Price
44 E 32nd St 12th Fl, New York, NY................212-213-1333

Big Fat TV/Phil Delbourgo
873 Broadway #500, New York, NY................212-420-0808

Bill Feigenbaum Designs/Joyce Feigenbaum
15 E 26th St #1615, New York, NY.................212-447-5550

Broad Street Productions/Tom Adler
10 W 19th St 11th Fl, New York, NY 212-924-4700

C & C Visual/Glen Botkin
1500 Broadway #400, New York, NY..............212-869-4900

Cacioppo Production Design/Tony Cacioppo
42 E 23rd St 5th Fl , New York, NY 212-777-1828

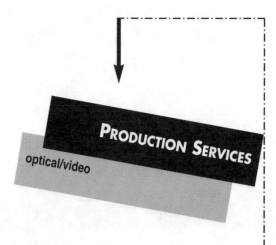

Caesar Video Graphics
137 E 25th St 2nd Fl, New York, NY.................212-684-7673

Capital Vectors/Lydia Rappold
826 Broadway 6th Fl, New York, NY 212-529-9696

Captain NY/Ed Sullivan Theater/David Niles
254 W 54th St, New York, NY........................212-307-4388

Caribiner Group
16 W 61st St, New York, NY 212-541-5300

CBS Inc
524 W 57th St , New York, NY 212-975-2318

Celefex/Dean DeCarlo
16 W 22nd St 4th Fl, New York, NY.................212-255-3470

Charlex Inc/Henry Frenzel
2 W 45th St 7th Fl, New York, NY 212-719-4600

Chelsea Television Studios/Eric Duke
221 W 26th St , New York, NY.......................212-727-1234

Chromavision/Robin Burkowitz
49 W 27th St 8th Fl, New York, NY 212-686-7366

Communications Plus/Geoffrey Fraize
102 Madison Ave, New York, NY.....................212-686-9570

Computer Graphic Resources/Michael Tarricone
8 W 40th St #1901, New York, NY..................212-764-3434

Creative Ways Inc/Stuart Sternbach
305 E 46th St, New York, NY 212-935-0145

Darino Films/Edward Darino
222 Park Ave S, New York, NY.......................212-228-4024

Data Motion Arts/Tony Caio
165 W 46th St #611, New York, NY................212-768-7411

DJM Films & Tape/Fran Drandoff
4 E 46th St, New York, NY............................212-687-0111

Dv8VIDEO
738 Broadway, New York, NY........................212-529-8204

EDEFX/Aran Friedman
219 E 44th St 9th Fl, New York, NY.................212-983-2686

Edit Decisions Inc/Gary Anthony
311 W 43rd St #701, New York, NY 212-757-4742

Editel
222 E 44th St 7th Fl, New York, NY 212-867-4600

Editvision/Michel Stylianou
265 Madison Ave 3rd Fl, New York, NY.............212-972-6677

Empire Video
216 E 45th St 11th Fl, New York, NY...............212-687-2060

Equitable Production Group
787 Seventh Ave, New York, NY 212-554-1389

FCL/Custom Lab
10 E 38th St 3rd Fl, New York, NY 212-679-9064

Fitz, Tracy
531-A Sixth Ave, Brooklyn, NY 718-768-8161

Frame Runner/Jackie Mauder
1995 Broadway #1100, New York, NY.............212-874-1730

Friedman Consortium, Harold
404 E 5th Ave #14A, New York, NY.................212-688-6434

Gramercy Broadcast Center
230 Park Ave S #1, New York, NY..................212-614-4184

GT Group/Greer Griffith
630 Ninth Ave #1000, New York, NY...............212-246-0154

HBO Studio Productions/Judy Glassman
120A E 23rd St, New York, NY.......................212-512-7800

HD/CG New York/Minako Sugiura
34-12 36th St/Landmark Front, Astoria, NY 718-361-1118

Horn/Eisenberg Film & Videotape
Editing/Mitch Garelick
16 W 46th St, New York, NY.........................212-391-8166

Ice Tea Productions/Rich Durkin
160 E 38th St #15B, New York, NY.................212-557-8185

PRODUCTION SERVICES

video

Image Group/Adriane Truex
305 E 46th St, New York, NY.....................212-752-3010
Joe Bevilacqua Stills/Film/Video/Joe Bevilacqua
202 E 42nd St, New York, NY.....................212-490-0355
Laser Edit East Inc/Gary Sharfin
304 E 45th St 2nd Fl, New York, NY.............212-983-3255
Lieberman Productions, Jerry
76 Laight St, New York, NY.......................212-431-3452
Lifetime Television/Stuart Lefkowitz
34-12 36th St 3rd Fl, Astoria, NY...............718-706-4222
Lightscape Productions
158 W 29th St 7th Fl, New York, NY.............212-695-6434
Little Caesar Productions
137 E 25th St 1st Fl, New York, NY..............212-779-0080
LRP Video/Shyamoli Pyne
3 Dag Hammarskjold Plaza, New York, NY........212-759-0822
Magno Sound & Video/Roland Blackway
729 Seventh Ave, New York, NY..................212-302-2505
Magnus, Mark
1666 Bell Blvd #739, Bayside, NY..............718-229-3268
Manhattan Transfer/Edit/Joanne Gross
545 Fifth Ave, New York, NY.....................212-687-4000
Mariah Productions Ltd/Ken Book
545 8th Ave 3rd Fl, New York, NY................212-947-0090
Marsden Reproductions/Steve Flores
30 E 33rd St, New York, NY......................212-725-9220
Merchandising Workshop/Roland Millman
550 W 43rd St, New York, NY....................212-239-4646
MHX Designs/Unitel Video
8 W 38th St 2nd Fl, New York, NY...............212-684-0508
MTI/Robert Marmiroli
885 Second Ave Level C, New York, NY..........212-355-0510
Multi Video Group Ltd/B Canavicks
Film Div/Jack Hubler
50 E 42nd St 11th Fl, New York, NY.............212-986-1577
Myrvik Productions, Ron
34 E 29th St, New York, NY......................212-685-0726
Napoleon Video Graphics
460 W 42nd St 2nd Fl, New York, NY............212-279-2000
National Video Industries
15 W 17th St, New York, NY......................212-691-1300
NBC Telesales/Elizabeth Davis
30 Rockefeller Plaza #601W, New York, NY......212-664-4444
Nexus Productions/Margaret Smagalski
10 E 40th St, New York, NY......................212-679-2180
Original Cinema Inc/Michael Bergman
130 W 57th St #12A, New York, NY..............212-545-0177
Otterson Television Video Inc/Bill Otterson
251 W 30th St #14W, New York, NY.............212-695-7417
Partners & Agostinelli Productions/Gregory Agostinelli
1123 Broadway #404, New York, NY.............212-697-7771
Perweiler & Assoc
107 E 16th St, New York, NY.....................212-925-8750
Post Perfect Inc/Dean Winkler
220 E 42nd St, New York, NY.....................212-972-3400
Premier Post/Keith Shapiro
630 Ninth Ave 14th Fl, New York, NY............212-757-1711
Primalux Video/Matt Clarke
30 W 26th St 7th Fl, New York, NY..............212-206-1402
Princzko Productions
9 E 38th St 3rd Fl, New York, NY................212-683-1300
Pyramid Recording
12 E 32nd St 3rd Fl, New York, NY..............212-686-8687
R/G Video/Cy Chang
21 W 46th St 4th Fl, New York, NY..............212-997-1464

Rebo Group/Clinton Powell
530 W 25th St 2nd Fl, New York, NY............212-989-9466
Ross-Gaffney Inc
21 W 46th St 9th Fl, New York, NY..............212-719-2744
Rutt Video Inc/Krystin Wagenberg
137 E 25th St, New York, NY.....................212-685-4000
Stark Studio
231 W 29th St #1005, New York, NY............212-868-5555
Stars Production Services/Richie Mahieu
30-30 Thomson Ave, Long Island City, NY........718-937-3510
Tapestry Productions/Nancy Walzog
920 Broadway, New York, NY....................212-677-6007
Teletechniques/Michael Temmer
1 W 19th St, New York, NY.......................212-206-1475
Telezign/Rob Wyatt
460 W 42nd St, New York, NY...................212-564-8888
Today Video/Beverly Seeger
45 W 45th St 12th Fl, New York, NY.............212-391-1020
Trans-Ocean Video Inc/Stephen, Marketing Dept
711 12th Av/Pass Ship Term 88, New York, NY.212-757-2707
Unitel Video/NY
515 W 57th St, New York, NY.....................212-265-3600
Video Dub Inc/Matthew Edelman
423 W 55th St, New York, NY.....................212-757-3300
Video Works/Harry Stroiber; Jim Riche
24 W 40th St, New York, NY......................212-869-2500
VSC/Shelley Riss
225 E 43rd St, New York, NY.....................212-599-1616
Wander Communications/Andy Wander
534 E 84th St #5W, New York, NY...............212-737-3058
Windsor Digital Graphics
8 W 38th St, New York, NY.......................212-944-9090
WNET/Net Telecon
356 W 58th St, New York, NY....................212-560-2068
Zander Productions, Mark
118 E 25th St, New York, NY.....................212-477-3900

EAST

Acme Recording Studios/Peter Denenberg
112 W Boston Post Rd, Mamaroneck, NY.........914-381-4142
Action Graphics/Tyrone Thompson
4725 Dorsey Hall Dr #A-402, Ellicott City, MD....410-992-0749
Active Video Inc/Tom Gittins; Chris Previte
265 Winter St, Waltham, MA......................617-890-6556
Advanced Video Services/Marty Pisano
1120 Bloomfield Ave #111, W Caldwell, NJ......201-882-6440
AdViz Advanced Visualizations/Allen H Cosgrove
709 Rt 206 N #333, Belle Mead, NJ...............908-281-0421
Adwar Video/Tom Miller
2370 Merrick Rd, Bellmore, NY..................516-785-1200
Amigo Business Computer/Bill Teller
192 Laurel Rd, E Northport, NY.................516-757-7334
Angelsea Productions
55 Russ St, Hartford, CT.........................203-241-8111
Arkay Productions Inc/Ralph Haselmann
25 S Main St, Manville, NJ.......................908-725-3003
Artemis Electronic Imaging Inc/George Otto
2285 Bristol Ave, State College, PA..............814-234-3165
Atlantic Video/Chris Cates
650 Massachusetts Ave NW, Washington, DC....202-408-0900
Audio Plus Video/Int Post/Beth Simon
235 Pegasus Ave, Northvale, NJ.................201-784-2190
AV Design/Joe Wall
1823 Silas Deane Hwy, Rocky Hill, CT...........203-529-2581
AVT/Edward Mctighe
26 W Highland Ave, Atlantic Highlands, NJ.......908-872-9090
Blanchard-Healy Video
Communications/Tom Kreuzberger
620 Sentry Parkway # 110, Blue Bell, PA..........215-834-5700
BOGH AV Productions
110 Jefferson Blvd, Warwick, RI.................401-737-1911
Cambridge Television Productions/Wilson Chao
67 Chapel St, Newton, MA.......................617-332-0084
Capital Presentations/Wolfgang Esh
10 Post Office Rd #2N, Silver Spring, MD.........301-588-9540
CCI COmmunications Inc/Virginia Frederick
1440 Phoenixville Pike, W Chester, PA...........215-296-7233
Center City Film & Video/John Gillespie
1503-05 Walnut St, Philadelphia, PA.............215-568-4134
Channel 3 Video/Jeffrey Page
PO Box 8781, Warwick, RI........................401-421-1616
Ciné-Med Inc/Kevin McGovern
127 Main St N/PO Box 1007, Woodbury, CT....203-263-0006
Cinecraft Audio Visual Services/Gary Brown
215B Central Avenue, Farmingdale, NY..........516-752-0700

Cineworks Productions
124 Great Bay Rd, Greenland, NH................603-431-4241
CIP/Provideo Inc
209 Butler St, York, PA..........................717-848-3655
Clark University
Dept of Visual/Performing Arts, Worcester, MA..508-793-7113
Computer Generated Imagery Inc/Rick Bowley
6270 Dean Pkwy, Ontario, NY...................716-265-1450
Computer Graphics Resources/
Computer Graphics Resources
One Dock St, Stamford, CT......................203-327-3635
Cramer Productions/Tom Martin
355 Wood Rd, Braintree, MA....................617-849-3350
Creative Video Design & Prod/Marie Leuchte
480 Neponset St, Canton, MA...................617-821-1110
Croma-Video, Inc./Lisa Warner
3801 Ridge Pike, Collegeville, PA................215-489-1070
CSR Productions Inc
One Bala Ave #G10, Bala Cynwyd, PA...........215-668-6353
D'Elia-Wittkofski Productions/Curt Powell
One Market St, Pittsburgh, PA....................412-391-2900
DAK Productions
46 Bayard St, New Brunswick, NJ................908-247-4740
DBF Media Company/Randy Runyon
PO Box 2458, Waldorf, MD......................301-843-7110
DCP Communications Grp Ltd/Alison Connors
301 Wall St, Princeton, NJ.......................609-921-3700
Devon Video Productions/John Christopher
800-C Lake St, Ramsey, NJ......................201-934-1250
DI Group/Sharon Lynch
651 Beacon St, Boston, MA......................617-267-6400
Doorbell Productions
370 Windsor Rd, Englewood, NJ.................201-586-6300
Dreamlight Inc/Michael Scaramozzino
50 Clifford St, Providence, RI.....................401-861-8002
Eaglevision Inc/Jeff Weil
880 Canal St, Stamford, CT......................203-359-8777
Edit One Video Productions/Fred Beckel
903 N Washington St, Wilmington, DE...........302-575-1022
Effective Communication Arts/David Jacobson
149 Dudley Rd/PO Box 250, Wilton, CT..........203-761-8787
Electro/Grafiks/Paul Wilson
1238 Callow Hill St #405, Philadelphia, PA........215-923-6440
ESPN Inc/Steve Ullman
1 ESPN Plaza, Bristol, CT........................203-585-2084
Expanded Video Inc/Ted Miles
477 Congress St, Portland, ME...................207-773-7005
Fairleigh Dickinson Univ/Madison/Harvey Flaxman
285 Madison Ave, Madison, NJ..................201-593-8500
Foremost Comm's/David Fox
360 Woodland St, Holliston, MA.................508-820-1130
Fox & Perla Ltd/Gene Perla
20 Martha St, Woodcliff Lake, NJ................908-604-6275
FSA Video Productions Inc/Andrew Henriques
42 S Village Dr, Bellport, NY.....................516-286-4241
GBH Productions/Jayne Pikor
125 Western Ave, Boston, MA...................617-492-9273
Gilmore Assoc Inc, Robert
360 Newbury St, Boston, MA....................617-536-0700
Girard Video/Jenny Gsell
1331 F St NW #250, Washington, DC............202-393-6666
Graphics 150/Van Smith
150 Speedwell Ave, Morris Place, NJ.............201-267-6446
Guymark Studios
3019 Dixwell Ave/PO Box 5037, Hamden, CT...203-248-9323
Half Moon Video Productions/Tom Horan
130 Central Ave, Jersey City, NJ.................201-792-1066
Hellawell, Dennis
201 Park Ave, Morristown, NJ...................201-644-4747
Hogpenny Productions Inc/Brian Treutlein
121 E 14th St/Ship Bottom,
Long Beach Island, NJ...........................609-494-6640
Idea Works
90 Bridge Dr, Newton, MA......................617-244-0101
Image Production/Jenny Coffelt
50 Water St, S Norwalk, CT......................203-853-3486
Image Recordings/Stephen Zelenko
5850 Ellsworth Ave #304, Pittsburgh, PA.........412-362-4050
Image Works/Dana Hutchins
537 Congress St #301, Portland, ME.............207-774-6399
In-House Video Editing/Tom Pernice
7 Walnut Tree Hill Rd, Shelton, CT...............203-929-0375
Intelligent Light
1099 Wall St W #387, Lyndhurst, NJ.............201-460-4700

Inter-Media Art Center/Michael Rothbard
370 New York Ave, Huntington, NY......516-549-9666
Interface Video Systems/Ken Maruyama
1233 20th St NW, Washington, DC........202-861-0500
Intl Media Services/Stuart Allen
718 Sherman Ave, Plainfield, NJ..........908-756-4060
Inverse Media/Chris Thomas
PO Box 1072, Southport, CT..............203-255-9620
John McKee Productions/John McKee
1111 Fawcett St, White Oak, PA..........412-664-7583
Keystone Media Group/Peter Putman
200D North St, Doylestown, PA...........215-345-8004
Lake Champlain Productions/Steve Murphy
4049 Williston Rd, S Burlington, VT......802-863-9780
Laurel Video Productions/Steve Tadzynski
1999 E Rte 70, Cherry Hill, NJ..........609-424-3300
Light & Power Productions/Bill Moffitt
26 N Broadway, Schenectady, NY..........518-381-6788
Long Island Video Entertainment/Peter Warzer
83 Hazel St, Glen Cove, NY..............516-759-2818
Media Mix/Joe Vargas
315 Rte 17 South, Paramus, NJ...........201-262-3700
Media Vision/James Lush
80 Great Hill Rd, Seymour, CT...........203-735-0002
Media Vision Productions/Janet May
240 New Britain Ave, Hartford, CT.......203-249-2424
Megcomm Film & Video Productions
226 Carey Ave, Wilkes-Barre, PA.........717-826-9805
Melovision Productions Inc/Mel Obst
190 Cold Soil Rd, Princeton, NJ.........609-895-1030
Mobile-Video Productions Inc/Stephen King
7315 Wisconsin Ave #325W, Bethesda, MD......301-656-2525
Modern Mass Media
PO Box 950, Chatham, NJ.................201-635-6000
Modern World Media Inc/Marketing Dept
6152 Encounter Row, Columbia, MD........410-944-4242
MPTV Video Production Unit/Mark Patterson
2116 Noble St, Pittsburgh, PA...........412-271-6788
Multivision
161 Highland Ave, Needham, MA...........617-449-5830
Nagy Films/Patrick Rafferty
9210 Corporate Blvd, Rockville, MD......301-258-1112
National Boston Video Center/Mark Jacques
115 Dummer St, Brookline, MA............617-734-4800
Oceana/Mark Rand
21 Princeton Pl, Orchard Park, NY.......716-662-8973
Optimedia Systems/George Hoffman
373 Rte 46 W, Fairfield, NJ.............201-227-8822
Optisonics Productions/Jim Brown
186 8th St, Cresskill, NJ...............201-871-4192
Our Town TV/Broadcast News Network
78 Church St, Saratoga, NY..............518-899-6989
Ovation Graphics/Kevin Dady
2 Graeme St, Pittsburgh, PA.............412-471-3350
PCI Recording/Caroline Maynard
737 Atlantic Ave, Rochester, NY.........716-288-5620
Penfield Productions Ltd
35 Springfield St, Agawam, MA...........413-786-4454
Pixel Light Communication/Carmine DeFalco
271 Rte 46 W # F205, Fairfield, NJ......201-808-1389
PREFIT Corporation/Joseph Chernicoff
PO Box 326, Willow Grove, PA............215-657-3976
Preston Productions
160 Locke Dr, Marlborough , MA..........508-624-4100
Producers East Media/Rosalyn Birnbaum
734 Rte 110 # 204, Melville, NY.........516-421-4800
Producers Video Corp
3700 Malden Ave, Baltimore, MD..........410-523-7520
Production House
PO Box 1076, Island Heights, NJ.........908-341-0689
Production House
327 N 17th St, Philadelphia, PA.........215-972-0934
Production Masters/Mike Killes
321 First Ave, Pittsburgh, PA...........412-281-8500
Production Works/Mark Phillips
360 East Ave, Rochester , NY............716-232-3709
PVS (Professional Video Services)/Liz Lokey
2030 M St NW, Washington, DC............202-775-0894
Q1 Productions
15 Campus Dr, Somerset, NJ..............908-563-2233
Quality Film & Video/Peter A Garey
232 Cockeysville Rd, Hunt Valley, MD....410-785-1920
Reider Video Productions /A Reider
2174 Morris Ave, Union , NJ.............908-688-8808

Resolution Inc/Linda Citro
19 Gregory Dr, S Burlington, VT.........802-862-8881
Robert Gilmore Associates/Tricia Turczynski
360 Newbury St, Boston, MA..............617-536-0700
Robert L Biel Associates
87 Greenwich Ave, Greenwich, CT.........203-622-6630
Schneider, Pat
Six Lincoln Gardens, Long Branch, NJ....908-222-6840
September Productions Inc
162 Columbus Ave, Boston, MA............617-482-9900
SEV/Sherwin-Greenberg Productions/Ron Ramos
117 W Chippewa St, Buffalo, NY..........716-856-7120
Shooters Inc/John Godley
2428 Rte 38 #302, Cherry Hill, NJ.......609-482-9090
Sound & Motion
180 Pool St, Biddeford, ME..............207-283-9191
Sound Image Inc/Jim Goodell
42 Richards St, Worcester, MA...........508-756-0673
Speaker Support Group/Jeff McPhie
19 Wall St, Princeton, NJ...............609-921-3400
Specter Video
4148 Windsor St, Pittsburgh, PA.........412-521-1444
Starfleet Productions
RD 5/Box 91/Avalon Rd, Altoona, PA......814-942-STAR
Stemrich, J David
1334 W Hamilton St, Allentown, PA.......215-776-0825
Steward Digital Video/Dave Bowers
525 Mildred Ave, Primos, PA.............215-626-6500
Studiolink/Andy Cordery
13 Roszel Rd, Princeton, NJ.............609-452-0846
Techniarts/Bill Moore
8555 Fenton St, Silver Spring, MD.......301-585-1118
Telesis Productions Inc/David Rose
277 Alexander St #600, Rochester, NY....716-546-5417
TRACES
10 N Presidential Blvd #115, Bala Cynwyd, PA...215-660-9699
Triad Media
PO Box 778, Frederick, MD...............301-663-1471
ULTITECH Inc/William J Comcowich
Foot of Broad St, Stratford, CT.........203-375-7300
Vidconn Productions Inc/Joe Seaton
360 Towlin Tpke #2E, Manchester, CT.....203-646-0660
Video Ad Services
244 Rutledge Dr, Bridgeville, PA........800-521-0994
Video Communication Services/Frank Siegel
523 Fellowship Rd #210, Mt Laurel, NJ...609-273-8800
Video Copy International/Hal Slifer
14 Lake St, Boston, MA..................617-787-7910
Video Editing Centers/Robin Gilmore
526-1 Lancaster Ave, Frazer, PA.........215-251-0465
Video Production Inc/Dan Miles
3 Crossgate Dr, Mechanicsburg, PA.......717-697-1506
Video Resource/Curtis Sink
140 Centennial Ave, Piscataway, NJ......908-457-7400
Video Services Unlimited/Paul Spencer
124 Canal St, Lewiston, ME..............207-782-5650
Video Techniques Productions Inc/Tom Mista
305 Regent St, Livingston, NJ...........201-992-0840
Video Works Productions/Ridge Amos
173 E Main St, Newark, DE...............302-454-1344
Video Workshop/Jay White
495 Forest Ave, Portland, ME............207-774-7798
Videocenter of New Jersey/Bob Schaffner
228 Park Ave, E Rutherford, NJ..........201-935-0900
Videosmith Inc/Post & Graphics/Thomas M Firchow
520 N Delaware Ave, Philadelphia, PA....215-238-5050
Videosmith of NJ/Rod Ammon
3 Independence Way, Princeton, NJ.......609-987-9099
Videotechniques Inc/Gene Wyckoff
313 Hope St, Stamford, CT...............203-323-7600
Vox-Cam Associates Ltd
813 Silver Spring Ave, Silver Spring, MD....301-589-5377
Wave Inc/Walter Henritze
11 California Ave, Framingham, MA........508-795-7100
WGBH Design
125 Western Ave, Boston, MA.............617-492-2777
WLNE-TV
10 Orme St, Providence, RI..............401-453-8000
Xerox Media Center/John Linn
780 Salt Rd/bldg 845, Webster, NY.......716-423-5090

SOUTH

Adco Productions/Earl Wainwright
7101 Biscayne Blvd, Miami, FL...........305-751-3118

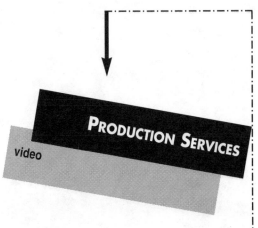
Advance Concepts Inc/Jim Brennsteiner
8229 Boone Blvd #102, Vienna, VA........703-448-0445
AIE Studios
3905 Braxton, Houston, TX...............713-781-2110
Allied Film and Video/Tony Bifano
6305 N O'Connor #111, Irving, TX........214-869-0100
American Media Productions/John Creech
PO Box 1722, Morehead City, NC..........919-247-5914
AMS Productions Inc/Suzanne Miller
13709 Gamma Rd, Dallas, TX..............214-701-0878
Ani-Majic Productions Inc/Alan Davidson
2699 Lee Rd #415, Winter Park, FL.......407-629-5757
Applied Video Communications/Mike Looper
719 Sawdust Rd, The Woodlands, TX.......713-364-1203
Art in Media Productions/Rick Sturnberg
7212 McNeil Dr #206, Austin, TX.........512-250-5535
Ashe/Bowie Productions/Ken Ashe
8531 Fairhaven, San Antonio, TX.........210-614-5678
Astro Audio-Visual
1336 W Clay, Houston, TX................713-528-7119
Automated Broadcast Consultants Inc
5401 Collins Ave Mezz, Miami Beach, FL...305-864-4800
AVID Inc/Leslie Draper
130 E Marks St, Orlando, FL.............407-423-9535
Baldridge Studios Inc/Donna Baldridge
347 N Main St, Memphis, TN..............901-522-1185
Banjo Video & Film Productions/Toby Jenkins
1188 St Andrews Rd, Columbia, SC........803-739-0336
Beachwood Productions/Peter Wilcox
696 S Yonge St #C, Ormond Beach, FL.....904-672-3698
BES Teleproductions/Ed Lazor
6829-E Atmore Rd, Richmond, VA..........804-276-5110
Bransby Productions/Sharon Grilliot
2124 Metro Cir, Huntsville, AL..........205-882-1161
Broadcast Resource Group/Southern Prods
255 French Landing Dr, Nashville, TN....615-248-1978
Broadcast Video Inc/Don Baret
20377 NE 15th Ct, Miami, FL.............305-653-7440
Caluger & Assocs/J Wayne Caluger
237 French Landing, Nashville, TN.......615-255-2792
Cannata Communications Corp/Jack Cannata
7031 Grand Blvd, Houston, TX............713-748-1684
Cat Walk/Richard Aldridge
418 W 5th St, Charlotte, NC.............704-342-3348
CEC Audio Visual/Video
PO Box 27109, Greenville, SC............803-288-0000
Century III at Universal Studios/Pam Lapp
2000 Universal Studios Plz, Orlando, FL...407-354-1000
Cinema Concepts/Stewart Harnell
2030 Powers Ferry Rd #214, Atlanta, GA...404-956-7460
Cinema East/Adam Rogers
5859 Biscayne Blvd, Miami, FL...........305-757-5859
Cinemasound Video Productions/Carla Thomas
1011 Arlington Blvd #320, Arlington, VA...703-524-1083
Cinetel Productions/Peter Franks
9701 Madison Ave, Knoxville, TN.........615-690-9950
Cinetron Computer Systems
4350 J Int'l Blvd, Norcross, GA.........404-925-4448
Circle Video Productions/Rich Sublett
631 Mainstream Dr, Nashville, TN........615-244-1717
Colordynamics/Ed Leach
150 E Bethany Rd, Allen, TX.............214-390-6500
Computer Studio, The/George Bowers
4000 Cumberland Pkwy 1400 #A1, Atlanta, GA 404-436-6092
Computerized Video Service/Johnnie Byers
Rte 2 Box 135, Comanche, OK.............405-439-5932

PRODUCTION SERVICES

video

Corporate Graphics Inc/Pat Murphy
3111 Monroe Rd, Charlotte, NC 704-335-0534
Corporate Media Services/Ralph Preston
632 W Summit Ave, Charlotte, NC 704-377-1601
Crawford Communication/Randy Bishop
535 Plasamour Dr NE, Atlanta, GA 404-876-7149
Creative Edge/Duncan Brown
808 Live Oak Dr #101, Chesapeake, VA......804-420-3605
Creative Post & Transfer/Chris John
377 Carowinds Blvd, Ft Mills, SC 803-548-7678
Creative Productions/WEAR-TV/Don Willis
PO Box 12278, Pensacola, FL 904-456-3333
Creative Video Inc/Guy Davidson/Armistead Whitney
1465 Northside Dr #110, Atlanta, GA 404-355-5800
Cypress Productions/George Cornelius
5410 Mariner St, Tampa, FL 813-289-6115
Delmar Communications Co/Gene Wash
2246 Dabney Rd #E, Richmond , VA 804-358-8880
Development Communications/Welby Smith
1800 N Beauregard St, Alexandria, VA.......703-845-7370
Diamond Video Productions Inc/Paul Lambert
129 Citation Ct, Birmingham, AL 205-942-8888
Digital Multi-Media Post Inc/Marie Hamlin
502 N Hudson St, Orlando, FL 407-293-3390
Dixieland Productions Inc/Bill Labord
3440 Oakcliff Rd #104, Atlanta, GA 404-458-1168
ECI Video/Shawn Sterling
2809 Ross Ave at Central, Dallas, TX.........214-969-6946
EDEFX/Obadia Marcos; Carol Enace
7355 NW 41st St, Miami, FL 305-593-6911
EFX Communications Inc/Jennifer Cortner
2300 S 9th St #136, Arlington, VA 703-486-2303
Electric Arts/Darryl Baird
2100 Highway 360 #700B, Grand Prairie, TX....214-988-3422
Englander Studios/Mel Englander
1931 NE 163rd St, N Miami Beach, FL 305-945-1700
F&F Productions/Bob Eisenstaedt
9675 4th St N, St Petersburg, FL 813-576-7676
Florida Film & Tape/Linda Fitzpatrick
3417 Lake Breeze Rd, Orlando, FL 407-297-0091
Florida Production Group Inc/Craig Meadows
913 Gulf Breeze Pkwy #16, Gulf Breeze, FL 904-934-5627
Franklin Video Inc/Mark Harmon
302 Jefferson St #300, Raleigh, NC 919-833-8888
Gannett Production Services/Nancy Putney
1611 W Peachtree St NE, Atlanta, GA 404-873-9182
Garman Audio/Video/Steve Garman
7701 N Broadway #A-6, Oklahoma City, OK.....405-842-3230
Georgia Pacific Television/Randy French
133 Peachtree St NE, Atlanta, GA 404-652-5690
GMG International/David Gardy
1950 Roland Clarke Pl, Reston, VA 703-339-8500
Greke Film/Video/Animation/Greg Leslie
7405 Kelley, Oklahoma City, OK 800-634-5045
Gulf Coast Audio Visual Producers/Richard Buehrig
19 W Garden St, Pensacola, FL 904-433-3016
Horizon Entertainment Inc/Gary Seline
7102 Grand, Houston, TX 713-747-6433
Hot Source Media/Michael Vaughan
1916 Wilson Blvd #304, Arlington, VA 703-527-5992
HP Productions/Heidi Pfisterer
8229 boone Blvd #390, Vienna, VA 703-356-3099
IBM Media/Mike Griner
3100 Windy Hill Rd, Marietta, GA 404-835-3426
Image Resources Inc
PO Box 616688, Orlando, FL 407-843-4200

Interactive Arts/Rick Ligas
501 E 4th St #511, Austin, TX..............512-469-0502
Island Video Productions/Tony Dabney
4211 Ave T 1/2 , Galveston, TX 409-762-2252
ITC/Joan Dasher
13515 Dulles Technology Dr, Herndon, VA........703-713-3335
JHT Multimedia/Gary Twitchell
5514 Lake Howell Rd, Winter Park, FL 407-657-2727
JIMANDI/James Spitler
734 Vosswood Dr, Nashville, TN 615-256-9887
Knowles Video Inc/Carl Knowles
2003 Apalachee Pkwy #204, Tallahassee, FL......904-878-2298
KXTX-TV/Dennis Brunn
PO Box 190307, Dallas, TX................214-521-3900
Longworth Communications/Jim Longworth
230 S Crater Rd, Petersburg, VA 804-862-9967
Marshall Productions Inc/Bill Hite
404 BNA Dr #102, Nashville, TN 615-399-8895
Match Frame Post Production/Don White
8531 Fairhaven, San Antonio, TX 210-614-5678
Max Place/Peter Sugarman
Rte 4 Box 54C, Louisa, VA.................703-894-0511
Media Events Concepts Inc
10139 Metropolitan Dr, Austin, TX 512-832-1142
Metropost/Vincent Hollister
501 N IH-35, Austin , TX 512-476-3876
Mind's Eye Graphics/Paul Lambert
2246 E Dabney Rd, Richmond, VA 804-353-7958
Mindflex/Steve Tanner
566 Dutch Valley RD, Atlanta, GA 404-892-6232
Moxie Media Inc/Martin Glenday
5734 Jefferson Hwy/POB 10203,
New Orleans, LA 504-733-6907
MPL Film & Video/Buddy Morgan
781 S Main St, Memphis, TN...............901-774-4944
Multi Image Group/Robert Silvey
1080 Holland Dr, Boca Raton, FL 407-994-3515
Multivision Video & Film/Bob Berkowitz
7000 SW 59th Pl, S Miami, FL..............305-662-6011
MVI/POST/Beck Lampert
2701-C Wilson Blvd, Arlington, VA 703-522-5335
Neo Geo/Diana Georgiou
130 E Marks St, Orlando, FL...............407-423-9524
Odyssey Communications Group/Rebecca Stevenson
6309 N O'Connor LB83, Irving, TX 214-432-9070
Omega Films/Cefus McRae
3100 Medlock Bridge Rd, Norcross, GA..........404-449-8870
On Video Inc/Jeff Schumm
2435 Southwell, Dallas, TX................214-406-9292
Park Avenue Teleproductions Inc/Melanie Cox
3500 Mayland Ct, Richmond, VA............804-346-3232
Pearlman Productions/Andrew Arten
2401 W Bellfort, Houston, TX...............713-668-3601
PhotoSonics/Audio Visual
1116 N Hudson St, Arlington, VA 703-522-1116
Pietrodangelo Prod Group Inc/Danny Pietrodangelo
216 E Oakland Ave, Tallahassee, FL..........904-681-2392
Pixel Graphics/Douglas H Parr
PO Box 63230, Pipe Creek, TX 210-535-9585
PO Box 5126/James L Webster
4220 Amnicola Hwy, Chattanooga, TN..........615-622-1193
Post Edge Inc/Mike Duncan
2040 Sherman St, Hollywood , FL 305-920-0800
Post Group At Disney/MGM Studios/Jim Dorriety
Roy O Disney Production Ctr,
Lake Buena Vista, FL 407-560-5600
Power Productions/John Mashburn
1967 Government St. #D, Mobile, AL 205-478-5530
Production Group, The/Bill Ford
510 A Leesville Rd, Lynchburg, VA...........804-237-0743
Pyramid Tele-Productions Corp
6305 N O'Connor Rd LB#6/#103, Irving, TX......214-869-3330
Quinn Associates Inc/Margaret Quinn
110 W Jefferson St, Falls Church, VA..........703-237-7222
Richard Kidd Productions Inc/John Kindervag
5610 Maple Ave, Dallas, TX................214-638-5433
Scene Three Inc/Greg Alldredge
1813 8th Ave S, Nashville, TN 615-385-2820
School of Architecture
Univ of Arkansas, Fayetteville, AR 501-575-3805
Selkirk Communications Video Services/Darice Lang
644 S Andrews Ave, Ft Lauderdale, FL 305-527-5007
Serious Robots/Rod Rich
1101 Capitol Blvd, Raleigh, NC.............919-821-0914

Smith/Taylor Productions/Wes Hoskins
2209 NPID #AB, Corpus Christi, TX...........512-289-0461
Southern Productions/David Deeb
255 French Landing Dr, Nashville, TN 615-248-1978
Southwest Teleproductions Inc/John Shives
2649 Tarna Dr, Dallas, TX 214-243-5719
Spectrum South Inc
15 S Main St #700, Greenville, SC 803-232-7369
Stuck, Jon D
6138 Edsall Rd #203, Alexandria, VA 703-354-5158
Summa Logics Corp./Karen Ceretto
8403 Golden Aspen Ct, Springfield, VA 703-912-7948
Sunbelt Video/Michael Poly
4205-K Stuart Andrew Blvd, Charlotte, NC 704-527-4152
Synthesis/Samuel Reynolds
PO Box 339, Conway, AR 501-327-2517
Take One Productions/Jere Sneider
101 Pheasant Wood Ct, Morrisville, NC 919-481-0000
Take Ten Teleproductions/Coby Cooper
1111 Bull St, Columbia , SC 803-758-1230
Teleproductions Unlimited/Mel Elzea
5820 S 129th East Ave, Tulsa, OK 918-252-2909
Texas Video & Post/Grant Guthrie
8964 Kirby Dr, Houston, TX 713-667-5000
TMP Video Communications/Howard Fox
PO Box 37364, Raleigh, NC 919-851-2202
Trimble Production Studios/Grady Trimble
612 Walnut, Little Rock, AR 501-666-8742
United States Video/Frank Garber
2070 Chain Bridge Rd #485, Vienna, VA.........703-848-1990
Univ of Texas Television/Greg West
7000 Fannin St #1111, Houston, TX 713-792-5017
Venture Productions
16505 NW 13th Ave, Miami, FL 305-621-5266
Video Co, The
9146 Jefferson Hwy, Baton Rouge, LA..........504-928-4814
Video Communications SE/Andy Bundschuh
311 N Adams, Tallahassee , FL 904-224-5420
Video Editing Services Inc/Arthur Rouse
215 E High St, Lexington, KY 606-255-9049
Video Eye/Bob Fitch
10960 Millridge N Dr #103, Houston, TX........713-890-8364
Video Image Productions/Charlie Case
2026 Powers Ferry Rd #100, Atlanta, GA........404-984-8288
Video One/James Kilpatrick
4912 Travis, Houston, TX.................713-524-8823
Video One Teleproductions/Tom Stout
220 St Michael St, Mobile, AL..............205-433-0013
Video Park/Charles Park Seward
11316 Pennywood Ave, Baton Rouge, LA........504-292-0840
Video Post & Transfer/Jack C. Bryan
2727 Inwood Rd, Dallas, TX 214-350-2676
Video Production Assocs/Lee Culpepper
2500 W Oakland Pk Blvd, Ft Lauderdale, FL.......305-731-3777
Video Workshop/Greg Cooper
1661 E Sample Rd, Pompano Beach, FL........305-942-3199
Videotexting/Jaf Fletcher
275 Commercial Blvd, Lauderdale-by-the-Sea, FL...305-771-5999
Videx Corp
944 Country Club Blvd, Cape Coral, FL 813-574-8999
Vidsat Communications
2313 Seven Springs Blvd, New Port Richey, FL 813-372-9005
Vision Design Teleproductions/Athena Schultz
5401 Corporate Woods Dr #500, Pensacola, FL .904-484-0655
Walton Electronics Inc
4324 SW 35th Terr, Gainesville, FL 904-376-5658
WDBJ-TV Inc
PO Box 7, Roanoke, VA 703-344-7000
West End Post/Lola Lott
2211 N Lamar #100, Dallas, TX.............214-871-3348
White Hawk Pictures/Michelle Barth
567 Bishopgate Ln, Jacksonville, FL..........904-634-0500
WOFL-Productions/Don Holt
35 Skyline Dr, Lake Mary, FL..............407-644-3535

MIDWEST

Admark Inc/Gary Piland
3630 SW Burlingame Rd, Topeka, KS 913-267-4712
Advanced Audio Visual/Ed Koskie
1212 W 96th St #C, Minneapolis, MN 612-881-4500
Advanced Video Communications/William Thermos
750 Pasquinelli Dr #244, Westmont, IL 708-323-7464
AGS&R Communications/IN
1835 S Calhoun, Fort Wayne, IN 219-744-4255

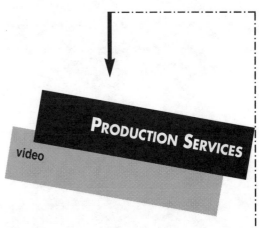
Allied Film & Video/Lew Wilson
1322 W Belmont Ave, Chicago, IL 312-348-0373
Allied Film and Video
7375 Woodward Ave, Detroit, MI.....................313-871-2222
**American Cablevision Teleproductions/
WTEN**/Tim Renshaw
3030 Roosevelt Ave, Indianapolis, IN 317-632-2288
Analog Video/Dick Madding
192 Walnut Grove Dr, Dayton, OH 513-299-3495
Animation Station Ltd/Terry Choate
633 S Plymouth Ct #1103, Chicago, IL.................312-939-8003
Associate Producers/Robert Hufstader
6545 Bloomfield Rd, Des Moines, IA 515-285-1209
Audio-Visual Associates/Bud Osborne
4760 E 65th St, Indianapolis, IN 317-255-6457
Avenue Edit/Benjamin Webber
625 North Michigan Ave, Chicago, IL 312-943-7100
AVP Inc/Verne Wandal
9136 Oakland, Portage, MI.................................616-324-3600
AVS Inc/Rob Ramseier
2109 Ward Ave, La Crosse, WI.......................608-787-1010
Badiyan Productions Inc/Dennis Gettler
720 W 94th St, Minneapolis, MN 612-888-5507
Barlow Productions Inc/Ron Barlow Sr
1115 Olivette Executive Pkwy, St Louis, MO 314-994-9990
Bauman Communications Inc/Chuck Joffe
602 Main St #300, Cincinnati, OH 513-621-6806
Beatty TeleVisual Productions/Dave Beatty
1287 Wabash Ave, Springfield , IL.......................800-777-2043
Blue Earth Picture Inc/James Ankeny
4808 Park Glen Rd, Minneapolis, MN.................612-922-3434
Blue Sky Productions/Peter Hickman
1020 S Creyts Rd #305, Lansing, MI.................517-886-2010
Busby Productions Inc/Don Flannery
1430 Locust St, Des Moines, IA.........................515-244-0404
Carlson Marketing Communications/Fred Carlson
Route 4/Box 384-B, Mora, MN........................612-679-4105
Charmed Productions Inc/Susan Arco
7535 Newburg Rd, Rockford, IL.......................815-332-9885
Cinecraft Productions Inc/Neil McCormick
2515 Franklin Blvd, Cleveland, OH 216-781-2300
Cinema Video Center
211 E Grand Ave, Chicago, IL 312-644-1650
City Animation Co/Emma Justice
57 Park St, Troy, MI...313-589-0600
Classic Video Inc/Jerry L Patton
5001 E Royalton, Cleveland, OH 216-838-5377
CMI Business Communications/Gary Brown
150 E Huron St, Chicago, IL...........................312-787-9040
Combined Services/David Patsy
9 W 14th St, Minneapolis, MN.......................612-871-5503
Communications Services/Gary Fisk
Governers State Univ, University Park, IL.............708-534-5000
Continental Cablevision of Cook Cnty/Terry Cantwell
688 Industrial Dr, Elmhurst, IL 708-530-4477
Continental Cablevision/Madison Hts/Jennifer Rossow
32090 John R, Madison Hts, MI 313-583-1354
Corplex Inc/Chris Blake
6444 N Ridgeway, Lincolnwood, IL 708-673-5400
Creative Concepts/Doug Crane
94-D Westpark Rd, Centerville, OH 513-436-2020
Creative Images/Steve Hendrickson
1400 Energy Park Dr #23, St Paul, MN 612-644-2157
Crystal Productions/Tim Dwyer
1024 Blouin Dr, Dolton, IL...............................708-841-2622
Curtis Inc/Dan Castagna
2025 Reading Rd # 130, Cincinnati, OH............513-621-8895
Cycle Sat/Jake Laate
John K Hanson Dr, Forest City, IA.....................515-582-6999
Daily Planet Ltd/Fred Berkover
455 Cityfrnt Plz/NBC Twr #2900, Chicago, IL 312-670-3766
Dakota Teleproductions/Curt Friesen
1134 S Sherman Ave, Sioux Falls, SD................605-331-5630
Darrell Brand/Moving Images Inc/Darrell J Brand
3308 St Paul Ave, Minneapolis, MN 612-922-3308
Dave Restuccia Productions/Dave Restuccia
2012 Queen Ave S, Minneapolis, MN................612-377-7560
Design Studios Teleproductions/David Harms
748 Ansborough, Waterloo, IA...........................319-232-4500
Dynaimages/Steve Hansen
3947 State Line, Kansas City, MO.....................816-531-3838
Edit Suite/Rich Melvin
1714 Boardman-Poland Rd # 15, Poland, OH.....216-757-1104
Editech Post Productions Inc/Bill Hartness
14344 Y St #200, Omaha, NE.........................402-896-9696

EDR Media/Niki Dias
23330 Commerce Park Rd, Beachwood, OH.......216-292-7300
Envision Inc/Jeff Blackwell
405 N Calhoun, Brookville, WI 414-789-8485
Film & Tape Works Inc/Daniel Robichud
211 E Ontario #300, Chicago, IL.....................312-280-2210
Film Craft Video/Sandy Ruby
37630 Interchange Dr, Farmington Hills, MI 313-474-3900
Finishing Group/Mark Place
312 E Buffalo St, Milwaukee, WI.....................414-277-7678
Frame One Inc/Paul Bernan
676 St Clair #425, Chicago, IL.........................312-440-1328
Fredrick Paul Productions Ltd/Jerry Jacobs
6547 W North Ave, Oak Park, IL.....................708-386-8055
Gannett Production Services/Bob Bye
8811 Olson Memorial Hwy, Golden Valley, MN.....612-541-8055
General Television Network/Sheila Minetola
13320 Northend, Oak Park, MI 313-548-2500
Grayson Media/Diane Kenna
70 W Hubbard #402, Chicago, IL 312-464-9852
Greatapes/Roger Rude
1523 Nicollet, Minneapolis, MN.......................612-872-8284
GRS Inc/Steve Andrews
13300 Broad St, Pataskala, OH 614-927-9566
GSI Ltd
9333 N Meridian St #102, Indianapolis, IN 317-575-0557
Harvest Communications/Marlin Davis
900 George Washington Blvd, Wichita, KS 316-652-9900
Hi-Tech Productions/Dave McClimans
33433 Boulevard, Eastlake, OH 216-953-0077
HVS Video/HVS Cable Adv/Patrick Krohlow
1100 Guns Rd, Green Bay , WI 414-468-3333
Image Producers/Mark Munroe
Box 130, Canfield, OH 216-533-0100
Image Video Teleproductions/Dean Marini
6755 Freedom Ave N W, N Canton, OH 216-494-9303
Innervision Studios/Bill Faris
11783 Borman Dr, St Louis, MO.......................314-569-2500
**Instant Replay Video & Film
Production Inc**/Stephanie Smith
1349 E McMillan St, Cincinnati, OH 513-569-8600
Inventive Images
150 Spring Dr #B220, Akron, OH 216-666-8838
Ipa/The Editing House/Maggie Magee
1208 W Webster, Chicago, IL.........................312-871-6033
IVL Post/Bob Rohde
Target Center/600 First Ave N, Minneapolis, MN ...612-673-1250
J Parker Ashley & Associates
119 N 4th #208, Minneapolis, MN 612-340-9807
Jones-Rasikas/Arnold Jones
800 Bond NW, Grand Rapids, MI.....................616-456-1855
Koplar Communications Ctr/Jim Wright
4935 Lindell Blvd, St Louis, MO.......................314-454-6320
Lamb & Co/Lawrence Lamb
650 Third Ave S, 17th Fl, Minneapolis, MN 612-333-8666
LaVoie's Photography/Rex LaVoie
01423 US 127 S, Bryan, OH 419-636-4602
Lightborne Communications/Bob Hanneken
632 Vine St #817, Cincinnati, OH 513-721-2272
Lighthouse Productions Inc/Rick Meredith/Chip Moore
1900 Hicks Rd, Rolling Meadows, IL.................708-506-1414
LTS Productions/David Lang
2810 Bennett Rd, Okemos, MI.........................517-332-1190
Mainsail Production Services/Ron Beam
521 Buyers Rd #109, Miamisburg, OH............513-866-7800
Mainstream Communications
9555 James Ave S #235, Minneapolis, MN........612-888-9000
Matrix Video Inc
6975 Washington Ave S, Edina, MN 612-944-9525
Media Consultants/Richard Wrather
201 N Stoddard/POB 130, Sikeston, MO 314-472-1116
Media Group Television/Curt Shaffer
7th Ave & 23rd St, Moline, IL...........................309-764-6411
Media Loft Inc/Pat Rousseau
333 N Washington Ave #210, Minneapolis, MN ...612-375-1086
Media Process Group Inc/Robert Hercules
770 N Halsted St #507, Chicago, IL.................312-850-1300
Metro Productions Inc/Tim McGovern
238 W 74th St, Kansas City, MO 816-444-7004
MGS Services/Viacom Div/Cliff Kroeter
65 E Wacker Pl #1900, Chicago, IL.................312-337-3761
Midwest Teleproductions/Tom Orrick
3947 State Line Rd, Kansas City, MO................816-531-3838
Motivation Media Inc/Rick Murray
1245 Milwaukee Ave, Glenview, IL 708-297-4740

Northwest Mobile Television/Michael Koenig
379 W 60th St, Minneapolis, MN.....................612-869-1119
**Northwest Teleproductions/
Chicago**/Carmen Trombetta; Amy Brierly
142 E Ontario 4th Fl, Chicago, IL.....................312-337-6000
Northwest Teleproductions/Edina/Bob Haak
4455 W 77th St, Edina, MN 612-835-4455
Nygard & Associates/Jim Nygard
400 N First St #120, Minneapolis, MN..............612-371-9228
On Line Video/Dennis Majewski
720 E Capitol Dr/PO Box 693, Milwaukee , WI..414-223-5254
**Ozam Film & Tape Productions, Imagination
Effects**/Dale Detoni
1228 S Fort, Springfield, MO 417-866-3232
Postique Inc/Mary Suzanne Patek
23475 Northwestern Hwy, Southfield, MI...........313-352-2610
Prairie Production Group
111 W Goose Alley, Urbana, IL 217-344-4675
Precision Tapes Inc/Daniel Piepho
2301 E Hennepin, Minneapolis, MN 612-379-7554
Promotovision Inc/Daniel G Endy
5929 Baker Rd #460, Minnetonka, MN 612-933-6445
Provideo/Jim Stiener
2302 W Badger Rd, Madison, WI 608-271-1226
Pulse Communications/Fred Monthey
211 N Broadway, Green Bay, WI.....................414-436-4777
Quicksilver Assoc Inc/Chuck Garasic
18 W Ontario St, Chicago, IL 312-943-7622
Rainbow Video Productions/Phil Troupe
Route 1/Box 82, Adams, NE.........................402-788-2556
RCA/Thompson Consumer Electronics
600 N Sherman Dr, Indianapolis, IN 317-267-5860
RDB Productions Inc/Russ Beckner
2100 Heatherwood Ct, Middletown, OH 513-422-9552
Rehabilitation Inst of Chicago/Craig Perryman
345 E Superior, Chicago, IL...........................312-908-6183
Renaissance Video Corp/John Feldbaille
130 S Jefferson, Chicago, IL...........................312-930-5000
Rivet Films, Inc.
748 Seminole, Detroit, MI...............................313-822-2700
Roscor Corp/Bennett Grossman
1061 Freehanville Dr, Mt Prospect , IL...............708-299-8080
Ross Roy Communications/Joey Silvian
100 Bloomfield Hills Pkwy, Bloomfield Hills, MI 313-433-6837
Schmitt & Co Prods Inc/Bob Schmitt
9840 Sharon Ln, Cleveland, OH 216-582-2282
Shaw Video Communications/Paul Quinn
3246 Henderson Rd, Columbus, OH.................614-457-4477
Skyview Film & Video/Vivian Craig
541 N Fairbanks, Chicago, IL.........................312-670-2020
Stribiak & Assoc/John Stribiak
506 N Clark St #2N, Chicago, IL.....................312-644-1285
Tech Pool Studios
1465 Warrensville Ctr Rd, Cleveland, OH 800-777-8930
Tele Edit Inc/Keith Pokorny
10 S 5th St #640, Minneapolis, MN.................612-333-5480
Total Video 3/Don Browers
10714 Mockingbird Dr, Omaha, NE.................402-592-3333
Transfer Zone/Roxane B Nusholtz
13251 Northend, Oak Park, MI 313-548-7580
Trillion Post Production/Ed Anderson
5989 Tahoe Dr S E, Grand Rapids, MI 616-940-9944
TV-5 WNEM/John Haupricht
107 N Franklin, Saginaw, MI 517-755-8191
Universal Images/Etta Menlo
26011 Evergreen #200, Southfield, MI..............313-357-4160

PRODUCTION SERVICES

video

Urban Communications/Kevin Hovey
14770 Martin Dr, Eden Prairie, MN 612-934-1100
Verba Communications Network
2001 S Hawley Rd, St Louis, MO 314-781-5252
VIDCAM Inc/Craig C Smith
10683 S Saginaw #E, Grand Blanc, MI.............313-694-0996
Video Arts/Art Phillips
1440 4th Ave N, Fargo, ND.........................701-232-3393
Video Features/Don Adamson
680 Northland Blvd Bldg C, Cincinnati, OH 513-742-6262
Video I-D Inc/Sam B Wagner
105 Muller Rd, Washington, IL......................309-444-4323
Video Impressions/Rick Roesing
1666 N Farnsworth, Aurora, IL.......................708-851-1663
Videographics Corp/Byr Shaw
441 E Erie #1804, Chicago, IL......................312-642-6652
Vision Communications Inc
7441 Second Blvd/Marketing, Detroit, MI...........313-873-7200
VPL Productions/Vincent Lucarelli
415 Assembly Dr, Bolingbrook, IL.................708-739-6439
Wadlund & Associates Inc/Jack Wadlund
2950 Howard Ct, St Paul, MN 612-770-1983
WFRV-TV
PO Box 19055, Green Bay, WI 414-437-5411
WKBN
3930 Sunset Blvd, Youngstown, OH 216-782-1144
Worthwhile Films
333 W Mifflin, Madison, WI 608-251-8855

WEST

10th St Video Productions/Jeff Bishop
140 N 16th St/PO Box 2, Boise, ID.................208-336-5222
3D Magic/FTI/Dave Altenau
55 Hawthorne St 10th Fl, San Francisco, CA 415-543-2517
4Media Company/Paul Sehenuk
2813 W Alameda Blvd, Burbank, CA.................818-840-7000
525 Post Production/Jerry Cancellieri
6424 Santa Monica Blvd, Hollywood, CA....213-466-3348
Acmetek/Ken Lau
1145 Hacienda Blvd #126,
Hacienda Heights, CA............................818-333-6161
Action Video/Alana Ireland
6616 Lexington Ave, Hollywood, CA...............213-461-3611
American Production Services/Ted Hill
2247 15th Ave W, Seattle, WA....................206-282-1776
Anderson Video/Frank Bluestein
100 Universal City Pl, Universal City, CA 818-777-7999
Angel Studios/Jill Hunt
5962 La Place Ct, Carlsbad, CA 619-929-0700
Annenberg Ctr at Eisenhower/TV Svcs/Karen Snyder
39000 Bob Hope Dr, Rancho Mirage, CA 619-773-4554
Applied Technical Video Inc/Dennis Kushner
1564 Commerical St SE, Salem, OR 503-371-9498
April Greiman Inc/Lorna Turner
620 Moulton Ave #211, Los Angeles, CA 213-227-1222
Armour Productions/Stephanie Tucker
306-C W Katella, Orange, CA.....................714-538-5811
Art F/X/Dawn Jones
3575 W Cahuenga Blvd, Los Angeles, CA.......213-876-9469
Artichoke Productions/Paul Kalbach
4114 Linden St, Oakland, CA.......................510-655-1283
Astavision Post Prod/Scott Kinnimon
910 N Citrus Ave, Hollywood, CA 213-465-3333
Avid Productions/Henry Bilbao
1300 Industrial Rd #1, San Carlos, CA..........415-593-2000

Bay Area Video Coalition/Luke Hones
1111 17th St, San Francisco, CA 415-861-3282
Big Zig Video/Elliot Porter
329 Bryant St #1B, San Francisco, CA............415-243-8880
Boyington Film Productions/Joel Newman
5875 Blackwelder, Culver City, CA 310-204-2200
Breene Kerr Production/David Richardson
1235 Pear Ave, Mt View, CA 415-965-3131
Bruce Hayes Productions/Bruce Hayes
959 Wisconsin St, San Francisco, CA..............415-282-2244
Button Interactive/Don Button
624 Parkman Ave, Los Angeles, CA..................213-483-6062
Calico/Jan Nagle
8843 Shirley Ave, Northridge, CA 818-701-5862
California Communications Inc/Hope Schenk
6900 Santa Monica Blvd, Los Angeles, CA ...213-466-8511
California Image Associates/Duane Thompson
3034 Gold Canal Dr #B, Rancho Cordova, CA...916-638-8383
California Video Center/Marian Stevens
5432 W 102nd St, Los Angeles, CA 310-216-5400
Catalyst Productions/Chris Negri
1431 Center St, Oakland, CA.......................510-836-1111
Central Image Video & Photography/Larry Gale
337 Majors St, Santa Cruz, CA......................408-454-0500
Character Shop, The
9033 Owensmouth Ave, Canoga Park, CA.........818-718-0094
Chesney Communications/Gay Kroll
2302 Martin St #125, Irvine, CA.....................714-263-5500
Cinema Network (Cinenet)/Jim Jarrard
2235 First St #111, Simi Valley, CA.................805-527-0093
Cinema Research Corp/Mary Stuart
6860 Lexington Ave, Hollywood, CA...............213-460-4111
CIS-Hollywood/Godfrey Pye
1144 N Las Palmas, Hollywood, CA...............213-463-8811
Commotion/Bob Robbins
3131 Cahuenga Blvd W, Los Angeles, CA.......213-882-6637
Commty TV of S Calif (KCET)/Stephen Kuczycki
4401 Sunset Blvd, Los Angeles, CA...............213-953-5269
Communications Concepts Co/Marijane Lynch
PO Box 3018, San Rafael, CA......................415-456-6495
Communications Video/Arthur Marchetti
PO Box 2125, Walnut Creek, CA 510-930-7788
Complete Post Inc/Neal Rydall; Bob Brian
6087 Sunset Blvd, Hollywood, CA..................213-467-1244
Computer Graphics Dept/University of Oregon
Deschutes Rm 120, Eugene, OR 503-346-4408
**Continental Productions, Continental TV
Network**/Cheryl Cordeiro
118 6th St S/PO Box 1219, Great Falls, MT 406-761-5536
Creative Alliance/David Ginsberg
6523-1/2 Leland Way, Los Angeles, CA...........213-463-6647
Creative Media Development/Doug Crane
710 SW Ninth Ave, Portland, OR....................503-223-6794
Creative Vision/Karen Chemiack
8745 E Orchard Rd #500,
Greenwood Village, CO............................303-688-5865
Crosspoint Productions/Peter Ten Eyck
940 Wadsworth Blvd, Lakewood, CO.............303-232-9572
Crowley & Associates/Paul Crowley
224 Mississippi St, San Francisco, CA 415-621-4477
CST Entertainment Imaging Inc/William A Schaeffer
5901 Green Valley Circle #400, Culver City, CA....310-417-3444
Digital Post & Graphics/Terri Williams
1921 Minor Ave, Seattle, WA 206-623-3444
Diversified Video Ind Inc/Sierra Video/Jerry Hennan
4216 N Maxxon, El Monte, CA 818-579-7023
Dollar & Associates/John Dollar
450 S 41st St, Boulder, CO........................303-494-8626
Dourmashkin Productions/Barbara Dourmashkin
3852 Camino de Solana, Sherman Oaks, CA 818-995-3997
Dub Masters/SP/Caroline Didiego
3923 S McCarran, Reno, NV 702-827-3821
EMA Video Productions/Ed Mellnik
3210 SW Dosch Rd, Portland, OR...................503-241-8663
Encore Video/Larry Chernoff
6344 Fountain Ave, Hollywood, CA 213-466-7663
Energy Productions' Timescape/Sue Kincade
12700 Ventura Blvd 4th Fl, Studio City, CA........818-508-1444
FilmCore/Valerie Petrusson
849 N Seward St, Hollywood, CA 213-464-7303
Flying Colours/Monica Fabian
499 Seaport Court #205, Redwood City, CA.....415-363-8767
Frame-by-Frame Media Services/Rick Sutter
5353 Randall Pl, Fremont, CA......................510-651-6330

Full Circle Production/Marigold Fine
1009 Third St, Santa Cruz, CA 408-459-8300
FX Plus Design/Elan Soltes
3025 W Olympic Blvd, Santa Monica, CA 310-315-2175
G.R.A.P.E. Video Recording/Sharon Robinson
89 W Neal St, Pleasanton, CA.....................510-462-1300
Gannett Productions/Jim Berger
500 Speer Blvd, Denver, CO 303-871-1899
Gerrard Associates/Cathy & Lewis Gerrard
5452 Evanwood Ave, Agoura , CA...............818-706-3959
Ginsburg Productions/Michael Ginsburg
2243 S Olive St, Denver, CO 303-757-2564
Glendale Studios/Egan Elledge
1239 S Glendale ave, Glendale, CA.................818-502-5300
Glenn Roland Films/Glenn Roland
10711 Wellworth Ave, Los Angeles, CA............310-475-0937
Golden Gaters Productions/Anita Norine
400 Tamal Plaza, Corte Madera , CA 415-924-7500
Graphic Evidence/Peter Rothenberg
200 Corporate Pt #300, Culver City, CA 310-568-1831
Gray Matter Advertising/Robert Gray
2434 Monoco St, Channel Islands, CA............818-884-7641
Greiman Inc/April Greiman
620 Moulton Ave #211, Los Angeles, CA 213-227-1222
Group Video Productions/Jerry Casey
10364 Rockingham Dr, Sacramento, CA............916-362-3964
Half-Inch Video/Sean Scott
185 Berry St #6501, San Francisco, CA 415-495-3477
Hawaii Production Center/Nick Carter
1534 Kapiolani Blvd, Honolulu, HI 808-973-5462
Helical Productions/Bob Brandon
3801 E Florida Ave # 613, Denver, CO 303-756-0044
Hollywood South Producers Centre/Jerry Kelley
2465 Campus Dr #A, Irvine, CA....................714-757-7588
I/O Productions Inc/Laurent Basset
7551 Sunset Blvd #202, Hollywood, CA 213-969-0113
Illusion Factory, The/Brian Weiner
23875 Ventura Blvd #104, Calabasas, CA..........818-223-8400
Imagic/Paul Elmi
1545 N Wilcox Ave, Los Angeles, CA...............213-461-7766
Impact Corporate Communications/Fred Gerhart
4601 Telephone Rd #117, Ventura, CA.............805-644-3645
Inland Audio Visual Co/Larry Bergman
27 W Indiana, Spokane, WA 800-584-8916
Inter Video/Chris Olson
10623 Riverside Dr, Toluca Lake, CA................818-VIDEO-24
Intermountain Studio/Robert Bell
1240 E 800 N, Orem, UT 801-225-8400
Intersound Inc/Garry Morris
8746 Sunset Blvd, Los Angeles, CA 310-652-3741
Jim Muse Presentation Technology/Jim Muse
10525 Lawson River, Fountain Valley, CA.......714-965-5205
KCFW Productions/Curt Smith
401 First Ave E, Kalispell, MT.......................406-755-5239
KGTV Channel 10/Jack Villarrubia
PO Box 85347, San Diego, CA 619-237-1010
Kirwin Communications/Alison Childs
1760 Prospector Ave, Park City, UT................801-649-8226
KNTV Production Services/Bill Glenn
645 Park Ave, San Jose, CA 408-286-1111
KTVU Retail Services/Rich Hartwit
2 Jack London Sq/PO Box 22222, Oakland, CA ...510-874-0228
LA Videograms/Larry Rosen
3203 Overland Ave #6157, Los Angeles, CA 310-836-9224
Laser Pacific Media Corp/Leon Silverman
809 N Cahuenga Blvd, Los Angeles, CA............213-462-6266
LateNite/Ken Jenkins
4630 E Elwood St #1, Phoenix, AZ 602-966-0465
Mchale Videofilm/Pat Walsh
1116 Auahi St, Honolulu, HI 808-545-4040
Media Enterprises West/John Sutphen
1220 Rosecrans St #259, San Diego, CA 619-238-2388
Media Staff/Jerry Maybrook
6926 Melrose Ave, Hollywood, CA 213-933-9406
Metrolight Studios/Dobbie Schiff
5724 W Third St #400, Los Angeles, CA 213-932-0400
Miller Design & Associates/Paul Miller
8151 N Northern Ave, Tucson, AZ 602-797-1775
Miller, Jack Paul
370 W Cedar Ave, Burbank, CA 818-841-4668
Mira Film & Video/Dave Sullivan
116 N Page St, Portland, OR 503-464-0630
Modern Videofilm Inc/Richard E Greenberg
4411 W Olive Ave, Burbank, CA....................818-840-1700

PRODUCTION SERVICES

video

Mr Film/Chris Walker
228 Main St #10, Venice, CA 310-396-0146
NBC Magic
3000 W Alameda Ave, Burbank, CA 818-840-4863
Norac Productions /Ashley Bittman
2300 15th St #100, Denver, CO 303-455-8200
North Country Media Group/Doug Blier
721 Second St S, Great Falls, MT......................406-761-7877
Northwest Media
PO Box 56, Eugene, OR................................503-343-6636
Oasis Studios
675 Holcomb Ave, Reno, NV...........................702-688-6262
On Tape Productions/Steve Zeifman
724 Battery St, San Francisco, CA....................415-421-5551
Pace Video Center/Romeo Carrera
2020 SW 4th Ave #700, Portland, OR..............503-226-7223
Pacific Focus Inc
1013 Kawaiahao St, Honolulu, HI....................808-536-3848
Pacific Media/Chris Largent
2600 W Olive Ave # 100, Burbank, CA.............818-841-1199
Pacific Ocean Post/Alan Kozlowski
730 Arizona Ave, Santa Monica, CA 310-458-3300
Pacific Video Resources/Dana Levy-Wendt
2331 Third St, San Francisco, CA 415-864-5679
Panorama Productions
2353 De la Cruz Blvd, Santa Clara, CA.............408-727-7500
Penrose Productions/Mickey Bernard
1042 Hamilton Ct, Menlo Park, CA 415-323-8273
Perception Labs/Rick English
1162 Bryant St, San Francisco, CA....................415-255-0751
Pettigrew, Stuart
1420 45th St/Annex, Emeryville, CA.................510-652-7932
Pinnacle Effects/Anthony Green
2334 Elliott Ave, Seattle, WA..........................206-441-9878
Planet Blue/Dick Voss
1040 N Las Palmas, Los Angeles, CA 213-871-8280
Playhouse Pictures
1401 N La Brea Ave, Hollywood, CA.................213-851-2112
Praxis Film Works Inc/Michael Fekete
6918 Tujunga Ave, N Hollywood, CA 818-508-0402
Precision Visual Communications/Herb Saunders
591 25th Rd, Grand Junction, CO......................303-242-1783
Production Co, The/Bob Patrick
601 S Rancho #C-26, Las Vegas, NV 702-263-0933
Production Masters Inc/Harry Tate
834 N 7th Ave, Phoenix, AZ.............................602-254-1600
Production West Inc/Jim Abel
2110 Overland Ave #120, Billings , MT 406-656-9417
PSI Inc/Jose Villanueva
18019-A Skypark Cir, Irvine, CA 714-261-6119
Realtime Video/Will Hoover
60 Broadway, San Francisco, CA 415-705-0188
REZ-N-8 Productions/Michele Ferrone
6430 Sunset Blvd #1000, Hollywood, CA...........213-550-8885
Rhythm & Hues/Suzanne Datz
910 N Sycamore Ave, Hollywood, CA 213-851-6500
Rick Resnick Audio Visual Design/Rick Resnick
5544 E Kings Ave, Scottsdale, AZ 602-482-3070
Rock Solid Productions/David Griffin
801 S Main St, Burbank, CA.............................818-841-8220
Rocky Mountain AV Productions/Terry Talley
4301 S Federal Blvd #101, Englewood, CO 303-730-1100
Rouda, Saul
PO Box 892, Sausalito, CA 415-927-7285
Ruby-Spears Productions
710 W Victory Blvd #201, Burbank, CA.............818-840-1234
San Diego Digital Post/Infovideo
9853 Pacific Hts Blvd #B, San Diego, CA...........619-452-9000
San Francisco Production Group/Debra Robins
550 Bryant St, San Francisco, CA.....................415-495-5595
Sidley Wright & Associates
953 N Highland, Los Angeles, CA 213-465-9527
Sight & Sound Co
4445 Shepherd St, Oakland, CA.......................510-530-3949
Slide Factory/Carlos Lara
300 Broadway #14, San Francisco, CA..............415-957-1369
Sound Photosynthesis/Faustin Bray
PO Box 2111, Mill Valley, CA 415-383-6712
South Bay Film & Video/Bill Buxton
20432 S Santa Fe Ave #K, Long Beach, CA........310-763-7788
**Southwest Television Production
Services**/Scott Tuchman
4633 E Van Buren, Phoenix, AZ 602-244-2982
Staging Techniques/WWBS
1921 Wilcox Ave, Hollywood, CA......................213-874-5106

Star Video Services
1020 Auahi St Bldg 7-3A, Honolulu, HI..............808-591-2298
Starlite Studios/Judy Darnell
1530 Old Oakland Rd #140, San Jose, CA........408-453-1166
Steve Michelson Productions/Steve Michelson
280 Utah St, San Francisco, CA........................415-626-3080
Stimulus/Grant Johnson
PO Box 170519, San Francisco, CA 415-558-8339
STS Production/Paul Roden
935 W Bullion St, Murry, UT............................801-263-3959
Sunset Post Inc/Bob Glassenberg
1813 Victory Blvd, Glendale, CA 818-956-7912
Tamagni, Charles
12616 Lakeshore Dr #46, Lakeside, CA.............619-443-1987
TeleScene Inc/Rick Larson
3487 W 2100 S, Salt Lake City, UT 801-973-3140
Tell-A-Vision/Dennis Hoe
729 Seward St 2nd Fl, Hollywood, CA 213-962-0298
Thirty Second St/Mitchell Kenison
1209 Mountain Road Pl, Albuquerque, NM........505-265-0224
THS Visuals/Todd Simon
PO Box 2192 , Lake Tahoe/Reno, NV 702-588-6976
Topix LA Inc/Chris Mitchell
1800 Vine St 3rd Fl, Hollywood, CA 213-962-4745
Total Video Co/Peter Roger
432 N Canal St #12, S San Francisco, CA 415-583-8236
Triad Media Services/Kirk Storms
3055 Triad Dr, Livermore, CA..........................510-449-0606
TVA/Television Associates Inc/Lauri A Clark
2410 Charleston Rd, Mt View, CA 415-967-6040
Unitel Video/Hollywood/Mark Brown
3330 Cahuenga Blvd W, Los Angeles, CA..........213-878-5800
Univ of Calif/San Diego/Kim MacConnel
Vis Arts 0327/9500 Gilman Dr, La Jolla, CA 619-534-5784
Valley Production Center/Frank Bronell
6633 Van Nuys Blvd, Van Nuys, CA 818-988-6601
Vanguard Video Productions/Christopher Robinson
PO Box 250, Ashland, OR................................503-488-0201
Varitel Video/Colleen Casey
1 Union St #rd Fl, San Francisco, CA................415-495-3328
Vegas Valley Prods/KVVU Bdcst Corp/Richard Walsh
25 TV5 Dr, Henderson, NV 702-435-5555
Video Arts/Kim Salyer
185 Berry St #5400, San Francisco, CA 415-546-0331
Video Data Services of Montana/Matthew Scotten
PO Box 1206, Whitefish, MT 406-862-0875
Video One Inc/Kevin Hamburger
10625 Chandler Blvd, N Hollywood, CA............818-980-0704
Video Transform/Joan Merrill
2450 Embarcadero Way, Palo Alto, CA.............415-494-1529
Video West Productions/Jim Yorgason
55 N 300 W, Salt Lake City, UT 801-575-7400
Videoplex, The/Katherine Cersosimo
3700 Woodland Dr #100, Anchorage, AK........907-248-9999
View Studio/Patty Moore
6715 Melrose Ave, Hollywood, CA....................213-965-1270
Vigon, Jay
11833 Brookdale Ln, Studio City, CA 213-650-0505
Virtual Images/Jane Bradley
3487 W 2100 S, Salt Lake City, UT...................801-977-0888
Visions Plus/Steve Dung
2223 N Main St, Walnut Creek, CA 510-256-9450
VSE Corp/Linda Marsh
760 Paseo Camarillo, Camarillo, CA.................805-987-7811
WalkerVision Interarts/Pat Walker
PO Box 22533, San Diego, CA 619-458-9038
Weinman, Lynda Animation
318 S Sparks St, Burbank, CA 818-843-5056
Western Images
600 Townsend #300W, San Francisco, CA 415-252-6000
Wickerworks Video Productions/Steve Hug
6020 Greenwood Plaza Blvd, Englewood, CO....303-741-3400
William Stetz Design/William Stetz
1108 Seward St, Los Angeles, CA......................213-461-4267
Winters Productions/Glenn Winters
1855 Foothill Blvd, La Canada, CA 213-682-1205
Woodholly Productions/Yvonne Parker
712 N Seward St, Hollywood, CA......................213-462-5330
Yurth Video Production Services/Les Yurth
PO Box 1305, Woodland Hills, CA.....................818-999-0080
Z-AXIS/Ray Hauschel
116 Inverness Dr E #110, Englewood, CO..........303-792-2400

CANADA

**07-TV Commercial Production Co
Animation**/Maryann Japhe
162 Parliament St, Toronto, ON.......................416-364-3556
ABS Productions Ltd/Robert Sandoz
196 Joseph Zatzman Dr, Dartmouth, NS 902-468-4336
ARCCA
67 Lombard St, Toronto, ON............................416-366-1662
Artray Ltd/Karen Pinnell
PO Box 4700, Vancouver, BC 604-420-2288
Atelier Infographe
860 St Zallier East, Quebec, QB 418-692-5407
Atlantic Television System (CHCH-TV)/John Jay
2885 Robie St, PO Box 1653, Halifax, NS 902-852-2860
Bradley, Shane
2596 Chaplot, Montreal, QB 514-270-1177
Broadcast Productions Media Image/Brian E Purdy
77 Huntley St #2522, Toronto, ON 416-961-1776
**CAP Comms (CKCO-TV)/
Electrohome Ltd**/Reg Salner
864 King St W, Kitchener, ON 519-578-1313
Carleton Productions, Inc./Dianne Van Velthoven
1500 Merivale Rd, Ottawa, ON 613-224-1313
Centre de Montage Electronique-TM/Pierre Potvin
1600 Maisonneuve Blvd E , Montreal, QB..........514-598-2938
Command Post & Transfer
179 John St, Toronto, ON................................416-585-9995
Creative Visual Associates/Glenn Rosin
413 Quebec St, Regina, SK 306-584-5423
Dan Krech Productions Inc/Daniel J Krech
48 River St, Toronto, ON 416-861-9269
Digital Magic
58 Fraser Ave, Toronto, ON.............................416-534-6154
Dignum Computer Graphics/Alain Leduc
1905 William St, Montreal, QB.........................514-931-4221
Dome Productions/Jens Olsen
300 Bremner Boulevard #3400, Toronto, ON ..416-341-2001
Don Krech Productions
48 River St, Toronto, ON.................................416-861-9269
DSC Laboratories
3565 Nashua Dr, Mississauga, ON 416-673-3211
Gastown Post & Transfer/Joe Fischer
50 West 2nd Ave, Vancouver, BC......................604-872-7000
Greenlight Corporation/Garth Douglas
70 Richmond St E #302, Toronto, ON 416-366-5444
Groupe Andre Perry/Mario Rachiele
1501 Parre St, Montreal Quebec, QB 514-933-0813
Icon Computer Graphic
225 Second Ave W, Vancouver, BC...................604-876-7678
Image Group Canada Ltd/Pat Cosway
26 Soho St, Toronto, ON.................................416-591-1400
Image Technix/Jim Marshall
76 Fraser Ave, Toronto, ON 416-975-1327
Image Ware
2 Berkeley #400, Toronto, ON 416-367-9632
Inter-Tel Image Ltd/Jacques Marcotte
1310 Lariviere, Montreal, QB 514-522-8599
ITV Productions/Hans J Dys
5325 104 St, Edmonton, AB............................403-436-1250
Le Studio Du Centre-Ville, Inc/Jean Raymond Bourque
1168 Bishop, Montreal, QB..............................514-878-3456
Magnetic Fax/Shelley Deslauruers
550 Queen St #205, Toronto, ON 416-367-8477
Magnetic North/Ann Cuthbertson
70 Richmond St E #100, Toronto, ON 416-365-7622

PRODUCTION SERVICES
video/video independent

Mirak Video Communications Services/Roz Royce
Jeff Sq, 64 Jefferson Ave #20, Toronto, ON.........416-531-1432
Mobil Image
26 Soho St, Toronto, ON...........................416-591-6854
Northwest Communications
2339 Columbia St #100, Vancouver, BC...........604-873-9330
Oasis Post Production Services/Don McMillan
340 Gerarrd St E, Toronto , ON....................416-466-5870
Omni Media Productions Ltd/Peter Murray
235 Martindale Rd #6, St Catharines, ON.........416-684-9455
Pacific Post Productions
770 Pacific Blvd S, Vancouver, BC................604-685-8686
Perkins & Associates/Kenn Perkins
7-1313 Border St, Winnipeg, MB..................204-633-4545
PFA Film & Video/Ed Higginson
111 Peter St 9th Fl, Toronto, ON.................416-593-0556
River Image/Robert Marinic
52 River St, Toronto, ON.........................416-862-7488
Shane Productions Inc, Lunny
305-560 Beatty St, Vancouver, BC................604-669-0333
Side Effects/Sonja Tessari
620 King St W, Toronto, ON......................416-367-8333
Studio Post & Transfer/Colin Minor
5305 104th St, Edmonton, AB.....................403-436-4444
Supersuite/François Garcia
4 Westmount Sq #200, Montreal, QB..............514-933-1161
Tele-4
1000 ave Myrand, Ste-Foy, QB....................418-688-9330
Tetes Video
5297 Bordeau, Montreal, QB......................514-521-4468
Tier One Communications
140 Renfrew Dr #101, Markham, ON...............416-470-0444
TOPIX
217 Richmond St W 2nd Fl, Toronto, ON..........416-971-7711
Utopix Infographie
2120 rue Sherbrooke est #1105, Montreal, QB...514-522-1254

INTERNATIONAL

Mediamix Limited/Lenny Little-White
16 South Ave, Kingston 10, JA.....................809-926-0917
Videomax Productions Ltd/Craig Pang-Sang
15 Haining Rd, Kingston 5, JA.....................809-929-6437

Video Independent
NEW YORK CITY

Atomique Film International/Alyce Wittenstein
110-20 71st Rd PH5, Forest Hills, NY 718-520-0354
Bakst, Edward
160 W 96th St, New York, NY 212-666-2579
Birmingham, Barbara
133 Barrow St #4A, New York, NY 212-691-5587
Britt, Ron
24 E24th St 3rd Fl, New York, NY 212-420-0686
BXB Inc/Henry Baker
532 La Guardia Pl #200, New York, NY 212-924-8654
Coppola, Richard
50-16 213th St, Bayside, NY 718-428-3636
Corporate Communication Group/Steve Rothman
7 W 30th St 12th Fl, New York, NY 212-268-2100
Critical Image/Michael Berger
4 W 37th St 2nd Fl, New York, NY 212-594-2586
Cynosure Films/Robert Luttrell
131 W 21st St, New York, NY 212-645-8216
Deas, D L
355 W 85th St #21, New York, NY 212-799-5149

Epure, Serban
60-11 Broadway #5L, Woodside, NY................718-335-7685
Fleishman, Carole
16 Abingdon Sq #1B, New York, NY 212-645-9071
Garrin Prods, Paul
252 E 7th St #17, New York, NY 212-260-7158
Goodman/Orlick Design/Judi Orlick
150 5th Ave #407, New York, NY.................212-620-9142
Gorewitz, Shalom
310 W 85th St #7C, New York, NY 212-724-2075
Hoberman, Perry
167 N 9th St #10, Brooklyn, NY 718-388-8241
Hudson Media/Martin Bluford
17 W 20th St, New York, NY 212-645-4445
Hunter, Marian
137 E 13th St, New York, NY 212-254-6620
Interface Arts/Robert Lyons
Liberty Studios/238 E 26th St, New York, NY......212-532-1865
John Cipoletti Productions/John Cipoletti
111 Senator St , Brooklyn, NY 718-680-3089
Lane, Morris
212-A E 26th St, New York, NY 212-696-0498
Lazar, Jeff
155 E 23rd St #406, New York, NY 212-533-0010
Lord, Rosalind
175 W 12th St #16N, New York, NY 212-807-7959
Mattlin, Brian
254 W 25th , New York, NY.....................212-627-4963
Maya Technology/Mark Gilliland
156 Chambers St, New York, NY 212-267-1195
Rogow, David
249 Eldridge St #23, New York, NY 212-254-0511
Rosebush Visions Corp/Judson Rosebush
154 W 57th St #826/New York, NY 212-398-6600
Schmidt, Mechthild
PO Box 453 Radio City Station, New York, NY ...212-245-5254
Schwartz, Roberta
22 Cheever Pl, Brooklyn, NY 718-624-4408
Severtson, Jeff
221 E 12th St #2, New York, NY 212-473-8086
Speer, Stephan
6105 80th St, Middle Village, NY................718-457-7641
Teletechniques/Michael Temmer
1 W 19th St, New York, NY 212-206-1475
Terra Incognita Prods/Dan Weissman
147 W 15th St #106, New York, NY 212-929-8277
Third World TV Exchange/
Nicola Katharina Wewer-Elejalde
30-33 32nd St #4H, Astoria, NY 718-626-6179
Vision Associates/Dita Domonkos
650 First Ave, New York, NY 212-481-7455

EAST

Artemis Electronic Imaging Inc/George Otto
2285 Bristol Ave, State College, PA..............814-234-3165
Domin, Jacqueline
26 Monroe St, Honeoye Falls, NY 716-624-3318
Doudoroff, Martin
504 Beacon St #2, Boston, MA 617-247-6145
Eagle, Steve
PO Box 1398, Cambridge, MA 617-864-5924
Fine Arts Video/Justin P West
203 Sylvester Rd, N Hampton, MA...............413-586-0730
Independent Media Corp
2655 Philmont Ave #203, Huntingdon Valley, PA 215-938-7710
Jackson, Crystal
14 Mary Ave, Union, NJ 908-964-5755
Laskey, Christopher
630 Weygadt Dr, Easton, PA....................215-252-6804
Leibowitz, David Scott
333 Vandelinda Ave, Teaneck, NJ 201-836-6712
Leonard Still Video/Barney Leonard
134 Cherry Ln, Wynnewood, PA 610-649-5588
Mistretta Illustration Studio/Andrea Mistretta
135 E Prospect St, Waldwick, NJ 201-652-7531
Mobius Artists Group/Nancy Adams
354 Congress St, Boston, MA 617-542-7416
Moser/Media, Michael
2000 P St NW #500, Washington, DC............202-293-1780
NY Inst of Technology/Peter Voci
School of Arch & Fine Arts , Old Westbury, NY...516-876-2752
Otitis Media Productions/Mary Scott
30 Thorne Ln, Port Jefferson, NY...............516-928-9645
Pocock-Williams, Lyn
37 Huemmer Terr, Clifton, NJ..................201-614-0314

Posch, Michael
716 Grand St, Hoboken, NJ 201-792-7003
Scion Corporation/Bonnie Polesky
152 W Patrick St, Frederick, MD 301-695-7870
Sloan, Richard C
242 Thunder Rd, Holbrook, NY 516-472-3898
Smalley, David
190 Old Colchester Rd, Quaker Hill, CT..........203-444-1150
Stroukoff, Ann
126-I The Orchard, Cranbury, NJ 609-448-5394
Taya, Hisao
720 Monroe St , Hoboken, NJ 201-659-2124
Wuilleumier Inc, Will
607 Boylston St, Boston, MA...................617-266-0103

SOUTH

Anello Technical Services/A J Anello
1915 W Waters Ave #1, Tampa, FL.............813-933-6009
Appalshop Inc/Dee Davis
306 Madison St, Whitesburg, KY 606-633-0108
Banchero, J
6506 Telegraph Rd, Alexandria, VA.............703-719-5654
Drabick, Matt
5909 N Hills Dr, Raleigh, NC 919-781-0158
Eales, Ray
PO Box 24691, Tampa, FL.....................813-237-0248
Ensemble Productions/Barry McConatha
1036 Cornstalk Ln, Auburn, AL.................205-826-3045
Hoit, Wayne
8430 Shady Glen Dr, Orlando, FL..............407-345-5273
JIMANDI/James Spitler
734 Vosswood Dr, Nashville, TN 615-256-9887
Media Projects Inc/Allen Mondell
5215 Homer St, Dallas, TX 214-826-3863
Simerman, Tony
Castle Harbor Way, Centreville, VA.............703-802-4950
Synthesis/Samuel Reynolds
PO Box 339, Conway, AR 501-327-2517
Young, Andrea
8430 Shady Glen Dr, Orlando, FL..............407-345-5273

MIDWEST

911 Gallery/Walter Wright & Mary A Kearns
911 E Main St, Indianapolis, IN 317-257-8350
● **BELDING, PAM (P 77)**
1601 Brightwood Dr, Minneapolis, MN 612-476-1338
BGN Unltd/Bryan Gerard Nelson
2420 31st Ave S, Minneapolis, MN 612-724-3430
Blue Sky Productions/Peter Hickman
1020 S Creyts Rd #305, Lansing, MI 517-886-2010
Channel 3 Productions/Michael Cooper
833 North Main, Wichita, KS 316-265-3333
Comware Inc/Wendy O'Neal
4225 Malsbary Rd, Cincinnati, OH 513-791-4224
Continental Cablevision/Madison Hts/Jennifer Rossow
32090 John R, Madison Hts, MI 313-583-1354
Eclipse/Michael Spaw
8117 Rosewood, Prairie Village, KS 913-864-2515
Goldsholl: Film Group/Deborah Goldsholl
420 Frontage Rd, Northfield, IL 708-446-8300
Hellman Animates/Jim Mathis
1225 W 4th St, Waterloo, IA 319-234-7055
J Parker Ashley Associates/Rick Benjamin
119 N Fourth St #208, Minneapolis, MN 612-340-9807
Jon Garon Creative Inc/Jon R Garon
3159 Lafayette Ridge Rd,
Minnetonka Beach, MN......................612-474-7000
LaVoie's Photography/Rex LaVoie
01423 US 127 S, Bryan, OH 419-636-4602
Lee DeForest Communications Ltd/Sherri Shallenberg
300 W Lake St, Elmhurst, IL 708-834-7855
Long Run Productions Ltd/Thomas Ingledew
219 Second St N, Minneapolis, MN............612-673-9729
Mallory Marketing Communications/Bob Mallory
1310 E Hwy 96, St Paul, MN 612-426-1875
Northwest Teleproductions/Edina/Bob Haak
4455 W 77th St, Edina, MN 612-835-4455
PC Pix/Peter Jacobs
596 Provident Ave, Winnetka, IL 708-446-8319
Prestige Production & Graphic/John Basso Jr
148 E Aurora Rd (Rte 82), Northfield, OH........216-467-8400
Shoulder-High Eye Productions/Ron Shook
4057 N Damen, Chicago, IL 312-883-1693
Siemer, Patrick
1809 W Division St, Chicago, IL 312-862-4244

Steve Blexrud Studio/Steve Blexrud
3722 22nd Ave S, Minneapolis, MN..............612-721-5325
Video Assist/Mark Adler
24265 Jamestown Dr, Novi, MI 313-349-2666

W E S T

Baird Productions/Doug Baird
578 Vermont St, San Francisco, CA 415-626-9494
Beckman Instruments Inc/Teri Beauchamp
2500 Harbor Blvd/MS B22D, Fullerton, CA 714-773-7682
Blume, Neil
10235 NE Prescott #16, Portland, OR 503-257-4560
C Design/Pat Coleman
915 Oak Ln #4, Menlo Park, CA...................415-321-2094
Call, Ed
757 Armada Terr, San Diego, CA 619-225-8569
Cheap Computer Graphics/Bruce Goren
28938 Morningside Dr, Val Verde, CA.............805-295-8487
Dimension Teleproductions/John Lunning
17602 N Black Canyon Hwy , Phoenix, AZ 602-866-0162
Direct Images/Bill Knowland
PO Box 29392, Oakland, CA.....................510-614-9783
Electric Zebra/Parker Solo/Gina Di Bari
1097 Woodland Ave, Menlo Park, CA 415-327-6574
EXP Graphics/Bruce Bennett
432 N Canal St #12, S San Francisco, CA 415-583-8236
Flom, Eric
829 Lincoln Ave, Alameda , CA 510-769-9391
Full Circle Production/Marigold Fine
1009 Third St, Santa Cruz, CA 408-459-8300
G-Nine Productions/Bill Groshelle
2030 Scott St, San Francisco, CA 415-922-0531
Goss, John C
1022 Keniston, Los Angeles, CA 213-938-9526
Graphic Design Group/Richard Browski
453 N Genesee Ave, Los Angeles, CA..............213-653-2454
Hanna Productions/Grant Hanna
PO Box 22938, Denver, CO 303-755-6686
House Film Design/James House
7033 W Sunset Blvd #301, Los Angeles, CA.......213-467-3429
How Studio/Howard Lim
1002 14th St #8, Santa Monica, CA 310-394-3005
InterVision/Will Doolittle
401 E Tenth Ave #160, Eugene, OR 503-343-2278
JAG Broadcast Video/John Gay
PO Box 12442, Reno, NV 702-688-6277
Johnson, Jay
2260 N Beachwood Dr, Hollywood, CA.............213-464-2606
Keller, Steve
11936 W Jefferson Blvd #C, Culver City, CA 310-390-8663
Laser Media Inc/Harvey Plotnick
6383 Arizona Cir, Los Angeles, CA 310-338-9200
MacNicol, Gregory
732 Chestnut St, Santa Cruz, CA 408-459-0880
Marcus & Assocs, Aaron/Aaron Marcus
1144 65th St #F, Emeryville, CA 510-601-0994
Mass Productions/Howard Petrick
1510 Guerrero St, San Francisco, CA 415-648-3789
McMillan Co Public Relations/Bill McMillan
1651 E Edinger Ave #106, Santa Ana, CA........714-953-7370
Monaci, Steven
1118 Abbot Ave, San Gabriel, CA 818-282-6376
Moojedi, Kamran
900 W Sierra Madre #122, Azusa, CA 818-969-5508
On-Q Productions/Vincent Quaranta
618 E Gutierrez St, Santa Barbara, CA...........805-963-1331
Palermo, David
3379 Calle Santiago, Carlsbad, CA..............619-944-9907
Phosphor Inc/Michael A Eddy
5951 Canterbury Dr Unit 7, Culver City, CA.......310-645-0122
Priestly, Joanna
1801 NW Upshur, Portland, OR 503-274-2158
PRINTZ/Charles Wyke-Smith
340 Townsend St, San Francisco, CA 415-543-5673
Production West Inc/Jim Abel
2110 Overland Ave #120, Billings , MT 406-656-9417
Raphael Film/Alexis Krasilovsky
1607 Lanta St, Los Angeles, CA 213-662-5746
Relf, Geoff Graphics Group
3511 Camino del Rio S, San Diego, CA 619-280-0922
Stanford University/Ed McGuigan
Meyer Library/Media Svcs, Palo Alto, CA 415-725-1176
stat™media/Gary W Birch
7077 E Shorecrest Dr, Anaheim Hills, CA..........714-280-0038

Studio M/Mike Michaels
8715 Waikiki Station, Honolulu, HI 800-453-3345
Taylor, Dawn Marie
AZ..602-973-1415
Total Vision Inc/Paul Rother
3015 Main St, Santa Monica, CA 310-450-1315
Unkeless, Charlie
5554 Village Green, Los Angeles, CA.............310-399-4072
Vanguard Video Productions/Christopher Robinson
PO Box 250, Ashland, OR......................503-488-0201
Video Pro of Wyoming/Eric Hamm
402 Broadway, Rock Springs, WY 307-382-8776
Video Professionals/Marjorie Franklin
1629 Telegraph Ave, Oakland, CA...............510-268-1266
View Studio/Patty Moore
6715 Melrose Ave, Hollywood, CA...............213-965-1270
Winchester, Shawn
422 E Huber St, Mesa, AZ 602-461-8424
Wise, Jennifer
4424 Murietta Ave #18, Sherman Oaks, CA.......818-981-4151
Worrell, Rand
1515 N Hayworth Ave, Los Angeles, CA 310-524-2521

C A N A D A

Bourgault, Jean
710 rue Bouvier #125, Quebec, QB.................418-628-4006
Struve-Dencher, Goesta
#303 750 E 7th Ave, Voncouver, BC.................604-872-3439

Video Animation

N E W   Y O R K   C I T Y

All American TV/Louise Perillo
1325 Ave of Americas, New York, NY 212-541-2800
Bernstein, Linda A
10-11 50th Ave, Long Island City, NY 718-784-1599
Cacioppo Production Design/Tony Cacioppo
42 E 23rd St 5th Fl , New York, NY 212-777-1828
Capital Vectors/Lydia Rappold
826 Broadway 6th Fl, New York, NY 212-529-9696
Celefex/Dean DeCarlo
16 W 22nd St 4th Fl, New York, NY 212-255-3470
Charlex Inc/Henry Frenzel
2 W 45th St 7th Fl, New York, NY 212-719-4600
Communications Plus/Geoffrey Fraize
102 Madison Ave, New York, NY 212-686-9570
Concrete Couple Productions/John Fekner
31-14 80th St, Jackson Heights, NY 718-651-3388
Creative Ways Inc/Stuart Sternbach
305 E 46th St, New York, NY....................212-935-0145
Data Motion Arts/Tony Caio
165 W 46th St #611, New York, NY 212-768-7411
Diamant Noir/Philippe Louis Houze
519 Broadway, New York, NY 212-226-9299
Doros Motion/Doros Evangeledes
40 W 21st St #1103, New York, NY 212-243-5520
Drucker Motion Pictures/Frank Drucker
166 E 34th St #5A, New York, NY 212-679-3569
EDEFX/Aran Friedman
219 E 44th St 9th Fl, New York, NY 212-983-2686
Empire Video
216 E 45th St 11th Fl, New York, NY 212-687-2060
Fitz, Tracy
531-A Sixth Ave, Brooklyn, NY 718-768-8161
Gadsden-Grant Graphic/Ingrid Williamson
676 E 57th St, Brooklyn, NY....................718-251-0577
ID Studios/Isauro De la Rosa
2 Jane St #4-A, New York, NY 212-243-5175
Ink Tank/B O'Connell
2 W 47th St 14th Fl, New York, NY 212-869-1630
Jiempreecha, Wichar
32-86-34th St #3A, Long Island City, NY...........718-721-6956
Klein, Josh
235 W 71st St #3, New York, NY 212-721-6294
Laser Edit East Inc/Gary Sharfin
304 E 45th St 2nd Fl, New York, NY 212-983-3255
Lieberman Productions, Jerry
76 Laight St, New York, NY.....................212-431-3452
Manhattan Transfer/Edit/Joanne Gross
545 Fifth Ave, New York, NY 212-687-4000
Marsden Reproductions/Steve Flores
30 E 33rd St, New York, NY 212-725-9220
Musivision/Fred Kessler
185 E 85th St, New York, NY 212-860-4420

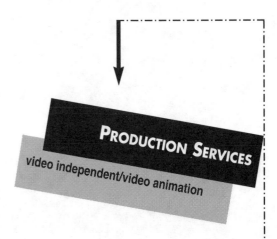

PRODUCTION SERVICES
video independent/video animation

NBC Telesales/Elizabeth Davis
30 Rockefeller Plaza #601W, New York, NY......212-664-4444
Nexus Productions/Margaret Smagalski
10 E 40th St, New York, NY.....................212-679-2180
Osama Ltd/Osama Hashem
50 Lexington Ave #105, New York, NY.............212-779-1923
Otterson Television Video Inc/Bill Otterson
251 W 30th St #14W, New York, NY 212-695-7417
Portfolio Artists Network/Marcus Nispel
601 W 26th St, New York, NY 212-633-6030
Post Perfect Inc/Dean Winkler
220 E 42nd St, New York, NY 212-972-3400
Premier Post/Keith Shapiro
630 Ninth Ave 14th Fl, New York, NY 212-757-1711
Princzko Productions
9 E 38th St 3rd Fl, New York, NY 212-683-1300
Rutt Video Inc/Krystin Wagenberg
137 E 25th St, New York, NY...................212-685-4000
Spano Design/Michael Spano
35 Regina Ln, Staten Island, NY 718-966-8979
Speer, Stephan
6105 80th St, Middle Village, NY................718-457-7641
Stark Studio
231 W 29th St #1005, New York, NY.............212-868-5555
Tapestry Productions/Nancy Walzog
920 Broadway, New York, NY 212-677-6007
Teletechniques/Michael Temmer
1 W 19th St, New York, NY 212-206-1475
Telezign/Rob Wyatt
460 W 42nd St, New York, NY 212-564-8888
Today Video/Beverly Seeger
45 W 45th St 12th Fl, New York, NY 212-391-1020
Trans-Ocean Video Inc/Stephen, Marketing Dept
711 12th Av/Pass Ship Term 88, New York, NY .212-757-2707
Video Works/Jim Riche
24 W 40th St, New York, NY 212-869-2500
Videograf/Michael Frenchman
144 W 27th St, New York, NY 212-206-8477
VSC/Shelley Riss
225 E 43rd St, New York, NY 212-599-1616
Wander Communications/Andy Wander
534 E 84th St #5W, New York, NY 212-737-3058
Windsor Digital Graphics
8 W 38th St, New York, NY 212-944-9090
Zen Over Zero/Helen Meschkow
249 E 48th St #15D, New York, NY 212-644-6364

E A S T

Advanced Video Services/Marty Pisano
1120 Bloomfield Ave #111, W Caldwell, NJ 201-882-6440
Animated Arts/Jeffrey Radden
265 Central Ave, Albany, NY 518-465-6448
Animated Design/Visual Communication
36 Smith Rd, Warwick, MA 508-544-6262
Artemis Electronic Imaging Inc/George Otto
2285 Bristol Ave, State College, PA..............814-234-3165
Atlantic Video/Chris Cates
650 Massachusetts Ave NW, Washington, DC....202-408-0900
AV Design/Joe Wall
1823 Silas Deane Hwy, Rocky Hill, CT 203-529-2581
AVT/Edward Mctighe
26 W Highland Ave, Atlantic Highlands, NJ........908-872-9090
Cambridge Television Productions/Wilson Chao
67 Chapel St, Newton, MA 617-332-0084
CCI COmmunications Inc/Virginia Frederick
1440 Phoenixville Pike, W Chester, PA 215-296-7233

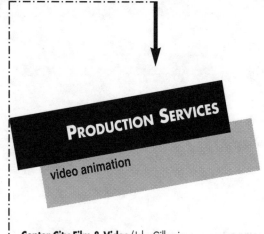

PRODUCTION SERVICES

video animation

Center City Film & Video/John Gillespie
1503-05 Walnut St, Philadelphia, PA 215-568-4134
Computer Creations/James Houghtaling
14 Balsam Cir, Whitesboro, NY 315-768-8471
Computer Generated Imagery Inc/Rick Bowley
6270 Dean Pkwy, Ontario, NY 716-265-1450
Cornell Theory Ctr/Cornell Univ/Wayne Lytle
621 Theory Ctr Bldg, Ithaca, NY 607-254-8793
Creative Video Design & Prod/Marie Leuchte
480 Neponset St, Canton, MA 617-821-1110
DI Group/Tricia Cadena
651 Beacon St, Boston, MA 617-267-6400
Flood, David Williams
43 Booraem Ave, Jersey City, NJ...................201-420-7475
GBH Productions/Jayne Pikor
125 Western Ave, Boston, MA 617-492-9273
Half Moon Video Productions/Tom Horan
130 Central Ave, Jersey City, NJ...................201-792-1066
Image Works/Dana Hutchins
537 Congress St #301, Portland, ME 207-774-6399
In-House Video Editing/Tom Pernice
7 Walnut Tree Hill Rd, Shelton, CT..................203-929-0375
Intelligent Light
1099 Wall St W #387, Lyndhurst, NJ................201-460-4700
Inter-Media Art Center/Michael Rothbard
370 New York Ave, Huntington, NY 516-549-9666
Interface Video Systems/Ken Maruyama
1233 20th St NW, Washington, DC 202-861-0500
Jackson, Crystal
14 Mary Ave, Union, NJ 908-964-5755
Jed Schwartz Productions/Jed Schwartz
RFD #2/Purlingbeck Rd, E Washington, NH 603-495-3125
Kleiser-Walczak Construction East/Jeff Kleiser
Riverview Rd, Lenox, MA 413-637-8944
Larsen & Ludwig/Matt Sell
2 Gateway Ctr 14th Fl, Pittsburgh, PA 412-338-0700
Lavin, Arnie
23 Glenlawn Ave, Seacliff, NY 516-676-1228
MB Productions/Ellen Brooks
17 Dorine Park, E Hanover, NJ 201-386-0044
MBTV Inc/Maureen Baeck
3006 Delaware River Rd, Frenchtown, NJ 908-996-6718
Media Mix/Joe Vargas
315 Rte 17 South, Paramus, NJ 201-262-3700
Megcomm Film & Video Productions
226 Carey Ave, Wilkes-Barre, PA...................717-826-9805
Melovision Productions Inc/Mel Obst
190 Cold Soil Rd, Princeton, NJ 609-895-1030
Mobile-Video Productions Inc/Stephen King
7315 Wisconsin Ave #325W, Bethesda, MD......301-656-2525
Modern World Media/Marketing Dept
6152 Encounter Row, Columbia, MD 410-944-4242
Motionart Studios/Pell Osborn
27 Common St, Boston, MA 617-242-1222
Nagy Films/Patrick Rafferty
9210 Corporate Blvd, Rockville, MD.................301-258-1112
Natl Boston Video Ctr/Mark Jacques
115 Dummer St, Brookline, MA 617-734-4800
NFL Films Video/Hal Lipman
330 Fellowship Rd, Mt Laurel, NJ 609-778-1600
Otitis Media Productions/Mary Scott
30 Thorne Ln, Port Jefferson, NY 516-928-9645
Our Town TV/Broadcast News Network
78 Church St, Saratoga, NY 518-899-6989
Ovation Graphics/Kevin Dady
2 Graeme St, Pittsburgh, PA........................412-471-3350

PCI Recording/Caroline Maynard
737 Atlantic Ave, Rochester, NY 716-288-5620
Penfield Productions Ltd
35 Springfield St, Agawam, MA....................413-786-4454
Producers Video Corp
3700 Malden Ave, Baltimore, MD 410-523-7520
Production House
327 N 17th St, Philadelphia, PA 215-972-0934
Production Masters/Mike Killes
321 First Ave, Pittsburgh, PA 412-281-8500
Production Works/Mark Phillips
360 East Ave, Rochester , NY 716-232-3709
Quality Film & Video/Peter A Garey
232 Cockeysville Rd, Hunt Valley, MD 410-785-1920
Rampion Visual Productions/Steven Tringali
316 Stuart St, Boston, MA 617-574-9601
Scabrini Design Inc, Janet
90 Maywood Rd, Norwalk, CT 203-853-6676
Smith, Mark
319 W Northfield Rd, Livingston, NJ................201-992-4213
Spicer Productions Inc/Sharon Jackson
1708 Whitehead Rd, Baltimore, MD 410-298-1200
Steward Digital Video/Dave Bowers
525 Mildred Ave, Primos, PA 215-626-6500
Studiolink/Andy Cordery
13 Roszel Rd, Princeton, NJ 609-452-0846
TDS Inc/Timothy Duffield
1551 Johnny's Way, West Chester, PA..............215-430-8557
Tech Graphics/Jim Sullivan
215 Salem St #3, Woburn, MA 617-933-6988
Telesis Productions Inc/David Rose
277 Alexander St #600, Rochester, NY 716-546-5417
Teletime Video Productions/Warren Manos
37-39 Watermill Ln, Great Neck, NY................516-466-3882
TRACES
10 N Presidential Blvd #115, Bala Cynwyd, PA...215-660-9699
Vidconn Productions Inc/Joe Seaton
360 Towlin Tpke #2E, Manchester, CT.............203-646-0660
Video Resource/Curtis Sink
140 Centennial Ave, Piscataway, NJ 908-457-7400
Video Services Unlimited/Paul Spencer
124 Canal St, Lewiston, ME.........................207-782-5650
Videocenter of New Jersey/Bob Schaffner
228 Park Ave, E Rutherford, NJ 201-935-0900
Videotroupe/Susanne Condon
3 Industrial Dr/POB 67, Windham, NH 603-893-4554
Viewpoint Computer Animation/Carlo Di Persio
13 Highland Circle, Needham, MA..................617-449-5858
Visual Logic/Jack Harris
724 Yorklyn Rd #150, Hockessin, DE 302-234-5707
Wave Inc/Walter Henritze
11 California Ave, Framingham, MA 508-795-7100
Xerox Media Center/John Linn
780 Salt Rd/Bldg 845, Webster, NY.................716-423-5090

SOUTH

American Media Productions/John Creech
PO Box 1722, Morehead City, NC 919-247-5914
AMS Productions Inc/Suzanne Miller
13709 Gamma Rd, Dallas, TX 214-701-0878
Ani-Majic Productions Inc/Alan Davidson
2699 Lee Rd #415, Winter Park, FL.................407-629-5757
Applied Video Communications/Mike Looper
719 Sawdust Rd, The Woodlands, TX..............713-364-1203
Arcimage/Richard Buday
4200 Montrose Blvd #330, Houston, TX 713-523-3425
Art in Media Productions/Rick Sturnberg
7212 McNeil Dr #206, Austin, TX...................512-250-5535
ASV Corp/Barbara Fishburn
2075 Liddell Dr NE, Atlanta, GA....................404-876-3445
Automated Broadcast Consultants Inc
5401 Collins Ave Mezz, Miami Beach, FL 305-864-4800
Baldridge Studios Inc/Donna Baldridge
347 N Main St, Memphis, TN.......................901-522-1185
Banjo Video & Film Productions/Toby Jenkins
1188 St Andrews Rd, Columbia, SC 803-739-0336
Beachwood Productions/Peter Wilcox
696 S Yonge St #C, Ormond Beach, FL 904-672-3698
BES Teleproductions/Ed Lazor
6829-E Atmore Rd, Richmond, VA.................804-276-5110
Caluger & Assocs/J Wayne Caluger
237 French Landing, Nashville, TN..................615-255-2792
Cannata Communications Corp/Jack Cannata
7031 Grand Blvd, Houston, TX 713-748-1684

Cat Walk/Richard Aldridge
418 W 5th St, Charlotte, NC 704-342-3348
Century III at Universal Studios/Pam Lapp
2000 Universal Studios Plz, Orlando, FL.............407-354-1000
Cinemasound Video Productions/Carla Thomas
1011 Arlington Blvd #320, Arlington, VA 703-524-1083
Cinetel Productions/Peter Franks
9701 Madison Ave, Knoxville, TN...................615-690-9950
Cinetron Computer Systems
4350 J Int'l Blvd, Norcross, GA 404-925-4448
Colordynamics/Ed Leach
150 E Bethany Rd, Allen, TX 214-390-6500
Computed Animation Technology/Trent DiGiulio
2200 N Lamar #301, Dallas, TX 214-871-5055
Computer Studio, The/George Bowers
4000 Cumberland Pkwy 1400 #A1, Atlanta, GA 404-436-6092
Corporate Graphics Inc/Pat Murphy
3111 Monroe Rd, Charlotte, NC 704-335-0534
Crawford Communication/Randy Bishop
535 Plasamour Dr NE, Atlanta, GA 404-876-7149
Creative Edge/Duncan Brown
808 Live Oak Dr #101, Chesapeake, VA...........804-420-3605
Creative Post & Transfer/Chris John
377 Carowinds Blvd, Ft Mills, SC 803-548-7678
Creative Video Inc/Armistead Whitney; Guy Davidson
1465 Northside Dr #110, Atlanta, GA 404-355-5800
Cyberiad Project/Dov Jacobson
6246 Columbia Pike, Falls Church, VA 703-642-8369
Cypress Productions/George Cornelius
5410 Mariner St, Tampa, FL.........................813-289-6115
Diamond Video Productions Inc/Paul Lambert
129 Citation Ct, Birmingham, AL 205-942-8888
Digital Multi-Media Post Inc/Marie Hamlin
502 N Hudson St, Orlando, FL.......................407-293-3390
Digital Wave/Mark Gould
2905 Coachman Ave, Tampa, FL 813-837-8014
ECI Video/Shawn Sterling
2809 Ross Ave at Central, Dallas, TX 214-969-6946
EDEFX/Obadia Marcos
7355 NW 41st St, Miami, FL........................305-593-6911
Editworks/Patrick Furlong
3399 Peachtree Rd NE #200, Atlanta, GA.........404-237-9977
Elite Post Nashville/George Betts
1025 16th Avenue S # 302, Nashville, TN 615-327-8797
Englander Studios/Mel Englander
1931 NE 163rd St, N Miami Beach, FL 305-945-1700
Ensemble Productions/Barry McConatha
1036 Cornstalk Ln, Auburn, AL......................205-826-3045
F&F Productions Inc/Bob Eisenstaedt
9675 4th St N, St Petersburg, FL....................813-576-7676
Florida Production Group Inc/Craig Meadows
913 Gulf Breeze Pkwy #16, Gulf Breeze, FL.......904-934-5627
Flying Foto Factory/Casey Herbert
107 Church St/PO Box 1166, Durham, NC........919-682-3411
Gannett Production Services/Nancy Putney
1611 W Peachtree St NE, Atlanta, GA 404-873-9182
Georgia Inst of Technology/Ray Haleblian
Multimedia Tech Lab/400 10 St, Atlanta, GA....404-894-4195
Georgia Pacific Television/Randy French
133 Peachtree St NE, Atlanta, GA...................404-652-5690
GMG International/David Gardy
1950 Roland Clarke Pl, Reston, VA 703-339-8500
Greke Film/Video/Animation/Greg Leslie
7405 Kelley, Oklahoma City, OK....................800-634-5045
Henninger Video Inc/Robert Henninger
2601-A Wilson Blvd, Arlington, VA 703-243-3444
Horizon Entertainment Inc/Gary Seline
7102 Grand, Houston, TX 713-747-6433
Hot Source Media/Michael Vaughan
1916 Wilson Blvd #304, Arlington, VA 703-527-5992
Imagic/Joe Huggins
1570 Northside Dr NW # 240, Atlanta, GA.........404-355-0755
JHT Multimedia/Gary Twitchell
5514 Lake Howell Rd, Winter Park, FL 407-657-2727
JIMANDI/James Spitler
734 Vosswood Dr, Nashville, TN 615-256-9887
Lee Martin Productions/Lee Martin
11250 Pagemill Rd, Dallas, TX......................214-556-1991
Longworth Communications/Jim Longworth
230 S Crater Rd, Petersburg, VA 804-862-9967
Match Frame Post Production/Don White
8531 Fairhaven, San Antonio, TX...................210-614-5678
Mind's Eye Graphics/Paul Lambert
2246 E Dabney Rd, Richmond, VA.................804-353-7958

MVI/POST/Beck Lampert
2701-C Wilson Blvd, Arlington, VA 703-522-5335
Neo Geo/Diana Georgiou
130 E Marks St, Orlando, FL.........................407-423-9524
Odyssey Communications Group/Rebecca Stevenson
6309 N O'Connor LB83, Irving, TX 214-432-9070
Omega Films/Cefus McRae
3100 Medlock Bridge Rd, Norcross, GA............404-449-8870
Pietrodangelo Prod Group Inc/Danny Pietrodangelo
216 E Oakland Ave, Tallahassee, FL..............904-681-2392
PO Box 5126/James L Webster
4220 Amnicola Hwy, Chattanooga, TN.............615-622-1193
Power Productions/John Mashburn
1967 Government St. #D, Mobile, AL...........205-478-5530
Production Group, The/Bill Ford
510 A Leesville Rd, Lynchburg, VA...................804-237-0743
Rencom/Roger Ennis
1333 Shady Pine Way #D, Tarpon Springs, FL....813-938-3172
Results Video/Larry Emerson
6515 Escondido #A-1, El Paso, TX 915-581-0184
Scene Three Inc/Greg Alldredge
1813 8th Ave S, Nashville, TN 615-385-2820
Selkirk Communications Video Services/Darice Lang
644 S Andrews Ave, Ft Lauderdale, FL............305-527-5007
Serious Robots/Rod Rich
1101 Capitol Blvd, Raleigh, NC.....................919-821-0914
Shooting Stars Post Inc/John Samaha
3106 W North "A" St, Tampa, FL 813-873-0100
Sirius Image/Larry Goddard
2062 Weems Rd, Tucker, GA.........................404-939-2004
Smart Concepts Ltd/Steve Sembritsky
Tulsa, OK..918-747-6006
Southern Productions/David Deeb
255 French Landing Dr, Nashville, TN 615-248-1978
Southwest Teleproductions Inc/John P Shives
2649 Tarna Dr, Dallas, TX.............................214-243-5719
Stokes Imaging Services/Steve Parker
7000 Cameron Rd, Austin , TX.......................512-458-2201
● **STRUTHERS, DOUG (P 52,53)**
6453 Southpoint Dr, Dallas, TX 214-931-0838
Summa Logics Corp./Karen Ceretto
8403 Golden Aspen Ct, Springfield, VA 703-912-7948
Sunbelt Video/Michael Poly
4205-K Stuart Andrew Blvd, Charlotte, NC.........704-527-4152
Synthesis/Samuel Reynolds
PO Box 339, Conway, AR..............................501-327-2517
Take One Productions/Jere Sneider
101 Pheasant Wood Ct, Morrisville, NC 919-481-0000
Video Co, The
9146 Jefferson Hwy, Baton Rouge, LA.............504-928-4814
Video Communications SE/Andy Bundschuh
311 N Adams, Tallahassee , FL.......................904-224-5420
Video Editing Services Inc/Arthur Rouse
215 E High St, Lexington, KY.........................606-255-9049
Video Image Productions/Charlie Case
2026 Powers Ferry Rd #100, Atlanta, GA........404-984-8288
Video One/James Kilpatrick
4912 Travis, Houston, TX..............................713-524-8823
Video One Teleproductions/Tom Stout
220 St Michael St, Mobile, AL........................205-433-0013
Video Park/Charles Park Seward
11316 Pennywood Ave, Baton Rouge, LA...........504-292-0840
Video Post & Transfer/Jack C. Bryan
2727 Inwood Rd, Dallas, TX..........................214-350-2676
Video Workshop/Greg Cooper
1661 E Sample Rd, Pompano Beach, FL.............305-942-3199
Videofonics Inc/Debbie Valentine
1101 Capital Blvd, Raleigh, NC.....................919-821-5614
Vision Design Teleproductions/Athena Schultz
5401 Corporate Woods Dr #500, Pensacola, FL .904-484-0655
Vogelei & Assocs
637-B Hoke St NW, Atlanta, GA......................404-352-1712
West End Post/Lola Lott
2211 N Lamar #100, Dallas, TX 214-871-3348
White Hawk Pictures/Michelle Barth
567 Bishopgate Ln, Jacksonville, FL.................904-634-0500
Willming Reams Animation/Doug Willming
900 Isom Rd #101, San Antonio, TX.................210-342-2141
ZFX/Dave Sieg
2645 Suffolk St, Kingsport, TN 800-452-4525

MIDWEST

Adpro Inc/Rob Robbins
331 N Main St, Sycamore, IL..........................815-895-2000

AGS&R Communications/IN
1835 S Calhoun, Fort Wayne, IN......................219-744-4255
Allied Film and Video
7375 Woodward Ave, Detroit, MI.....................313-871-2222
Analog Video/Dick Madding
192 Walnut Grove Dr, Dayton, OH....................513-299-3495
Animation Station Ltd/Terry Choate
633 S Plymouth Ct #1103, Chicago, IL.............312-939-8003
Associate Producers/Robert Hufstader
6545 Bloomfield Rd, Des Moines, IA................515-285-1209
Avenue Edit/Benjamin Webber
625 North Michigan Ave, Chicago, IL 312-943-7100
Bazooka Graphics/Div Swell Pictures/Dave Mueller
233 E Wacker Dr, Chicago, IL.........................312-649-9000
Blue Sky Productions/Peter Hickman
1020 S Creyts Rd #305, Lansing, MI................517-886-2010
Bromley Chapin Designs/Al Chapin
400 Renaissance Ctr #2250, Detroit, MI..........313-259-2661
Classic Animation/David Baker
1800 S 35th St, Galesburg, MI.......................616-665-4800
Classic Video Inc/Jerry L Patton
5001 E Royalton, Cleveland, OH......................216-838-5377
Corplex Inc/Chris Blake
6444 N Ridgeway, Lincolnwood, IL.................708-673-5400
Crystal Productions/Tim Dwyer
1024 Blouin Dr, Dolton, IL............................708-841-2622
Curtis Inc/Dan Castagna
2025 Reading Rd # 130, Cincinnati, OH..........513-621-8895
Daily Planet Ltd/Fred Berkover
455 Cityfrnt Plz/NBC Twr #2900, Chicago, IL ...312-670-3766
Digital Creations/Jeff McFall
5380 North Carrolton, Indianapolis, IN 317-257-8685
Dynaimages/Steve Hansen
3947 State Line, Kansas City, MO....................816-531-3838
Edge Multimedia/John Sink
804 Burr Oak Dr, Westmont, IL 708-920-1005
Editech Post Productions Inc/Bill Hartness
14344 Y St #200, Omaha, NE.........................402-896-9696
Editel Design/Glen Noren
301 E Erie St, Chicago, IL.............................312-440-2360
EDR Media/Niki Dias
23330 Commerce Park Rd, Beachwood, OH.......216-292-7300
Emerald City Animation/Post Prod/Robert Southerly
2660 Horizon Dr, Grand Rapids, MI 616-949-9283
Envision Inc/Jeff Blackwell
405 N Calhoun, Brookville, WI 414-789-8485
Film & Tape Works Inc/Daniel Robichud
211 E Ontario #300, Chicago, IL 312-280-2210
Film Craft Video/Sandy Ruby
37630 Interchange Dr, Farmington Hills, MI 313-474-3900
Filmack Studios/Robert Mack
1327 S Wabash Ave, Chicago, IL.....................312-427-3395
Forcade & Associates/Tim Forcade
1440 Lawrence Ave, Lawrence, KS...................913-843-1605
Forcade & Associates/Mark Levine
1213 Mulford St, Evanston, IL........................708-328-1318
Fredrick Paul Productions Ltd/Jery Jacobs
6547 W North Ave, Oak Park, IL 708-386-8055
Golan Productions/Ari Golan
507 W North Ave, Chicago, IL........................312-274-FIL
Golden Dome Prdctns/WNDU-TV/Chuck Huffman
PO Box 1616, South Bend, IN 219-631-1616
Goldsholl: Film Group/Deborah Goldsholl
420 Frontage Rd, Northfield, IL......................708-446-8300
Grace & Wild Studios/Steven Wild
23689 Industrial Pk Dr, Farmington Hills, MI........313-471-6010
Grayson Media/Diane Kenna
70 W Hubbard #402, Chicago, IL 312-464-9852
Greatapes/Roger Rude
1523 Nicollet, Minneapolis, MN......................612-872-8284
GRS Inc/Steve Andrews
13300 Broad St, Pataskala, OH.......................614-927-9566
● **H-GUN/DIGITAL DIV**/Robert Bial **(P 43)**
2024 S Wabash St 7th Fl, Chicago, IL...............312-808-0134
Hartley Metzmer Huenink Comm/John Huenink
3076 S Calhoun Rd, New Berlin, WI................414-784-1010
Harvest Communications/Marlin Davis
900 George Washington Blvd, Wichita, KS........316-652-9900
Hellman Animates/Jim Mathis
1225 W 4th St, Waterloo, IA 319-234-7055
Hellman Associates/Kathy Forslund
400 First Ave N #218, Minneapolis, MN...........612-375-9598
Hi-Tech Productions/Dave McClimans
33433 Boulevard, Eastlake, OH......................216-953-0077

PRODUCTION SERVICES
video animation

HVS Video/HVS Cable Adv/Patrick Krohlow
1100 Guns Rd, Green Bay , WI.......................414-468-3333
Image Assoc/Ann de Lodder
1625 W Big Beaver Rd, Troy , MI....................313-649-2200
Image Producers/Mark Munroe
Box 130, Canfield, OH.................................216-533-0100
Industrial Video/Ray Grubic
1601 N Ridge Rd, Lorain, OH.........................216-233-4000
**Information Technology Design
Assoc Inc**/Daniel Klassen
7156 Shady Oak Rd, Eden Prairie, MN............612-944-8244
**Instant Replay Video & Film
Production Inc**/Stephanie Smith
1349 E McMillan St, Cincinnati, OH 513-569-8600
Ipa/The Editing House/Maggie Magee
1208 W Webster, Chicago, IL.........................312-871-6033
IVL Post/Bob Rohde
Target Center/600 First Ave N, Minneapolis, MN ...612-673-1250
J Parker Ashley Associates/Rick Benjamin
119 N Fourth St #208, Minneapolis, MN 612-340-9807
Kasper Studio/Walt Kasper
17903 Huron River Dr, New Boston, MI.............313-753-9100
Lamb & Co/Lawrence Lamb
650 Third Ave S 17th Fl, Minneapolis, MN........612-333-8666
Lightborne Communications/Bob Hanneken
632 Vine St #817, Cincinnati, OH 513-721-2272
Lighthouse Productions Inc/Rick Meredith/Chip Moore
1900 Hicks Rd, Rolling Meadows, IL................708-506-1414
LTS Productions/David Lang
2810 Bennett Rd, Okemos, MI 517-332-1190
Mainsail Production Services/Ron Beam
521 Buyers Rd #109, Miamisburg, OH............513-866-7800
Mainstream Communications Inc/Becky Beck
955 James Ave S #235, Bloomington, MN..........612-888-9000
Marx Production Ctr/Tom Deming
3100 W Vera Ave, Glendale, WI 414-351-5060
Media Consultants/Richard Wrather
201 N Stoddard/POB 130, Sikeston, MO 314-472-1116
Media Group Inc/Gregory Woods
1480 Dublin Rd, Columbus, OH......................614-488-0621
Metro Productions Inc/Tim McGovern
238 W 74th St, Kansas City, MO 816-444-7004
Micrographics
550 W Jackson Blvd #340, Chicago, IL.............312-648-1000
Mike Jones Film Corp/David Jones
5250 W 74th St, Minneapolis, MN...................612-835-4490
**Mills-James Prdctns/Teleproduction
Center**/Jim Zangmeister
3545 Fishinger Blvd, Columbus, OH.................614-777-9933
Motivation Media Inc/Rick Murray
1245 Milwaukee Ave, Glenview, IL 708-297-4740
**Northwest Teleproductions/
Chicago**/Carmen Trombetta/Amy Brierly
142 E Ontario 4th Fl, Chicago, IL....................312-337-6000
Northwestern Mutual Life/Betty Hoff
720 E Wisconsin Ave, Milwaukee, WI...............414-271-1444
Omni Productions/W H Long
655 W Carmel Dr, Carmel, IN.........................317-844-6664
On Line Video/Dennis Majewski
720 E Capitol Dr/PO Box 693, Milwaukee , WI..414-223-5254
Optimus Inc/Cat Gulick
161 E Grand Ave, Chicago, IL........................312-321-0880
**Ozam Film & Tape Productions,
Imagination Effects**/Dale Detoni
1228 S Fort, Springfield, MO 417-866-3232

Paradigm Communication Group/Keith Meyer
250 Production Plz, Cincinnati, OH 513-381-7100
Phar-Mor Productions/Diane Olenik
20 Federal Plaza W, Youngstown, OH 216-746-6641
Post Effects/Jacqui Jones
400 W Erie St #101, Chicago, IL 312-944-1690
Post Production Services Inc/Dave Dittgen
602 Main St #900, Cincinnati, OH 513-621-6677
Postique Inc/Mary Suzanne Patek
23475 Northwestern Hwy, Southfield, MI...........313-352-2610
Precision Tapes Inc/Daniel Piepho
2301 E Hennepin, Minneapolis, MN 612-379-7554
Prism Studios/Jack Graff
2412 Nicollet Ave, Minneapolis, MN 800-659-2001
Producers Color Service Inc/Susan Flynn
24242 Northwestern Hwy, Southfield , MI...........313-352-5353
Production House/ImageLab
811 St John S, Highland Park, IL 708-433-3172
Promedia Productions/Steve Keller
2593 Hamline Ave, Roseville, MN.....................612-631-3681
Provideo/Jim Stiener
2302 W Badger Rd, Madison, WI 608-271-1226
Pulse Communications/Scott Koffarnus
211 N Broadway, Green Bay, WI 414-436-4777
Quantum Group/Janice Glassfort
101 W 10th St, Indianapolis, IN 317-639-6001
Quicksilver Assoc Inc/Chuck Garasic
18 W Ontario St, Chicago, IL 312-943-7622
Rainbow Video Productions/Phil Troupe
Route 1/Box 82, Adams, NE 402-788-2556
Real Productions/Robert J Schuster
1821 University Ave #N-153, St Paul, MN 612-646-9472
Reelworks Animation Studio/Peter Lang
318 Cedar Ave S #300, Minneapolis, MN 612-333-5063
Reitz Data Communications/Larry Reitz
8225 Farnsworth Rd #A, Waterville, OH 419-878-3334
Renaissance Video Corp/John Feldbaille
130 S Jefferson, Chicago, IL 312-930-5000
Rivet Films, Inc.
748 Seminole, Detroit, MI 313-822-2700
RW Video Inc/Gerrit Marshall
4902 Hammersley Rd, Madison, WI..................608-274-4000
Sacks, Ron/Ron Sacks
1189 Rosebank Dr, Worthington, OH................614-846-1921
Sanders & Co/Denita Clifford
3610 N Meridian, Indianapolis, IN 317-926-2841
Second Sight Productions/Thom Bixbe
950 N Meridian St # 33, Indianapolis, IN 317-237-7888
Shaw Video Communications/Paul Quinn
3246 Henderson Rd, Columbus, OH 614-457-4477
Sinnott & Associates/Cynthia Neal
676 N LaSalle St, Chicago, IL 312-440-1875
SOS Productions/Mark Braver
753 Harmon Ave, Columbus, OH......................614-221-0966
Sound/Video Impressions/Bill Holtane
110 River Rd, Des Plaines, IL 708-297-4360
Soundlight Productions/Terry Luke
1915 Webster, Birmingham, MI 313-642-3502
Still Life & Kicking/Martha Coleman
6573 City West Pkwy, Minneapolis, MN 612-943-1029
Take 1 Productions Ltd/Teri Murphy
5325 W 74th St, Minneapolis, MN 612-831-7757
TechPool Studios Inc/TechPool Studios
1463 Warrensville Ctr Rd, Cleveland , OH 216-382-1234
Telemation Productions/John Dussling
100 S Sangamon, Chicago, IL 312-421-4111

TeleVideo Productions/Mike Sopa
611 S Farwell St, Eau Claire, WI......................715-833-9269
TGA Recording Co/Thomas G Alti
295 Urbandale Dr, Benton Harbor, MI...............616-926-7581
Thirteen Fifty-one Productions/Tom Kwilosz
1351 W Grand, Chicago, IL..............................312-421-0400
Thomas EFX Group/Tom Olson
7436 W 78th St, Minneapolis, MN 612-942-8702
USA Teleproductions/Jamie Burns
1440 N Meridian, Indianapolis, IN 317-632-5900
Vanguard Productions Inc/Jerry Beck
7020 Huntley Rd, Columbus, OH......................614-436-4610
VGI Productions/Video Genesis Inc/Howard Schwartz
4949 Galaxy Pkwy, Cleveland, OH 216-464-3635
Vidcam Productions Inc/Mary Detomaso
7150 Hart St/Unit A-1, Mentor, OH 216-255-5050
Video Genesis/Jim Gerber
4949 Galaxy Pkwy, Cleveland, OH 216-464-3635
Video Post/M Kappleman
4600 Madison #120, Kansas City, MO..............816-531-1225
Video Post & Graphics Inc/Claude Kennedy
14 Cambridge Ave, Dayton, OH 513-276-3113
Video Wisconsin/Jeff Utschig
18110 W Bluemound Rd, Brookfield, WI 414-785-1110
Visual Motivation Inc/Robert Jackson
10901 Valley View Rd, Eden Prairie, MN 612-943-0060
Wayne Boyer Studio/Wayne Boyer
1324 Greenleaf St, Evanston, IL 708-491-6363
Xenas Communications/Greg Peck
354 Gest St, Cincinnati, OH 513-621-2729

WEST

10th St Video Productions/Jeff Bishop
140 N 16th St/PO Box 2, Boise , ID...................208-336-5222
22 Product/Terry Green
440 Davis Ct #509, San Francisco, CA 415-399-9744
3D Magic/FTI/Dave Altenau/David Muir
55 Hawthorne St 10th Fl, San Francisco, CA 415-543-2517
Acmetek/Ken Lau
1145 Hacienda Blvd #126,
Hacienda Heights, CA 818-333-6161
Action Video/Alana Ireland
6616 Lexington Ave, Hollywood, CA.................213-461-3611
Advanced Digital Imaging/Cliff Auchmoody
1250 N Lakeview Ave, Anaheim, CA 714-779-7772
American Film Technology/Susan Crampton
11585 Sorrento Valley Rd #104, San Diego, CA...619-259-8112
American Production Services/Ted Hill
2247 15th Ave W, Seattle, WA.........................206-282-1776
Angel Studios/Jill Hunt
5962 La Place Ct, Carlsbad, CA........................619-929-0700
Animation Farm/Don Smith
21850 Poppy Lane, Nuevo, CA 909-928-1577
Antigravity Workshop/Dan Lutz
456 Lincoln Blvd, Santa Monica, CA 310-393-9747
Applied Technical Video Inc/Dennis Kushner
1564 Commerical St SE, Salem, OR..................503-371-9498
Armour Productions/Stephanie Tucker
306-C W Katella, Orange, CA 714-538-5811
Art F/X/Dawn Jones
3575 W Cahuenga Blvd, Los Angeles, CA..........213-876-9469
Artichoke Productions/Paul Kalbach
4114 Linden St, Oakland, CA 510-655-1283
Astavision Post Prod/Scott Kinnimon
910 N Citrus Ave, Hollywood, CA 213-465-3333
Atomic Media Group/Carol Leigh
1933 Davis St #213, San Leandro, CA...............510-430-8012
Avid Productions/Henry Bilbao
1300 Industrial Rd #1, San Carlos, CA 415-593-2000
Barking Art Productions/Allen Battino
126 1/2 S Flores St, Los Angeles, CA.................213-852-1937
Bechtold Studios/Bob Bechtold
430 S Niagara St, Burbank, CA.........................818-562-1751
Beckman Instruments Inc/Teri Beauchamp
2500 Harbor Blvd/MS B22D, Fullerton, CA 714-773-7682
Big Zig Video/Elliot Porter
329 Bryant St #1B, San Francisco, CA 415-243-8880
Blume, Neil
10235 NE Prescott #16, Portland, OR 503-257-4560
Bookmakers Ltd/David Bolinsky
PO Box 1086, Taos, NM 505-776-5435
Boss Film Studios/Bob Mazza
13335 Maxella Ave, Marina Del Rey, CA...........310-823-0433
Boyington Film Productions/Joel Newman
5875 Blackwelder, Culver City, CA 310-204-2200

Calico/Jan Nagle
8843 Shirley Ave, Northridge, CA 818-701-5862
California Communications Inc/Hope Schenk
6900 Santa Monica Blvd, Los Angeles, CA 213-466-8511
California Image Associates/Duane Thompson
3034 Gold Canal Dr #B, Rancho Cordova, CA...916-638-8383
Catalyst Productions/Chris Negri
1431 Center St, Oakland, CA 510-836-1111
Central Image Video & Photography/Larry Gale
337 Majors St, Santa Cruz, CA.........................408-454-0500
Cimity Arts/Special Visual Effects/Barbara Cimity
800 S Robertson Blvd, Los Angeles, CA.............310-659-4504
Cinema Engineering Co/Richard L Bennett
7243 Atoll Ave, N Hollywood, CA.....................818-765-5340
Commotion/Bob Robbins
3131 Cahuenga Blvd W, Los Angeles, CA 213-882-6637
Communications Video/Arthur Marchetti
PO Box 2125, Walnut Creek, CA 510-930-7788
Complete Post/Bob Brian
6087 Sunset Blvd, Hollywood, CA.....................213-467-1244
Continental Productions, Continental TV Network/Cheryl Cordeiro
118 6th St S/PO Box 1219, Great Falls, MT 406-761-5536
Creative License–Film & Video/Bonnie MacBird
131 S Barrington Pl #200, Los Angeles, CA 310-476-9725
Creative Media Development/Doug Crane
710 SW Ninth Ave, Portland, OR......................503-223-6794
Crosspoint Productions/Peter Ten Eyck
940 Wadsworth Blvd, Lakewood, CO 303-232-9572
DeGraf/Wahrman/Michael Wahrman
1845 N Curson Ave, Los Angeles, CA 213-876-2341
Depixion
27125 Wapiti Dr, Evergreen, CO 303-670-7822
Digidesign Inc/Peter Gotcher
1360 Willow Rd #101, Menlo Park, CA..............415-688-0600
Digital Artworks/Paul Scott
2295 Coburg Rd #104, Eugene, OR 503-344-6541
Digital Design Simulations/Dan Biggs
PO Box 565, Bonita, CA..................................619-421-2107
Digital Post & Graphics/Terri Williams
1921 Minor Ave, Seattle, WA...........................206-623-3444
Diversified Video Ind Inc/Sierra Video/Jerry Hennan
4216 N Maxxon, El Monte, CA 818-579-7023
DLH Studios/Dennis Hocking
2900 Adams #B-29, Riverside, CA....................909-687-6654
Dolphin Multi Media/Cynthia Kondratieff
2440 Embarcadero Way, Palo Alto, CA.............415-962-8310
Don Pierce Productions/Don Pierce
236 Lupe Ave, Newbury Park, CA 805-498-2776
Double Vision Productions/Betty Vivado
1955 Railroad Dr #A, Sacramento, CA..............916-921-6986
Dourmashkin Productions/Barbara Dourmashkin
3852 Camino de Solana, Sherman Oaks, CA 818-995-3997
Dowlen Artworks/James Dowlen
2129 Grahn Dr, Santa Rosa, CA........................707-579-1535
Dream Quest Images/Greg Pappas
2635 Park Center Dr, Simi Valley, CA 805-581-2671
DrewPictures Inc/Drew Huffman
246 First St #402, San Francisco, CA.................415-247-7600
Duffus, Bill
1745 Wagner, Pasadena, CA............................818-577-7531
Dynamic Perspectives/Linda Deschambault
2066 Donald Dr, Moraga, CA...........................510-631-9017
Editel/Los Angeles/Michael Moreale
729 N Highland Ave, Hollywood, CA 213-931-1821
Effective/Visual Imagery/Mia Corrales
151 Kalmus #K1, Costa Mesa, CA 714-641-5222
Electric Paint & Design/Jon Paul Davidson
PO Box 8822, Incline Village, NV 702-831-3490
Elson, Matt
11901 Sunset Blvd #207, Los Angeles, CA 310-471-4511
EP Graphics/Burbank/Eddie Pong
2525 N Naomi, Burbank, CA............................818-953-9375
Flying Colours/Monica Fabian
499 Seaport Court #205, Redwood City, CA 415-363-8767
Frame-by-Frame Media Services/Rick Sutter
5353 Randall Pl, Fremont, CA...........................510-651-6330
Fred Wolf Films/Fred Wolf
4222 W Burbank Blvd, Burbank, CA 818-846-0611
FX Plus Design/Elan Soltes
3025 W Olympic Blvd, Santa Monica, CA 310-315-2175
G.R.A.P.E. Video Recording/Sharon Robinson
89 W Neal St, Pleasanton, CA..........................510-462-1300
Gentry, David
1145 Wisconsin St, San Francisco, CA...............415-824-2920

Ginsburg Productions/Michael Ginsburg
2243 S Olive St, Denver, CO 303-757-2564
Glenn Roland Films/Glenn Roland
10711 Wellworth Ave, Los Angeles, CA 310-475-0937
Global One Design & Innovation/Drew Cronk
Box 2218, Healdsburg, CA 707-431-7664
Global Vision/Homoud B Alkouh
2082 Business Center Dr #175, Irvine, CA.........714-252-1106
Gomes, Steve
1901 S Bascom Ave #1190, Campbell, CA.......408-559-6369
GR Graphics Audio Visual/Guillermo Rodriguez
1850 N 15th Ave, Phoenix, AZ....................602-252-6525
Graphix Zone/Angela Aber
38 Corporate Pk #100, Irvine, CA................714-833-3838
Gray Matter Advertising/Robert Gray
2434 Monoco St, Channel Islands, CA...........818-884-7641
Half-Inch Video/Sean Scott
185 Berry St #6501, San Francisco, CA 415-495-3477
Hamm & Assocs/Gene Hamm
1644 Moclips Dr, Petaluma, CA...................707-765-2198
Helical Productions/Bob Brandon
3801 E Florida Ave # 613, Denver, CO 303-756-0044
Herigstad, Dale
3112 Ledgewood Dr, Los Angeles, CA 213-463-1339
Hollywood South Producers Centre/Jerry Kelley
2465 Campus Dr #A, Irvine, CA 714-757-7588
I/O Productions Inc/Laurent Basset
7551 Sunset Blvd #202, Hollywood, CA............213-969-0113
ILM (Industrial Light & Magic)/Andrea Merrim
PO Box 2459, San Rafael, CA.....................415-258-2000
Image Corp/Dave Hoffman
1617 Wazee St, Denver, CO 303-573-6206
Impact Corporate Communications/Fred Gerhart
4601 Telephone Rd #117, Ventura, CA............805-644-3645
International Corporate Video Inc/James Fasso
1020 Serpentine Ln #114, Pleasanton, CA.........510-426-8230
Intersound Inc/Garry Morris
8746 Sunset Blvd, Los Angeles, CA 310-652-3741
Invisions Inc/Scott Tuckman
3641 N 52nd St, Phoenix, AZ 602-840-1090
Jane Lily Design/Mason Lyte
610 Anacapa St, Santa Barbara, CA...............805-683-4884
JMTV/John Thill
2808 Roosevelt St, Carlsbad, CA..................619-434-3363
KCFW Productions/Curt Smith
401 First Ave E, Kalispell, MT.....................406-755-5239
Killingsworth Presentations/Penny Engel
3834 Long Beach Blvd, Long Beach, CA............310-595-7796
Kirwin Communications/Alison Childs
1760 Prospector Ave, Park City, UT 801-649-8226
KNTV Production Services/Bill Glenn
645 Park Ave, San Jose, CA 408-286-1111
Kroyer Films Inc/Bill Kroyer
12517 Chandler Blvd, North Hollywood, CA 818-755-0280
LA Videograms/Larry Rosen
3203 Overland Ave #6157, Los Angeles, CA 310-836-9224
Laguna Productions/Charlie King
2708 S Highland Dr, Las Vegas, NV 702-731-5600
LePrevost Corporation/John LePrevost
29350 W Pacific Coast Hwy, Malibu, CA...........310-457-3742
Lightform Design/Bill Lae
1771 Bel Air Rd, Bel Air, CA 310-476-8617
Lumeni Productions/Tony Valdez
1362 Flower, Glendale, CA 213-462-2110
Master Communication/Robert Masters
3429 Kerckhoff Ave, San Pedro, CA................310-832-3303
Master Videoworks/Jeff Killian
3611 S Harbor Blvd #150, Santa Ana, CA.......714-241-7724
McCamant Productions Animation/David McCamant
1520 Akard Dr, Reno , NV 702-746-8276
MDC Teleproductions/McDonnell
Douglas/Milton Moline
4000 Lakewood Blvd, Long Beach, CA 310-496-9040
Media Lab/Jim Turrentine
400 McCaslin Blvd #211, Louisville, CO............303-499-5411
Media Staff/Jerry Maybrook
6926 Melrose Ave, Hollywood, CA.................213-933-9406
Metrolight Studios/Dobbie Schiff
5724 W Third St #400, Los Angeles, CA...........213-932-0400
Metropolitan Entertainment/Andy Ullman
1680 N Vine St #600, Hollywood, CA.............213-856-7060
Midland Production Corp/Yas Takata
435 S Second St, Richmond, CA...................510-233-4236
Miller Design & Associates/Paul Miller
8151 N Northern Ave, Tucson, AZ..................602-797-1775

Moving Pixels/Dave Thomas
1833 Verdugo Vista Dr, Glendale, CA..............818-247-9508
Mr Film/Chris Walker
228 Main St #10, Venice, CA 310-396-0146
Murphy, Alan
133 Sutphen St, Santa Cruz, CA..................408-423-4770
NBC Magic
3000 W Alameda Ave, Burbank, CA 818-840-4863
North Country Media Group/Doug Bliler
721 Second St S, Great Falls, MT..................406-761-7877
Northwest Media
PO Box 56, Eugene, OR...........................503-343-6636
Novocom/GRFX Productions/Jack Schaeffer
6314 Santa Monica Blvd, Hollywood, CA..........213-461-3688
Nutopia Digital Video/Eric Myers
300 Valley St #301, Sausalito, CA.................415-331-0714
On Tape Productions/Steve Zeifman
724 Battery St, San Francisco, CA.................415-421-5551
Pace Video Center/Romeo Carrera
2020 SW 4th Ave #700, Portland, OR..............503-226-7223
Pacific Data Images/Brad Lewis
1111 Karlstad St, Sunnyvale, CA..................408-745-6755
Pacific Interactive Design Corp/Scott Palamar
1460 Fourth St #200, Santa Monica, CA..........310-458-1898
Pacific Ocean Post/Alan Kozlowski
730 Arizona Ave, Santa Monica, CA 310-458-3300
Penrose Productions/Mickey Bernard
1042 Hamilton Ct, Menlo Park, CA................415-323-8273
Pettigrew, Stuart
1420 45th St Annex, Emeryville, CA...............510-652-7932
Pinnacle Effects
2334 Elliott Ave, Seattle, WA 206-441-9878
Pinnacle Post/Vance Martin
2334 Elliot Ave, Seattle, WA 206-443-1000
Pittard Sullivan Design/Bill Pittard
6430 Sunset Blvd #200, Hollywood, CA...........213-462-1190
Post Group/Rich Ellis
6335 Homewood Ave, Los Angeles, CA 213-462-2300
Precision Visual Communications/Herb Saunders
591 25th Rd, Grand Junction, CO.................303-242-1783
Production Co, The/Bob Patrick
601 S Rancho #C-26, Las Vegas, NV..............702-263-0933
Production Masters Inc/Harry Tate
834 N 7th Ave, Phoenix, AZ.......................602-254-1600
Rase Productions, Bill
955 Venture Ct, Sacramento, CA 916-929-9181
Realtime Video/Will Hoover
60 Broadway, San Francisco, CA..................415-705-0188
Results! Computers & Video
12600 Morrow Ave NE, Albuquerque, NM 505-291-8998
ReVerb/Lorraine Wild
5514 Wilshire Blvd #900, Los Angeles, CA 213-936-7305
REZ-N-8 Productions/Michele Ferrone
6430 Sunset Blvd #1000, Hollywood, CA..........213-550-8885
Rhythm & Hues/Suzanne Datz
910 N Sycamore Ave, Hollywood, CA 213-851-6500
Rick Resnick Audio Visual Design/Rick Resnick
5544 E Kings Ave, Scottsdale, AZ 602-482-3070
Rock Solid Productions/David Griffin
801 S Main St, Burbank, CA.......................818-841-8220
Rocky Mountain AV Productions/Terry Talley
4301 S Federal Blvd #101, Englewood, CO 303-730-1100
S.L.A.D.E. Corporation/Stephanie Slade
PO Box 10176, Beverly Hills, CA..................213-466-7171
San Francisco Production Group/Debra Robins
550 Bryant St, San Francisco, CA.................415-495-5595
Sidley Wright & Associates
953 N Highland, Los Angeles, CA 213-465-9527
Steve Michelson Productions/Steve Michelson
280 Utah St, San Francisco, CA...................415-626-3080
Stimulus/Grant Johnson
PO Box 170519, San Francisco, CA...............415-558-8339
Studio Productions Inc
650 N Bronson Ave #223, Hollywood, CA.........213-856-8048
Syd Mead Inc/Roger Servick
1716 N Gardner St, Los Angeles, CA 213-850-5225
Telemation Denver/Dan Boyd
8745 E Orchard Rd #500,
Greenwood Village, CO...........................303-290-8000
TeleScene Inc/Rick Larson
3487 W 2100 S, Salt Lake City, UT 801-973-3140
Thirty Second St/Mitchell Kenison
1209 Mountain Road Pl, Albuquerque, NM........505-265-0224
THS Visuals/Todd Simon
PO Box 2192 , Lake Tahoe, Reno, NV.............702-588-6976

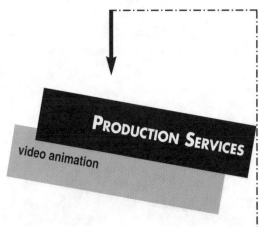

Topix LA Inc/Chris Mitchell
1800 Vine St 3rd Fl, Hollywood, CA 213-962-4745
Total Video Co/Peter Roger
432 N Canal St #12S, San Francisco, CA 415-583-8236
● **TRACER DESIGN**/Chad Little (P 42)
4206 N Central Ave, Phoenix, AZ.................602-265-9030
Triad Media Services/Kirk Storms
3055 Triad Dr, Livermore, CA.....................510-449-0606
Unitel Video/Hollywood/Mark Brown
3330 Cahuenga Blvd W, Los Angeles, CA.........213-878-5800
Univ of Calif at Santa Barbara/Peter Allen
Coll Engnr/Office of Dean#1001,
Santa Barbara, CA................................805-893-4803
US Animation/David Lipman
818 N La Brea Ave, Hollywood, CA...............213-465-2200
US Animation/John Whitney
818 N LaBrea Ave, Hollywood , CA...............213-465-2200
Van der Veer Photo Effects/Tom Anderson
724 S Victory Blvd, Burbank, CA..................818-841-2512
Varitel Video/Colleen Casey
1 Union St, San Francisco, CA 415-495-3328
Veridian/Ronnie Sampson
268 Ninth Ave, San Francisco, CA................415-979-4980
Video Arts/Kim Salyer
185 Berry St #5400, San Francisco, CA 415-546-0331
Video Media Productions/Ann Bonanno
2727 W Southern Ave, Tempe, AZ................602-966-6545
Video West Productions/Jim Yorgason
55 N 300 W, Salt Lake City, UT 801-575-7400
Virtual Images/Jane Bradley
3487 W 2100 S, Salt Lake City, UT 801-977-0888
Vision Productions/KSAZ-TV/Bill Lucas
511 W Adams, Phoenix, AZ.......................602-262-5131
Visions Plus/Steve Dung
2223 N Main St, Walnut Creek, CA 510-256-9450
Visual Effects Lab/Bob Durrenberger
8815 Friant St, San Diego, CA....................619-578-3363
Visual Promotions/Paul Tokarski
1905 N Edgemont St, Los Angeles, CA.............213-661-9560
West Coast Projections Inc/David Gibbs
11245 W Bernardo Ct #100, San Diego, CA......619-674-7334
Westcom Creative Group/Steve Ogle
2295 Coburg Rd #105, Eugene, OR..............503-484-4314
Wickerworks Video Productions/Steve Hug
6020 Greenwood Plaza Blvd, Englewood, CO....303-741-3400
Windstar Studios Inc/Dale Mitchell
525 Communications Cir, Colorado Springs, CO.719-635-0422
Woodholly Productions/Yvonne Parker
712 N Seward St, Hollywood, CA 213-462-5330
Z-AXIS/Ray Hauschel
116 Inverness Dr E #110, Englewood, CO.........303-792-2400
Zilberts & Assocs, Ed
5690 DTC Blvd #190, Englewood, CO 303-220-5040

CANADA

Animatics Inc/Alfredo Coppola
126 York St #207, Ottawa, ON 613-235-9000
Ann-imation Video
179 John St, Toronto, ON..........................416-971-9762
Artray Ltd/Karen Pinnell
PO Box 4700, Vancouver, BC.....................604-420-2288
Calibre Digital Design/Cindy Cosenzo
2 Melanie Dr Unit 5, Bramalea, Toronto, ON 416-792-7611
Canadian Film Aid/James Eng
782 King St W, Toronto Canada, ON 416-363-5071

Command Post & Transfer
179 John St, Toronto, ON..................................416-585-9995
DHD Post Image/Angus MacKay
6265 St Jacques #200, Montreal, QB.................514-489-8989
Digital Magic
58 Fraser Ave, Toronto, ON 416-534-6154
Dignum Computer Graphics/Alain Leduc
1905 William St, Montreal, QB.......................514-931-4221
Groupe Andre Perry/Mario Rachiele
1501 Parre St, Montreal, QB.........................514-933-0813
Les Animation Drouin, Inc/Andre Drouin
19 Le Royer O #401, Montreal, QB..................514-849-0548
Oasis Post Production Services/Don McMillan
340 Gerarrd St E, Toronto , ON......................416-466-5870
Pyrate Animation
451 St-Sulpice #L, Montreal, QB.....................514-284-0761
Side Effects/Sonja Tessari
620 King St W, Toronto, ON 416-367-8333
Soft Image Inc/Elizabeth Jones
3510 Blvd St Laurent #214, Montreal, QB..........514-845-1636
Studio Post & Transfer/Colin Minor
5305 104th St, Edmonton, AB 403-436-4444
TOPIX
217 Richmond St W 2nd fl, Toronto, ON 416-971-7711

Video Sound Integration

NEW YORK CITY

Aesthetic Engineering/Laurie Speigel
175 Duane St, New York, NY 212-925-7049
Amphion Music/Stephanie Martini
60 W 87th St, New York, NY 212-873-9715
Anne Phillips Productions/Anne Phillips
170 West End Ave, New York, NY 212-580-2349
Arnold Black Productions/Ruth Black
895 West End Ave, New York, NY 212-865-5933
Audio Directors
325 W 19th St, New York, NY 212-924-5850
Big Time Productions/Paul Conti
100 Riverside Dr, New York, NY......................212-626-6544
Bob Gerardi Music/Bob Gerardi
160 W 73rd St, New York, NY 212-874-6436
Chromavision Corp/Bruce Testa
49 W 27th St, New York, NY 212-686-7366
City College of NY/Communications
Film & Video/Eileen Gilmartin
138th St & Convent Ave, New York, NY 212-650-7421
Corelli Jacobs Recording/Andy Jacobs
25 W 45th St, New York, NY 212-382-0220
Crushing Enterprises
157 W 57th St #1200, New York, NY 212-265-0741
David Horowitz Music Assoc/Patty Forbes
373 Park Ave S 2nd Fl, New York, NY 212-779-3030
DSM Producers
161 W 54th St #803, New York, NY 212-245-0006
Electra Nova Studios
342 Madison Ave, New York, NY 212-687-5838
Elias Associates/Terry O'Gara
6 W 20th St, New York, NY 212-807-6151
Fitz, Tracy
531-A Sixth Ave, Brooklyn, NY 718-768-8161
Four/Four Productions/Frank Di Minno
630 Ninth Ave #5, New York, NY 212-581-4444
Freedom Sound/Tallman Music/Roger Tallman
226 E 54th St 7th Fl, New York, NY 212-688-5120

Gerald Alters Inc/Gerald Alters
201 E 87th St PHJ, New York, NY 212-369-5454
Harmonic Ranch/Brook Williams
59 Franklin St, New York, NY 212-966-3141
Jonathan Helfand Music
35 W 45th St, New York, NY 212-921-0555
Josell Communications/Jessica Josell
185 West End Ave #22C, New York, NY 212-877-5560
JPM Enterprises/Jessie Miller
PO Box 937, New York, NY 212-724-6467
Kamen Audio Productions/Roy Kamen
701 Seventh Ave, New York, NY 212-575-4660
Katz, Marco
79th St Boat Basin #53, New York, NY...............212-769-4836
Lee Gagnon Music Production/Lee Gagnon
240 W 73rd St #1212, New York, NY 212-496-9245
Look Inc/Jeanne Neary
170 Fifth Ave, New York, NY 212-627-3500
M & I Recording/Mitch Yuspeh
630 Ninth Ave, New York, NY 212-582-0210
Magic Venture Studios/Bill Prickett
111 W 24th St, New York, NY 212-645-4747
Magnetic Music Publishing Co/Reynold Weidenaar
5 Jones St #4, New York, NY 212-255-8527
Mambo Music Ltd/Megan Cariola
233 W 26th St #2W, New York, NY 212-691-0303
McBrien Productions Inc
PO Box 7070/FDR Station, New York, NY.........212-758-0505
Mega Music Corp
345 E 62nd St, New York, NY 212-838-3212
Michael Dansicker Music/Michael Dansicker
353 W 56th St #8C, New York, NY 212-757-3841
New York Audio Productions
140 W 22nd St, New York, NY 212-243-6826
Newfound Music/Bob Montero
15 W 26th St 4th Fl, New York, NY 212-889-8899
North Forty Productions/Ric Kallaher
30 W 21st St 12th Fl, New York, NY................212-243-4040
O'Farrill Music Ltd/Chico O'Farrill
574 West End Ave, New York, NY 212-799-3297
Otterson Television Video Inc/Bill Otterson
251 W 30th St #14W, New York, NY 212-695-7417
Partners & Agostinelli Productions/Gregory Agostinelli
1123 Broadway #404, New York, NY 212-697-7771
Petersen Music, John
205 E 16th St, New York, NY 212-477-6124
PMC Intl/J C Addison
250 W 49th St #804, New York, NY 212-541-4620
Pyramid Recording
12 E 32nd St 3rd Fl, New York, NY 212-686-8687
Radio Band of America/Marie Lenz
1350 Ave of Americas 12th Fl, New York, NY 212-687-4800
Rick Ulfik Productions/Rick Ulfik
130 W 42nd St #904, New York, NY 212-704-0888
RK Music Production/Robert Kahn
54 W 21st St #410, New York, NY 212-229-2279
Rocky Road Music Productions/Jeff DeLinko
142 West End Ave, New York, NY 212-873-9481
Schwartz Recording Inc
420 Lexington Ave #1934, New York, NY..........212-687-4180
Shelton Leigh Palmer & Co/Lon Palmer
19 W 36th St, New York, NY 212-714-1710
Sicurella & Assoc/Gary Schwartzman
62 W 45th St, New York, NY 212-398-1211
Smythe & Co/Leslie Baum
16 W 19th St, New York, NY 212-645-1166
Soundtrack/NY/Mike Korash
936 Broadway 4th Fl, New York, NY 212-420-6010
Sun Group/Arthur Custer
1133 Broadway #1527, New York, NY 212-255-1000
Sutcliffe Music Inc
8 W 19th St 2nd Fl, New York, NY 212-989-9292
Terry Waldo Enterprises/Terry Waldo
335 W 88th St, New York, NY 212-362-5922
The Music Picture/David Forrest
147 E 37th St, New York, NY 212-683-4600
TRF Production Music Libraries/Michael Nurko
747 Chestnut Ridge Rd, Chestnut Ridge, NY 800-899-6874
Two Pie Are Music/Tom DeSisto
18 E 16th St, New York, NY 212-645-9290
Videomix/Aquarius/Ira Taxin
123 W 18th St, New York, NY 212-627-7700
Vilinsky/Snyder Music/Patrick Thouron
1776 Broadway, New York, NY 212-757-7003

VSC/Shelley Riss
225 E 43rd St, New York, NY 212-599-1616
Woloshin Inc/Sid Woloshin
95 Madison Ave, New York, NY 212-684-7222

EAST

Acme Recording Studios/Peter Denenberg
112 W Boston Post Rd, Mamaroneck, NY 914-381-4142
Art Institute of Philadelphia/Michael Santaspirt
1622 Chestnut St/Visual Comm, Philadelphia, PA....215-567-7080
Audio Plus Video/Int Post/Beth Simon
235 Pegasus Ave, Northvale, NJ 201-784-2190
D'Elia-Wittkofski Productions/Curt Powell
One Market St, Pittsburgh, PA.......................412-391-2900
Dr T's Music Software Inc/Cheryl Brown
124 Crescent Rd #3, Needham, MA 617-455-1454
Fox & Perla Ltd/Gene Perla
20 Martha St, Woodcliff Lake, NJ....................908-604-6275
Hogpenny Productions Inc/Brian Treutlein
121 E 14th St/Ship Bottom,
Long Beach Island, NJ...............................609-494-6640
Modern Video Productions Inc/Jim Burt
1650 Market St 3rd Fl, Philadelphia, PA 215-569-4100
Oasis Music Inc/Steve Horelick
577 Warburton Ave, Hastings-on-Hudson, NY 914-478-0400
Omnimusic/Laura Johnson
52 Main St, Port Washington, NY 516-883-0121
PCI Recording/Caroline Maynard
737 Atlantic Ave, Rochester, NY 716-288-5620
PG Music
266 Elmwood Ave #111, Buffalo, NY...............416-528-2368
Plotkin Music Assoc/Jerry Plotkin
PO Box 660, Goldens Bridge, NY 212-246-8400
Robilotto, Philip
777 Stone Rd Route 9W, Glenmont, NY.............518-767-3196
Sheffield Audio Video Productions/Nancy Riskin
13816 Sunnybrook Rd, Phoenix, MD 410-628-7260
Slide Center/Lorraine Walsh
186 South St 6th Fl, Boston, MA....................617-451-9902
Sound & Motion
180 Pool St, Biddeford, ME 207-283-9191
Spicer Productions Inc/Sharon Jackson
1708 Whitehead Rd, Baltimore, MD.................410-298-1200
Starfleet Productions
RD 5/Box 91/Avalon Rd, Altoona, PA 814-942-STA
Studiolink/Andy Cordery
13 Roszel Rd, Princeton, NJ.........................609-452-0846
Teletime Productions/Warren Manos
37-39 Watermill Ln, Great Neck, NY................516-466-3882
Xerox Media Center/John Linn
780 Salt Rd/bldg 845, Webster, NY 716-423-5090

SOUTH

615 Music Productions/Randy Wachtler
1030 16th Ave S, Nashville, TN 615-244-6515
Banjo Video & Film Productions/Toby Jenkins
1188 St Andrews Rd, Columbia, SC.................803-739-0336
Beachwood Productions/Peter Wilcox
696 S Yonge St #C, Ormond Beach, FL 904-672-3698
Circle Video Productions/Rich Sublett
631 Mainstream Dr, Nashville, TN 615-244-1717
Crawford Communication/Randy Bishop
535 Plasamour Dr NE, Atlanta, GA 404-876-7149
Cypress Productions/George Cornelius
5410 Mariner St, Tampa, FL 813-289-6115
Dallas Sound Lab/Johnny Marshall
6305 N O'Connor #119, Irving, TX 214-869-1122
Development Communications/Welby Smith
1800 N Beauregard St, Alexandria, VA..............703-845-7370
Diamond Video Productions Inc/Paul Lambert
129 Citation Ct, Birmingham, AL 205-942-8888
Digital Multi-Media Post Inc/Marie Hamlin
502 N Hudson St, Orlando, FL.......................407-293-3390
Disk Productions/Joey Decker
1100 Perkins Rd, Baton Rouge, LA..................504-343-5438
EDEFX/Obadia Marcos
7355 NW 41st St, Miami, FL........................305-593-6911
Editworks/Patrick Furlong
3399 Peachtree Rd NE #200, Atlanta, GA....404-237-9977
Firstcom/Music House/Chappel/Andrea Bergeron
13747 Montfort Dr #220, Dallas, TX 214-934-2222
Florida Film & Tape/Linda Fitzpatrick
3417 Lake Breeze Rd, Orlando, FL 407-297-0091
Gannett Production Services/Nancy Putney
1611 W Peachtree St NE, Atlanta, GA 404-873-9182

Garman Audio/Video/Steve Garman
7701 N Broadway #A-6, Oklahoma City, OK.....405-842-3230
Gulf Coast Audio Visual Producers/Richard Buehrig
19 W Garden St, Pensacola, FL 904-433-3016
Henninger Video Inc/Robert Henninger
2601-A Wilson Blvd, Arlington, VA 703-243-3444
Hummingbird Productions/Susie Judkins
7 Music Sq W, Nashville, TN............................615-254-2200
Knowles Video Inc/Carl Knowles
2003 Apalachee Pkwy #204, Tallahassee, FL......904-878-2298
KXTX-TV/Dennis Brunn
PO Box 190307, Dallas, TX..............................214-521-3900
Loyola College/Silvia Pengilly
6363 St Charles Ave, New Orleans, LA 504-865-2595
MPL Film & Video/Buddy Morgan
781 S Main St, Memphis, TN...........................901-774-4944
Multivision Video & Film/Bob Berkowitz
7000 SW 59th Pl, S Miami, FL.........................305-662-6011
Music Works/Linda Thornberg
2719 Polk St #3, Hollywood, FL 305-920-4418
MVI/POST/Beck Lampert
2701-C Wilson Blvd, Arlington, VA 703-522-5335
Omega Films/Cefus McRae
3100 Medlock Bridge Rd, Norcross, GA.............404-449-8870
Park Avenue Teleproductions Inc/Melanie Cox
3500 Mayland Ct, Richmond, VA......................804-346-3232
Post Group At Disney/MGM Studios/Jim Dorriety
Roy O Disney Production Ctr, Lake Buena Vista, FL ..407-560-5600
Power Productions/John Mashburn
1967 Government St. #D, Mobile, AL 205-478-5530
Scene Three Inc/Greg Alldredge
1813 8th Ave S, Nashville, TN 615-385-2820
Selkirk Communications Video Services/Darice Lang
644 S Andrews Ave, Ft Lauderdale, FL 305-527-5007
Southwest Teleproductions Inc/John P Shives
2649 Tarna Dr, Dallas, TX...............................214-243-5719
Take One Productions/Jere Sneider
101 Pheasant Wood Ct, Morrisville, NC 919-481-0000
Take Ten Teleproductions/Coby Cooper
1111 Bull St, Columbia , SC 803-758-1230
Texas Video & Post/Grant Guthrie
8964 Kirby Dr, Houston, TX.............................713-667-5000
TMP Video Communications/Howard Fox
PO Box 37364, Raleigh, NC 919-851-2202
Trans-Delta Productions/Steve Candelera
3346 Walnut Bend Ln, Houston, TX 800-544-1681
Trimble Production Studios/Grady Trimble
612 Walnut, Little Rock, AR 501-666-8742
Video One/James Kilpatrick
4912 Travis, Houston, TX................................713-524-8823
Video Park/Charles Park Seward
11316 Pennywood Ave, Baton Rouge, LA 504-292-0840
Video Post & Transfer/Jack C. Bryan
2727 Inwood Rd, Dallas, TX 214-350-2676
Videofonics Inc/Debbie Valentine
1101 Capital Blvd, Raleigh, NC 919-821-5614
Vision Design Teleproductions/Athena Schultz
5401 Corporate Woods Dr #500, Pensacola, FL....904-484-0655
White Hawk Pictures/Michelle Barth
567 Bishopgate Ln, Jacksonville, FL..................904-634-0500

MIDWEST

Absolute Music/Johnny Hagen
420 N Fifth St #664, Minneapolis , MN 612-339-6758
AIGA/Minnesota/John DuFresne
275 Market St #54, Minneapolis, MN 612-339-6904
Allied Film and Video
7375 Woodward Ave, Detroit, MI......................313-871-2222
Alpha Video & Audio/Carlie Madison
7711 Computer Ave, Edina, MN 612-881-2055
**American Cablevision Teleproductions/
WTEN**/Tim Renshaw
3030 Roosevelt Ave, Indianapolis, IN 317-632-2288
Anderson Audio Light/Kurt Alan Anderson
7825 Wayzata Blvd, St Louis Park, MN 612-544-2001
ASCAP/Ken Clauses
3459 Washington Dr #202, Eagan, MN 612-456-9495
Asche & Spencer Music/Camille Benoit
322 1st Ave N Ste 600, Minneapolis, MN........612-338-0032
Associate Producers/Robert Hufstader
6545 Bloomfield Rd, Des Moines, IA 515-285-1209
Audio-Visual Associates/Bud Osborne
4760 E 65th St, Indianapolis, IN 317-255-6457
Avenue Edit/Benjamin Webber
625 North Michigan Ave, Chicago, IL 312-943-7100

Avsense Productions/Rich Jamieson
4133 W 45 St, Minneapolis, MN......................612-920-3770
Banghaus Music Productions/John Mannella
205 E Grand, Chicago, IL................................312-467-5441
Barlow Productions Inc/Ron Barlow Sr
1115 Olivette Executive Pkwy, St Louis, MO 314-994-9990
Bauman Communications Inc/Chuck Joffe
602 Main St #300, Cincinnati, OH 513-621-6806
Bazooka Graphics/Div Swell Pictures/Dave Mueller
233 E Wacker Dr, Chicago, IL 312-649-9000
Big City Productions/Greg Hazard
4040 Nicols Rd, Eagan, MN 612-452-8210
Big City Productions/Jim Fuller
4040 Nichols Rd, Eagan, MN 612-452-1108
Blumberg Communications Inc/Chris Hoy
525 N Washington Ave, Minneapolis, MN.........612-333-1271
Brad R Callahan & Co/Brad R Callahan
4920 Logan Ave S, Minneapolis, MN 612-929-3136
Bradley Johnson Productions/Bradley Johnson
8500 Vagabond Ln N, Maplegroove, MN 612-494-4020
Brown, Jensen & Garloff/Nils Jensen
3534 Architect Ave, Minneapolis, MN...............612-788-6588
Calliope/Michael Loftus
1034 Summit Ave, St Paul, MN........................612-222-0166
Cameron & Co/Jim Cameron
2233 University Ave #150, St Paul, MN 612-645-4002
Caryn Professional Modeling School & Agency
63 S Ninth St #200, Minneapolis, MN 612-337-8400
Chris Jones & Associates/Chris Jones
5841 Covington Ln, Minnetonka, MN 612-934-9288
Cine Service/Matthew Quastt
4015 Auburn Dr, Minneapolis, MN 612-929-0029
City Animation Co/Emma Justice
57 Park St, Troy, MI......................................313-589-0600
Classic Video Inc/Jerry L Patton
5001 E Royalton, Cleveland, OH 216-838-5377
CMI Business Communications/Gary Brown
150 E Huron St, Chicago, IL............................312-787-9040
Coast, The/Rob Barrett
1844 W Wayzata Blvd, Long Lake, MN 612-476-2204
Communications Services/Gary Fisk
Governers State Univ, University Park, IL 708-534-5000
Continental Cablevision/Madison Hts/Jenifer Rossow
32090 John R, Madison Hts, MI 313-583-1354
Control Sound Inc/Scott Bauer
3819 42nd Ave S, Minneapolis, MN.................612-724-6579
Corbett Communications/Sandy Corbett
5129 Valley View Rd, Minnetonka, MN..............612-470-4103
Corplex Inc/Chris Blake
6444 N Ridgeway, Lincolnwood, IL 708-673-5400
Creative Visuals Inc/Lois Gower
731 Harding St NE, Minneapolis, MN 612-378-1621
Crystal Productions/Tim Dwyer
1024 Blouin Dr, Dolton, IL..............................708-841-2622
Curtis Inc/Dan Castagna
2025 Reading Rd #130, Cincinnati, OH 513-621-8895
Custom Recording Studios/Jim Reynolds
4800 Drake Rd, Golden Valley, MN...................612-521-2950
Cycle Sat/Jake Laate
John K Hanson Dr, Forest City, IA 515-582-6999
Dakota Teleproductions/Curt Friesen
1134 S Sherman Ave, Sioux Falls, SD................605-331-5630
Danger Studios/C David Erbele
33 S Sixth St #4130 , Minneapolis , MN 612-338-2510
Darrell Brand/Moving Images Inc/Darrell J Brand
3308 St Paul Ave, Minneapolis, MN 612-922-3308
David A Roth Music/David A Roth
PO Box 1082, Westerville, OH 614-890-7684
David Tremblay Productions/David Tremblay
210 King Creek Rd, Minneapolis, MN 612-546-8416
Design Studios Teleproductions/David Harms
748 Ansborough, Waterloo, IA.........................319-232-4500
Doane-Tone Recorders/Christopher Doane
12809 Woodview Ln, Burnsville, MN 612-894-8387
Edcom Productions/Raymond Zalokar
26991 Tungsten Rd, Cleveland, OH 216-261-3222
Edge Multimedia/John Sink
804 Burr Oak Dr, Westmont , IL 708-920-1005
Edit Suite/Rich Melvin
1714 Boardman-Poland Rd # 15, Poland, OH.....216-757-1104
Editech Post Productions Inc/Bill Hartness
14344 Y St #200, Omaha, NE..........................402-896-9696
Editel Design/Glen Noren
301 E Erie St, Chicago, IL................................312-440-2360

PRODUCTION SERVICES
video sound integration

Elsberry Production Services/Richard Elsberry
13718 Country Pl, Burnsville, MN 612-435-7591
Elsner, John P
PO Box 8176, Minneapolis, MN........................612-377-9244
EMC Productions/Dick Stevens
300 York Ave, St Paul, MN...............................612-771-1555
Emerald City Animation/Post Prod/Robert Southerly
2660 Horizon Dr, Grand Rapids, MI 616-949-9283
Everyday Music Co/Jerry Brunskill
510 First Ave N #305, Minneapolis, MN............612-333-0502
Flying Colors Inc/Rick Cornish
1128 Harmon Pl #10, Minneapolis, MN............612-338-3405
General Television Network/Sheila Minetola
13320 Northend, Oak Park, MI 313-548-2500
**Golden Dome Productions/
WNDU-TV**/Chuck Huffman
PO Box 1616, South Bend, IN 219-631-1616
GR Barron & Co/Greg R Barron
2303 Wycliff St #200, St Paul, MN 612-645-8569
Grace & Wild Studios/Steven Wild
23689 Industrial Pk Dr, Farmington Hills, MI........313-471-6010
Greatapes/Roger Rude
1523 Nicollet, Minneapolis, MN........................612-872-8284
GRS Inc/Steve Andrews
13300 Broad St, Pataskala, OH 614-927-9566
Hartley Metzmer Huenink Comm/John Huenink
3076 S Calhoun Rd, New Berlin, WI 414-784-1010
Hartwig Music/Paul W Hartwig
123 N Third St, Minneapolis, MN 612-375-9578
Harvest Communications/Marlin Davis
900 George Washington Blvd, Wichita, KS 316-652-9900
Hellman Animates/Jim Mathis
1225 W 4th St, Waterloo, IA 319-234-7055
Herschel Commercial Inc/Donna Norwich
333 E Ontario #1802B, Chicago , IL.................312-664-4354
Hi-Tech Productions/Dave McClimans
33433 Boulevard, Eastlake, OH 216-953-0077
Hoffman Communications/Mark Hoffman
160 Glenwood Ave N #100, Minneapolis, MN ..612-673-9630
Hunt Productions Inc/John Hunt
1500 E 79th St, Bloomington, MN 612-854-5044
HVS Video/HVS Cable Adv/Patrick Krohlow
1100 Guns Rd, Green Bay, WI 414-468-3333
Image Assoc/Ann de Lodder
1625 W Big Beaver Rd, Troy , MI......................313-649-2200
Image Producers/Mark Munroe
Box 130, Canfield, OH 216-533-0100
Image Video Teleproductions/Dean Marini
6755 Freedom Ave N W, N Canton, OH 216-494-9303
In Sync/Scott Constans
5604 Sheldon St, Shoreview, MN 612-786-2334
Industrial Video/Ray Grubic
1601 N Ridge Rd, Lorain, OH 216-233-4000
Innervision Studios/Bill Faris
11783 Borman Dr, St Louis, MO 314-569-2500
**Instant Replay Video & Film
Production Inc**/Stephanie Smith
1349 E McMillan St, Cincinnati, OH 513-569-8600
Iowa Teleproduction Center Inc/Brad Morford
4800 Corporate Drive, West Des Moines, IA........515-225-7800
J Parker Ashley Associates/Rick Benjamin
119 N Fourth St #208, Minneapolis, MN 612-340-9807
John Flomer's Primal Cinema/John Flomer
PO Box 80514, Minneapolis, MN.......................612-822-6732
Jones-Rasikas/Arnold Jones
800 Bond NW, Grand Rapids, MI 616-456-1855

PRODUCTION SERVICES

video sound integration

Joy Art Music/Brenda Stewart
   247 E Ontario, Chicago, IL............312-642-7100
Juntunen Video/Field & Post Production/Diane De Lisi
   708 N 1st St #131, Minneapolis, MN....612-341-3348
Kasper Studio/Walt Kasper
   17903 Huron River Dr, New Boston, MI......313-753-9100
Keynote Productions Inc/Richard Hahn
   4322 Mahoning Ave, Youngstown, OH........216-793-7295
Koplar Communications Ctr/Jim Wright
   4935 Lindell Blvd, St Louis, MO........314-454-6320
Kuether Music Productions/Steven Kuether
   2206 Sheridan Ave N, Minneapolis, MN....612-521-1581
L-E-O Systems Inc/Jill O'Neil
   1505 E David Rd, Dayton, OH........513-298-1503
Lamb & Co/Lawrence Lamb
   650 Third Ave S 17th Fl, Minneapolis, MN......612-333-8666
Learn-PC Video System
   10729 Bren Rd E, Minneapolis, MN........800-532-7672
Lighthouse Productions Inc/Rick Meredith/Chip Moore
   1900 Hicks Rd, Rolling Meadows, IL......708-506-1414
Location & Back Ltd/Randy Berglin
   15381 Trillium Cir, Eden Prairie, MN......612-937-2985
LTS Productions/David Lang
   2810 Bennett Rd, Okemos, MI........517-332-1190
LVI Video Productions/Dirk Dantuma
   975 Grand Ave, St Paul, MN........612-292-9325
Mallory Marketing Communications/Bob Mallory
   1310 E Hwy 96, St Paul, MN........612-426-1875
Marketing By Video Inc/Bill Hagen
   6440 Flying Cloud Dr #101, Eden Prairie, MN....612-944-3323
Martin/Bastian Communications/Steve Martin
   105 Fifth Ave S, Minneapolis, MN........612-375-0055
Marx Production Ctr/Tom Deming
   3100 W Vera Ave, Glendale, WI........414-351-5060
MAS Services/Loren Deutz
   4151 Wentworth Ave S, Minneapolis, MN......612-823-3333
Master Video Productions Inc/Bill Dean
   7322 Ohms Ln, Minneapolis, MN........612-831-0877
Matney & Associates Inc/Ed Matney
   8651 Spring Creek Rd, Northfield, MN......507-663-1048
Media Consultants/Richard Wrather
   201 N Stoddard/PO Box 130, Sikeston, MO......314-472-1116
Media Group Inc/Gregory Woods
   1480 Dublin Rd, Columbus, OH........614-488-0621
Media Group Television/Curt Shaffer
   7th Ave & 23rd St, Moline, IL........309-764-6411
Media Loft Inc/Pat Rousseau
   333 N Washington Ave #210, Minneapolis, MN ...612-375-1086
Media Productions Inc/Judith Kessel
   119 N Fourth St #205, Minneapolis, MN......612-339-1355
Media Team/Jeff Sylvestre
   25 University Ave SE #407, Minneapolis, MN......612-379-0898
Menten Music Inc/Dale Menten
   10 S Fifth St #450, Minneapolis, MN......612-333-4650
Metro Productions Inc/Tim McGovern
   238 W 74th St, Kansas City, MO........816-444-7004
MID-CO TV Systems Inc
   3440 Kilmer Ln N, Plymouth, MN........612-544-3375
Midwest Teleproductions/Tom Orrick
   3947 State Line Rd, Kansas City, MO......816-531-3838
Mill City Records/Al Heigl
   PO Box 177, Northfield, MN........507-663-6090
Minnesota Video Productions/Tim O'Phelan
   11358 Viking Dr, Eden Prairie, MN......612-721-7100
Multimedia
   5610 Rowland Rd, Minnetonka, MN........612-931-2112

Music Advantage/Paul Gerike
   112328 Chatfield Ct, Chaska, MN........612-448-7544
Music Staff Inc/Decker Velie
   121 W Franklin, Minneapolis, MN........612-870-7430
Northwestern Mutual Life/Betty Hoff
   720 E Wisconsin Ave, Milwaukee, WI........414-271-1444
Nygard & Associates/Jim Nygard
   400 N First St #120, Minneapolis, MN........612-371-9228
Omni Productions/W H Long
   655 W Carmel Dr, Carmel, IN........317-844-6664
On Line Video/Dennis Majewski
   720 E Capitol Dr, PO Box 693, Milwaukee , WI..414-223-5254
Optimus Inc/Cat Gulick
   161 E Grand Ave, Chicago, IL........312-321-0880
Ozam Film & Tape Productions, Imagination Effects/Dale Detoni
   1228 S Fort, Springfield, MO........417-866-3232
Paradigm Communication Group/Keith Meyer
   250 Production Plz, Cincinnati, OH........513-381-7100
Phar-Mor Productions/Diane Olenik
   20 Federal Plaza W, Youngstown, OH........216-746-6641
Platinum Productions/Rob Gibbons
   8100 N High St, Columbus, OH........614-888-4181
Post Effects/Jacqui Jones
   400 W Erie St #101, Chicago, IL........312-944-1690
Post Production Services Inc/Dave Dittgen
   602 Main St #900, Cincinnati, OH........513-621-6677
Postique Inc/Mary Suzanne Patek
   23475 Northwestern Hwy, Southfield, MI....313-352-2610
Prairie Production Group
   111 W Goose Alley, Urbana, IL........217-344-4675
Precision Tapes Inc/Daniel Piepho
   2301 E Hennepin, Minneapolis, MN........612-379-7554
Producers Color Service Inc/Susan Flynn
   24242 Northwestern Hwy, Southfield , MI.......313-352-5353
Production House/ImageLab
   811 St John St. Marketing Dept, Highland Park, IL708-433-3172
Project Four Communications/Jack Marr
   4536 Garfield Ave S, Minneapolis, MN........612-824-3881
Promedia Productions/Steve Keller
   2593 Hamline Ave, Roseville, MN........612-631-3681
Pulse Communications/Fred Monthey
   211 N Broadway, Green Bay, WI........414-436-4777
RDB Productions Inc/Russ Beckner
   2100 Heatherwood Ct, Middletown, OH........513-422-9552
RE Miller Communications/Roger E Miller
   3520 W 21st St, Minneapolis, MN........612-925-0781
Real Productions/Robert J Schuster
   1821 University Ave #N-153, St Paul, MN......612-646-9472
Reel Good Productions/Kyrl Henderson
   PO Box 6, Excelsior, MN........612-470-0232
Reelworks Animation Studio/Peter Lang
   318 Cedar Ave S #300, Minneapolis, MN........612-333-5063
Renaissance Video Corp/John Feldbaille
   130 S Jefferson, Chicago, IL........312-930-5000
Revelation Video/Paul A Tatge
   7085 Shady Oak Rd, Eden Prairie, MN........612-437-6604
Roaring Fork Productions/Clay Nicles
   11720 28th Ave N, Plymouth,, MN........612-559-1198
Russell-Manning Productions/Bruce Clark
   18 N 4th St, Minneapolis, MN........612-338-7761
RW Video Inc/Gerrit Marshall
   4902 Hammersley Rd, Madison, WI........608-274-4000
Sanders & Co/Denita Clifford
   3610 N Meridian, Indianapolis, IN........317-926-2841
Second Sight Productions/Thom Bixbe
   950 N Meridian St # 33, Indianapolis, IN........317-237-7888
Semantic Music/C Lynn Seacord
   PO Box 284, Hopkins, MN........612-935-6976
Seraphim Communications Inc/Bill Huff
   1568 Eustis St, St Paul, MN........612-645-9173
Severson, Michael R
   17343 Ithaca Ln, Lakeville, MN........612-435-8520
Shaw Video Communications/Paul Quinn
   3246 Henderson Rd, Columbus, OH........614-457-4477
Sims, John L
   2428 Dupont Ave S, Minneapolis, MN........612-377-9911
Skyview Film & Video/Vivian Craig
   541 N Fairbanks, Chicago, IL........312-670-2020
SOS Productions/Mark Braver
   753 Harmon Ave, Columbus, OH........614-221-0966
Sound Decisions Inc/Brian Basilico
   10 E 22nd St #108, Lombard, IL........708-916-7070
Sound Resources/Michele Jansen
   1400 Energy Park Dr #21, St Paul, MN........612-644-3660

Sound/Video Impressions/Bill Holtane
   110 River Rd, Des Plaines, IL........708-297-4360
Steve Blexrud Studio/Steve Blexrud
   3722 22nd Ave S, Minneapolis, MN........612-721-5325
Steve Ford Music/Ken Kolasny
   506 N Clark, Chicago, IL........312-828-0556
Tagteam Film & Video/Jeff Byers
   2525 Franklin Ave E #203, Minneapolis, MN....612-338-3360
Take 1 Productions Ltd/Teri Murphy
   5325 W 74th St, Minneapolis, MN........612-831-7757
Talent*Media Presents/Don Cosgrove
   1846 Summit Ave, St Paul, MN........612-698-7141
Tele Edit Inc/Keith Pokorny
   10 S 5th St #640, Minneapolis, MN........612-333-5480
Tele-Producers Inc/Harlan Meyer
   7085 Shady Oak Rd, Eden Prairie, MN........612-941-2988
Telemation Productions/John Dussling
   100 S Sangamon, Chicago, IL........312-421-4111
TeleVideo Productions/Mike Sopa
   611 S Farwell St, Eau Claire, WI........715-833-9269
Terry Fryer Music/Terry Fryer
   1334 Asbury, Evanston, IL........312-440-0944
TGA Recording Co/Thomas G Alti
   295 Urbandale Dr, Benton Harbor, MI........616-926-7581
Tharnstrom Music/John Tharnstrom
   610 N Fairbanks, Chicago, IL........312-266-7773
The Roland Co/Will Sullivan
   420 Summit Ave 3rd Fl, St Paul, MN........612-222-8716
Thirteen Fifty-one Prdctns/Tom Kwilosz
   1351 W Grand, Chicago, IL........312-421-0400
Toby's Tunes Inc/Harley Toberman
   2325 Girard Ave S, Minneapolis, MN........612-377-0690
Total Video 3/Don Browers
   10714 Mockingbird Dr, Omaha, NE........402-592-3333
Track Record Studios Inc/Norton Lawellin
   1561 Sherburne Ave, St Paul, MN........612-645-9281
Track Seventeen Productions/Curtis R Olson
   7500 Noble Ave N, Minneapolis, MN........612-566-4859
Tracking Station/Brad Stokes
   3504 44th Ave S, Minneapolis, MN........612-729-8712
Tullio & Ruppert/Ann Ryan
   211 E Ohio #1909, Chicago, IL........312-744-0197
TV-5 WNEM/John Haupricht
   107 N Franklin, Saginaw, MI........517-755-8191
Twenty-Nine Point Five Music/John Calder
   3712 Garfield Ave S, Minneapolis, MN........612-825-2900
Urban Communications/Kevin Hovey
   14770 Martin Dr, Eden Prairie, MN........612-934-1100
USA Teleproductions/Jamie Burns
   1440 N Meridian, Indianapolis, IN........317-632-5900
Vanguard Productions Inc/Jerry Beck
   7020 Huntley Rd, Columbus, OH........614-436-4610
Vaughn Communications Inc/Don Drapeau
   7951 Computer Ave S, Minneapolis, MN........612-832-3100
VGI Productions/Video Genesis Inc/Howard Schwartz
   4949 Galaxy Pkwy, Cleveland, OH........216-464-3635
Vidcam Productions Inc/Mary Detomaso
   7150 Hart St/Unit A-1, Mentor, OH........216-255-5050
Video Duplication Inc/Bill Nord
   7322 Ohms Ln, Edina, MN........612-831-2011
Video Features/Don Adamson
   346 Gest St, Cincinnati, OH........513-241-6262
Video Genesis/Jim Gerber
   4949 Galaxy Pkwy, Cleveland, OH........216-464-3635
Video Impressions/Rick Roesing
   1666 N Farnsworth, Aurora, IL........708-851-1663
Video Post/M Kappleman
   4600 Madison #120, Kansas City, MO........816-531-1225
Video Post & Graphics Inc/Claude Kennedy
   14 Cambridge Ave, Dayton, OH........513-276-3113
Video Wisconsin/Jeff Utschig
   18110 W Bluemound Rd, Brookfield, WI........414-785-1110
Visions/Bruce Baldwin
   6729 Amherst Ln, Eden Prairie, MN........612-934-2120
VPL Productions/Vincent Lucarelli
   415 Assembly Dr, Bolingbrook, IL........708-739-6439
Wizardly Teleproductions Inc/John R Sparr
   5705 44th Ave S, Minneapolis, MN........612-727-2009
Xenas Communications/Greg Peck
   354 Gest St, Cincinnati, OH........513-621-2729

W E S T

3D Magic/FTI/Dave Altenau
   55 Hawthorne St 10th Fl, San Francisco, CA........415-543-2517

4Media Company/Paul Sehenuk
2813 W Alameda Blvd, Burbank, CA...............818-840-7000
525 Post Production/Jerry Cancellieri
6424 Santa Monica Blvd, Hollywood, CA...........213-466-3348
Acmetek/Ken Lau
1145 Hacienda Blvd #126,
Hacienda Heights, CA.....................................818-333-6161
Anderson Video/Frank Bluestein
100 Universal City Pl, Universal City, CA 818-777-7999
Annenberg Ctr at Eisenhower/TV Svcs/Karen Snyder
39000 Bob Hope Dr, Rancho Mirage, CA.............619-773-4554
Applied Technical Video Inc/Dennis Kushner
1564 Commerical St SE, Salem, OR 503-371-9498
Aristosoft
7040 Koll Ctr Pkwy #160, Pleasanton, CA..........510-426-5355
Armock, Amy
492 N 2nd St #2, San Jose, CA.....................408-294-0664
Armour Productions/Stephanie Tucker
306-C W Katella, Orange, CA.........................714-538-5811
Artichoke Productions/Paul Kalbach
4114 Linden St, Oakland, CA.........................510-655-1283
Associated Production Music/Cassie Lord
6255 Sunset Blvd #820, Hollywood, CA............800-543-4276
Atomic Media Group/Carol Leigh
1933 Davis St #213, San Leandro, CA..............510-430-8012
Avid Productions/Henry Bilbao
1300 Industrial Rd #1, San Carlos, CA.............415-593-2000
Bimstein, Phillip Kent
1942 Zion Park Blvd, Springdale, UT...............801-772-3839
Bob Ohlsson Digital Audio/Bob Olhsson
PO Box 555, Novato, CA.............................415-898-2981
Bruce Hayes Productions/Bruce Hayes
959 Wisconsin St, San Francisco, CA................415-282-2244
California Communications Inc/Hope Schenk
6900 Santa Monica Blvd, Los Angeles, CA 213-466-8511
California Image Associates/Duane Thompson
3034 Gold Canal Dr #B, Rancho Cordova, CA...916-638-8383
Cassette Productions Unlimited/Joel V Brooks
5796 Martin Rd, Irwindale, CA......................818-969-6881
Catalyst Productions/Chris Negri
1431 Center St, Oakland, CA.........................510-836-1111
Central Image Video & Photography/Larry Gale
337 Majors St, Santa Cruz, CA......................408-454-0500
Cinema Engineering Co/Richard L Bennett
7243 Atoll Ave, N Hollywood, CA 818-765-5340
CommuniCreations
2130 S Bellaire, Denver, CO 303-759-1155
Complete Post/Bob Brian
6087 Sunset Blvd, Hollywood, CA 213-467-1244
**Continental Productions, Continental TV
Network**/Cheryl Cordeiro
118 6th St S/PO Box 1219, Great Falls, MT 406-761-5536
Crosspoint Productions/Peter Ten Eyck
940 Wadsworth Blvd, Lakewood, CO 303-232-9572
Crowley & Associates/Paul Crowley
224 Mississippi St, San Francisco, CA 415-621-4477
Dial Music Productions/David Dial
PO Box 2043, Burbank, CA..........................818-841-3301
Digidesign Inc/Peter Gotcher
1360 Willow Rd #101, Menlo Park, CA.............415-688-0600
DLH Studios/Dennis Hocking
2900 Adams #B-29, Riverside, CA...................909-687-6654
Double Vision Productions/Betty Vivado
1955 Railroad Dr #A, Sacramento, CA..............916-921-6986
Dream Quest Images/Greg Pappas
2635 Park Center Dr, Simi Valley, CA 805-581-2671
Dub Masters/SP/Caroline Didiego
3923 S McCarran, Reno, NV 702-827-3821
Editel/Los Angeles/Michael Moreale
729 N Highland Ave, Hollywood, CA 213-931-1821
Effective/Visual Imagery/Mia Corrales
151 Kalmus #K1, Costa Mesa, CA 714-641-5222
Encore Video/Larry Chernoff
6344 Fountain Ave, Hollywood, CA 213-466-7663
EP Graphics/Burbank/Eddie Pong
2525 N Naomi, Burbank, CA.........................818-953-9375
FilmCore/Valerie Petrusson
849 N Seward St, Hollywood, CA 213-464-7303
Full Spectrum Productions/John McCauley
150 E Dana St, Mountain View, CA 415-967-1883
G.R.A.P.E. Video Recording/Sharon Robinson
89 W Neal St, Pleasanton, CA.......................510-462-1300
Gannett Productions/Jim Berger
500 Speer Blvd, Denver, CO 303-871-1899

Glenn Roland Films/Glenn Roland
10711 Wellworth Ave, Los Angeles, CA 310-475-0937
Gomes, Steve
1901 S Bascom Ave #1190, Campbell, CA........408-559-6369
Goss, John C
1022 Keniston, Los Angeles, CA 213-938-9526
Graphic Media Inc/Doug New
411 SW Second, Portland, OR.......................503-223-2262
Half-Inch Video/Sean Scott
185 Berry St #6501, San Francisco, CA 415-495-3477
Helical Productions/Bob Brandon
3801 E Florida Ave # 613, Denver, CO 303-756-0044
Hollywood South Producers Centre/Jerry Kelley
2465 Campus Dr #A, Irvine, CA.....................714-757-7588
I/O Productions Inc/Laurent Basset
7551 Sunset Blvd #202, Hollywood, CA............213-969-0113
Inland Audio Visual Co/Larry Bergman
27 W Indiana, Spokane, WA 800-584-8916
Inter Video/Chris Olson
10623 Riverside Dr, Toluca Lake, CA.................818-VIDEO-24
Interactive Audio/Gary Levenberg
544 Natoma St, San Francisco, CA..................415-431-0778
Interactive Records/Steven Rappaport
921 Church St, San Francisco, CA 415-285-8650
International Corporate Video Inc/James Fasso
1020 Serpentine Ln #114, Pleasanton, CA..........510-426-8230
Intersound Inc/Garry Morris
8746 Sunset Blvd, Los Angeles, CA 310-652-3741
JRA Interactive/Jane McLaughlin
3647 Sunset Beach Dr NW, Olympia, WA........206-866-0533
KCFW Productions/Curt Smith
401 First Ave E, Kalispell, MT........................406-755-5239
Keller & Cohen/Beth Fenn
1315 Bridgeway, Sausalito, CA 415-331-6377
KGTV Channel 10/Jack Villarrubia
PO Box 85347, San Diego, CA 619-237-1010
Killer Tracks/Phil Spieller
6534 Sunset Blvd, Hollywood, CA 213-957-4455
Killingsworth Presentations/Penny Engel
3834 Long Beach Blvd, Long Beach, CA 310-595-7796
Klang Sound
75 Wood St, San Francisco, CA 415-931-3382
KTVU Retail Services/Rich Hartwit
2 Jack London Sq/PO Box 22222, Oakland, CA.510-874-0228
Laguna Productions/Charlie King
2708 S Highland Dr, Las Vegas, NV 702-731-5600
Laser Pacific Corp /Steve Mitchell
540 Hollywood Way, Burbank, CA...................818-842-0777
Main Street Video Productions Inc/Mark Tyson
119 N Wahsatch Ave, Colorado Springs, CO.....719-520-9969
Master Communication/Robert Masters
3429 Kerckhoff Ave, San Pedro, CA.................310-832-3303
Master Videoworks/Jeff Killian
3611 S Harbor Blvd #150, Santa Ana, CA.........714-241-7724
Matthews/Griffith Music/Kathy Matthews
6565 Sunset Blvd, Hollywood , CA 213-993-2700
**MDC Teleproductions/McDonnell
Douglas**/Milton Moline
4000 Lakewood Blvd, Long Beach, CA 310-496-9040
Media Services–SoundMax/Eric Shed Brandon
PO Box 179, Cameron Park, CA.....................916-676-3342
Media Staff/Jerry Maybrook
6926 Melrose Ave, Hollywood, CA..................213-933-9406
Metropolitan Entertainment/Andy Ullman
1680 N Vine St #600, Hollywood, CA..............213-856-7060
Moving Pixels/Dave Thomas
1833 Verdugo Vista Dr, Glendale, CA...............818-247-9508
Network Music Inc/Ken Berkowitz
15150 Ave of Science, San Diego, CA..............800-854-2075
Northwest Media
PO Box 56, Eugene, OR..............................503-343-6636
Omnimusic/Jerry Burnham
6255 Sunset Blvd #803, Hollywood, CA...........800-828-6664
Opcode Systems Inc/Chris Halaby
3950 Fabian Way #100, Palo Alto, CA 415-856-3333
Pace Video Center/Romeo Carrera
2020 SW 4th Ave #700, Portland, OR.............503-226-7223
Pacific Ocean Post/Alan Kozlowski
730 Arizona Ave, Santa Monica, CA................310-458-3300
Pacific Rim Video/Roger Proulx
663 Maulhardt Ave, Oxnard, CA.....................805-485-9930
Passport Designs Inc/Steve Bertges
100 Stone Pine Rd, Half Moon Bay, CA.............415-726-0280
Penrose Productions/Mickey Bernard
1042 Hamilton Ct, Menlo Park, CA 415-323-8273

PRODUCTION SERVICES
video sound integration

Piece of Cake/Sandro Lobinger
4425 Clybourn Ave, N Hollywood, CA 818-763-2087
Post Group/Rich Ellis
6335 Homewood Ave, Los Angeles, CA 213-462-2300
Precision Visual Communications/Herb Saunders
591 25th Rd, Grand Junction, CO 303-242-1783
Premore Inc
5130 Klump Ave , N Hollywood, CA.................818-506-7714
Production West Inc/Jim Abel
2110 Overland Ave #120, Billings , MT 406-656-9417
Rase Productions, Bill
955 Venture Ct, Sacramento, CA 916-929-9181
Rick Resnick Audio Visual Design/Rick Resnick
5544 E Kings Ave, Scottsdale, AZ 602-482-3070
Rocky Mountain AV Productions/Terry Talley
4301 S Federal Blvd #101, Englewood, CO 303-730-1100
San Francisco Production Group/Debra Robins
550 Bryant St, San Francisco, CA.....................415-495-5595
Signet Sound Studio/David Dubow
7317 Romaine St, Los Angeles, CA 213-850-1515
Slide Factory/Carlos Lara
300 Broadway #14, San Francisco, CA..............415-957-1369
Sound Photosynthesis/Faustin Bray
PO Box 2111, Mill Valley, CA 415-383-6712
Starlite Studios/Judy Darnell
1530 Old Oakland Rd #140, San Jose, CA........408-453-1166
Steve Michelson Productions/Steve Michelson
280 Utah St, San Francisco, CA......................415-626-3080
Stimulus/Grant Johnson
PO Box 170519, San Francisco, CA 415-558-8339
Sunset Post Inc/Bob Glassenberg
1813 Victory Blvd, Glendale, CA 818-956-7912
Tanglewood Productions/Michael Eardly
1105 Terminal Way #217, Reno, NV................702-688-6282
TeleScene Inc/Rick Larson
3487 W 2100 S, Salt Lake City, UT 801-973-3140
THS Visuals/Todd Simon
PO Box 2192, Lake Tahoe, Reno, NV 702-588-6976
Total Video Co/Peter Roger
432 N Canal St #12S San Francisco, CA 415-583-8236
**University of North Colorado/
School of Music**/Gene Aitken
Greeley, CO 303-351-2577
Valentine Productions at Omega/Laurie Gordon
725 Mariposa, Denver, CO...........................303-825-1900
Valley Production Center/Frank Bronell
6633 Van Nuys Blvd, Van Nuys, CA.................818-988-6601
Vegas Valley Prods/KVVU Bdcst Corp/Richard Walsh
25 TV5 Dr, Henderson, NV...........................702-435-5555
Video One Inc/Kevin Hamburger
10625 Chandler Blvd, N Hollywood, CA............818-980-0704
Video West Productions/Jim Yorgason
55 N 300 W, Salt Lake City, UT 801-575-7400
Vision Productions/KSAZ-TV/Bill Lucas
511 W Adams, Phoenix, AZ...........................602-262-5131
WalkerVision Interarts/Pat Walker
PO Box 22533, San Diego, CA 619-458-9038
West Coast Projections Inc/David Gibbs
11245 W Bernardo Ct #100, San Diego, CA......619-674-7334
Westcom Creative Group/Steve Ogle
2295 Coburg Rd #105, Eugene, OR................503-484-4314
Western Images
600 Townsend #300W, San Francisco, CA 415-252-6000
Wickerworks Video Productions/Steve Hug
6020 Greenwood Plaza Blvd, Englewood, CO....303-741-3400

Windstar Studios Inc/Dale Mitchell
525 Communications Cir, Colorado Springs, CO.719-635-0422
Woodholly Productions/Yvonne Parker
712 N Seward St, Hollywood, CA 213-462-5330
Yamaha Corp of America/Joan McKabin
6600 Orangethorpe Ave, Buena Park, CA 714-522-9011
Yurth Video Production Services/Les Yurth
PO Box 1305, Woodland Hills, CA 818-999-0080
Z-AXIS/Ray Hauschel
116 Inverness Dr E #110, Englewood, CO..........303-792-2400

CANADA

Command Post & Transfer
179 John St, Toronto, ON............................416-585-9995
Groupe Andre Perry/Mario Rachiele
1501 Parre St, Montreal , QB 514-933-0813
Hoffert Communications/Paul Hoffert
73 Brookview Dr, Toronto, ON 416-781-4191
Oasis Post Production Services/Don McMillan
340 Gerarrd St E, Toronto , ON 416-466-5870
Pacific Post Productions
770 Pacific Blvd S, Vancouver, BC 604-685-8686

Prepress

NEW YORK CITY

● **ACE GROUP, THE (P 101)**
149 W 27th St, New York, NY.......................212-255-7846
AMKO Color USA (Pixel Andot Inc)/Howard Kwak
301 Madison Ave, New York, NY....................212-687-0400
Bengal Graphics Inc/Teresa Powers
175 Varick St, New York, NY......................212-924-1762
Callaway Editions/Nicholas Callaway
54 Seventh Ave S, New York, NY...................212-929-5212
● **COPYTONE**/Mark Bevilacque (P 103)
115 W 45th St 10th Fl, New York, NY 212-575-0235
Cotton Communications, Jim
417 Canal St, New York, NY 212-431-1923
Designers Atelier/Dave Perskie
45 W 45th St #808, New York, NY..................212-221-8585
Dots/Steve Ku
31 E 28th St 5th Fl, New York, NY..................212-779-4888
Elite Color Inc/Tom Tivo
127 W 25th St 8th Fl, New York, NY...............212-989-3289
EMR Systems Comm/Frank Santisi
134 W 26th St, New York, NY......................212-989-4442
Image Axis Inc/Kevin O'Neill
38 W 21st St, New York, NY.......................212-989-5000
Kwik Color Ltd/Richard Sirota
229 W 28th St, New York, NY......................212-643-0200
Lloyd & Germain/Stuart Germain
31 E 28th St 6th Fl, New York, NY................212-686-8181
Micropage/Microtype/Michael Comodo
900 Broadway, New York, NY......................212-533-9180
Pergament Graphics
38 E 30th St, New York, NY.......................212-213-8310
Potomac Graphic Industries Inc/Tony Giordano
508 W 26th St, New York, NY.....................212-924-4880
Speed Graphics
342 Madison Ave , New York, NY..................212-682-6520

EAST

Aero Graphics
15 Hudson Ave, Rochester, NY 716-232-4848
Baltimore Color Plate
9006 Yellow Brick Rd, Baltimore, MD 410-391-8855

CCS Princeton Inc/Michael Burns
521 Cornwall Ave, Buffalo, NY......................716-837-0800
Color Logic/Jim Longo
299 Webro Rd, Parsippany, NJ 201-515-0099
Color Professionals Inc
217 Brook Ave, Passaic Park, NJ....................201-228-2058
ColorScan Services Inc
241 Stuyvesant Ave, Lyndhurst , NJ................201-438-6729
Colotone Inc
260 Branford Rd, PO Box 97, N Branford, CT.....203-481-6190
CompuColor Inc
108 Ridgedale Ave, Morristown, NJ 201-538-9696
CrownGraphics Inc
828 Evarts St NE, Washington, DC 202-832-6767
Design Imaging/Joan Baird
1730 M Street NW #505, Washington, DC......202-296-3799
Design Imaging
1730 M St NW #505, Washington, DC............202-296-3797
E Lambert Inc
5721 Rte 130 S, Pennsauken, NJ...................609-665-3680
Expert Imaging/Slide Service
60 Farmington Ave, Longmeadow, MA 413-567-0574
Fidelity Color Inc
PO Box 100, Brownstown , PA......................717-627-1311
Gamma One Inc/Kieram Doherty
12 Corporate Dr, North Haven, CT..................203-234-0440
Graphic Color Image
1 Rex Dr, Braintree, MA 617-849-1706
Graphics Trade Services Inc
1027 Commerical Ave, E Petersburg, PA.........717-569-8751
Graphtec Inc/Marriott Jr Winchester
1724 Whitehead Rd, Baltimore, MD.................410-298-6100
GS Imaging Services
701 Ashland, Folcroft, PA.........................215-532-2170
Hahn Graphics Inc/Helene Hahn
201 W Padonia Rd #400, Timonium, MD..........410-252-8888
HK Graphics/Elizabeth Disessa
82 Spring St, Everett, MA 617-387-3301
Image Foundry/Tim Fields
3100 Elm Ave, Baltimore, MD.....................410-889-6660
Image Press/Bob Sollish
439 Hawley Ave, Syracuse, NY 315-422-8984
Infinity USA Inc/Harvey Gelber
141 Eileen Way, Syosset, NY......................516-496-3900
K&L Color Graphics/Dave Langlois
9 Currier St, White River Junction, VT...............802-295-1661
Lanman Companies/Michael Wight
120 Q St NE, Washington, DC.....................202-269-5400
Laser Images Inc/Jim Condio
704 Second Ave, Pittsburgh, PA 412-471-2882
Media Works/Jay M Arancio
185 Hickory Corner Rd, Milford, NJ................908-996-7855
Mirage Visuals/Curt Brooks
206 Monroe St, Monroeville, PA 412-372-4181
Pace Color Separation Inc/Charles Travis
84 Liberty St, Quincy, MA 617-773-1972
PH Zimmerman Multiples
8 Wildey Rd, Barrytown, NY.......................914-758-4488
Phototype Color Graphics/Joel Rubin
7890 Airport Hwy, Pennsauken, NJ................609-663-4100
Prepworks Inc/Bill Manning
1050 Commonwealth Ave, Boston, MA 617-232-3010
Prestige Color/Jim Dommel
19 Prestige Lane, Lancaster, PA....................717-392-1711
Spectrum/Tina Gagnon
4 Raymond Ave, Salem, NH 603-893-9294
Swan Engraving/Stuart Swan
31 Wordin Ave, Bridgeport , CT 203-366-4308
TruColor Labs/David Calhoun
20 Powers Dr, Paramus, NJ........................201-261-2107
Wilson, Barry
654-1/2 Morton Pl NE, Washington, DC.........703-276-5386
Wrap-Arounds Inc
80 Windsor St, W Springfield, MA..................413-781-5440

SOUTH

American Color
3874 Fiscal Ct, W Palm Beach, FL..................407-842-7201
Blackhawk Color Corp/Leann Sanderson
13900 49th St N, Clearwater, FL...................813-535-4641
Blanks Color Imaging
2343 N Beckley Ave, Dallas, TX....................214-741-3905
Color Trend Corp/David Reierson
1141 Holland Dr #9, Boca Raton, FL 407-994-2657

Comp-U-Type/Chris Cochran
1920 Tenth Ave S, Birmingham, AL 205-323-8898
Dimension Inc
1507 W Cass St, Tampa, FL........................813-251-0244
Ellison Graphics Inc/Clarke Blacker
1300 S 14th St, Lantana, FL........................407-746-9256
Eyebeam/Miles Wright
2500 Gateway Center Blvd #600, Morrisville, NC ...919-469-3859
Graphic Enterprises Inc/Laura Rogers
316 E Abram St, Arlington, TX 817-277-9442
Graphics Group/Vivian Darby
2820 Taylor St, Dallas, TX 214-749-2222
Graphtec Inc/Frank Mauro
505 NW 65th Court, Ft Lauderdale, FL 800-444-1931
Holmes, Michael
409A N Tyler, Dallas, TX...........................214-942-4061
International Imaging Inc/Red Maxwell
1531-J Westbrook Plaza Dr, WInston-Salem, NC .919-760-0770
Laser Tech Color Inc/Damien Gough
1505 Luna Rd Bldg 2 #202, Carrollton, TX.........800-365-8957
Magna IV Color Imaging/Gary Middleton
701 W 7th , Little Rock, AR........................501-376-2397
Martin/Greater Film Graphics Inc/Dan Bontempo
611 Julia St, New Orleans, LA 504-524-1741
Moberley, Connie
215 Asbury, Houston, TX..........................713-864-3638
NEC Inc/Roy Luckett
1504 Elm Hill Pike, Nashville, TN 615-367-9110
Peerless Engravers/Jerry Masters
823 Main, Little Rock, AR 501-375-8266
Photo Offset Plate Service Inc
706 E Chestnut, Louisville, KY.....................502-584-3363
PhotoEngraving Inc
502 N Willow Ave, Tampa, FL......................813-253-3427
Precision Color Inc
905 Gleaves St, Nashville , TN.....................615-254-1496
Presentation Resource/Dee Hardie
5 W Cary St, Richmond, VA 804-648-7854
Professional Publications/Ann McMahon
1646 Belmont Ave, Baton Rouge, LA...............504-346-0707
Quebecor Printing Prepress Group/Lew Gravis
111 Mackenan Dr, Cary, NC 919-481-0627
Savannah Color Separations Inc
417 E Broughton St, Savannah, GA................912-233-8053
Technigraphics Inc
1709-1721 Western Ave, Knoxville, TN............615-637-6811
Vista Color Corp/Henry Serrano
3401 NW 36th St, Miami, FL.......................305-635-2000
Wallace Engraving Co/John McGlothlin
Box 17606, Austin, TX.............................512-444-2244
Wilson Engraving Co Inc/Fred Fard
PO Box 655591, Dallas, TX........................214-565-9000
Words & Pictures/Diane Maldin
4372 Spring Valley Rd, Dallas, TX..................214-702-9119
WRE/Colortech Prepress/John Comerford
533 Banner Ave, Greensboro, NC 919-275-9821

MIDWEST

ASG Inc
3365 N Drake, Chicago, IL 312-604-4333
Bass Litho Color/Dick Bass
3015 E Cairo, Springfield, MO 417-866-4929
Blackbourn Inc/Tom Scherkenbach
5270 W 84 St Ste 200, Bloomington, MN........612-835-9040
Brown Graphics/Bob Smith
2320 Pine, St Louis, MO 314-241-2214
**Central Apex Engraving Co–
Reproductions Inc**/Tom Kletzker
1307 Washington, St Louis, MO...................314-421-4436
Color Associates Inc/Ted Pech
10818 Midwest Ind Blvd, St Louis, MO............314-423-9300
Color Image/Tom O'Donnell
461 N Milwaukee Ave, Chicago, IL................312-666-2844
Color Systems
1514 Springfield St, Box 1333, Dayton, OH.......513-253-8959
Color-Imetry Inc
5557 W Howard St, Skokie, IL.....................708-679-5757
Colorbrite Inc
1001 Plymouth Ave N, Minneapolis, MN...........612-522-6711
Colorcraft Inc
3765 N 35th St, Milwaukee, WI....................414-442-1344
Crown Color Corp/Nelson Flackus
7 S Rte 12, Fox Lake, IL 708-587-2177
Dakota Photographics Inc/Steve Berglund
669 Fourth Ave N, Fargo, ND......................701-232-8808

El Dorado Graphics/George Gaulke
  15791 W Ryerson Rd, New Berlin, WI 414-784-3520
Entercor Studio/Stuart Bodin
  4330 Nicollet Ave S, Minneapolis, MN 612-825-6620
Enteron Group/John Reilly
  815 S Jefferson, Chicago, IL............................312-922-8816
ETS Graphics Inc/Bruce Leisy
  340 Pattie Ave/PO Box 606, Wichita, KS..........316-265-4621
Flying Color Graphics Inc
  1001 W North St, Pontiac, IL..........................815-842-2811
Graphic Services/Steve Kelly Sr
  1520 Second Ave, Des Moines, IA 515-244-0112
Holland Litho Service/Brian Baarman
  101 W 17th St, Holland, MI 616-392-4644
IPP LithoColor
  1313 W Randolph St, Chicago, IL 312-243-0465
Jim Walters Color Separation/Jim Walters
  2836 Bartells Dr, Beloit, WI 608-365-8634
Kayenay Graphics
  715 N Federal, Mason City, IA 515-424-2535
Kibby/Dick Kemp
  25235 Dequindre, Madison Heights, MI 313-542-1223
King Graphics Inc/Nancy Gorski
  571 W Polk St, Chicago, IL..............................312-943-2202
Kreber Graphics Inc/Joe Kreber
  670 Harmon Ave, Columbus, OH......................614-228-3501
Litho Arts Inc/Tom Horn
  1200 W 6th St, Mishawaka, IN........................219-259-9991
LithoCraft Color Service Inc
  305 W Erie St, Springfield, MO........................417-883-7144
Metzger's Prepress/Tom Metzger
  305 Tenth St, Toledo, OH................................419-241-7195
Noral Color Corp/Bob Petty
  5560 N Northwest Hwy, Chicago, IL................312-775-0991
Nordell Graphic Communications Inc
  PO Box 96, 102 N 6th St, Staples, MN 218-894-3591
OEC Graphics Inc
  555 W Waukau Ave, Oshkosh, WI....................414-235-7770
Omaha Graphics Inc
  3115 S 61st Ave, Omaha, NE..........................402-556-0335
Orent Graphic Arts/James Bennett
  4805 G St, Omaha, NE....................................402-733-6400
Photo Lith
  948 Scribner NW, Grand Rapids, MI 616-459-3586
Photo Specialties Inc
  6939 W 59th St, Chicago, IL............................312-586-9577
Phototype
  2144 Florence Ave, Cincinnati, OH..................513-281-0999
Phototype Engraving Co/Steve Bell
  2144 Florence Ave, Cincinnati, OH..................513-281-0999
Precision Color Plate Inc
  9200 General Dr, Plymouth , MI 313-459-5640
Pro Litho Plate Inc
  2235 Rockwell Ave, Cleveland, OH 216-861-4668
Professional Litho Art Inc
  807 13th Ave S, Minneapolis, MN 612-338-0400
Quality Color Process
  215 E Ninth St, Cincinnati, OH 513-241-6644
Quebecor Printing Prepress Group/Phil Dickert
  59 W Seegers Rd, Arlington Hts, IL..................708-593-2878
Rayson Films Inc
  W 222 N 600 Cheaney Rd, Waukesha, WI 414-544-6699
Riverside Color Corp
  3000 84th Ln NE, Minneapolis, MN..................612-784-5808
Ropkey Graphics
  117 N East St, Indianapolis, IN........................317-632-5446
Salem Graphics
  1705 S Ellsworth Ave, Salem , OH 216-332-1500
Schawk Inc
  1695 River Rd, Des Plaines, IL 708-827-9494
Schmidt Printing/Nola Hoyt
  1101 Frontage Rd NW, Byron, MN....................507-775-6400
Stevenson Photo Color Co
  5325 Ridge Ave, Cincinnati, OH......................513-351-5100
SuperS●n Div/Seller Photo
  2121 S Imboden Ct, Decatur, IL 217-429-7700
Techtron Imaging Center/Shelly Slimak
  160 E Illinois St, Chicago, IL............................312-644-4999
Terre Haute Engraving Co Inc
  648 Walnut, Terre Haute, IN 812-232-2151
Trico Graphics
  1642 N Besly Ct 2nd Fl, Chicago, IL................312-489-7181
Tru-Color Inc/Donald Harvey
  3803 Ford Cir, Cincinnati, OH..........................513-272-1470

Tukaiz Litho/Sam Bianco
  2917 N Latoria Ln, Franklin Park, IL........708-455-1588
Wace USA/Graphic Warehouse/Bob Prange
  2 N Riverside Plz #1400, Chicago, IL............312-876-0533
Watt/Peterson Inc/David Peterson
  15020 27th Ave N, Plymouth, MN..................612-553-1617
Widen Colourgraphics Ltd
  2614 Industrial Dr, Madison, WI 608-222-1296

WEST

Agency Images/Jack Mauck
  3900 W Alameda 17th Fl, Burbank, CA............818-972-1887
American Color/Melissa Trevino
  3511 Hancock St, San Diego, CA....................619-296-6211
American Color/Pete Rosales
  2840 Howe Rd #B, Martinez, CA....................510-370-8100
American Color/Paul Haase
  3900 E Timrod, Tucson, AZ 602-881-5060
American Color/Scott Lewison
  3700 Havana #100, Denver, CO......................303-371-2010
American Color/Bob Day
  2340 E Polk St, Phoenix, AZ..........................602-275-4347
Arizona Four Color Corp/Cliff Lindstrom
  2967 W Fairmont, Phoenix, AZ 602-241-1485
Balzer/Shopes Inc
  175 S Hill Dr, Brisbane, CA............................415-468-6550
Bow-Haus Inc/Charles James
  8017 Melrose Ave, Los Angeles, CA................213-658-5129
Capitol Color Imaging Inc/John Matuze
  47 W 200 South, Salt Lake City, UT................801-363-8801
Capitol Color Imaging Inc/Greg Schriener
  1463 W Alameda Ave, Denver, CO..................800-634-1564
CD-ROM Strategies/Dr Ash Pahwa
  Six Venture #208, Irvine, CA 714-453-1702
Color Image Inc
  2343 Miramar Ave, Long Beach, CA................310-498-3731
Color Masters Inc
  2926 W Virginia Ave, Phoenix, AZ 602-269-3245
ColorBus Software/Reuben Gooch
  18271 McDermott West St #E, Irvine , CA........800-742-2695
Colorgraphics/Sherri Gillenwater
  3707 E Broadway Rd, Phoenix, AZ 602-437-2720
ColorType/Mike McCarty
  2192 Kurtz St, San Diego, CA 619-295-0910
Creative Color Service Corp
  294 Martin Ave, Santa Clara, CA....................408-727-0674
Effective Graphics Inc
  1821 Kona Dr, Compton, CA 310-604-0952
Filmworks, The/Don Lee
  1011 Armando St, Anaheim, CA......................714-632-7775
Fox Colour/Tom Fox
  290 N Benson Ave Bldg 1, Upland, CA............909-981-5050
Graphic Design Services/Sharon Noller
  515 W Lambert, Brea, CA................................714-529-7003
Graphics Plus/Steve Wheeler
  6020 Washington Blvd, Culver City, CA............310-559-3732
Honolulu Graphics Art Inc/Edith Kawashima
  849 Halekauwila St, Honolulu, HI....................808-591-8228
Intersep/Tom Pettinger
  2345 S 2700 West, Salt Lake City, UT 801-973-6720
Laserscan Inc/Gary Harris
  10220 S 51 St, Phoenix, AZ............................602-893-7777
● **LAX SYNTAX DESIGN**/Lance Jackson (P 61)
  1790 5th St, Berkeley, CA..............................510-849-4313
O.N.E. Color Communications/Gary Leach
  1001 42nd St, Oakland, CA 510-652-9005
Pacific Digital Image/Jeremy Smith
  1050 Sansome St #100, San Francisco, CA........415-274-7234
Phaedrus Productions/Dennis Dunbar
  714 Hill St #2, Santa Monica, CA....................310-450-9319
PrePress Studio/Terry Rorie
  5657 Wilshire Blvd #350, Los Angeles, CA 213-938-3956
Professional Printing Svcs/Everett Jess
  2225-A Hancock St, San Diego, CA..................619-297-1188
Quality Graphics/Max Loftin
  502 Goetz Ave, Santa Ana, CA........................714-546-2923
Raging Fingers/Jason Clark
  6301 Sunset Blvd, Hollywood, CA....................213-462-0575
Scan-Graphics Inc/Jerry Seehof
  PO Box 3179, Huntington Beach, CA................714-840-5380
SelectGraphics/Lynn Stewart
  7105 Havenhurst Ave, Van Nuys, CA................818-787-5600
Solzer, Wolfgang
  515 Folsom St, San Francisco, CA....................415-495-8440

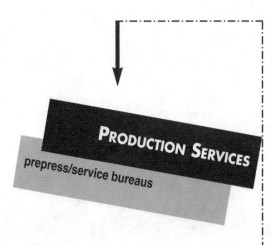

PRODUCTION SERVICES
prepress/service bureaus

Spectrum Inc
  6275 Joyce Dr, Golden, CO 800-426-5677
Star Graphic Art/Crissy Moore/Bruce Hillman
  919 E Hillsdale Blvd #100, Foster City, CA.........415-345-3321
Summerfield Graphics/David Lewis
  860 Piner Rd #17, Santa Rosa, CA 707-526-1515
Trade Litho Inc/Clyde Thornburg
  2187 NW Reed, Portland, OR..........................503-228-8458
Typesetting Room, The/Renee Gauthier
  150 Paularino #150, Costa Mesa, CA..............714-751-3131
Veritel/Lori Anderson
  1 Union St, San Francisco, CA........................415-495-3328
Wescan Color Corp
  15305 NE 40th St #100, Redmond , WA..........206-883-7020
Western Imaging/David Krauss/Dan Goldstein
  115 Constitution Dr, Menlo Park, CA................415-325-0300
William Stetz Design/William Stetz
  1108 Seward St, Los Angeles, CA....................213-461-4267

CANADA

H&S Reliance Ltd/William Sanham
  42 Industrial St, Toronto, ON 416-425-5750
Passage Productions Inc/Alan Dew
  720 King St W #806, Toronto, ON....................416-361-0989

Service Bureaus

NEW YORK CITY

Admaster
  104 E 25th St, New York, NY..........................212-673-8100
Anagraphics Inc/John Lanahan
  39 W 38th St 10th Fl, New York, NY................212-840-2912
AVI Visual Productions/Barney Bloom
  915 Broadway 12th Fl, New York, NY 212-505-9155
Axiom Design Systems/Howard Greenberg
  6 W 18th St, New York, NY............................212-989-1100
Brilliant Image Inc/Jerry Kahn
  7 Pennsylvania Plz 11th Fl, New York, NY 212-736-9661
Brilliant Image Inc/David Lapidus
  7 Pennsylvania Plz 11th Fl, New York, NY 212-736-9661
Business Link
  312 Fifth Ave 3rd Fl, New York, NY 212-268-0777
Cohen Design Inc, Hayes
  133 Cedar Rd, E Northport, NY........................516-368-2031
● **COPYTONE**/Mark Bevilacque (P 103)
  115 W 45th St 10th Fl, New York, NY 212-575-0235
Designers Atelier/Dave Perskie
  45 W 45th St #808, New York, NY 212-221-8585
Digital Exchange/Curt Steyer
  1 W 20th St, New York, NY............................212-929-0566
Duggal Color Projects Inc
  9 W 20th St, New York, NY............................212-924-6363
Elite Color Inc/Tom Tivo
  127 W 25th St 8th Fl, New York, NY 212-989-3289
FCL/Colorspace/Terry Chamberlain
  10 E 38th St 3rd Fl, New York, NY 212-889-7787
Hi-res Slide Graphics
  305 Seventh Ave 19th Fl, New York, NY 212-366-4860
Imaging Consortium Inc/Jack Yeh
  1200 Broadway #2E, New York, NY 212-689-2008
JCH Group Ltd/Michael Wineglass
  44 W 28th St, New York, NY..........................212-532-4000
Line & Tone/Paul Delnunzio
  246 W 38th St, New York, NY........................212-921-8333

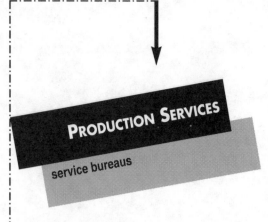

Magic Graphics/Carmine Corinello
424 W 33rd St 12th Fl, New York, NY..............212-947-1942
MetroGrafik/John Cavanagh
330 Seventh Ave, New York, NY..................212-695-7778
Microcomputer Publishing Center/Melanie Rosenzweig
4 W 20th St, New York, NY.......................212-463-8585
Micropage/Microtype/Michael Comodo
900 Broadway, New York, NY.....................212-533-9180
Mirenburg & Co/Barry L Mirenburg
301 E 38th St, New York, NY.....................212-573-9200
Morton Graphics/Stewart Feuer
19 W 21st St #1104, New York, NY................212-620-5944
Next Wave Productions/Juan Moreiras
134 Tenth Ave/Grnd Lev, New York, NY...........212-989-2727
Peter X (+C) Ltd
627 Greenwich St 12th Fl, New York, NY.........212-366-6600
Pintura Graphics/Angelo Gomez
1641 Third Ave #31F, New York, NY...............212-534-3102
Presentation Graphics/Jeff Genshaft
37 W 17th St 7th Fl, New York, NY...............212-255-5355
Presentation Source/Peter Roberto
230 Park Ave S, New York, NY....................212-614-4111
Quad Right/Dennis Shedd
711 Amsterdam Ave, New York, NY................212-222-1220
R C Communications Inc
1633 Broadway, New York, NY....................212-621-5080
Renaissance Art Center/Andrew Edwards
29 W 38th St 3rd Fl, New York, NY...............212-575-6100
Sam Flax
12 W 20th St, New York, NY......................212-620-3038
Show & Tell
39 W 38th St, New York, NY......................212-840-2912
Slide House, The/Tyrone Rice
200 Park Ave S, New York, NY....................212-505-5321
Sprintout New York/Dan Martin
406 W 31st St, New York, NY.....................212-239-2714
The Big Picture
37 West 17th St , New York, NY..................212-255-5355
Time Life Photo Lab/Hanns Kohl
Time/Life Bldg Rockefller Ctr, New York, NY 212-522-2127
Total Visuals
110 W 40th St 20th Fl, New York, NY.............212-944-8788
Typogram
560 Broadway, New York, NY.....................212-219-9770
Typogram/Sarah Riegelman
900 Broadway, New York, NY.....................212-505-1640
Waltzer Digital Service, Carl/Carl Waltzer
873 Broadway #412, New York, NY................212-475-8748
Wander Communications/Andy Wander
534 E 84th St #5W, New York, NY.................212-737-3058
Zero One
33 W 17th St, New York, NY......................212-255-0011

EAST

Advanced Laser Graphics
1101 30th St NW #100, Washington, DC...202-342-2100
Amanuensis/Diane Romagnoli
144 W Merrimack St, Manchester, NH.........603-624-2704
American Color/Jeff Waterhouse
1211 Washington St, Allentown, PA...................215-437-6505
American Color/Jim Bravosky
275 Hudson St, Hackensack , NJ.....................201-488-4676
American Color/John Rebman
2500 Walden Ave Bldg 1 #A, Buffalo, NY 716-684-8100
Aquidneck Graphics Inc/Stuart MacNaught
28 Jacome Way, Middletown, RI.....................401-849-9930

Artco Graphics Inc
271-79 Westfield Ave, Elizabeth, NJ 908-352-7207
Associated Graphic Services Ltd/Andy Schoenfeld
8 Corporate Circle, Albany, NY 518-456-6789
Atlantic Group/Stephanie Nizolek
10 Fairfield Ave, Stamford, CT 203-359-4228
Auras Design/Robert Sugar
1746 Kalorama Rd NW, Washington, DC...........202-745-0088
Autograph/John Pecora
21 Hyde Rd, Farmington, CT.......................203-677-9323
Berry & Homer/Harry Hollingsworth
2035 Richmond St, Philadelphia, PA...............800-522-0888
Blazing Graphics Inc/Loribeth Holland
25 Amflex Dr/PO Box 20377, Cranston, RI.........401-946-6100
Bokland Custom Visuals/Renee Guzek
122 Industrial Park Rd, Albany, NY 518-482-3137
Boston Business Graphics Inc/Joe Nickleson
300 Wildwood St #20, Woburn, MA...............617-938-6525
CA/GRAFX/Michael Moreau
433 Ft Pitt Blvd, Pittsburgh, PA....................412-471-6231
Chromagen Digital Imaging/Philip Bryan
1100 University Ave #141, Rochester, NY..........716-256-6218
Chromakers/Richard Lepkowski
880 Canal St, Stamford, CT........................203-323-7277
Color Group/Reubens Color/Mark Weinstein
168 Saw Mill River Rd/Rte 9A, Hawthorne, NY...914-769-8484
Commonwealth Creative Group/Bob Fields
27 Strathmore Rd, Natick, MA.....................508-651-7556
Copy Print Co/Andrew Cohen
50 Rte 5, Warwick, RI..............................401-739-2300
CR Waldman Graphics Communications/Rich Greene
9100 Pennsaucken Hwy, Pennsaucken, NJ..........800-820-3000
Creative Typesetting Inc/Craig Newell
1800 Post Rd, Warwick, RI..........................401-737-1766
Dots Enuff!/David Geiser
100 N Patterson St, State College, PA..................814-238-4527
Everett Studios/Robbie Everett
22 Barker Ave, White Plains, NY....................914-997-2200
Expresslides Inc/Bob Seiple
324 Chestnut St, Union, NJ.........................908-964-3933
FEA Laser Service/Louis Peddicord
2404 Ravenview Rd, Baltimore, MD 410-252-8910
Filmet Color Labs/Rick Bachelder
7436 Washington St, Pittsburgh, PA................412-351-3510
G R Waldman Graphics/George Gunzelman
9100 Pennsauken Hwy, Pennsauken, NJ 609-662-9111
G&G Laser Typesetting/Don Gallant
1030 Congress St, Portland, ME.....................207-774-7338
General Graphics Co
3651 Hill Rd, Parsippany, NJ........................201-829-1320
Generated Image/Susan Martin
164 Middle St, Portland, ME........................207-774-4455
George Monagle/Graphic Partners/George Monagle
445 Hawley Ave, Syracuse, NY......................315-426-0513
GIST/John Robinson
700 State St, New Haven, CT.......................203-495-7500
Graphic Accent/Bruce Alexander
446 Main St, PO Box 243, Wilmington, MA.......508-658-7602
Graphic Art & Design/Andre Herviou
4 Hendrickson Ave & Hwy 35, Red Bank, NJ.......908-741-1160
Graphic Arts Service Bureau/Rowland Brandwein
159 Coburn Woods, Nashua, NH.....................603-880-4844
Graphic Connexions Inc/Ken Silver
10 Abeel Rd, Cranbury, NJ.........................609-655-8970
Graphic Designers/Kenneth Shafton
20 Summer St, Stamford, CT.......................203-357-7370
Graphic Hearts/Adam Bell
255 Cherry St, Milford, CT.........................203-874-1305
Graphics 150/Van Smith
150 Speedwell Ave, Morris Place, NJ 201-267-6446
Graphics Workshop/Sam Mason
1005 W Fayette St, Syracuse, NY....................315-422-8115
Graphique/Thomas Garen
306 Warren St, N Babylon, NY......................516-321-4907
Guymark Studios
3019 Dixwell Ave/PO Box 5037, Hamden, CT ...203-248-9323
Harmony Press/Fred Grotenhuis
717 W Berwick St, Easton, PA.......................215-559-9800
High Resolution Inc/Peter Koons
PO Box 397, Camden, ME..........................207-236-3777
Image Resolutions/William Wenzel
105 Webster St #6, Hanover, MA...................617-871-8300
Imageset Design/Mark Beale
470 Forest Ave, Portland, ME.......................207-775-4738

Imagesetter/Alison Poor
440 Humphrey St, Swampscott, MA 617-592-1234
Imageworks
3679 Concord Rd/PO Box 3325, York, PA........717-757-5710
Information International Inc/Gary Moore
10 Commerce Way, Woburn, MA 617-937-9400
Inspired Solutions/John Santore
270 Farmington Ave, Farmington, CT...............203-828-9862
Intl Colour Service/Michael Degus
911 E Main St, Rochester, NY......................716-442-7601
Kalnin Graphics/Michael Malloy
Benjamin Fox Pavillion #A8, Jenkintown, PA........215-887-6970
Laser Image Reprographics/Stephen Hall
129 Main St/POB 23, Binghamton, NY.............607-723-2145
Media Power/Robert Pena
41 Wellman St 3rd Fl, Lowell, MA 508-458-7000
Media Works/Jay M Arancio
185 Hickory Corner Rd, Milford, NJ 908-996-7855
Microprint/Terry Moore
240 Bear Hill Rd, Waltham, MA 617-890-7500
Needham Graphics Inc/Jim Friend
633 Highland Ave, Needham, MA...................617-449-8404
Optigraphix
138 Simsbury Rd, Avon, CT.........................203-677-6569
Pageworks Inc/Mark Prevost
501 Cambridge St, Cambridge, MA.................617-374-6000
Photo Communications Corp
815 Greenwood Ave, Jenkintown, PA...............215-572-5900
PhotoSep Inc/James Boyle
809 Hylton Rd Unit 4 #12, Pennsauken, NJ........609-663-1444
PrePress Computer Works/Bob Hunt
486 New Park Ave, W Hartford, CT.................203-232-1595
Prestige Color/Jim Dommel
19 Prestige Lane, Lancaster, PA.....................717-392-1711
Putnam Photo Lab/William McCann
23 Sugar Hollow Rd, Danbury, CT...................203-790-9650
Regester Photo Services Inc/Eugene Regester
50 Kane St, Baltimore, MD 410-633-7600
Regional Typographers Inc/Alan Coopersmith
131 Henry St, Freeport, NY 516-378-4422
Sentry Color Labs Ltd/Arline Green
571 South Ave, Rochester, NY......................716-262-2030
Slide Center/Lorraine Walsh
186 South St 6th Fl, Boston, MA....................617-451-9902
Slide Sense
360-1 Knickerbocker Ave, Bohemia, NY 516-563-3072
Spectracomp/Jeff Fackler
5170 E Trindle Rd, Mechanicsburg, PA...............717-697-8600
Spectratone
230 Ferris Ave, White Plains, NY....................914-946-3336
Spectrum/Tina Gagnon
4 Raymond Ave, Salem, NH.........................603-893-9294
Spectrum Arts Ltd./Alan Chantker
1823 Eutaw Pl, Baltimore, MD 410-462-6900
Sprintout Corp/J Higgins
50 Clifford St, Providence, RI........................401-421-2264
Sukolsky-Brunelle Inc/David Kull
908 Penn Ave 2 nd Fl, Pittsburgh, PA................412-391-6440
Tele-Measurements Inc/David Bastadi
145 Main Ave, Clifton, NJ...........................201-473-8822
Today's Graphics Inc/Tim Canny
2130 Arch St 3rd Fl, Philadelphia, PA................215-567-0332
TruColor Labs/David Calhoun
20 Powers Dr, Paramus, NJ.........................201-261-2107
Tyava Productions/Mary Ross
259 Oak St, Binghamton, NY 607-722-1457
Typeline/Bob Badaracco
668 American Legion Dr, Teaneck, NJ...............201-836-2300
Typeworks Plus/Tom Voight
One Dupont Circle #126, Washington, DC 202-857-0775
Typographic House/Peter Brotman
63 Melcher St, Boston, MA..........................617-482-1719
Vision Graphics/Dave Monette
100 Moody St, Ludlow, MA..........................413-589-9551
Visual In-Seitz/Bill Kelley
225 Oak St, Rochester, NY..........................716-454-4350
Winchell Company, The/George Finley
1315 Cherry St, Philadelphia, PA....................215-568-1770
Wordwrap/Bob Thurber
10 Church St, N Attleboro, MA......................508-695-8066
Wrap-Around Inc/Bob Cheney
80 Windsor St, W Springfield, MA...................413-781-5440

Adron Graphics Inc/Wayne Canoy
300 Poplar Ave, Memphis , TN...........901-521-0625
Adv Computer Imaging/Skip Griswold
2185 Northlake Bldg 8 #3, Atlanta, GA 404-934-5220
Adv Computer Imaging/George Powell
2359 Windy Hill Rd, Marietta, GA...................404-952-6294
American Color/Tom Otte
7108 Crossroads Blvd #308, Brentwood, TN 615-371-1060
American Color/Syd Beeston
16178 W Hardy, Houston, TX......................713-876-2170
American Color/Alayne Jurgens
14320 Midway Rd, Dallas , TX....................214-392-4540
AV COM Productions/Doug Myer
3158 S 108th East Ave #275, Tulsa, OK...........918-627-2301
Award Publications/Roger Modey
3040 Williams Dr #200, Fairfax, VA.............703-641-9400
BC Printing/Paul Moncman
532 E Main St, Norfolk, VA.....................804-627-8020
Boca Laser Inc/Sally Stewart
750 South Dixie Hwy, Boca Raton, FL 407-393-1994
Cobb Inc/Tom Bell
901 Monmouth St, Newport, KY...................606-291-1146
Color Place, The/Jim Moss
1330 Conant St, Dallas, TX.....................214-631-7174
ColorCopy Inc/Mitch Powers
1602 21st Ave S, Nashville, TN.................615-321-5740
Comp-U-Type/Chris Cochran
1920 Tenth Ave S, Birmingham, AL 205-323-8898
Cumberland Graphics/Fred Von Collin
118 20th Ave S, Nashville, TN 615-320-1195
Data Slide
510 14th St, Atlanta, GA 404-881-8844
Datastream/Amy Salmons
585 W Main St, Lexington, KY...................606-255-6686
Davidson & Co/Ken Davidson
1750 Lower Roswell Rd, Marietta, GA 404-973-9637
● **DIGITAL IMAGE**/Ray Parrott **(P 90)**
1758 Tullie Circle, Atlanta, GA...............404-325-6955
Digital Photographic Imaging/Kent Cranor
6085 Barfield Rd #206, Sandy Springs, GA 404-843-2050
Douglas Graphics/Peggy Hooper
1701 Pinecroft Rd #105, Greensboro, NC..........910-299-4621
Eyebeam/Miles Wright
2500 Gateway Center Blvd #600,
Morrisville, NC................................919-469-3859
Final Proof Inc/Kathy Weaber
1750 N Florida Mango Rd #303,
W Palm Beach, FL..............................407-684-9434
Graphic Zone, The/David Brown
10 Office Park Cir #100, Birmingham, AL 205-870-5300
Graphics Group/Vivian Darby
2820 Taylor St, Dallas, TX.....................214-749-2222
Graphtec Inc/Frank Mauro
505 NW 65th Court, Ft Lauderdale, FL 800-444-1931
Hanson Graphics of Memphis/Marty Guyse
3086 Bellbrook Dr, Memphis, TN 901-396-4350
Hot Graphics & Printing Inc/Robert Dawson
5241 Elmore Rd, Memphis, TN 901-387-1717
Houston Photolab
5250 Gulfton #3B, Houston, TX.....................713-666-0282
Icca Graphic Services/Craig Hurt
429 Briabend Dr, Charlotte, NC...................704-523-7219
Image Center/Lisa Phillips
1011 Second St SW, Roanoke, VA................703-343-8243
Inkspot Printing Media Service/Jim Lanham
1212 Yarborough, El Paso, TX...................915-598-1138
International Imaging
650 Tanners Ln, Earlysville, VA................804-978-1111
Laser Tech Color Inc/Damien Gough
1505 Luna Rd Bldg 2 #202, Carrollton, TX.......800-365-8957
Lewis Creative Technologies/Stephen Jones
900 W Leigh St, Richmond, VA...................804-648-2000
Lino Graphics Services/Stacey Williams
610 S Jennings Ave, Ft Worth, TX...............817-332-4070
Lithographic Services Inc
1065 W Adams St, Jacksonville, FL 904-353-8566
Lunar Graphics/Layne St Julien
655 St Ferdinand St, Baton Rouge, LA............504-383-7652
Macfactory/Eric VonColln
118 20th Ave S, Nashville, TN..................615-327-3437
Meteor Photo & Imaging Co/Steve Carroll
680 14th St NW, Atlanta, GA 404-892-1688
Micron/Green/Millard Pate
1240 NW 21st Ave, Gainesville, FL 904-376-1529

MMI/Maxine Mooney
11660 Alpharetta Hwy #540, Roswell, GA.........404-664-9665
National Color/Greg Warrick
341 Victory Dr, Herndon, VA....................703-834-0100
National Photographic Labs/Mike Krodel
1926 W Gray, Houston, TX.......................713-527-9300
NEC Inc/Roy Luckett
1504 Elm Hill Pike, Nashville, TN...............615-367-9110
Norling Studios Inc/Glenn McVicker
221 Swathmore Ave, High Point, NC..............919-434-3151
Noumenon Labs Inc/Elles Gallion
1349 Empire/Central #310, Dallas, TX 214-688-4100
● **OLDROYD LTD, BOB (P 100)**
4 Points Studio Box 5599, Atlanta, GA 404-581-0653
Patrick Graphics/Pat Curran
702-B Seaboard St, Myrtle Beach, SC................803-448-7777
● **PHOTOEFFECTS**/Trevor Donovan **(P 87)**
3350 Peachtree NE #1130, Atlanta, GA 404-816-8000
Picture Conversion Inc/Go Babaoglu
5109 Leesburg Pike #212, Falls Church, VA........703-998-5777
Presentation Resource/Dee Hardie
5 W Cary St, Richmond, VA 804-648-7854
PressLink/Don Kent
11800 Sunrise Valley Dr #1130, Reston, VA.......703-758-1740
Prototype Graphics/Joe True
9391 Grogans Mill, The Woodlands, TX 713-367-8973
Publication Technology Corp/Cindy Garza
11200 Waples Mill Rd #310, Fairfax, VA 703-591-0687
Quality Chrome/Steve Pepper
5300 Old Pineville Rd #108, Charlotte, NC.......704-527-5194
Quebecor Printing/Robert DeAngelo
6300 Hazeltine Natl Dr #108, Orlando, FL.........407-851-3681
Riddick Advertising Art/Mike Bamrath
401 E Main St, Richmond, VA....................804-780-0006
Ringier America/John Forgach
8700 J Red Oak Blvd, Charlotte, NC 704-527-6865
Roland House Inc/Cosby/Williams
2020 N 14th St, Arlington, VA..................703-525-7000
Royal Graphics/Jay Agee
523 Ringo St, Little Rock, AR..................501-375-8255
Slide Step/Fred Lines
22 Seventh St NE, Atlanta, GA..................404-873-5353
Slidemasters/Derek Bevan
1100 Business Pkwy, Richardson, TX..............214-437-0542
Stokes Imaging Services/Steve Parker
7000 Cameron Rd, Austin , TX 512-458-2201
Storter Childs Printing Company/Kevin McCuller
1540 NE Waldo Rd, Gainesville, FL 904-376-2658
Texas Photocomp/Bill Minter
1482 N First St, Abilene, TX...................915-673-6725
Type Case Inc/Linda Jennings
611 Allston Ave, Ft Worth, TX..................817-332-7563
Typografiks Inc/Eli Huffman
4701 Nett, Houston, TX.........................713-861-2290
Web Co/Dave Wedel
401 N Weatherford, Midland, TX 915-570-4747
Words & Pictures/Diane Maldin
4372 Spring Valley Rd, Dallas, TX..............214-702-9119
World Ventures Corp/Barbara Herring
5231 S Nova Rd, Daytona Beach, FL.................800-552-5353
WRE/Colortech Prepress/John Comerford
533 Banner Ave, Greensboro, NC 919-275-9821

MIDWEST

A1 Printers Inc/Nick Meyer
5616 Lincoln Dr, Edina, MN 612-933-3852
Action Nicholson Color Inc/John Haugn
6519 Eastland Rd, Brook Park, OH................216-234-5370
AGS & R Communications/Mike Foster
314 W Superior St, Chicago, IL 312-649-4500
Alley Newspaper, Steve
PO Box 7599, Minneapolis, MN 612-721-6051
Alpha Channel
770 N Halsted St #300, Chicago, IL..................312-421-8000
Alternative Type & Graphics/Dan Kauffman
9700 W Foster Ave, Chicago, IL.................312-992-2050
American Color
3740 Stern Ave, St Charles , IL 708-377-8660
American Litho Graphic Corp/Bob Gittins
1107 N 19th St, Omaha, NE 402-346-3377
Anzographics Computer Typographers/Chuck Anzilotti
770 La Salle St #800, Chicago, IL................312-642-8973
Applied Communications/Cathy Wissler
3380 Successful Way, Dayton, OH 513-233-0070

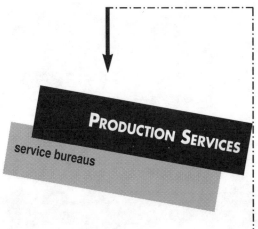

PRODUCTION SERVICES
service bureaus

Art (to the Nth) Lab/Ellen Sandor
III Inst of Tech Rm 319 Wshnk , Chicago , IL 312-567-3762
As Soon As Possible Inc/George West
3000 France Ave, St Louis Park, MN 612-926-4735
ASI Image Studios/Krys Ciaciarulo
10 Second St NE #214, Minneapolis, MN..........612-379-7117
Bolger Publications Creative Printing/Mike Oslund
3301 Como Ave SE, Minneapolis, MN...............800-999-6311
Butler Graphics Inc/Jack Butler
294 Town Center, Troy, MI 313-528-2808
Carey Color/Bill Smith
775 S Progress Dr, Medina , OH 216-725-5637
Clicks Inc/Susan Vince
3363 Commercial Ave, Northbrook, IL.............708-291-1020
CMI Business Communications/Gary Brown
150 E Huron St, Chicago, IL 312-787-9040
CN Graphics Inc/Corry J Walbridge
3300 University Ave SE #203, Minneapolis, MN.612-378-0808
Color Graphics Inc/Jim Petrole
7660 West Industrial Dr, Forest Park, IL.......708-771-7660
Color Perfect Inc/Bob Zbik
7450 Woodward, Detroit, MI 313-872-5115
ColorCraft Inc/Dick Johnson
3765 N 35th St, Milwaukee, WI..................414-442-1344
Commercial Art Service/Steve Dittbenner
9112 Constance St, Lenexa, KS 913-894-9391
Communitronics Corp/Rita Anderson
1907 S Kings Hwy, St Louis, MO 314-771-7160
Computer Chrome Inc/Tony Anderson
803 Transfer Rd, St Paul, MN...................612-646-2442
Computer Graphic Center/Drew Waddell
1125 High St, Des Moines, IA 515-282-0000
Computer Graphic Group/Dave Bitner
230 Findlay, Cincinnati, OH 513-241-8300
Computype/Darlene Nelson
2285 W Country Rd County, St Paul, MN 612-633-0633
Creative Photo/Mike Dieter
1515 Linden, Des Moines, IA 515-243-4010
Creative Visuals Inc/Lois Gower
731 Harding St NE, Minneapolis, MN..............612-378-1621
● **CUSTOM COLOR CORP**/Guy Clark **(P 102)**
300 W 19th Terrace, Kansas City, MO 800-821-5623
Dahl + Curry/Ray Johnson
1320 Yale Pl, Minneapolis, MN..................612-338-7171
David Tremblay Productions/David Tremblay
210 King Creek Rd, Minneapolis, MN.............612-546-8416
Digicliques/Gerry Moore
4320 High View Pl, Minnetonka, MN 612-377-8202
Dots Per Inch/Meyer Associates/Sandra Muselter
14 N Seventh Ave, St Cloud, MN.................612-259-4000
Dwight Yaeger Typographer/Jon Yaeger
935 W Third Ave, Columbus, OH 614-294-6326
El Dorado Graphics/George Gaulke
15791 W Ryerson Rd, New Berlin, WI.............414-784-3520
Electronic Easel/David Alsbury
2124 University Ave, St Paul, MN 612-659-2424
Electronic Imaging Inc/John Callahan
515 N Neil St, Champaign, IL...................217-351-1550
Electronic Publishing Ctr/Greg Lindhout
6532 Clay Ave SW, Grand Rapids, MI.............616-698-9890
Enterprise Information Service/Dave Gronauer
520 S Main #2413, Akron, OH 216-762-2222
Express Slide/Tom Zurowski
11605 Manda Dr, Huntley, IL 708-669-0033
Faxgraphix/Charlie Schrandt
200 12th Ave S, Minneapolis, MN................612-341-0884

PRODUCTION SERVICES

service bureaus

Fontastik Inc/Cliff Ping
911 Main St #1717, Kansas City, MO 816-474-4366
Graphic Arts Inc/Julie Tucker
631 SW Harrison St, Topeka, KS 913-354-8596
Graphics Unlimited Inc/Lee Marwede
3000 Second St N, Minneapolis, MN...............612-588-7571
Harlan Type/Steve Bravard
1107 Dublin Rd, Columbus, OH 614-486-9641
Henderson Typography/Kathi Wilson
222 W Huron 2nd Fl, Chicago, IL 312-951-8973
Holland Printing/Dave Hargens
1007 E 162nd St, S Holland, IL...............708-596-9000
House of Graphics/Jim Hadlich
1991 Stowater Ave, St Paul, MN...............612-738-1143
Imagesetter Inc/Mary Howards
2423 American Ln, Madison, WI 608-244-6243
Intran Electronic Publ Svcs/
Xerox Corp/Debbie Brusehaver
7900 Xerxes Ave S #400, Bloomington, MN.......612-831-0342
Jodee Kulp Graphic Arts Services/John Theis
119 N 4th St #401, Minneapolis, MN...............612-341-9870
Kalamazoo Color Lab/Heinze Jahnke
1326 Portage St, Kalamazoo, MI 616-344-6000
Kalamazoo Label Co/Russ Wilson
321 W Ransom, Kalamazoo, MI 616-381-5820
Kibby/Dick Kemp
25235 Dequindre, Madison Heights, MI 313-542-1223
Kingery Printing/John Kingery
3012 S Banker St, Effingham, IL 217-347-5151
Kreber Graphics Inc/Joe Kreber
670 Harmon Ave, Columbus, OH 614-228-3501
Lasercom Inc/George Lowe
1701 E Lake Ave, Glenview, IL 708-724-2490
Lattice Inc/Liz Dunham
3010 Woodcreek Dr #A, Downers Grove, IL.......708-769-4060
Lighthouse Productions Inc
1900 Hicks St, Rolling Meadows, IL 708-506-1414
Linhoff Corporate Color/Greg Linhoff
4400 France Ave S, Minneapolis, MN...............612-927-7333
Martin/Bastian Communications/Steve Martin
105 Fifth Ave S, Minneapolis, MN 612-375-0055
Max PostScript/Gary Schlueter
613 Main St, Cincinnati, OH 513-421-5500
Mead Paper/Shawn Palmer
Courthouse Plz NE, Dayton, OH 513-495-3308
Metzger's Prepress/Tom Metzger
305 Tenth St, Toledo, OH 419-241-7195
MGS Services/Viacom Div/Cliff Kroeter
65 E Wacker Pl #1900, Chicago, IL 312-337-3761
Midtown Imaging/Ralph Green
3420 Carnegie Ave, Cleveland, OH 216-431-6500
Midwest Litho Arts/Rudy Galfi
125 E Oakton St, Des Plaines, IL...............708-296-2000
N2 Communications Inc/Randall Powell
817 W 2nd St, Wichita, KS 316-265-1630
Noral Color Corp/Bob Petty
5560 N Northwest Hwy, Chicago, IL 312-775-0991
Novachrome/Jim Varney
710 N Tucker #303, St Louis, MO 314-241-9150
Number One Graphics Inc/Thomas Drolett
325 N Clippert, Lansing, MI 517-332-6231
NW Printcrafters/Stephen Mike
120 W Plato Blvd, St Paul, MN 612-227-7721
Oms Business Services/Terese Aldrich
2785 White Bear Ave #210, Maplewood, MN...612-779-2207

Photo Lab Inc/Joy Smith
1026 Redna Terr, Cincinnati, OH...............513-771-4400
Photographic Specialties Inc
1718 Washington Ave N, Minneapolis, MN.......612-332-6303
Picas & Points Inc/Doris Brennan
8070 Morgan Cir, Minneapolis, MN...............612-881-2626
Port-To-Print/Jim Devine
2851 Index Rd, Madison, WI 608-273-4887
Preplink/Dean Hanisco
4005 Clark Ave, Cleveland, OH 216-651-5465
Print Craft Inc/Tommy Merickel
315 Fifth Ave NW, St Paul, MN...............612-633-8122
Printing Station/Toni McCormick
1420 Locust Ave, Des Moines, IA 515-243-8144
Pro Graphics Inc/Tony Brinza
2077 S 116th St, W Allis, WI...............414-543-0200
Pro-Color/Terry Nelson
909 Hennepin S Ave, Minneapolis, MN...............612-673-8900
Quicksilver Assoc Inc/Chuck Garasic
18 W Ontario St, Chicago, IL 312-943-7622
Robin Color Imaging/Chuck Eckart
2106 Central Pkwy, Cincinnati, OH...............513-381-5116
Ross Ehlert Photolabs/Gary Riecke
225 W Illinois, Chicago, IL 312-644-0244
Sauer & Associates/Christian Sauer
2844 Arsenal, St Louis, MO...............314-664-4646
Sinkler Or Sinkler/Jim Wilieko
1214 N LaSalle, Chicago, IL...............312-649-9504
Skyline Displays Inc/Gary Eittreim
12345 Portland Ave S, Burnsville, MN...............612-895-6360
Slide Services Inc/Ronald Lovoll
2537 25th Ave S, Minneapolis, MN...............612-721-2434
Smart Set Inc/Kevin Brown
210 N Second St #102, Minneapolis, MN...........612-339-7725
Spectrum Image Group/Byron Sletton
10835 Midwest Ind Blvd, St Louis, MO...............314-423-8111
Square One Desktop Publishing/Pati Anderson
2208 W 21st St, Minneapolis, MN...............612-377-1152
Superior Color Graphics/Virgil Jones
411 W 7th St, Kansas City, MO 816-221-4117
TDS/William McKinley
113 North May St, Chicago, IL 312-733-4464
TriTel Productions Inc/Ted Traiforos
53 W Seegers Rd, Arlington Hts, IL 708-952-0020
Typetronics/Eric Landmann
301 S Bedford #7A, Madison, WI...............608-257-2939
Universal Color Inc/Paul Anderson
7701 42nd Ave N, Minneapolis, MN...............612-535-6435
Video Arts/Art Phillips
1440 4th Ave N, Fargo, ND...............701-232-3393
Villager Graphics/Tom Hamlin
757 S Snelling Ave, St Paul, MN...............612-699-1462
Visual Motivation Inc/Robert Jackson
10901 Valley View Rd, Eden Prairie, MN...........612-943-0060
Watt/Peterson Inc/David Peterson
15020 27th Ave N, Plymouth, MN...............612-553-1617
Web Tech Inc/James J Matthias
4560 N 127th St, Butler, WI...............414-781-8805
Weimer Typesetting/Frank Morton Jr
111 E McCarty St, Indianapolis, IN...............317-267-0565
Winsted Corp/Judy Pahl
10901 Hampshire Ave S, Minneapolis, MN.......612-944-8556
Xpress Graphics/Tony Munno
137 N Oak Park Ave #200, Oak Park, IL...........708-848-8651
Zimmer Inc/Brad Bishop
Boggs Indstrl Park/PO Box 708, Warsaw, IN......219-372-4310

WEST

Abracadabra Slide Production
1331 E Edinger Ave, Santa Ana, CA...............714-667-1010
Adage Graphics/Larry Christian
8939 S Sepulveda Blvd #424, Los Angeles, CA...310-216-2828
AGS
1153 Lincoln Ave #A, San Jose, CA 408-287-5811
Alphagraphics Printshops of the Future/Patel Kanu
10910 Lindbrook, Los Angeles, CA 310-208-2679
American Color
3310 W MacArthur Blvd, Santa Ana, CA...............714-545-7622
American Translation Services
245 Loma Corta Dr, Solana Beach, CA...............619-481-6046
Andre's Photo Lab/Andre Schellenberg
7686 Miramar Rd, San Diego, CA 619-549-3900
Andresen Typographics/Sandi Perri
1900 E Warner Ave Bldg 1-A, Santa Ana, CA 714-250-4450

Andresen Typographics
2639 29th St, Santa Monica, CA 310-452-5521
Aptos Post Inc/John Andrew
1119 Pacific Ave #202, Santa Cruz, CA...........408-459-6888
Aquila Technologies Group Inc/Mark Bronson
8401 Washington Pl NE, Albuquerque, NM........505-828-9100
Art Director, The/Steve Helali
5512 Wilshire Blvd, Los Angeles, CA 213-933-9668
Artistic Typesetting & Graphic/Tammy Gortmaker
2748 S Santa Fe, San Marcos, CA...............619-599-9060
Associated Publications & Graphics
3825 W 226th St , Torrance, CA...............310-373-8406
Blazing Pages/Frank Iachelli
7755 Center Ave #290, Huntington Beach, CA...714-891-8786
Bow-Haus Inc/Charles James
8017 Melrose Ave, Los Angeles, CA 213-658-5129
Burian Imagesetting/Ron Burian
423 SE 13th St, Portland, OR...............503-238-6712
Capitol Color Imaging Inc/John Matuze
47 W 200 South, Salt Lake City, UT 801-363-8801
Captured Images/Bill Grimm
31316 Via Colenas #117, Westlake Village, CA...818-707-9491
Cbm Type/Jay McKendry
624 East Evelyn Ave #G, Sunnyvale, CA...........408-739-0460
Central Graphics/Chuck Surprise
725 13th St, San Diego, CA...............619-234-6633
CIS-Hollywood/Godfrey Pye
1144 N Las Palmas, Hollywood, CA...............213-463-8811
Citizen's Graphic Arts
709 SE 7th , Portland, OR...............503-232-8501
Color Service/Barbara DiGuiseppe
509 Fairview Ave N, Seattle, WA...............206-587-0278
ColorType/Dale Stauffer/Mike McCarty
2192 Kurtz St, San Diego, CA 619-299-8411
Columbia Lithograph Inc
12078 E Florence Ave, Santa Fe Springs, CA......310-944-9486
Communicate/Emilie Caruso
3201 Wilshire Blvd, Santa Monica, CA 310-998-9228
Comp-U-Copy Inc/Marcie Miller
801-B 14th St, Golden, CO...............303-278-1271
Compographics
517 Mercury Ln, Brea, CA...............714-990-3878
Composition Type/Gordon MacIver
6076 Bristol Pkwy #203, Culver City, CA...........310-410-0410
Computer Typesetting & Graphics/Lacey La Moire
2636 Churn Creek Rd, Redding, CA...............916-223-3444
Computer Graphics Services
1245 Old Tale Rd, Boulder, CO...............303-440-4400
Computer Imaging/Jake Koebrich
1617 Wazee St Unit D, Denver, CO...............303-573-7812
Copy Spot Inc
3130 Wilshire Blvd, Santa Monica, CA 310-393-0693
Custom Color
1236 S Central Ave, Glendale, CA 818-240-6100
Custom Composition
7241 Owensmouth Ave, Canoga Park, CA......818-883-6544
Dee Typographers Inc/Larry DiPasquale
3407 W 6th St #520, Los Angeles, CA 213-389-0608
Design & Type Inc/Lori Beckerman
739 Bryant St, San Francisco, CA...............415-495-6280
Deskprint
7352 Beverly Blvd, Los Angeles, CA 213-933-9138
Desktop Graphics & Design/Marion Wilson
17921 #H Sky Park Cir, Irvine, CA...............714-261-1881
Desktop Publishing Inc/Bill Wertz
2015 Larkspur Landing Cir, Larkspur, CA...........415-925-2030
Desktop Studio
1254 Westwood Blvd, Los Angeles, CA...............310-474-9544
Digital Pond/Chris McKenney
50 Minna St, San Francisco, CA...............415-495-7663
DP&C
220 S Kenwood #100, Glendale, CA...............818-507-4433
DuPuis Design/Steven DuPuis
29800 Agoura Rd #103, Agoura Hills, CA........818-706-2864
Duraset Typesetting
1139 S Julian, Los Angeles, CA...............213-749-9199
DynaMedia Design & Graphics Inc/Kathryn A Smith
501 E Harvard, Glendale, CA...............818-243-1114
E&J Advertising/William Flint
1400 N Dutton #9, Santa Rosa, CA...............707-527-5195
Electric Page, The/Sam Toll
1915 21st St, Sacramento, CA 916-737-3900
Electric Pencil/Samantha Seplar
7809 Melrose Ave, Los Angeles, CA 213-852-9665

Electronic Publishing Network/Steve Walsten
  PO Box 3001, San Clemente, CA.......................714-248-9025
Electronic Sweat Shop
  990 N Hill St #200, Los Angeles, CA 213-225-1918
Energy Productions' Timescape/Sue Kincade
  12700 Ventura Blvd 4th Fl, Studio City, CA.........818-508-1444
Ferrari Color Digital Imaging/Scott Powell
  2574 21st St, Sacramento, CA 800-533-6333
Filmworks, The/Don Lee
  1011 Armando St, Anaheim, CA 714-632-7775
Fontographics/Hossein Farmani
  998 S Robertson #202, Los Angeles, CA 310-659-0122
Fotocraft/Allen Shores
  769 N Orange Grove Blvd, Pasadena, CA 818-796-5307
Gibson Studio Inc
  954 San Rafael Ave, Mt View, CA 415-961-9534
Gonzauras/Howard Robbins
  5757 Wilshire M101, Los Angeles, CA 213-749-4800
GP Color Inc/John Portaro
  201 S Oxford Ave, Los Angeles, CA 213-386-7901
Grafica/Carol Girardi
  7053 Owensmouth Ave, Canoga Park, CA 818-712-0071
Graphic Design Services/Sharon Noller
  515 W Lambert, Brea, CA 714-529-7003
Graphic Express/Robin Donofrio
  2150 Old Middlefield Way, Mt View, CA...........415-962-9900
Graphic Presentation Services/Alan Halpern
  1001-A Colorado Ave, Santa Monica, CA 310-451-1307
Graphic Traffic/Evelyn Bradley
  1528 State St, Santa Barbara, CA 805-965-2372
Graphics 4 Typography Inc/Paul Fitterer
  300 NW 14th Ave, Portland, OR 503-225-0148
Graphis Type
  2400 W Ninth St, Los Angeles, CA 213-383-0751
Grayphics Type & Design/Tim Gray
  29 W Anapamu St, Santa Barbara, CA 805-899-2387
GTS Graphics/Elliott Derman
  5650 Jillson St, Los Angeles, CA....................213-888-8889
Hallas Photo Lab/Jim Davis
  4532 Telephone Rd #11, Ventura, CA.............805-642-8063
Hannaway & Assoc, G W/Joe Pezzillo
  839 Pearl St, Boulder, CO 303-440-9631
HB Type & Graphics/Glenn Byron
  1615 Alabama St, Huntington Beach, CA 714-536-3939
Headline Typography
  627 Encinitas Blvd, Encinitas, CA 619-436-0133
Hi-Rez Graphics/Frances Carrasco
  1610 S La Cienega #201, Los Angeles, CA 310-858-5711
Hollywood Vaults Inc/David Wexler
  742 N Seward St, Hollywood, CA 805-569-5336
House Graphics/John Rickey
  3511 Camino del Rio S #200, San Diego, CA 619-563-5752
Hunza Graphics/Paul Marcus
  2527 Dwight Way, Berkeley, CA 510-549-1766
Icon West/Reginald Green
  7961 West 3rd St, Los Angeles, CA 213-938-3822
Image Communications/Glenn Michel
  2908 Oregon Ct #I-8, Torrance, CA 310-533-8911
Image Source/Eli Kozak
  2239 Omega Rd #5, San Ramon, CA 510-837-9015
ImageSetters
  11418-B Moorpark St, N Hollywood, CA 818-760-6595
Indian Rock Image Setting/Jay Nitschke
  1794 Fifth St, Berkeley, CA...........................510-843-8973
Infomania/Tom Mornini
  11492 Sunrise Gold Circle, Rancho Cordova, CA..916-852-5900
Inland Imaging Service/Via Type Corp/Kip Kiphart
  150 N Oberlin Ave, Claremont, CA 909-626-8973
Integrated Graphic Media/Kirk Keller
  5700 Ayala Ave, Irwindale, CA......................818-815-2775
Intersep/Tom Pettinger
  2345 S 2700 West, Salt Lake City, UT 801-973-6720
Interset/John Marsh
  3407 W 6th St #520, Los Angeles, CA 213-666-1000
IPG Digital Imaging Ctr/Joe Maguire
  19 Corporate Park, Irvine, CA 714-261-8777
Joe's Type Shop/Larry Bruynincks
  727 W 7th St #1145, Los Angeles, CA 213-622-1449
Kedie Image Systems/Dave Kedie
  744 San Aleso, Sunnyvale, CA 408-734-9005
Klein Design/Richard Klein
  6290 Sunset Blvd #505, Hollywood, CA............213-466-0633
Krishna Copy Center/Singh Garcha
  1221 Broadway, Oakland, CA........................510-763-0193

Krishna Copy Center/Amrik Singh Bullar
  1676 N California Blvd #119, Walnut Creek, CA..510-256-6898
Krishna Copy Center/Anil
  1785 Paloverde Ave, Long Beach, CA 310-431-9974
Krishna Copy Center/Sanjeeb Rai
  551 Mission Street, San Francisco, CA...............415-543-3688
Krishna Copy Center/Vinod Kumar
  2501 Telegraph Ave, Berkeley, CA 510-549-0506
Krishna Copy Center/Bill Row
  66 Kearney Street, San Francisco, CA 415-986-6161
Krishna Copy Center/Ted Espejo
  2111 University Ave, Berkeley, CA 510-540-5959
L Graphix/Dan Meireis
  267 Commercial St SE, Salem, OR 503-399-7798
L Graphix/Rick Sanders
  108 NW 9th St #201, Portland, OR 503-248-9713
Laser Express/Amir Kaholi
  7822 Convoy Ct, San Diego, CA 619-694-0204
Laser Writing Inc/Ed Bell
  100 W 11th Ave, Denver, CO 303-592-1144
Laser Writing Inc/Tom Samuelson
  8480 E Orchard Rd, Englewood, CO 303-741-4100
Lazer Graphix/Jerry Stephens
  3021 Valley View #209, Las Vegas, NV............702-871-5511
Lazertouch/Patty Olivera
  44 East 4th Ave, San Mateo, CA 415-348-7010
Lee's Copy Network/Chris Lee
  4410 Easton Dr, Bakersfield, CA 805-322-3450
Lithographics/Robert Seastrom
  1616 J Street, Sacramento, CA 916-447-3219
Loral Ads
  13810 SE Eastgate #500, Bellevue, WA............206-746-6800
Marinstat/Von Wall
  352 Miller Ave, Mill Valley, CA.......................415-381-1188
Marketing Resources
  716 Ventura, Oxnard, CA 805-984-7622
Masters Presentation Graphics/Julie Bradley
  2405 Lake View Ave, Los Angeles, CA 213-660-7010
Matrix Communication Arts & Svcs/Daniel Hayashi
  4646 Manhattan Beach Blvd #C, Longdale, CA...310-542-9410
Matrix Communications/Laura Manss
  229 Pajaro #201, Salinas, CA........................408-757-4164
Max Loftins Quality Graphics Inc/Lance Brush
  502 Goetz Ave, Santa Ana, CA 714-546-2923
McIntyre Advertising/Jean Haley
  11650 Iberia Pl #130, San Diego, CA 619-485-6852
Medior
  2800 Campus Dr #150 , San Mateo, CA 415-525-4000
Mel Typesetting/Ed Nies
  1519 S Pearl St, Denver, CO 303-777-5571
MMZ Graphics/Mike Howard
  4672 Southeastern Ave, City of Commerce, CA...213-724-7225
Modern Graphics/Theresa Do
  7676 Claremont Mesa Blvd, San Diego, CA........619-560-5383
MSI Marketing/Communications/Stan Barbarich
  3 Main Dock, Sausalito, CA............................415-331-6400
National Teleprinting Inc/Jim Howe
  1420 Blake St, Denver, CO 303-623-2800
Omniprint/Randi Putz
  5670 El Camino Real #F, Carlsbad, CA.............619-931-6664
On The Ball
  1884 S Santa Cruz, Anaheim, CA....................714-978-9057
On Word Inc/Mark Ross
  1608 Pacific Ave #204, Venice, CA 310-399-7733
Pacific Data Images/Brad Lewis
  1111 Karlstad Dr, Sunnyvale, CA.....................408-745-6755
Pacific Digital Image/Jeremy Smith
  1050 Sansome St #100, San Francisco, CA........415-274-7234
Pacific Title & Art/Peter Hubbard
  6350 Santa Monica Blvd, Hollywood, CA..........213-464-0121
Page One/Doug Lax
  2402 Michelson #100, Irvine, CA 714-851-1530
Page Printing/Gil Zazquez
  21541 Nordhoff St, Chatsworth , CA.................818-341-6033
Phoenix Press/Diane Biggerstaff
  2772 Main St, Irvine, CA 714-261-0333
Photovault
  1045 17th St, San Francisco, CA 415-552-9682
Pinnacle Publishing/Toni Cardoza
  55 Osgood Pl, San Francisco, CA 415-989-TYPE
Pip Printing/Bob Dodson
  1500 S 336th St #15, Federal Way, WA...........206-838-9150
PIX Imaging/Jim Cox
  6100 Melrose Ave, Los Angeles, CA 213-462-2255

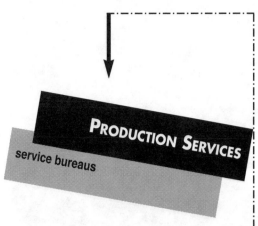
Pixmil Digital Imaging Service/Don Redding
  6150 Lusk Blvd #B102, San Diego, CA 619-597-1177
PrePress Studio/Terry Rorie
  5657 Wilshire Blvd #350, Los Angeles, CA 213-938-3956
Primary Color/Romero Flores
  16841 Milliken Ave, Irvine, CA 714-660-7080
Pro Lab/John Bigelow
  123 NW 36th St, Seattle, WA 206-547-5470
Pro Type Computer Graphics/Jim Price
  7108 De Soto Ave #203, Canoga Park, CA 818-710-8973
Professional Printing Svcs/Everett Jess
  2225-A Hancock St, San Diego, CA..................619-297-1188
ProLab Imaging & Graphics/Steve McCullum
  1285 W Buers, Denver, CO 303-733-2200
Quest/Robert Schlaff
  432 S Broadway, Denver, CO 303-722-5965
Quicktype & Design/Rick Valasek
  2001 Miraloma, Placentia, CA 714-579-8800
R B Images/Roseanne Brown
  723 N Cahuenga, Hollywood , CA 213-962-6660
Raging Fingers/Jason Clark
  6301 Sunset Blvd, Hollywood, CA...................213-462-0575
**Rapid Lasergraphics Photography &
Design**/Brian Blackwelder
  633 Battery St 2nd Fl, San Francisco, CA...........415-291-8166
Repli Color/Ramon Mangelson
  238 Edison St, Salt Lake City, UT....................801-328-0271
RFX Inc/Ray Feeney
  710 Seward St, Hollywood, CA 213-851-2100
Rhythm & Hues/Suzanne Datz
  910 N Sycamore Ave, Hollywood, CA 213-851-6500
Ross Typographic Service Inc
  1000 Fremont Ave #130, S Pasadena, CA 818-799-1610
Scan-Graphics Inc/Jerry Seehof
  PO Box 3179, Huntington Beach, CA 714-840-5380
Scarlett Letters/Tom Tetrault
  6430 Sunset Blvd, Hollywood, CA 213-461-5959
Second Original Transparencies/Doug Mitchell
  17705 NE 29th St, Redmond, WA....................206-885-1238
Sid's Typographers
  3386 S Robertson Blvd, Los Angeles, CA............310-839-2491
Skill Set Graphix/Manny Pavia
  5757 Wilshire Blvd # M101, Los Angeles, CA 213-937-5757
Slide House/Stephen Suga
  3737 Torrance Blvd #105, Torrance, CA 310-792-2092
Southern California PrintCorp/Mike Flanagan
  1010 E Union St #205, Pasadena, CA 818-398-3501
Subia Corporation/Steve Barberio
  2450 S 24th St, Phoenix, AZ 602-275-6565
Superslides/Jane Wallace
  1900 Wazee St #209, Denver, CO...................303-295-7131
Taft Printing/Deskprint Graphics/David Klass
  7352 Beverly Blvd, Los Angeles, CA.................213-933-9138
Thompson Type Inc/John Pierce
  3687 Voltaire St, San Diego, CA......................619-224-3137
Thurston Productions/Chuck Thurston
  485 Third St, San Francisco, CA 415-882-4800
Tom Buhl Imaging Ctr/Tom Buhl
  621 Chapala St, Santa Barbara, CA 805-963-8841
Truline Graphics/Mark Allen
  738 N Cahuenga Blvd, Los Angeles, CA............213-469-1611
Trunk Line/Debra Hughes
  8800 Venice Blvd, Los Angeles, CA 310-204-2882
Type Factory Inc/Eileen LaRue
  380 E First S #A, Tustin, CA...........................714-730-0990

PRODUCTION SERVICES
service bureaus/integrated printing/ cd-rom publishers

Type Inc/Dale Robertson
5900 Wilshire Blvd #460, Los Angeles, CA 213-931-1068
Type Plus/Byron King
222-W Mission St #118, Spokane, WA.............509-328-6102
Typesetting Room, The/Renee Gauthier
150 Paularino #150, Costa Mesa, CA................714-751-3131
Ultratype Inc/Gerald Phillips
342 W 200 S #150, Salt Lake City, UT 801-521-5914
Valley Printers/Sally Barton
2180 Iowa Ave, Riverside, CA.......................909-682-5539
Veritel/Lori Anderson
1 Union St, San Francisco, CA 415-495-3328
Via Type/Clarence Kiphardt
150 N Oberlin Ave, Claremont, CA 909-626-8351
Visicom Design Group/Mark Wood
2625 Park Blvd, Palo Alto, CA 415-327-6280
Vivid Graphics & Typesetting/Ken Ormelinda
9235 Activity Rd #108, San Diego, CA 619-578-4843
Western Imaging/David Krauss/Dan Goldstein
115 Constitution Dr, Menlo Park, CA 415-325-0300
Western Type & Printing/Warren Funnell
2218 Fifth Ave, Seattle, WA 206-728-6700
Wizywig/Lynn Christensen
3151 Elliot Ave #310, Seattle, WA 206-283-3069
WP Plus/Steve Hirsh
2858 El Roble Dr, Los Angeles, CA.................213-255-5515
Wright Lithography/Ken Wright
125 W LaPalma Ave, Anaheim , CA 714-778-4499
Wy'east Color/Dwight Cummings
1427 116th Ave NE, Bellevue, WA..................206-454-8006
Wy'east Color Inc/Dean Cummings
4200 SW Corbett Ave, Portland , OR 503-228-7053
ZZYZX Visual Systems/Kim Kapin
1011 N Orange Dr, Los Angeles, CA 213-883-1060

CANADA

Ainsworth Group, The/Lori Klooster
65 Hanson Ave, Kitchner, ON 519-578-0530
Bergman Graphics Limited/GE Bergman
66 Nuggett Court, Brampton, ON 416-792-8385
GM&A Advertising/Joe Costa
1785 Matheson Blvd E, Mississauga, ON 416-625-1734
Toronto Image Works/Ed Burtynski
80 Spadina Ave #207, Toronto, ON 416-363-1999

Integrated Printing

Daniels Printing Co/Esther Fisher
40 Commercial St, Everett, MA.......................617-389-7900
Dupont Imaging Systems
1007 Market St, Wilmington, DE.....................302-774-1000
Graphic Enterprises Inc/Laura Rogers
316 E Abram St, Arlington, TX........................817-277-9442
Graphics Illustrated/Kelly Arnold
1500 Australian Ave, Riviera Bch, FL 407-848-8989
Graphtec Inc/Frank Mauro
505 NW 65th Court, Ft Lauderdale, FL 800-444-1931
Harvest Productions/Maryann Doe
6312 E Santa Ana Canyon Rd 350,
Anaheim, CA 714-281-0844
Image Transform Ltd/Ron Cherkas
106 SW Second Ave, Des Moines, IA................515-288-0000
King Graphics Inc/Nancy Gorski
571 W Polk St, Chicago, IL............................312-943-2202
LaserTools Corp/Sales Line
1250 45th St #100, Emeryville, CA..................800-767-8004

Washburn Graphics
801 S McDowell, Charlotte, NC.........................800-438-0378

CD-ROM Publishers

NEW YORK CITY

ABC News
47 W 66th St 6th Fl, New York, NY 212-456-3749
Bantam Books Inc
666 Fifth Ave, New York, NY 212-765-6500
Byron Preiss Multimedia
24 W 25th St, New York, NY 212-645-9870
Facts On File Inc
460 Park Ave S, New York, NY 212-683-2244
McGraw Hill
11 W 19th St, New York, NY 212-337-5916
Sony Electronic Publishing
711 Fifth Ave, New York, NY 212-702-6273
St Martin's Press
175 Fifth Ave, New York, NY 212-674-5151

EAST

Agfa Compugraphic
90 Industrial Way, Wilmington, MA 800-424-8973
Agfa Division/Miles Inc
200 Ballardvale St, Wilmington, MA 508-658-5600
Applied Optical Media/Ted Winsor
1450 Boot Rd/Bldg 400, West Chester, PA 215-429-3701
Bureau of Elec Publishing
141 New Rd, Parsippany, NJ..........................201-808-2700
Compact Publishing
5141 MacArthur Blvd, Washington, DC.............800-964-1518
Computerized Bizmart Systems
451 Moody St #206, Waltham, MA 615-894-4452
DC Heath & Co
125 Spring St, Lexington, MA 617-862-6650
Deep River Publishing
565 Congress St, Portland, ME.......................207-871-1683
DRI/McGraw Hill
24 Hartwell Ave, Lexington, MA......................716-863-5100
Eastman Kodak
100 Carlson Rd, Rochester, NY.......................716-724-4000
Free Spirit Software
58 Noble St/PO Box 128, Kutztown, PA............215-683-5609
Fujitsu Imaging Systems
404 Wyman St #300, Waltham, MA 617-890-4833
G K Hall & Co
70 Lincoln St, Boston, MA.............................617-423-3990
Grolier Electronic Publishing Inc
Sherman Turnpike, Danbury, CT 203-797-3530
INTEL/Princeton Operation
DVI Technology, CN 5325, Princeton, NJ...........609-936-7619
Internet Start Up/Editorial Inc
PO Box 2647/Whistlestop Mall, Rockport, MA....508-546-7346
Little, Brown & Co
34 Beacon St, Boston, MA.............................617-859-5549
Macmillan New Media
124 Mount Auburn St, Cambridge, MA 617-661-2955
Multimedia Products Corp
300 Airport Executive Park, Spring Valley, NY....914-426-0400
National Geographic Society
PO Box 98018, Washington, DC 800-368-2728
Pilgrim New Media
955 Massachusetts Ave, Cambridge, MA...........617-491-7660
Syracuse Language Systems
719 E Genesee St, Syracuse, NY 315-478-6729

SOUTH

Digital Wisdom Inc
444 Water Ln, Box 2070, Tappahannock, VA.....804-758-0670
Hyperglot Software Co Inc
518-D Kngstn Pke, PO Box 10746, Knoxville, TN.615-558-8270
Intechnica Int'l Inc
PO Box 3877, Midwest City, OK 405-732-0138
Jane's Information Group
1340 Braddock Pl #300, Alexandria , VA............703-683-3700
Multimedia Publishing Studio
4111 Northside Prkwy #L05S1, Atlanta, GA 404-238-4304
Nimbus Info Systems
State Rd 629/Guildford Farm, Ruckersville, VA....804-985-1100
Nimbus Records
PO Box 7427, Charlottesville, VA.....................804-985-1100

MIDWEST

Alde Publishing
6520 Edenvale Blvd, Eden Prairie, MN 612-934-4239
American Express/Info Svcs
1825 Old Mill Rd, Omaha, NE 800-338-6073
Compton's Learning
310 S Michigan/Britannica Ctr, Chicago, IL........312-347-7128
Digital Publishing
8100 Wayzata Blvd, Golden Valley, MN 612-595-0801
NEC Home Electronics Computer
1255 Michael Dr, Wood Dale, IL 312-860-9500
NTC Publishing Group
4255 W Touhy Ave, Lincolnwood, IL 708-679-5500
Quanta Press Inc
1313 Fifth St SE #208C, Minneapolis, MN 612-379-3956
Wayzata Technology Inc
412 Pokegama Ave N, Box 807,
Grand Rapids, MN 218-326-0597
Wheeler Arts/Paula Wheeler
66 Lake Park, Champaign, IL 217-359-6816

WEST

7th Level
5225 San Fernando Rd W, Los Angeles, CA 818-547-1955
Activision/Howard Marks
11601 Wilshire Blvd 10th Fl, Los Angeles, CA....415-329-0800
Addison Wesley/Jack Hankin
2725 Sandhill Rd, Menlo Park, CA 415-854-0300
Adobe Systems
1585 Charleston Rd, Mountain View, CA...........415-961-4400
Artspace/AIM
11111 Santa Monica Blvd #1000,
Los Angeles, CA 213-473-4136
Atari Computer Corp
1196 Borregas Ave/PO Box 3427,
Sunnyvale, CA 408-745-2000
Clarinet Communications
4880 Stevens Creek Blvd #206, San Jose, CA....408-296-0366
Claris Corp
5201 Patrick Henry Dr, Santa Clara, CA............800-325-2747
Compton's New Media/Carlsbad
2320 Camino Vida Roble, Carlsbad, CA 619-929-2500
Compton's New Media/Solana Beach
722 Genevieve #M, Solana Beach, CA 619-259-0444
Conter Software
6170 Cornerstone CT E, San Diego, CA.............800-CD-TEAC
Creative Multimedia Corp
514 NW 11th Ave #203, Portland, OR............503-241-4351
Cyan Inc
PO Box 28096, Spokane, WA.........................509-468-0807
East West Communications
1631 Woods Dr, Los Angeles, CA....................213-848-8436
Ebook Inc
1009 Pectin Ct, Milpitas, CA 408-262-0502
Educorp
7434 Trade St, San Diego, CA 619-536-9999
Gazelle Technologies
San Diego, CA...619-693-4030
Hewlett Packard Co
19310 Pruneridge Ave, Cupertino, CA 415-691-5805
Interactive Music Co, The
921 Church St , San Francisco, CA 415-285-8650
Interoptica Publishing
300 Montgomery St #201, San Francisco, CA.....415-788-8788
J & D Publishing
Orem, UT...800-937-6493
Jasmine Multimedia Publishing
64746 Valjean Ave #100, Van Nuys, CA...........800-798-7535
Masque Publishing Inc
PO Box 5223, Englewood, CO 800-290-9853
Media Vision
47300 Bayside Prkwy, Freemont, CA................510-770-8600
Medio Multimedia
Redman, WA...206-867-5500
Multicom Publishing
1100 Olive Way #1250, Seattle, WA................206-622-5530
Opcode Interactive/Tracy Robinson
39950 Fabian Way #100, Palo Alto, CA 415-494-1112
PC Comix
Ashland, OR..503-488-3727
Philips Interactive Media Systems
11111 Santa Monica Blvd #1000,
Los Angeles, CA 310-473-4136
Reflective Arts International
180 Knowles Way, Los Gatos, CA 408-395-4332

Substance Interactive Media/Nick Roberts
    444 Grove St, San Francisco, CA......................415-626-2147
Sumeria Inc
    329 Bryant St #3D, San Francisco, CA 415-904-0800
Supermac Technology
    485 Potrero Ave, Sunnyvale, CA 408-245-2202
Uni Disc Inc
    3941 Cherryvale Ave #1, Soquel, CA 408-464-0707
Voyager Co
    1315 Pacific Coast Hwy 3rd Fl,
    Santa Monica, CA 310-451-1383
Warner New Media
    3500 West Olive Ave #1050, Burbank, CA........818-955-6485
Xiphias
    Helms Hall/8758 Venice Blvd, Los Angeles, CA...213-841-2790

C A N A D A

Discis Knowledge Research Inc
    45 Sheppard Ave #410, Toronto, ON 416-250-6537

PRODUCTION SERVICES
cd-rom publishers

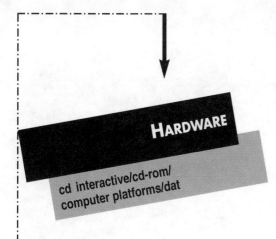

CD Interactive

Apple Computer Inc/Tara Griffin
135 E 57th St 17th Fl, New York, NY 212-339-3700
Ayala Computer Imaging
631 W Stratford Pl, Chicago, IL.......................312-327-1156
BetaCorp Technologies Inc/John J Bekto
6770 Davand Dr #40, Mississauga, ON 416-564-2424
**Blanchard-Healy Video
Communications**/Tom Kreuzberger
620 Sentry Parkway # 110, Blue Bell, PA...........215-834-5700
CD Folios/CD Publishing/Randy Fugate
22647 Ventura Blvd #199, Woodland Hills, CA.805-527-9227
Current Designs Corp/David Rapkin
163 W 23rd St #5, New York, NY 212-463-0795
Kodak
343 State St, Rochester, NY 800-445-6325
Optimage Interactive Services LP/
Lynn Goodman
1501 50th St #100, W Des Moines, IA 515-225-7000
Philips Consumer Electronics Co/Paul Fredrickson
One Philips Dr, Knoxville, TN 615-521-4316
Telerobotics Intl Inc/Daniel Kuban
7325 Oak Ridge Hwy #104, Knoxville , TN........615-690-5600
Viper Optics/JoAnn Gillerman
950 61st St, Oakland, CA.............................510-654-2880

CD-ROM

Apple Computer Inc
20525 Mariani Ave, Cupertino, CA.................408-996-1010
BetaCorp Technologies Inc/John J Bekto
6770 Davand Dr #40, Mississauga, ON 416-564-2424
CD Technology/Bill Liu
766 San Aleso, Sunnyvale, CA 408-752-8500
CD-ROM Strategies/Dr Ash Pahwa
Six Venture #208, Irvine, CA 714-453-1702
Chinon America Inc/Scott Elrich
615 Hawaii Ave, Torrance, CA.......................310-533-0274
Commodore Business Machines/Bill Sacko
1200 Wilson Dr, West Chester, PA.................215-431-9100
Crown & Shield Software Inc/Mary Lynn Davis
29 Crafts St #200, Newton, MA 617-965-3383
Dataware Technology Inc
5775 Flatiron Pkwy #220, Boulder , CO...........303-449-4157
Digipress/Denis Oudard
2016 Bainbridge Row Dr, Louisville, KY...........502-895-0565
Eagle Eye Publishers Inc/Paul Murphy
115 Park St SE #300, Vienna, VA....................703-242-4201
Educorp/Technical Support Dept/Suzi Nawabi
7434 Trade St, San Diego, CA.......................619-536-9999
Executive Technologies Inc/Jim Geer
2120 16th Ave S, Birmingham, AL...................205-933-5494
Follett Software Co/Lisa McManaman
809 N Front St, McHenry, IL...........................815-344-8700
Geovision Inc/Kenneth Shain
PO Box 921266, Norcross, GA 404-448-8224
**Hewlett Packard Co/Personal Info
Products Group**/Bob Frankenberg
5301 Stevens Creek Blvd, Santa Clara, CA........408-553-7585
Hitachi Sales Corp of America/Eric Kamayatsu
401 W Artesia Blvd, Compton, CA...................310-537-8383
IBM Marketing Svcs/Carol Quinn
Rte 9W, Palisades, NY 914-732-6000
IBM/Educational Systems/Tom Wall
Box 2150/Northside Pkwy, Atlanta, GA 404-238-2000

Images Inc/Warren Lieb
216 N May #101, Chicago, IL 312-226-7660
Info Systems/David Ujita
9740 Irvine Blvd, Irvine, CA..........................714-583-3000
IS Grupe Inc/Peter Schipma
948 Springer Dr, Lombard, IL.......................708-627-0550
JVC Co of America/Harry Elias
41 Slater Dr, Elmwood Park, NJ 201-794-3900
Knowledge Access Intl/Alan Guzzetti
2685 Marine Way #1305, Mt View, CA............415-969-0606
Kodak
343 State St, Rochester, NY 800-445-6325
Laser Magnetic Storage International Co/Rob van Eijk
4425 Arrowswest Dr, Colorado Springs, CO.....719-593-7900
Mammoth Microproductions Inc/Robert Odgon
1536 Cole Blvd #165, Golden, CO 303-237-5776
Meridian Data Inc/Richard Krueger
5615 Scotts Valley Dr, Scotts Valley, CA............408-438-3100
Micro Design International/Mark Gillmore
6985 University Blvd, Winter Park, FL 407-677-8333
MicroSearch/Lon Ketcham
9000 Southwest Freeway #330, Houston, TX......713-988-2818
Microtech International Inc/Elaine Deleo
158 Commerce St, East Haven, CT.................203-468-6223
National Teleprinting Inc/Jim Howe
1420 Blake St, Denver, CO...........................303-623-2800
NEC Technologies (USA) Inc/Peter Gion
1414 Massachusetts Ave, Boxboro, MA............508-264-8000
Nimbus Info Systems/Bob Hedrick
PO Box 7427, Charlottesville, VA...................800-782-0778
OnLine Computer Systems Inc/Lisa Wolin
20251 Century Blvd, Germantown, MD 301-428-3700
Optical Access Intl/Mike Maloney
500 W Cummings Park, Woburn, MA...............617-937-3910
Optical Media International
180 Knowles Dr, Los Gatos, CA.....................408-376-3511
Panasonic
50 Meadowlands Pkwy, Secaucus, NJ...............201-348-7000
Philips Consumer Electronics Co/Paul Fredrickson
One Philips Dr, Knoxville, TN 615-521-4316
Relax Technology/Frank Gabrielli
3101 Whipple Rd, Union City , CA...................510-471-6112
Roland Corp US/Scott Summers
7200 Dominion Cir, Los Angeles, CA 213-685-5141
Sony Corp of America
3 Paragon Dr, Montvale, NJ..........................800-635-7669
Sun Moon Star/Personal Computer Div
/Michael Wu
1941 Ringwood Ave, San Jose, CA.................408-452-7811
Todd Enterprises Inc/Michael Scheibel
224-49 67th Ave, Bayside, NY 718-343-1040
Trillium Computer Resources/Clare Snyder
450 Phillip St #A4, Waterloo, ON 519-886-4404

Computer Platforms

AGFA/Div of Miles/Seamus Carroll
One Ramland Rd, Orangeburg, NY 914-365-0190
Altronix Systems/Allen Cowie
100 Ford Rd, Denville, NJ.............................201-625-8220
Apple Computer Inc/Tara Griffin
135 E 57th St 17th Fl, New York, NY 212-339-3700
Apple Computer Inc
20525 Mariani Ave, Cupertino, CA.................408-996-1010
Art Machines Inc/Hugh Smyser
594 Broadway #307, New York, NY 212-431-4400
AST Research Inc/Grant Johnson
16215 Alton Pkwy, Irvine, CA........................714-727-7901
BetaCorp Technologies Inc/John J Bekto
6770 Davand Dr #40, Mississauga, ON 416-564-2424
CD Technology/Bill Liu
766 San Aleso, Sunnyvale, CA 408-752-8500
Colby Systems Corp/Charles Edwards
2991 Alexis Dr, Palo Alto, CA 415-941-9090
Commodore Business Machines/Bill Sacko
1200 Wilson Dr, West Chester, PA 215-344-3040
Commodore Business Machines
1200 Wilson Dr, West Chester, PA 215-431-9100
Computer Friends/Scott Huish
14250 NW Science Park Dr, Portland, OR..........503-626-2291
Computer Group/Bob Beecher
14 Ellis Potter Ct, Madison, WI.......................608-273-1803
Dicomed/Tim Whalen
11401 Rupp Dr, Burnville, MN........................800-888-7979

Everex/Ned Nevels
5020 Brandin Ct, Fremont, CA........................510-438-5783
Farallon Computing/Georgeann Benesch
2470 Mariner Square Loop, Alameda, CA.........510-814-5100
Future Video Products Inc/Steven Godfrey
28 Argonaut #150, Aliso Viejo, CA 714-770-4416
**Hewlett Packard Co/Personal Info
Products Group**/Bob Frankenberg
5301 Stevens Creek Blvd, Santa Clara, CA........408-553-7585
HSD Microcomputer US Inc/Dave Peter
345 1st St #0, Encinedas, CA 619-632-9700
Hyperspeed Technologies/Rod Davis
10225 Barnes Canyon Rd #A206,
San Diego, CA 619-578-4893
IBM Marketing Svcs/Carol Quinn
Rte 9W, Palisades, NY 914-732-6000
Imapro Corp/Fred Androni
2400 Saint Laurent Blvd, Ottawa, ON 613-738-3000
Info Systems/David Ujita
9740 Irvine Blvd, Irvine, CA..........................714-583-3000
Julie Research Laboratories/Loebe Julie
508 W 26th St, New York, NY 212-633-6625
Kodak
343 State St, Rochester, NY 800-445-6325
Kubota Pacific Computers/Michael Keeler
2630 Walsh Ave, Santa Clara, CA 408-727-8100
Magni Systems Inc/David Jurgensen
9500 SW Gemini Dr, Beaverton, OR 503-626-8400
MicroSearch/Lon Ketcham
9000 Southwest Freeway #330, Houston, TX......713-988-2818
NEC Technologies Inc
1414 Massachusetts Ave, Boxborough, MA 508-264-8970
NeXT Inc/Steve Jobs
900 Chesapeake Dr, Redwood City, CA.............415-366-0900
Nth Graphics/Tom Hall
11500 Metric Blvd #210, Austin, TX.................512-832-1944
Panasonic
50 Meadowlands Pkwy, Secaucus, NJ 201-348-7000
Philips Consumer Electronics Co/Paul Fredrickson
One Philips Dr, Knoxville, TN 615-521-4316
Quantum Computers/Grant Asplund
5003 Tacoma Mall Blvd, Tacoma, WA...............206-475-7000
RasterOps Corp/Marketing Dept
2500 Walsh Ave, Santa Clara, CA 408-562-4200
● **SILICON GRAPHICS INC**/
Deborah Miller (**P 106,107,BACK COVER**)
2011 N Shoreline Blvd, Mountain View, CA........415-960-1980
Sun Microsystems Inc
2550 Garcia Ave, Mt View, CA.......................415-960-1300
Sundance Technology Group/Rush Beesly
6309 N O'Connor #111 LB/128, Irving, TX 214-869-1002
Symbolics
555 Virginia Rd, Concord, MA 508-287-1000
Telerobotics Intl Inc/Daniel Kuban
7325 Oak Ridge Hwy #104, Knoxville , TN........615-690-5600
Trident Microsystems Inc/Michael Maia
205 Ravendale Dr, Mt View, CA 415-691-9211
Videk Corp/Thomas Braun
1100 Corporate Dr, Canandaigua, NY..............716-924-6100
Videologic Inc/Lynn Dacy
245 First St, Cambridge , MA 617-494-0530
Workstation Technologies Inc/Christopher Miner
18010 Sky Park Cir #155, Irvine, CA................714-250-8983

DAT

ADIC
PO Box 97057, Redmond, WA.......................206-881-8004
Alliance Peripheral Systems
6131 Deramus, Kansas City, MO.....................800-235-2752
Custom Recording Studios/Jim Reynolds
4800 Drake Rd, Golden Valley, MN...................612-521-2950
Denon America/Steve Baker
222 New Rd, Parsippany, NJ..........................201-575-7810
FWB Inc/David Lamont Booth
240 Polk #215, San Francisco, CA 415-474-8055
Hard Drives International
1912 W 4th St, Tempe, AZ 800-776-3475
MacProducts
608 W 22nd St, Austin, TX.............................800-622-8721
Memory Technology Texas
3007 N Lamar, Austin, TX 512-451-2600
MicroNet Technology Inc/Exec Committee
80 Technology, Irvine, CA.............................714-453-6000

Microtech International Inc/Elaine Deleo
158 Commerce St, East Haven, CT 203-468-6223
PCPC/Robert Leeds
34 Jerome Ave #214, Bloomfield, CT..............203-243-5320
Peripheral Land Inc/Frank Jaramillo
47421 Bayside Pkwy, Fremont, CA 510-657-2211

Digital Editing

Analog Devices Inc/Vicki Chase
One Technology Way, Norwood, MA 617-329-4700
Avid Technology Inc/Mark Overington
One Park W 2nd Fl, Tewksbury, MA..................508-640-6789
CD Folios/CD Publishing/Randy Fugate
22647 Ventura Blvd #199, Woodland Hills, CA.805-527-9227
Digital F/X Inc/Barbara Koalkin
755 Ravendale Dr, Mt View, CA 415-961-2800
Epix Inc/Charles Dijak
381 Lexington Dr, Buffalo Grove, IL 708-465-1818
IBM/Educational Systems/Tom Wall
Box 2150/Northside Pkwy, Atlanta, GA 404-238-2000
Imapro Corp/Fred Androni
2400 Saint Laurent Blvd, Ottawa, ON 613-738-3000
Jonathan Helfand Music
35 W 45th St, New York, NY 212-921-0555
JVC Co of America/Harry Elias
41 Slater Dr, Elmwood Park, NJ 201-794-3900
Knox Video/Roland Blood
8547 Grovemont Cir, Gaithersburg, MD 301-840-5805
Kodak
343 State St, Rochester, NY 800-445-6325
MicroSearch/Lon Ketcham
9000 Southwest Freeway #330, Houston, TX......713-988-2818
Mitsubishi Electronics America Inc/Doug Mackeroy
800 Cottontail Ln, Somerset, NJ.......................908-563-9889
Results! Computers & Video
12600 Morrow Ave NE, Albuquerque, NM 505-291-8998
Selectra/Richard Comfort
PO Box 5497, Walnut Creek, CA 800-874-9889
Sundance Technology Group/Rush Beesly
6309 N O'Connor #111 LB/128, Irving, TX 214-869-1002
Video Snapshots
2295 Stepstone Ln, Hanover Park, IL 708-213-2227
Videomedia Inc/Amy Gomersall
175 Lewis Rd, San Jose, CA............................408-227-9977

Digitizers

ASC Systems/R Martin
PO Box 566, St Clair Shores, MI 313-882-1133
Aspex Inc/Gerald Henrici
536 Broadway, New York, NY 212-966-0410
Digital Vision/John Pratt
270 Bridge St, Dedham, MA............................617-329-5400
Epix Inc/Charles Dijak
381 Lexington Dr, Buffalo Grove, IL 708-465-1818
GTCO Corp/Harry Stolte
7125 Riverwood Dr, Columbia , MD 410-381-6688
Koala Acquisitions/Earl Liebich
PO Box 1924, Morgan Hill, CA 408-776-8181
Kodak
343 State St, Rochester, NY 800-445-6325
Matrox Electronic Systems Ltd/Andree DeLaurier
1055 St Regis Blvd, Doval, QB.........................514-685-2630
New Media Graphics/Marty Duhms
780 Boston Rd, Billerica , MA 508-663-0666
Orange Micro Inc/Greg Petersen
1400 N Lakeview Ave, Anaheim, CA 714-779-2772
Recognition Concepts Inc/Ron Earwood
5200 Convair Dr , Carson City, NV 702-882-7817
Summagraphics Houston Instrument
8500 Cameron Rd, Austin, TX..........................800-776-9989
Videk Corp/Thomas Braun
1100 Corporate Dr, Canandaigua, NY...............716-924-6100
Videologic Inc/Lynn Dacy
245 First St, Cambridge , MA 617-494-0530
Willow Peripherals/Elucian Royall
190 Willow Ave, Bronx, NY 718-402-9500

Display Systems

Apple Computer Inc/Tara Griffin
135 E 57th St 17th Fl, New York, NY 212-339-3700
Aspex Inc/Gerald Henrici
536 Broadway, New York, NY 212-966-0410

Atari Corp/Greg LaBrec
1196 Borregas Ave, Sunnyvale, CA 408-744-0880
Boxlight Corp
17771 Fjord Dr NE , Poulsbo, WA.....................206-697-4008
Colby Systems Corp/Charles Edwards
2991 Alexis Dr, Palo Alto, CA 415-941-9090
Elographics Inc/Debbie Maxey
105 Randolph Rd, Oak Ridge, TN 615-482-4100
Focus Enhancements/Susan Tetrault
800 W Cummings Pk #4500, Woburn, MA.......800-538-8865
Hard Drives International
1912 W 4th St, Tempe, AZ 800-776-3475
Hyperspeed Technologies/Rod Davis
10225 Barnes Canyon Rd #A206,
San Diego, CA 619-578-4893
IBM Marketing Svcs/Carol Quinn
Rte 9W, Palisades, NY 914-732-6000
IBM/Educational Systems/Tom Wall
Box 2150/Northside Pkwy, Atlanta, GA 404-238-2000
● **IKEGAMI**/Joe C Guzman **(P 108)**
37 Brook Ave, Maywood, NJ...........................201-368-9171
JVC Co of America/Harry Elias
41 Slater Dr, Elmwood Park, NJ 201-794-3900
Kodak
343 State St, Rochester, NY 800-445-6325
Magni Systems Inc/David Jurgensen
9500 SW Gemini Dr, Beaverton, OR.................503-626-8400
Mitsubishi Electronics America Inc/
Doug Mackeroy
800 Cottontail Ln, Somerset, NJ.......................908-563-9889
Mitsubishi Electronics America Inc/Mike Foster
5665 Plaza Dr, Cypress, CA............................714-220-2500
NEC Technologies (USA) Inc/Peter Gion
1414 Massachusetts Ave, Boxboro, MA.............508-264-8000
Nth Graphics/Tom Hall
11500 Metric Blvd #210, Austin, TX..................512-832-1944
Panasonic
50 Meadowlands Pkwy, Secaucus, NJ 201-348-7000
Pioneer New Media Technology/Henry Ebers
Sherbrooke Pl/600 E Crescent,
Upper Saddle River, NJ 201-327-6400
Radius Inc/Carol Johnson
1710 Fortune Dr, San Jose, CA 408-434-1010
RasterOps Corp/Marketing Dept
2500 Walsh Ave, Santa Clara, CA 408-562-4200
Reflection Technology Inc/Liz Goldstein
230 Second Ave, Waltham, MA 617-890-5905
Relax Technology/Frank Gabrielli
3101 Whipple Rd, Union City , CA....................510-471-6112
RGB Spectrum/Robert Markus
950 Marina Village Pkwy, Alameda, CA............510-814-7000
Supermac Technology/Margaret Hughes
215 Moffett Pk Dr, Sunnyvale, CA 408-541-6100
Tektronix Inc/Display Products/Sat Narayanan
PO Box 500/MS 46-943, Beaverton, OR............503-627-5000
Telephoto Communications Inc/Leonard G Roberts
11722-D Sorrento Valley Rd, San Diego, CA.......619-452-0903

Electronic Photography

Altronix Systems/Allen Cowie
100 Ford Rd, Denville, NJ...............................201-625-8220
Analog Devices Inc/Vicki Chase
One Technology Way, Norwood, MA 617-329-4700
Andrews, Jackie
CA 415-546-5666
Apple Computer Inc
20525 Mariani Ave, Cupertino, CA 408-996-1010
Beckman Instruments Inc/Teri Beauchamp
2500 Harbor Blvd/MS B22D, Fullerton, CA 714-773-7682
Boxlight Corp
17771 Fjord Dr NE , Poulsbo, WA.....................206-697-4008
Dicomed/Tim Whalen
11401 Rupp Dr, Burnville, MN.........................800-888-7979
Digital Vision/John Pratt
270 Bridge St, Dedham, MA............................617-329-5400
Dycam Inc/Debbie Bedford
9588 Topanga Canyon Blvd, Chatsworth, CA 818-998-8008
Epix Inc/Charles Dijak
381 Lexington Dr, Buffalo Grove, IL 708-465-1818
Imapro Corp/Fred Androni
2400 Saint Laurent Blvd, Ottawa, ON 613-738-3000
Kodak
343 State St, Rochester, NY 800-445-6325

HARDWARE

digital editing/digitizers/display systems/
electronic photography/graphic input
devices/hdtv

Leaf Systems/Patty Foley
250 Turnpike Rd, Southboro, MA 508-460-8300
Megavision/Ken Boydston
5765 Thorwood Dr, Santa Barbara, CA.............805-964-1400
Nikon Inc/EID/Joe Carfora
1300 Walt Whitman Rd, Melville, NY 516-547-4200
Polaroid Resource Center
784 Memorial Dr, Cambridge, MA 800-225-1618
Presentation Technologies/Jim Creede
779 Palomar Ave, Sunnyvale, CA 408-730-3700
Prime Option Inc/Nick Vasilikis
2341 W 205th St #116, Torrance, CA................310-618-0274
Todd Enterprises Inc/Michael Scheibel
224-49 67th Ave, Bayside, NY 718-343-1040

Graphic Input Devices

Apple Computer Inc
20525 Mariani Ave, Cupertino, CA 408-996-1010
Artist Graphics/Jim Triggs
2675 Patton Rd, St Paul, MN 612-631-7800
Caere Corporation/Larry Lunetta
100 Cooper Ct, Los Gatos, CA 408-395-7000
Chisholm/David Dicklich
910 Campisi Way, Campbell, CA......................408-559-1111
Digital Vision/John Pratt
270 Bridge St, Dedham, MA............................617-329-5400
Everex/Ned Nevels
5020 Brandin Ct, Fremont, CA.........................510-438-5783
Imapro Corp/Fred Androni
2400 Saint Laurent Blvd, Ottawa, ON 613-738-3000
Kensington Microware/Mary Shank
2855 Campus Dr, San Mateo, CA 415-572-2700
Kodak
343 State St, Rochester, NY 800-445-6325
Law Cypress Distributing
5883 Eden Park Pl, San Jose, CA......................408-363-4700
New Media Graphics/Marty Duhms
780 Boston Rd, Billerica , MA 508-663-0666
Nikon Inc/EID/Joe Carfora
1300 Walt Whitman Rd, Melville, NY 516-547-4200
nView Corp/Alicia Blanchard
860 Omni Blvd, Newport News, VA...................804-873-1354
Simgraphics Engineering Corp/Dave Verso
1137 Huntington Dr #A-1, S Pasadena, CA.........213-255-0900
Summagraphics Houston Instrument
8500 Cameron Rd, Austin, TX..........................800-776-9989
Television Laboratories Inc/Maureen Kappenman
30 Hwy 36, Atlantic Highlands, NJ 908-291-8600
Vermont Microsystems Inc/Diane Magnusson
11 Tigan St, Winooski, VT...............................802-655-2860
Willow Peripherals/Elucian Royall
190 Willow Ave, Bronx, NY 718-402-9500

HDTV

Kodak
343 State St, Rochester, NY 800-445-6325
Panasonic
50 Meadowlands Pkwy, Secaucus, NJ 201-348-7000
RCA/Thompson Consumer Electronics
600 N Sherman Dr, Indianapolis, IN 317-267-5860
Rebo Group/Clinton Powell
530 W 25th St 2nd Fl, New York, NY.................212-989-9466
Sony Corp of America
3 Paragon Dr, Montvale, NJ............................800-635-7669

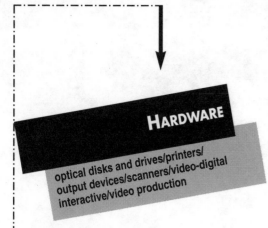

HARDWARE

optical disks and drives/printers/
output devices/scanners/video-digital
interactive/video production

Optical Disks & Drives

Adaptec/Jim Switz
691 Milpitas Blvd, Milpitas, CA......408-945-8600
Advanced Retrieval Systems/Doug Strickland
1 Appletree Sq #801, Bloomington, MN......612-854-1937
Advanced Storage Concepts Inc/Bill Casey
10713 Ranch Rd 620 N #601, Austin, TX...512-335-1077
Apple Computer Inc
20525 Mariani Ave, Cupertino, CA......408-996-1010
Applied Memory Technology Inc/Michael Harvey
2822 Walnut Ave, Tustin, CA......714-838-1860
BetaCorp Technologies Inc/John J Bekto
6770 Davand Dr #40, Mississauga, ON......416-564-2424
CD Technology/Bill Liu
766 San Aleso, Sunnyvale, CA......408-752-8500
Chinon America Inc/Anne Cousino
615 Hawaii Ave, Torrance, CA......310-533-0274
Current Designs Corp/David Rapkin
163 W 23rd St #5, New York, NY......212-463-0795
Dataware Technology Inc
5775 Flatiron Pkwy #220, Boulder , CO......303-449-4157
Denon America/Steve Baker
222 New Rd, Parsippany, NJ......201-575-7810
Digital Equipment Corp/Alan Goldman
334 South St, Shrewsbury, MA......508-841-6625
FWB Inc/David Lamont Booth
240 Polk #215, San Francisco, CA......415-474-8055
Galic Maus Ventures/George Galic
5140 St Moritz Dr NE, Columbia Heights, MN....612-571-7961
GEOVISION Inc/Kenneth Shain
PO Box 921266, Norcross, GA......404-448-8224
Hard Drives International
1912 W 4th St, Tempe, AZ......800-776-3475
**Hewlett Packard Co/Personal Info
Products Group**/Bob Frankenberg
5301 Stevens Creek Blvd, Santa Clara, CA......408-553-7585
Ilomega Corp/Iomega Corp
1821 W 4000 South, Roy, UT......800-456-5522
JVC Disc America/Al Crawford
2 JVC Rd, Tuscaloosa, AL......205-556-7111
Kodak
343 State St, Rochester, NY......800-445-6325
Laser Magnetic Storage International/Rob van Eijk
4425 Arrowswest Dr, Colorado Springs, CO......719-593-7900
Micro Design International/Mark Gillmore
6985 University Blvd, Winter Park, FL......407-677-8333
MicroSearch/Lon Ketcham
9000 Southwest Freeway #330, Houston, TX......713-988-2818
Microtech International Inc/Elaine Deleo
158 Commerce St, East Haven, CT......203-468-6223
Mirror Technologies/John Norberg
5198 W 76th St, Edina, MN......800-643-2680
NEC Technologies (USA) Inc/Peter Gion
1414 Massachusetts Ave, Boxboro, MA......508-264-8000
Optical Access Intl/Mike Maloney
500 W Cummings Park, Woburn, MA......617-937-3910
Panasonic
50 Meadowlands Pkwy, Secaucus, NJ......201-348-7000
Peripheral Land Inc/Frank Jaramillo
47421 Bayside Pkwy, Fremont, CA......510-657-2211
Personics Corp/Greg Ballard
981 Bing St, San Carlos, CA......415-592-1700
Pioneer New Media Technology/Henry Ebers
Sherbrooke Pl/600 E Crescent,
Upper Saddle River, NJ......201-327-6400

Rancho Technology Inc/Steve Bierwirth
10783 Bell Court, El Rancho Cucamonga, CA.....909-987-3966
Relax Technology/Frank Gabrielli
3101 Whipple Rd, Union City , CA......510-471-6112
Ricoh Corp/Dell Glover
3001 Orchard Pkwy, San Jose, CA......408-432-8800
Sony Corp of America
3 Paragon Dr, Montvale, NJ......800-635-7669
Sony Corp of America/Lisa Becker
3300 Zanker Rd, San Jose, CA......408-955-4217
Sun Microsystems Inc
2550 Garcia Ave, Mt View, CA......415-960-1300
Sun Moon Star/Personal Computer Div/Michael Wu
1941 Ringwood Ave, San Jose, CA......408-452-7811
Sundance Technology Group/Rush Beesly
6309 N O'Connor #111 LB/128, Irving, TX......214-869-1002
TEAC America Inc/Mike Wynn
7733 Telegraph Rd, Montebello, CA......213-726-0303
Todd Enterprises Inc/Michael Scheibel
224-49 67th Ave, Bayside, NY......718-343-1040
Zoran Corp/Michael Nell
1705 Wyatt Dr, Santa Clara, CA......408-986-1314

Printers/Output Devices

3M Image Reprographic Systems/Steve Fullerton
Bldg 223-2N-01, St Paul, MN......612-736-0801
Apple Computer Inc
20525 Mariani Ave, Cupertino, CA......408-996-1010
BGL Technology Corp/LaserLeaders/Alan Lang
451 Constitution Ave, Camarillo, CA......805-987-7305
Brother International Corp/Bill Hall
200 Cottontail Ln/S Vantage, Somerset, NJ......908-356-8880
Commodore Business Machines
1200 Wilson Dr, West Chester, PA......215-431-9100
Hitachi Sales Corp of America/Eric Kamayatsu
401 W Artesia Blvd, Compton, CA......310-537-8383
HSD Microcomputer US Inc/Dave Peter
345 1st St #0, Encinedas, CA......619-632-9700
Ilford Photo/Patricia Harmon
W 70 Century Rd, Paramus, NJ......201-265-6000
Imapro Corp/Fred Androni
2400 Saint Laurent Blvd, Ottawa, ON......613-738-3000
Information International Inc/Bruce Chew
5933 Slauson Ave, Culver City, CA......310-390-8611
Kodak
343 State St, Rochester, NY......800-445-6325
Kyocera Electronics Inc/Robert Donnelly
100 Randolph Rd, Somerset, NJ......800-245-8979
● **LVT-KODAK COMPANY**/Gail Corey
(P 117)
100 Kings Highway #1400, Rochester, NY......716-724-1111
Mannesmann Tally Corp/Bruce Ridley
8301 S 180th St, Kent , WA......206-251-5524
Mitsubishi Electronics America Inc/Mike Foster
5665 Plaza Dr, Cypress, CA......714-220-2500
Mitsubishi Electronics America Inc/Doug Mackeroy
800 Cottontail Ln, Somerset, NJ......908-563-9889
Mitsubishi Intl Corp/Mike Kurahashi
701 Westchester Ave #101 W, White Plains, NY......914-997-4999
NewGen Systems Corp/Meryl Cook
17550 Newhope St, Fountain Valley, CA......714-641-8600
Panasonic
50 Meadowlands Pkwy, Secaucus, NJ......201-348-7000
Polaroid Resource Ctr
784 Memorial Dr, Cambridge, MA......800-225-1618
Prime Option Inc/Nick Vasilikis
2341 W 205th St #116, Torrance, CA......310-618-0274
Ricoh Corp/Dell Glover
3001 Orchard Pkwy, San Jose, CA......408-432-8800
Summagraphics Houston Instrument
8500 Cameron Rd, Austin, TX......800-776-9989
Talaris Systems Inc/Kevin King
PO Box 261580, San Diego, CA......619-587-0787
Tall Tree Systems/Rick McFiren
2585 E Bayshore Rd, Palo Alto, CA......415-493-1980
Texas Instruments Inc/Tom Miller
5701 Airport Rd, Temple, TX......817-774-6102

Scanners

AGFA/Div of Miles/Seamus Carroll
One Ramland Rd, Orangeburg, NY......914-365-0190
Black Box Corp/Dave Kenney
PO Box 12800, Pittsburgh, PA......412-745-5565

Chinon America Inc/Anne Cousino
615 Hawaii Ave, Torrance, CA......310-533-0274
DFI Inc
135 Main Ave, Sacramento, CA......916-568-1234
Dicomed/Tim Whalen
11401 Rupp Dr, Burnville, MN......800-888-7979
Everex/Ned Nevels
5020 Brandin Ct, Fremont, CA......510-438-5783
Follett Software Co/Lisa McManaman
809 N Front St, McHenry, IL......815-344-8700
Fujitsu Computer Product of America/Steve Butterfield
2904 Orchard Pkwy, San Jose, CA......408-894-3512
HSD Microcomputer US Inc/Dave Peter
345 1st St #0, Encinitas, CA......619-632-9700
Imapro Corp/Fred Androni
2400 Saint Laurent Blvd, Ottawa, ON......613-738-3000
Information International Inc/Bruce Chew
5933 Slauson Ave, Culver City, CA......310-390-8611
Kodak
343 State St, Rochester, NY......800-445-6325
Leaf Systems/Patty Foley
250 Turnpike Rd, Southboro, MA......508-460-8300
Microtek Lab Inc/Gina Scalese
3715 Doolittle Dr, Redondo Beach, CA......310-297-5000
Mirror Technologies/John Norberg
5198 W 76th St, Edina, MN......800-643-2680
NISCA Inc/Julia Maxey
1919 Old Denton Rd #104, Carrollton, TX......214-242-9696
Ocron Inc
3350 Scott Blvd/Bldg 36/Mrktng,
Santa Clara, CA......800-933-1399
Panasonic
50 Meadowlands Pkwy, Secaucus, NJ......201-348-7000
Polaroid Resource Ctr
784 Memorial Dr, Cambridge, MA......800-225-1618
Ricoh Corp/Dell Glover
3001 Orchard Pkwy, San Jose, CA......408-432-8800

Video-Digital Interactive

Andrews, Jackie
CA......415-546-5666
Artificial Reality/Myron Krueger
Box 786, Vernon, CT......203-871-1375
ASC Systems/R Martin
PO Box 566, St Clair Shores, MI......313-882-1133
BetaCorp Technologies Inc/John J Bekto
6770 Davand Dr #40, Mississauga, ON......416-564-2424
Colby Systems Corp/Charles Edwards
2991 Alexis Dr, Palo Alto, CA......415-941-9090
Current Designs Corp/David Rapkin
163 W 23rd St #5, New York, NY......212-463-0795
IBM Multimedia/Information Center
4111 Northside Pkwy, Atlanta, GA......800-426-9402
Kodak
343 State St, Rochester, NY......800-445-6325
National Teleprinting Inc/Jim Howe
1420 Blake St, Denver, CO......303-623-2800
Pioneer New Media Technology/Henry Ebers
Sherbrooke Pl/600 E Crescent ,
Upper Saddle River, NJ......201-327-6400

Video Production

Abekas/Pete Mountanos
101 Galveston Dr, Redwood City, CA......415-369-5111
Analog Devices Inc/Vicki Chase
One Technology Way, Norwood, MA......617-329-4700
Arkay Technologies/Ray Khorram
5 Tsienneto Rd #2, Derry, NH......603-425-2149
Cinetron Computer Systems
4350 J Intl Blvd, Norcross, GA......404-925-4448
Computer Prompting & Captioning Co/Sidney Hoffman
3408 Wisconsin Ave NW, Washington, DC......202-966-0980
Cornerstone Associates/Suzanne Barrett/Edward Sayles
211 Second Ave, Waltham, MA......617-890-3773
Diaquest Inc/Dan Lindheim
1440 San Pablo Ave, Berkeley, CA......510-526-7167
Digital Pictures Inc/Tom Grotting
1212 N Second S, Minneapolis, MN......612-371-4515
Folsom Research Inc/Ed Hart
526 E Bidwell St, Folsom, CA......916-983-1500
Grass Valley Group/Janice Haighley
6 Forest Ave, Paramus , NJ......201-845-8900

Intelligent Resources Integrated Systems, Inc./Barbara Maes
3030 Salt Creek Ln #10G, Arlington Hghts, IL 708-670-9388

Leitch Inc
920 Corporate Ln, Chesapeake, VA 804-548-2300

LSI Logic/Philip Pfeifer
1501 McCarthy Blvd, Milpitas, CA 408-433-7635

Lyon Lamb Video Animation Systems Inc
4531 Empire Avenue, Burbank, CA 818-843-4831

Mainstream Communications Inc/Becky Beck
955 James Ave S #235, Bloomington, MN..........612-888-9000

Mammoth Microproductions Inc/Robert Odgon
1536 Cole Blvd #165, Golden, CO 303-237-5776

New Microtime/David Acker
1280 Blue Hills Ave, Bloomfield, CT 203-242-4242

P E Photron/Yuki Fujikawa
4030 Moor Park Ave #108, San Jose, CA 408-261-3613

Ray, Al
2304 Houston Street, San Angelo, TX 915-949-2716

Real Productions/Robert J Schuster
1821 University Ave #N-153, St Paul, MN..........612-646-9472

Russell-Manning Productions/Bruce Clark
18 N 4th St, Minneapolis, MN 612-338-7761

Selectra/Richard Comfort
PO Box 5497, Walnut Creek, CA 800-874-9889

Supermac Technology/Margaret Hughes
215 Moffett Pk Dr, Sunnyvale, CA 408-541-6100

Videofax Company/Bruce Biegel
60 Madison Ave #903, New York, NY...............212-689-3440

Visionetics International Corp/Jay Jedankes
21311 Hawthorne Blvd #235, Torrance, CA.......310-316-7940

Virtual Reality

Advanced Gravis Computer Technologies Ltd
7400 MacPherson Ave #111, Burnaby , BC........604-434-7274

Altronix Systems/Allen Cowie
100 Ford Rd, Denville, NJ................................201-625-8220

Artificial Reality/Myron Krueger
Box 786, Vernon, CT 203-871-1375

AutoDesk Inc/Multimedia Div/Anna Melillo
2320 Marinship Way, Sausalito, CA..................415-331-0356

CiS Graphics
1 Stiles Rd #305, Salem, NH 603-894-5999

Computed Animation Technology/Trent DiGiulio
2200 N Lamar #301, Dallas, TX 214-871-5055

Crystal River Engineering
12350 Wards Ferry Rd, Groveland, CA 209-962-6382

Division Ltd
Quarry Rd/Chipping Sodbury, Bristol, UK 045-432-4527

Exos 8
2A Gill St, Woburn, MA...................................617-933-0022

Focal Point 3D Audio
1402 Pine Ave #127, Niagara Falls, NY 716-285-3930

Gyration Inc
12930 Saratoga Ave Bldg C, Saratoga, CA........408-255-3016

Hyperspeed Technologies/Rod Davis
10225 Barnes Canyon Rd #A206,
San Diego, CA 619-578-4893

Imapro Corp/Fred Androni
2400 Saint Laurent Blvd, Ottawa, ON 613-738-3000

Immersive Technologies Inc/Martin Perlmutter
2866 McKillop Rd, Oakland, CA 510-261-0128

Latent Image Development Corp/David Geshwind
2 Lincoln Sq, New York, NY 212-873-5487

Law Holographics/Linda E Law
234 Main st, Huntington Station, NY 516-673-3138

Leep Systems Inc
241 Crescent St, Waltham, MA.........................617-647-1395

Logitech Inc
6505 Kaiser Dr, Fremont, CA............................510-795-8500

Man/Environment Inc
2251 Federal Ave, Los Angeles, CA 310-477-7922

Metaware Inc
2161 Delaware Ave, Santa Cruz, CA.................408-429-6382

Ono-Sendai Corp/Tim Wood
2830 Alameda St, San Francisco, CA................415-487-2040

Phar Lap Software Inc
60 Aberdeen Ave, Cambridge, MA....................617-661-1510

Polhemus/Ed Costello
1 Hercules Dr/PO Box 560, Colchester, VT 802-655-3159

S E Tice Consulting/Steve Tice
15860 Dartford Way, Sherman Oaks, CA 818-906-3322

SimGraphics Engineering Corp
1137 Huntington Dr, S Pasadena, CA................213-255-0900

Spaceball Technologies Inc
600 Suffolk St, Lowell, MA................................508-970-0330

StereoGraphics
2171 E Francisco Blvd, San Rafael, CA..............415-459-4500

Straylight
150 Mt Bethel Rd, Warren, NJ 908-580-0086

Telerobotics Intl Inc/Daniel Kuban
7325 Oak Ridge Hwy #104, Knoxville , TN........615-690-5600

TiNi Alloy Co
1621 Neptune Dr, San Leandro, CA..................510-483-9676

Virtual Research
3193 Bellick St #2, Santa Clara, CA 408-748-8712

Virtual Technologies
2175 Park Blvd, Palo Alto, CA 415-321-4900

Vivid Group
317 Adelaide St W #302, Toronto, ON 416-340-9290

Vivid Group/Vincent John Vincent
317 Adelaide St W #302, Toronto, ON 416-340-9290

VREAM
2568 N Clark St #250, Chicago, IL...................312-477-0425

Xtensory Inc
140 Sunridge Dr, Scotts Valley, CA 408-439-0600

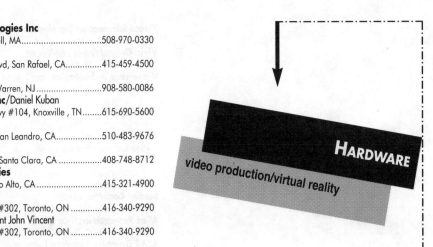

HARDWARE

video production/virtual reality

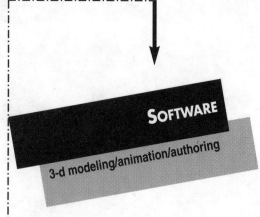

3-D Modeling

Acme Animotion Group Ltd/Evan Hirsch
40 Newport Parkway, Jersey City, NJ 201-420-6440
Acuris Inc/Eduardo Llach
931 Hamilton Ave, Menlo Park, CA...................415-329-1920
Alias Research Inc/Peter Goldie
110 Richmond St E, Toronto, ON 416-362-9181
AT&T Graphics Software Labs/Laura Van Den Dries
8888 Keystone Crossing #1490,
Indianapolis, IN 317-844-4364
AutoDesk Inc/Multimedia Div/Anna Melillo
2320 Marinship Way, Sausalito, CA..................415-331-0356
AutoDesk Retail Products/Sue Wickham
11911 N Creek Pkwy S, Bothell, WA 800-228-3601
Byte By Byte Corp/Scott Peterson
9442A Capital of TX Hwy N #650, Austin, TX.....512-795-0150
Computer Putty/David Poole
408 West Portola, Los Angeles, CA...................415-948-3319
Do While Studio/Jennifer Hall
273 Summer St 7th Fl, Boston, MA 617-338-9129
Dynaware USA Inc/Sho Hayami
950 Tower Ln #1150, Foster City, CA.................415-349-5700
Frame-by-Frame Media Services/Rick Sutter
5353 Randall Pl, Fremont, CA...........................510-651-6330
Graphisoft/Bruce Rienhart
400 Oyster Point Blvd #429,
S San Francisco, CA 800-344-3468
Hammes, Alan
65 Downing St #1B, New York, NY 212-691-6387
● **HERBERT, JONATHAN (P 63)**
501 Fifth Ave #1407, New York, NY..................212-490-2450
ILM (Industrial Light & Magic)/Andrea Merrim
PO Box 2459, San Rafael, CA...........................415-258-2000
Imagetects/Verdonna Ahrens
PO Box 4, Saratoga, CA...................................408-252-5487
Intex Solutions Inc/Jim Wilner
35 Highland Cir, Needham, MA.........................617-449-6222
Latent Image Development Corp/David Geshwind
2 Lincoln Sq, New York, NY 212-873-5487
ModaCAD/Linda Freedman
1954 Cotner Ave, Los Angeles, CA...................310-312-6632
New England Software Inc/Bjorn Nordwall
Greenwich Office Park 3, Greenwich, CT...........203-625-0062
PM Source/Jim Newton
1500 Laurel St, San Carlos, CA.........................415-591-8978
Ray Dream Inc/Eric Hautemont
1804 N Shoreline Blvd #240, Mt View, CA 415-960-0765
SAS Institute/O B Barrett
SAS Campus Dr, Cary, NC.................................919-677-8000
● **SILICON GRAPHICS INC**/Deborah Miller
(P 106,107,BACK COVER)
2011 N Shoreline Blvd, Mountain View, CA.......415-960-1980
● **SPECULAR INTERNATIONAL**/Dave Trescot
(P 114,115)
479 West St, Amherst, MA 413-253-3100
Stone Design Group/Andrew Stone
3725 Rio Grande NW, Albuquerque, NM 505-345-4800
● **STRATA STUDIO**/Reed Terry
(P 110,111)
2 W St George Blvd #2100, St George, UT 801-628-5218
Virtus Corp
117 Edinburgh S #204, Cary, NC.......................919-467-9700
Visual Business Systems Inc/Carl Hire
380 Interstate N Pkwy #190, Atlanta, GA...........404-956-0325

Visual Information Development Inc/Eric Popejoy
136 W Olive Ave, Monrovia, CA......................818-358-3936
Vivid Group/Vincent John Vincent
317 Adelaide St W #302, Toronto, ON 416-340-9290
Wavefront Technologies/Tom Sullivan
530 E Montecito St, Santa Barbara, CA.............805-962-8117

Animation

Accent Software
4546 El Camino Real #S, Los Altos, CA 415-949-2711
Adobe
1585 Charleston Rd, Mt View, CA 415-324-8080
Aimtech (Advanced Interactive Media) Corp/Elaine LeBlanc
20 Trafalgar Sq, Nashua, NH............................603-883-0220
Alias Research Inc/Peter Goldie
110 Richmond St E, Toronto, ON 416-362-9181
AT&T Graphics Software Labs/Laura Van Den Dries
8888 Keystone Crossing #1490,
Indianapolis, IN 317-844-4364
AutoDesk Inc/Multimedia Div/Anna Melillo
2320 Marinship Way, Sausalito, CA..................415-331-0356
AutoDesk Retail Products/Sue Wickham
11911 N Creek Pkwy S, Bothell, WA 800-228-3601
BetaCorp Technologies Inc/John J Bekto
6770 Davand Dr #40, Mississauga, ON 416-564-2424
Bill Feigenbaum Designs/Joyce Feigenbaum
15 E 26th St #1615, New York, NY....................212-447-5550
Bridgestone Multimedia/Bill Lamphear
1979 Palomar Oaks Way, Carlsbad, CA............619-431-9888
Bright Star Technology Inc/Alan Higginson
40 Lake Bellevue #350, Bellevue, WA 206-451-3697
Byte By Byte Corp/Scott Peterson
9442A Capital of TX Hwy N #650, Austin, TX.....512-795-0150
Cornell/Abood/Cheryl Abood
4121-1/2 Rodford Ave, Studio City, CA..............818-505-6655
Diaquest Inc/Dan Lindheim
1440 San Pablo Ave, Berkeley, CA 510-526-7167
Digital Arts/Frank Hartman
4531 Empire Ave #229, Burbank, CA................818-972-2112
Do While Studio/Jennifer Hall
273 Summer St 7th Fl, Boston, MA 617-338-9129
East Design/Roger East
600 Townsend St #415W, San Francisco, CA.....415-552-2300
Equilibrium Technologies/Frank Colin
475 Gate Five Rd #225, Sausalito, CA...............415-332-4343
Graphisoft/Bruce Rienhart
400 Oyster Point Blvd #429,
S San Francisco, CA 800-344-3468
Hammes, Alan
65 Downing St #1B, New York, NY 212-691-6387
Informatics Group Inc/Michael Ganci
100 Shield St, W Hartford, CT...........................203-953-4040
Linker Systems/Sheldon Linker
13612 Onkayha Cir, Irvine, CA..........................714-552-1904
Microware Systems Corp
1900 NW 114th St, Des Moines, IA....................515-224-1929
NTERGAID/Laura Schrader
2490 Black Rock Tpke #337, Fairfield, CT 203-380-1280
Octree Software Inc/Eric Denny
1955 Landings Dr, Mt View, CA 415-390-9600
Paul Mace Software/Sheila Watson
400 Williamson Way, Ashland, OR 503-488-2322
PM Source/Jim Newton
1500 Laurel St, San Carlos, CA 415-591-8978
Presentation Works/Carol Greenberg
1541 Yardley St, Santa Rosa, CA......................707-545-1400
SAS Institute/O B Barrett
SAS Campus Dr, Cary, NC.................................919-677-8000
● **SILICON GRAPHICS INC**/Deborah Miller
(P 106,107,BACK COVER)
2011 N Shoreline Blvd, Mountain View, CA........415-960-1980
Simgraphics Engineering Corp/Dave Verso
1137 Huntington Dr #A-1, S Pasadena, CA.........213-255-0900
Softsync Inc/Rod Campbell
SW 37th Ave #765, Coral Gables, FL.................305-444-0080
● **SPECULAR INTERNATIONAL**/Dave Trescot
(P 114,115)
479 West St, Amherst, MA 413-253-3100
● **STRATA STUDIO**/Reed Terry **(P 110,111)**
2 W St George Blvd #2100, St George, UT 801-628-5218
Sun Moon Star/Personal Computer Div/Michael Wu
1941 Ringwood Ave, San Jose, CA....................408-452-7811

Vermont Microsystems Inc/Diane Magnusson
11 Tigan St, Winooski, VT 802-655-2860
Viacom New Media/Michelle Boeding
648 S Wheeling Rd, Wheeling , IL 800-877-4266
Vision Database Systems Inc/Emil Bonaduce
853 Don Ross Rd, Juno Beach, FL 407-694-2211
Visual Information Development Inc/Eric Popejoy
136 W Olive Ave, Monrovia, CA......................818-358-3936
Vivid Group/Vincent John Vincent
317 Adelaide St W #302, Toronto, ON 416-340-9290
Wavefront Technologies/Tom Sullivan
530 E Montecito St, Santa Barbara, CA.............805-962-8117

Authoring

Adobe
1585 Charleston Rd, Mt View, CA 415-324-8080
Advanced IDAS/Henry Luong
9550 Firestone Blvd #210, Downey, CA.............310-861-4245
Aimtech (Advanced Interactive Media) Corp/Elaine LeBlanc
20 Trafalgar Sq, Nashua, NH............................603-883-0220
American Computer Imaging/Bill Glassgold
351 Hiatt Dr, Palm Beach Gardens, FL 407-624-6820
BetaCorp Technologies Inc/John J Bekto
6770 Davand Dr #40, Mississauga, ON 416-564-2424
Big Noise Software/David Lee
PO Box 23740, Jacksonville, FL.........................904-730-0754
Bourbaki Inc/Chris Severud
PO Box 2867, Boise, ID 208-342-5849
Bovey & Assocs, Ruth
1443 W Altgeld, Chicago, IL.............................312-327-2886
Bridgestone Multimedia/Bill Lamphear
1979 Palomar Oaks Way, Carlsbad, CA............619-431-9888
Cabimat/David B Kelley
88 W Broadway 2nd Fl, New York, NY...............212-385-8191
Cochenille Computer/Susan Lazear
PO Box 4276, Encinitas, CA..............................619-259-1698
Comware Inc/Wendy O'Neal
4225 Malsbary Rd, Cincinnati, OH 513-791-4224
Dataware Technology Inc
5775 Flatiron Pkwy #220, Boulder , CO............303-449-4157
Deneba Software/Doug Levy
7400 SW 87th Ave, Miami, FL 305-596-5644
East Design/Roger East
600 Townsend St #415W, San Francisco, CA.....415-552-2300
Elmsoft Inc/Frank Elmore
7954 Helmart Dr, Laurel, MD 301-470-3451
Executive Technologies Inc/Jim Geer
2120 16th Ave S, Birmingham, AL 205-933-5494
IBM Multimedia/Informations Cntr
4111 Northside Pkwy, Atlanta, GA.....................800-426-9402
Image Technology/Vance Gayle
5605 Roanne Way, Greensboro, NC...................919-299-2452
InfoAccess/Frances Bigley
2800 156th Ave SE, Bellevue, WA......................206-747-3203
Informatics Group Inc/Michael Ganci
100 Shield St, W Hartford, CT 203-953-4040
Information Navigation Inc/Leslie Killeen
4201 University Dr #102, Durham, NC 919-493-4390
IS Grupe Inc/Peter Schipma
948 Springer Dr, Lombard, IL 708-627-0550
Kapili, Jesse V
324 Paxson Ave, Hamilton Sq, NJ 609-584-1770
Knowledge Garden Inc/R John Slade
12-8 Technology Dr, Setauket, NY 516-246-5400
KnowledgeSet Corp/Marty Miller
250 Sobrante Way, Sunnyvale, CA....................408-738-3400
KWA Inc/Ken Wood
6004 S Kipling #208, Littleton, CO.....................303-972-9206
Lightscape Graphics Software/Rod Recker
4040 Moor Park Ave #108, San Jose, CA 408-246-1155
Media Management Systems Inc/Wynne Ragland
One Meca Way #600, Norcross , GA 404-564-5606
NTERGAID/Laura Schrader
2490 Black Rock Tpke #337, Fairfield, CT 203-380-1280
Odom, Laddie Scott
2135 W Giddings #3W, Chicago, IL 312-728-4942
OnLine Computer Systems Inc/Lisa Wolin
20251 Century Blvd, Germantown, MD 301-428-3700
Optimage Interactive Services LP/Lynn Goodman
1501 50th St #100, W Des Moines, IA 515-225-7000
Paradise Software/John Malleo-Roach
7 Centre Dr #9, Jamesburg, NJ 609-275-4475

Paul Hertz Media Arts/Paul Hertz
  2215 W Fletcher St, Chicago, IL.........................312-975-9153
Paul Mace Software/Sheila Watson
  400 Williamson Way, Ashland, OR................503-488-2322
Pixel Productions/Rachel McAfee
  67 Mowat Ave #547, Toronto, ON....................416-535-3058
Presentation Works/Carol Greenberg
  1541 Yardley St, Santa Rosa, CA....................707-545-1400
Reactor/Mike Saenz
  445 W Erie #208, Chicago, IL.........................312-573-0800
Realta/Kimble Jenkins
  2000 Madison Ave, Memphis, TN 901-725-0855
Right Brain Inc/John Nelson
  4263 Brigadoon Dr, Shoreview, MN 612-229-6299
RIX Softworks Inc/Paul Harker
  17811 Sky Park Cir #B, Irvine, CA..................714-476-8266
Simgraphics Engineering Corp/Dave Verso
  1137 Huntington Dr #A-1, S Pasadena, CA.........213-255-0900
TMS Inc/Jana Otto
  110 W 3rd St, Stillwater, OK............................405-377-0880
Tumble Interactive Media/Cal Vornberger
  910 West End Ave #3D, New York, NY 212-316-0200
Viacom New Media/Michelle Boeding
  648 S Wheeling Rd, Wheeling , IL...................800-877-4266
Videofax Company/Bruce Biegel
  60 Madison Ave #903, New York, NY...............212-689-3440
Vision Database Systems Inc/Emil Bonaduce
  853 Don Ross Rd, Juno Beach, FL....................407-694-2211
Visual Access Technology Inc/Linda Berger
  39 Old Ridgebury RD, Danbury, CT..................203-794-2424
Vivid Group/Vincent John Vincent
  317 Adelaide St W #302, Toronto, ON 416-340-9290
Vortex Interactive/Richard Currier
  5962 La Place Ct #225, Carlsbad, CA 619-929-8144
Voyager/Jane Wheeler
  578 Broadway #406, New York, NY...............212-431-5199
zpwk/Ikar Kozak
  161 S First St, Brooklyn, NY...............................718-963-2925

Digital Editing

Adobe
  1585 Charleston Rd, Mt View, CA 415-324-8080
Avid Technology Inc/Mark Overington
  One Park W 2nd Fl, Tewksbury, MA.................508-640-6789
BetaCorp Technologies Inc/John J Bekto
  6770 Davand Dr #40, Mississauga, ON 416-564-2424
Bob Olhsson Digital Audio/Bob Olhsson
  PO Box 555, Novato, CA..................................415-898-2981
Coda Music Technology/Peggy Wagner
  6210 Bury Dr, Eden Prairie, MN 800-843-2066
Delta Tao Software/Joe Williams
  760 Harvard Ave, Sunnyvale, CA 408-730-9336
Digital F/X Inc/Barbara Koalkin
  755 Ravendale Dr, Mt View, CA 415-961-2800
Informatics Group Inc/Michael Ganci
  100 Shield St, W Hartford, CT 203-953-4040
Radius Inc/Carol Johnson
  1710 Fortune Dr, San Jose, CA.........................408-434-1010
RIX Softworks Inc/Paul Harker
  17811 Sky Park Cir #B, Irvine , CA...................714-476-8266
Selectra/Richard Comfort
  PO Box 5497, Walnut Creek, CA 800-874-9889
System Generation Assoc Inc/Dick Lewis
  122 N Cortez #305, Prescott, AZ......................602-778-4840
Vivid Group/Vincent John Vincent
  317 Adelaide St W #302, Toronto, ON 416-340-9290
White Pine Software/Andrew Nilssen
  120 Flanders Rd, Westboro , MA......................508-836-4400

Imaging/Digital Effects

Alvuso Corp/Steve Koschman
  9770 Carroll Ctr Rd #J, San Diego, CA 619-695-6956
American Computer Imaging/Bill Glassgold
  351 Hiatt Dr, Palm Beach Gardens, FL 407-624-6820
Amitech Corp/Jack Frost
  2721-E Merrilee Dr, Fairfax, VA.......................703-698-5057
Amtronics Inc/Ron Schulingkamp
  PO Box 24190, New Orleans, LA......................504-831-0691
ASDG Inc/Jolaine Benson
  925 Stewart St, Madison, WI...........................608-273-6585
AT&T Graphics Software Labs/Laura Van Den Dries
  8888 Keystone Crossing #1490,
  Indianapolis, IN...317-844-4364

AXS/Optical Technology Resource/Tod Carter
  2560 9th St #219, Berkeley, CA....................510-540-5232
BetaCorp Technologies Inc/John J Bekto
  6770 Davand Dr #40, Mississauga, ON 416-564-2424
Bourbaki Inc/Chris Severud
  PO Box 2867, Boise, ID 208-342-5849
BTS/Broadcast TV Systems/Skip Ferderber
  94 W Cochran St, Simi Valley, CA 805-584-4700
Business & Professional Software Inc/Judy Farrell
  139 Main St, Cambridge, MA 800-342-5277
Byte By Byte Corp/Scott Peterson
  9442A Capital of TX Hwy N #650, Austin, TX.....512-795-0150
Cambridge Media/David Titus
  71 E Wyoming Ave, Melrose, MA.....................617-665-3053
Claris Corp/Customer Svc
  5201 Patrick Henry Dr, Santa Clara, CA.............800-325-2747
Color Age
  900 Technology Pk Dr Bldg 8, Billerica, MA 508-667-8585
Computer Aided Technology Inc/Suzanne Fletcher
  10132 Monroe, Dallas, TX 214-350-0888
Computer Associates International Inc/Susan Scanlon
  1 Computer Assocs Plz, Islandia, NY...............800-477-5880
Computer Friends/Scott Huish
  14250 NW Science Park Dr, Portland, OR.........503-626-2291
Crystal Graphics/Bob Howard
  3110 Patrick Henry Dr, Santa Clara, CA.............408-496-6175
Delta Tao Software/Joe Williams
  760 Harvard Ave, Sunnyvale, CA 408-730-9336
Diaquest Inc/Dan Lindheim
  1440 San Pablo Ave, Berkeley, CA 510-526-7167
Digital F/X Inc/Barbara Koalkin
  755 Ravendale Dr, Mt View, CA 415-961-2800
● **DIGITAL STOCK PROFESSIONAL**/Charles Smith **(P 99)**
  400 S Sierra Ave #100, Solana Beach, CA.........619-794-4040
Direct Images/Bill Knowland
  PO Box 29392, Oakland, CA............................510-614-9783
Electronic Imagery Inc/Bob Heinken
  1100 Park Central Blvd S #3400,
  Pompano Beach, FL 305-968-7100
Elixir Technologies Corp/Kevin Laracey
  302 N Montgomery, Ojai, CA 805-640-8054
Equilibrium Technologies/Frank Colin
  475 Gate Five Rd #225, Sausalito, CA..............415-332-4343
● **FARR PHOTOGRAPHY**/Lori Farr **(P 36,37)**
  12 Walnut St, Rochester, NY...........................716-235-1479
Filmworks, The/Don Lee
  1011 Armando St, Anaheim, CA 714-632-7775
G W Hannaway/George Hannaway
  839 Pearl St, Boulder, CO 303-440-9696
Graphisoft/Bruce Rienhart
  400 Oyster Point Blvd #429, S San Francisco, CA800-344-3468
Haynes, Barry
  820 Memory Ln, Boulder Creek, CA 408-338-4569
Image Concepts Technologies Inc/Cliff Nickerson
  33 Boston Post Rd W, Marlborough , MA............508-481-6882
Imaging Technology Inc/Steve Silver
  55 Middlesex Tpke, Bedford, MA 617-275-2700
Informatics Group Inc/Michael Ganci
  100 Shield St, W Hartford, CT 203-953-4040
INSET Systems Inc/Daryl Squires
  71 Commerce Dr, Brookfield, CT......................203-740-2400
Intel Corporation/Stewart Rosenberg
  2200 Mission College Blvd, Santa Clara, CA 408-765-8080
Iterated Systems Inc/Rick Darby
  5550 Peachtree Pkwy #545, Norcross, GA.........404-840-0728
Julie Research Laboratories/Loebe Julie
  508 W 26th St, New York, NY 212-633-6625
Kleiser-Walczak Construction East/Jeff Kleiser
  Riverview Rd, Lenox, MA.................................413-637-8944
Knowledge Access Intl/Alan Guzzetti
  2685 Marine Way #1305, Mt View, CA 415-969-0606
KWA Inc/Ken Wood
  6004 S Kipling #208, Littleton, CO 303-972-9206
Laserscan Systems Inc/Leonard Fox
  5310 NW 33rd Ave #115, Ft Lauderdale, FL......305-739-7715
Latent Image Development Corp/David Geshwind
  2 Lincoln Sq, New York, NY 212-873-5487
● **LETRASET USA**/Alex Kalfatides **(P 113)**
  40 Eisenhower Dr, Paramus, NJ.......................201-845-6100
Lightpath Imageworks/Barry Pribula
  59 First Ave, New York, NY..............................212-777-7612
Linker Systems/Sheldon Linker
  13612 Onkayha Cir, Irvine, CA.........................714-552-1904
Medialogic Inc/Linda Lynam
  17351 Sunset Blvd #304, Pacific Palisades, CA ..310-573-7575

SOFTWARE

authoring/digital editing/imaging-digital
effects/raytracing-rendering/video
production

Micro Alliance Inc/David Pomerantz
  107 Greenwood St, Wakefield, MA 617-245-8879
Paul Hertz Media Arts/Paul Hertz
  2215 W Fletcher St, Chicago, IL........................312-975-9153
● **PIXAR**/Pam Kerwin **(P 116)**
  1001 W Cutting Blvd, Richmond, CA 510-236-4000
Red Window Studio/Les Reiss
  4921 Jefferson, New Orleans, LA......................504-736-0465
Scott, Ron
  1000 Jackson Blvd, Houston, TX.......................713-529-5868
Software Publishing Corp/Helen Sales Dept
  3165 Kifer Rd, Santa Clara, CA........................408-986-8000
Telephoto Communications Inc/Leonard G Roberts
  11722-D Sorrento Valley Rd, San Diego, CA.......619-452-0903
Truevision/Kathleen Asch
  7340 Shadeland Station, Indianapolis, IN...........317-841-0332
Unisys
  Box 1110, Princeton, NJ.................................908-329-4071
Vivid Group/Vincent John Vincent
  317 Adelaide St W #302, Toronto, ON 416-340-9290
Zoran Corp/Michael Nell
  1705 Wyatt Dr, Santa Clara, CA.......................408-986-1314

Raytracing/Rendering

Alias Research Inc/Peter Goldie
  110 Richmond St E, Toronto, ON 416-362-9181
Artbeats/Phil Bates
  2611 S Myrtle Rd, Myrtle Creek, OR 503-863-4429
AutoDesk Retail Products/Sue Wickham
  11911 N Creek Pkwy S, Bothell, WA 800-228-3601
Comware Inc/Wendy O'Neal
  4225 Malsbary Rd, Cincinnati, OH 513-791-4224
Cross, Lloyd
  PO Box 672, Gualala, CA.................................707-884-9139
Engineered Software/Susan Stanley
  PO Box 18344, Greensboro, NC 919-299-4843
Graphisoft/Bruce Rienhart
  400 Oyster Point Blvd #429,
  S San Francisco, CA 800-344-3468
Kapili, Jesse V
  324 Paxson Ave, Hamilton Sq, NJ 609-584-1770
● **PIXAR**/Pam Kerwin **(P 116)**
  1001 W Cutting Blvd, Richmond, CA 510-236-4000
Ray Dream Inc/Eric Hautemont
  1804 N Shoreline Blvd #240, Mt View, CA 415-960-0765
Stone Design Group/Andrew Stone
  3725 Rio Grande NW, Albuquerque, NM 505-345-4800
System Generation Assoc Inc/Dick Lewis
  122 N Cortez #305, Prescott, AZ......................602-778-4840
Virtus Corp
  117 Edinburgh S #204, Cary, NC 919-467-9700
Visual Information Development Inc/Eric Popejoy
  136 W Olive Ave, Monrovia, CA.......................818-358-3936
Vital Images Inc
  505 N Fourth St, Fairfield, IA 515-472-7726
Wavefront Technologies/Tom Sullivan
  530 E Montecito St, Santa Barbara, CA..............805-962-8117

Video Production

Byte By Byte Corp/Scott Peterson
  9442A Capital of TX Hwy N #650, Austin, TX.....512-795-0150
Cognetics Corp
  51 Everett Dr #103B/POB 386,
  Princeton Junction, NJ 609-799-5005

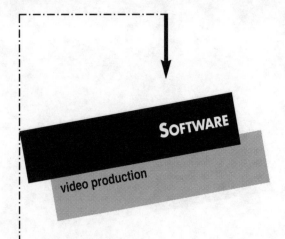

SOFTWARE

video production

Data Translation/Patrick Rafter
100 Locke Dr, Marlborough , MA.......................508-481-3700

Diaquest Inc/Dan Lindheim
1440 San Pablo Ave, Berkeley, CA....................510-526-7167

Digital Arts/Frank Hartman
4531 Empire Ave #229, Burbank, CA...............818-972-2112

East Design/Roger East
600 Townsend St #415W, San Francisco, CA.....415-552-2300

EZTV Video Ctr/Michael Masucci
7280 Melrose Ave, Los Angeles, CA 213-939-7100

Grass Valley Group/Janice Haighley
6 Forest Ave, Paramus , NJ 201-845-8900

**Information Technology Design
Assoc Inc**/Daniel Klassen
7156 Shady Oak Rd, Eden Prairie, MN 612-944-8244

Lake Compuframes Inc/Laurie Kaplan
PO Box 890, Briarcliff Manor, NY 914-941-1998

● **LETRASET USA**/Alex Kalfatides **(P 113)**
40 Eisenhower Dr, Paramus, NJ 201-845-6100

Parallax Graphics/Lee Caswell
2500 Condensa St, Santa Clara, CA..................408-727-2220

**Rochester Institute of Technology/
Printing**/Frank Romano
Box 9887, Rochester, NY 716-475-7023

Sayett Technology/Randy Carson
17 Toby Village, Pittsford, NY 716-264-9250

Time Arts Inc/Pat Radnich
1425 Corporate Ctr Pkwy, Santa Rosa, CA 707-576-7722

Triad Media
PO Box 778, Frederick, MD...............................301-663-1471

Truevision/Kathleen Asch
7340 Shadeland Station, Indianapolis, IN...........317-841-0332

Viacom New Media/Michelle Boeding
648 S Wheeling Rd, Wheeling , IL 800-877-4266

Videofax Company/Bruce Biegel
60 Madison Ave #903, New York, NY.............212-689-3440

Videotex Systems Inc/Bob Gillman
8499 Greenville Ave #205, Dallas, TX 214-343-4500

VMI Inc/Gretchen Schott
211 Weddell Dr, Sunnyvale, CA 408-745-1700

Publications

NEW YORK CITY

Act III Publishing/Marketing Dept
110 E 59th 6th Fl, New York, NY......................212-909-0430
Art Direction/Advertising Trade Publications Inc
10 E 39th St 6th Fl, New York, NY 212-889-6500
Back Stage/Back Stage Publications
1515 Broadway, New York, NY 212-764-7300
BackStage/SHOOT/BPI Communications Inc
1515 Broadway, New York, NY 212-947-0020
**Cahners Publishing Co/Graphic Arts
Monthly**/Karen Lehrer AD
249 W 17th St, New York, NY 212-463-6834
Electric Pages/Jack Powers
405 Fourth St, Brooklyn, NY 718-499-1884
Graphic Design: USA/Kaye Publishing Corp
1556 Third Ave #405, New York, NY 212-534-5500
Microcomputer Publishing Center/Melanie Rosenzweig
4 W 20th St, New York, NY 212-463-8585
RC Publications Inc/Print
104 Fifth Ave 19th Fl, New York, NY..................212-463-0600
Rosebush Visions Corp/Judson Rosebush
154 W 57th St #826/New York, NY 212-398-6600
Television Broadcast/PSN Publications Inc
2 Park Ave #1820, New York, NY 212-779-1919
U&Lc/International Typeface Co
2 Hammarskjold Plaza, New York, NY 212-371-0699
● **VAN NOSTRAND REINHOLD (P 120)**
115 Fifth Ave, New York, NY 800-544-0550

EAST

Cahners Publishing/Datamation/Steve Twombly
275 Washington St, Newton, MA.......................617-964-3030
Cahners Publishing/Digital Review/Gina Biancucci Sr Ad
275 Washington St, Newton, MA.......................617-964-3030
Computer Reseller News/CMP
600 Community Dr, Manhasset, NY 516-562-5000
Electronic Publishing/Paul McPherson
10 Tara Blvd 5th Fl, Nashua, NH 603-891-0123
IDG Communications/Amigaworld/Howard Happ
80 Elm St, Peterborough, NH 603-924-0100
Knowledge Industry Pubs/Computer Pictures
701 Westchester Ave #109W, White Plains, NY.914-328-9157
Montage Publ Inc/Computer Pictures
701 Westchester Ave #109W, White Plains, NY.914-328-9157
Multimedia Review, The Journal of
Meckler 11 Ferry Lane W, Westport, CT 203-226-6967
NAB Multimdedia News/NAB/Interactive MM Assoc
1771 N St NW, Washington, DC 202-429-5300
**PennWell Publishing Co/Computer Graphics
World**/Jan Horner AD
10 Tara Blvd, Nashua, NH 603-891-0123
Post/Producers Quarterly/Ken McGorry
25 Willowdale Ave, Port Washington, NY 516-767-2500
**PTN Publishing/Advanced Imaging
Magazine**/Charles Grecky, Publisher
445 Broad Hollow Rd #21, Melville, NY..............516-845-2700
S Klein Newsletter on Computer Grphcs/Stanley Klein
730 Boston Post Rd, Sudbury, MA.....................508-443-4671
Scientific Computing & Automation/Calvin Carr
301 Gibraltar Dr, Morris Plains, NJ 201-292-5100
Seybold Reports/Seybold Publications
428 E Baltimore Pike, Media, PA........................610-565-2480

SOUTH

IBM Multimedia Solutions/Larry Baratto
4111 Northside Prkwy, Atlanta, GA 800-426-9402
Macintosh Product Registry/Redgate
Communications Corp
660 Beachland Blvd, Vero Beach, FL..................407-231-6904
Multimedia & VideoDisc Monitor/Future Systems Inc
PO Box 26, Falls Church, VA 703-241-1799
Professional Publications/Ann McMahon
1646 Belmont Ave, Baton Rouge, LA..................504-346-0707
Weigand Report/Rick Doucette
PO Box 690, Cedar Hill, TX 214-291-4005

MIDWEST

Business Publishing/Hitchcock Chilton Publishing Co.
191 S Gary Ave, Carol Stream, IL 708-665-1000
Dynamic Graphics Inc/Peter Force
6000 N Forest Park Dr, Peoria, IL 309-688-8800
How/F&W Publications
PO Box 12575, Cincinnati, OH 800-365-0963

Hunter Publishing
25 N Westpoint Blvd #800, Elk Grove Village, IL.708-427-9512
Metatec Corp/Nautilus
7001 Discovery Blvd, Dublin, OH......................800-637-3472
Peed Corp/PC Today/Lisa AD Scarboro
120 W Harvest Dr/PO Box 85380, Lincoln, NE ..800-424-7900
PH Brink International/Jeff Brink
6100 Golden Valley Rd, Minneapolis, MN..........612-591-1977
Publishers Group/Wendell Stark
1022 W 80th St, Minneapolis, MN 612-881-3183
Southwind Publishing Co
8340 Mission Rd #106, Prairie Village, KS 913-642-6611
St Louis Magazine/Barbara Lutman
305 N Broadway, St Louis, MO 314-231-4185
Video Disc System Sound & Video/Intertec Publishing
PO Box 12901, Overland Park, KS 913-341-1300
Video Systems Magazine/Tom Brick
9800 Metcalf Ave, Overland Park, KS 913-967-1834

WEST

Advanced Systems/IDG Communications/SF
501 Second St, San Francisco, CA 415-243-0500
Aldus Corp/Aldus Magazine/Allan Routh
411 First Ave S, Seattle, WA 206-343-4200
Autologic Inc/Ruta Medina
1050 Rancho Conejo Blvd, Thousand Oaks, CA..805-498-9611
Before & After/Page Lab Inc
PO Box 418252, Sacramento, CA 916-784-3880
CADalyst/CADalyst Publications Ltd
PO Box 10460, Eugene, OR.............................503-343-1200
Calvert, Carl
119 Alton #D, Santa Ana, CA 714-262-1546
Film & Video/Steven Rich
8455 Beverly Blvd #508, Los Angeles, CA..........213-653-8053
Frank Relations/Nancy Frank
1342 Stevenson, San Francisco, CA 415-626-5742
IDG/MacWorld/Joanne Hoffman
501 Second St #500, San Francisco, CA 415-243-0505
Invisions Inc/Scott Tuckman
3641 N 52nd St, Phoenix, AZ 602-840-1090
Journal, THE/Edward W Warnshuis
150 El Camino Real #112, Tustin, CA................714-730-4011
Los Angeles Macintosh Group/Suzy Prieto
12021 Wilshire Blvd #349, Los Angeles, CA 310-278-LAMG
MacArtist/Fritz Richard
119 E Alton Ave #D, Santa Ana, CA 714-262-1546
MacWeek/Ellie Leishman
301 Howard St 15th Fl, San Francisco, CA 415-243-3500
Media Culture/Frank Wiley
215 Fair Oaks, San Francisco, CA 415-824-3993
MicroPublishing News
21150 Hawthorne Blvd #104, Torrance, CA.......310-371-5787
MicroTimes/Bam Publications Inc
3470 Buskirk, Pleasant Hill, CA........................510-934-3700
Mondo 2000/Bert Nagel
PO Box 10171, Berkeley, CA 510-845-9018
New Media/Lydia Lee
901 Mariners Island Blvd #365, San Mateo, CA .415-573-5170
Next World/PCW Communications Inc/Laura Stoll
501 Second St, San Francisco, CA 415-243-0500
Open Computing
1900 O'Farrell St 2nd Fl, San Mateo, CA 415-513-6800
Pacific Media Associates/William Coggshall
1121 Clark Ave, Mountain View, CA 415-948-3080
PC Graphics & Video/Michael Frocillo
201 E Sand Pointe Ave #600, Santa Ana, CA.....714-513-8400
PC World/PCW Communications Inc/Greg Silva
501 Second St, San Francisco, CA 415-243-0500
Presentation Magazine
23410 Civic Ctr Way #E10, Malibu, CA............310-456-2283
Publish/PCW Communications Inc
501 Second St #310, San Francisco, CA 415-243-0600
The Future Image Rep/Alexis J Gerard
1020 Parrott Dr, Burlingame, CA......................415-579-0493
Verbum Magazine/Michael Gosney
2187-C San Elijo Ave, Cardiff, CA 619-944-9977
Windows Watcher/Tina Gregg
15127 NE 24th #344, Redmond, WA 206-881-7354
Ziff-Davis/MacUser/Jeff Cohen
950 Tower Ln 18th Fl, Foster City, CA................415-378-5600

CANADA

Pixel–The Computer Animation News People Inc
217 George St, Toronto, ON 416-367-0088

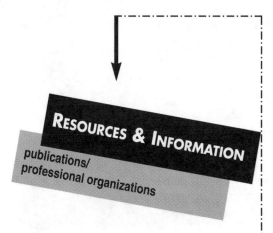

Professional Organizations

NEW YORK CITY

ACM–Assn for Computing Machinery Inc/Lillian Israel
1515 Broadway 17th Fl, New York, NY 212-869-7440
Advertising Club of New York/Peder Saxer
235 Park Ave S 6th Fl, New York, NY 212-533-8080
Advertising Council Inc/Ruth Wooden
261 Madison Ave 11th Fl, New York, NY...........212-922-1500
**Advertising Photographers of
New York Inc**/Marilyn Wallen
27 W 20th St #601, New York, NY 212-807-0399
American Assn of Adv Agencies/Karen Wagner
666 Third Ave 13th Fl, New York, NY 212-682-2500
American Institute of Graphic Arts/
Caroline Hightower
1059 Third Ave, New York, NY 212-752-0813
Art Directors Club of New York/Myrna Davis
250 Park Ave S, New York, NY 212-674-0500
Association of American Publishers/Tom McKee
220 E 23rd St 2nd Fl, New York, NY 212-689-8920
Association of the Graphic Arts/William Dirzulaitis
330 Seventh Ave 9th Fl, New York, NY 212-279-2100
Audio Engineering Society/Donald Plunkett
60 E 42nd St #2520, New York, NY 212-661-8528
Childrens Book Council Inc/Paula Quint
568 Broadway, New York, NY 212-966-1990
Electronic Arts Intermix/Stephen Vitiello
536 Broadway 9th Fl, New York, NY 212-966-4605
Experimental Intermedia/Phill Niblock
224 Centre St, New York, NY............................212-431-6430
Guild of Bookworkers/Louise Kuflik
521 Fifth Ave, New York, NY 212-757-6454
International Advertising Association Inc/Ellen Correy
342 Madison Ave #2000, New York, NY...........212-557-1133
SIGGRAPH/Lois Blankstein
1515 Broadway 17th FL, New York, NY.............212-869-7440
Society of American Graphic Artists/Michael DiCirbo
32 Union Sq E #1214, New York, NY 212-260-5706
Society of Illustrators/Terry Browne
128 E 63rd St, New York, NY 212-838-2560
Society of Publication Designers/Bride Whelan
60 E 42nd St #721, New York, NY....................212-983-8585
**Special Interest Group for Computers &
Physically Handicapped**/Diane Darrow
1515 Broadway 17th Fl, New York, NY 212-869-7440
**Special Interest Group for Software
Engineers**/Patrick McCarren
1515 Broadway 17th Fl, New York, NY 212-869-7440
Type Directors Club of New York/Carol Wahler
60 E 42nd St #721, New York, NY....................212-983-6042
Volunteer Lawyers for the Arts/Daniel Mayer
1 E 53rd St 6th Fl, New York, NY......................212-319-2910
Women in the Arts/Roberta Crown
1175 York Ave #2G, New York, NY 212-751-1915

EAST

American Advertising Federation/Wally Snyder
1101 Vermont Ave #500, Washington, DC 202-898-0089
American Assn for the Advancement of Science
1333 H St NW, Washington, DC 202-326-6400
American Booksellers Association Inc/Bernie Rath
828 S Broadway, Tarrytown, NY.......................914-591-2665
American Holographic Society
2018 R Street NW, Washington, DC..................202-667-6322

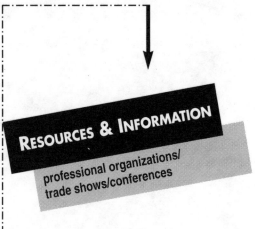

Art Directors Club of Philadelphia/Bob Ecker
2017 Walnut St, Philadelphia, PA 215-569-3650
Art Institute of Philadelphia/Michael Santaspirt
1622 Chestnut St/Visual Comm, Philadelphia, PA215-567-7080
Assn of Professional Design Firms/Sarah Vance
1 Story St, Cambridge, MA.............................617-864-7474
Boston Computer Society/Mike Costello
1972 Massachusetts Ave, Cambridge, MA.........617-864-1700
Boston Visual Artists Union/Diane Edstrom
33 Harrison Ave 7th Fl, Boston, MA...................617-695-1266
CD-CINC/Fred Durr
3100 St Paul St, Baltimore, MD 410-243-0797
Connecticut Art Directors Club/Elizabeth Smith
PO Box 639, Avon, CT...................................203-651-0886
Electronic Frontier Foundation/Mitchell Kapor
238 Main St #400, Cambridge, MA 617-576-4590
Graphic Artists Guild/Marilyn Masterson
14 Eaton Sq #8, Needham, MA........................617-455-0363
Guild of Natl Science Illustrators/Leslie Becker
PO Box 652/Ben Franklin Sta, Washington, DC ..301-309-1514
**Institute of Electrical & Electronic Engineering
Computer Soc**/Dr T Michael Elliott
1730 Massachusetts Ave NW, Washington, DC..202-371-0101
Interactive Multimedia Assoc/Susan Dodds
3 Church Circle #800, Annapolis, MD...............410-626-1380
Interactive Services Association/Robert Smith Jr
8403 Colesville Rd #865, Silver Spring, MD 301-495-4955
Intl Design by Electronics Assn/Don Sparkman
1120 Connecticut Ave #270, Washington, DC....202-785-2414
Intl Graphic Arts Education Assn/Dr Lenore Collins
4615 Forbes Ave, Pittsburgh, PA 412-682-5170
Myers, Barry
407 Thayer Ave, Silver Spring, MD 301-585-8617
National Assn of Artists' Organization/Helen Brunner
918 F St NW, Washington, DC 202-347-6350
National Assn of Desktop Publishing
462 Old Boston St, Topsfield, MA 508-887-7900
**National Association of Printers &
Lithographers**/Greg Van Wert
780 Palisade Ave, Teaneck, NJ 201-342-0700
Natl Assn of Broadcasters/Edward Fritts
1771 N St NW, Washington, DC 202-429-5300
New England Computer Arts Assn/William Eldridge
24 Harold St, Somerville, MA...........................617-776-8682
PageMaker User Group/NADTP/Tom Tetreault
462 Old Boston St, Topsfield, MA 800-874-4113
Point-of-Purchase Advertising Institute/Richard K Blatt
66 N Van Brunt St, Englewood, NJ...................201-894-8899
**Research & Eng Council of the Graphic Arts
Indus**/Fred Rogers
PO Box 639, Chadds Ford, PA..........................610-388-7394
**Society for Environmental Graphic
Design**/Sarah Speare
1 Story St, Cambridge, MA...............................617-868-3381
Software Publishers Assn/David Tremblay
1730 M St NW #700, Washington, DC............202-452-1600

SOUTH

Advertising Club of Ft Worth/D'Ann Dagen
PO Box 820376, Ft Worth, TX..........................817-283-3615
Art Directors Club of Houston/Susan Strickland
PO Box 271137, Houston, TX...........................713-961-3434
Association for Multi-Image/Marilyn Kulp
10008 N Dale Mabry Hwy #113, Tampa, FL.......813-960-1692
Brave Young Artists, The/Rachel Jackson
205 Smyer Terrace, Birmingham, AL...................205-871-4083

Creative Club of Atlanta/Sal Kibler
PO Box 77244, Atlanta, GA.............................404-881-9991
Dallas Society of Visual Communications/Sue Reynolds
3530 High Mesa Dr, Dallas, TX.........................214-241-2017
**Graphic Art Marketing Information
Services**/Jacqueline Bland
100 Dangerfield Rd, Alexandria , VA 703-519-8179
Graphic Arts Employers of America/William Solomon
100 Dangerfield Rd, Alexandria, VA 703-519-8150
Graphic Arts Mrktng Info Svc
100 Dangerfield Rd, Alexandria, VA 703-519-8179
Information Technology Assn of America/ITAA/Luanne James
1616 N Fort Myer Dr #1300, Arlington, VA........703-522-5055
Natl Assn of Schools of Art & Design/Samuel Hope
11250 Roger Bacon Dr #21, Reston, VA............703-437-0700
Natl Computer Graphics Assn/Irene Cahill
2722 Merrilee Dr #200, Fairfax, VA....................703-698-9600
**NPES/Assn for Suppliers of Print & Publshng
Techn**/Regis DelMontagne
1899 Preston White Dr, Reston, VA 703-264-7200
Printing Ind Assoc of the South/Dave Bacon
305 Plus Park Blvd, Nashville, TN......................615-366-1094
Scitex Graphic Arts Users Assn/Don Dunham
PO Box 290249, Nashville, TN 615-366-1798
Society for Technical Communications/William Stolgitis
901 N Stuart St, Arlington, VA...........................703-522-4114
World Computer Graphics Association/Caby Smith
5201 Leesburg Pike, Falls Church, VA................703-578-0301

MIDWEST

ACM/SIGGRAPH/Cindy Stark
401 N Michigan Ave #2300, Chicago, IL...........312-644-6610
Advertising Club of Cincinnati/Betty Worthington
PO Box 43252, Cincinnati, OH 513-576-6068
Advertising Club of Indianapolis/Leo Demos
3833 N Meridian #305B, Indianapolis, IN 317-924-5577
Advertising Club of Kansas City/Whitey Kuhn
9229 Ward Pkwy #260, Kansas City, MO 816-822-0300
Advertising Federation of Minnesota/Jennifer Velasco
275 Market St #C-19, Minneapolis, MN............612-339-5470
AIGA/Minnesota/John DuFresne
275 Market St #54, Minneapolis, MN 612-339-6904
American Center for Design/Mary Davis
233 E Ontario #500, Chicago, IL.......................312-787-2018
ASCAP/Ken Clauses
3459 Washington Dr #202, Eagan, MN 612-456-9495
**Assn for Dev of Computer
Instruction Systems**/Howard Troutman
1121 Worthington Woods #199,
Worthington, OH...614-847-0859
Computing Technology Industry Assoc/John Venator
450 E 22nd St #230, Lombard, IL 708-268-1818
Design Collective/DF Cooke
130 E Chestnut St, Columbus, OH 614-464-2880
Electronic Artists Group/Jim Brod
PO Box 580783, Minneapolis, MN....................612-649-4641
Institute of Business Designers/Terri Carlton
341 Merchandise Mart, Chicago, IL...................312-467-1950
Michigan Guild of Artists & Artisans/Mary Strope
118 N Fourth Ave, Ann Arbor, MI.....................313-662-3382
Midwest Independent Publishers Association/Pat Bell
9561 Woodridge, Eden Prairie, MN 612-941-5053
Milwaukee Advertising Club/Marilyn Brocker
231 W Wisconsin Ave #500, Milwaukee, WI.....414-271-7351
Minnesota Broadcasters Assn/Jim Wychor
3517 Raleigh Ave S/Box 16030,
St Louis Park, MN 612-926-8123
**Minnesota Commercial Indstrl Photographers
Assn**/Terry Jacobson
3340 Rhode Island Ave S, Minneapolis, MN.......612-938-0646
Minnesota Film Board/Kelly Pratt
401 N 3rd St #460, Minneapolis, MN 612-332-6493
**National Center for Super Computing
Applications**/Colleen Bushell
237 Com App Bdg/605 E Sprfld, Champaign, IL 217-244-6830
Optical Publishing Assn/Richard Bowers
PO Box 21268, Columbus, OH.........................614-442-8805
Printing Industry of Minnesota Inc/Kelvin Johnson
450 N Syndicate #200, St Paul, MN..................612-646-4826
School of Communications Arts Inc/Kathy Dale
2526 27th Ave S, Minneapolis, MN...................612-721-5357

WEST

Advertising Club of Los Angeles/Kerrie Berchon
11340 W Olympic #352, Los Angeles, CA.........310-914-7711
Advertising Production Association/LA/Veronica Urias
444N Larchmont Blvd #200, Los Angeles, CA 213-463-7851
Allied Arts of Seattle Inc/Alf Collins
105 S Main St #201, Seattle, WA 206-624-0432
**AM/FM Int Automated Mapping & Facilities
Mgmt**/James Black
14456 E Evans Ave, Aurora, CO.......................303-337-0513
Art Directors and Artists Club of Sacramento
2791 24th St, Sacramento, CA.........................916-731-8802
Art Directors Club of Denver/Mary Hall
1900 Grant St #805, Denver, CO......................303-830-7888
Art Directors Guild/Gene Allen
11365 Ventura Blvd #315, Studio City, CA.........818-762-9995
ArtCom/Carl Loeffler
PO Box 193123/Rincon Ctr, San Francisco, CA..415-431-7524
Association of Desktop Publishers/N E Paddock
4677 30th St #800, San Diego, CA...................619-563-9714
Black Creative Professionals/Kidogo
PO Box 34272, Los Angeles, CA 213-964-3550
Briley, Mike
370 Altair Way #190, Sunnyvale, CA................408-730-0341
**Communicating Arts Group of
San Diego**/Sidney Cleaver
3108 Fifth Ave #F, San Diego, CA 619-295-5082
Computer Consultants/Randy Tobin
1309 Riverside Dr, Burbank, CA 818-955-5830
Conejo Valley Mac Users Group/Lewis Kane
2509 Thousand Oaks Blvd #147,
Thousand Oaks, CA.......................................818-706-8545
Directors Guild Of America
7920 Sunset Blvd, Los Angeles, CA 213-851-3671
Frank Relations/Nancy Frank
1342 Stevenson, San Francisco, CA..................415-626-5742
International Interactive Comms Society/Debra Palm
14657 SW Teal Blvd #119, Beaverton, OR........503-579-4427
Intl Assn of Busn Communicators/LA
PO Box 2727, El Segundo, CA 213-892-6392
Los Angeles Creative Club, The/Carol Golden
6404 Wilshire Blvd #111, Los Angeles, CA 213-387-7432
Los Angeles Macintosh Group/Suzy Prieto
12021 Wilshire Blvd #349, Los Angeles, CA 310-278-LAMG
Mac Valley Users Group/Robert Wright
PO Box 4297, Burbank, CA..............................818-241-8479
Pete Bleyer Studio Inc/Pete Bleyer
807 N Sierra Bonita Ave, Los Angeles, CA..........213-653-6567
Printing Industries Assn/So Cal/Bob Lindgren
5800 S Eastern Ave #400, Los Angeles, CA........213-728-9500
Public Relations Society of America/Barbara Shore
7060 Hollywood Blvd #614, Los Angeles, CA.....213-461-4595
Publishers Marketing Assn/Jan Nathan
2401 Pacific Coast Hwy #102,
Hermosa Beach, CA.......................................310-372-2732
Randy and Cindy Tobin/PC Consultant
1309 Riverside Dr, Burbank, CA 818-955-5830
San Francisco Creative Alliance/Teresa Rodriguez
PO Box 410387, San Francisco, CA 415-387-4040
Seattle Ad Federation/Becky McLaughlin
800 Fifth Ave #274, Seattle, WA......................206-448-4481
Society for Computer Simulation/Brian O'Neill
PO Box 17900, San Diego, CA 619-277-3888
Society of Illustrators of Los Angeles/Monica Heath
11480 Burbank Blvd, N Hollywood, CA.............818-952-7452
Society of Illustrators/San Francisco
690 Market St #920, San Francisco, CA.............415-399-1681
Ventura Publisher Users Group/Bob Moody
7502 Aaron Pl, San Jose, CA 408-227-5068
Video Electronics Standards Assn/Roger McKee
2150 N First #440, San Jose, CA......................408-435-0333
Visual Arts Assn
1420 N Mckinley Ave, Los Angeles, CA 310-898-6090
WICI/Women in Communications Inc
8306 Wilshire Blvd #635, Beverly Hills, CA........310-288-7177
YLEM
Box 749, Orinda, CA 510-482-2483

Trade Shows/Conferences

NEW YORK CITY

**Conference Management Group/Oct 28-30,
1994**/VISCOMM 94
Jacob Javits Center, New York, NY....................203-852-0500

Digital Video NY/April 19-21 1994/Victor Harwood
NY Sheraton, New York, NY 212-226-4141
Image World/Video Expo NY/Sept 11-13
1994/Knowledge Industry Publications Inc
Jacob Javits Center, New York, NY...................914-328-9157
Showbiz Expo East/Jan 6-8, 1994/Live Time
NY Hilton, New York, NY................................213-668-1811
UNIX EXPO/Oct 4-6, 1994/Bruno Blenheim Inc
Jacob Javits Center, New York, NY....................201-346-1400

EAST

GRAPH EXPO East/Sept 25-28, 1994/
Patrick LaFramboise Inc
Pennsylvania Conv Ctr, Philadelphia, PA 703-264-7200
Interactive Multimedia 94/Aug 17-19, 1994/
Soc for Applied Learning Technology
J W Marriott Hotel, Washington, DC...........800-457-6812
Intermedia 94/San Jose, CA/ March 1-3, 1994/
Reed Exhibtion Co
999 Summer St, Stamford, CT 203-964-0000
MACWORLD Expo Boston/Aug 2-5, 1994/
Mitch Hall Associates
Bayside & World Trade Center, Boston, MA 617-361-8000
Mitch Hull & Assocs/Oct 5-7, 1994/CD-ROM Expo
World Trade Ctr, Boston, MA............................617-361-2001
NAB/Multimedia World, Las Vegas, March 21-24,
1994/National Assoc of Broadcasters
1771 N Street NW, Washington, DC.................202-429-5336

SOUTH

ACM-SIGCSE/Computer Science in Ed/Mar 9-10,
1995/ACM
Opryland, Nashville, TN.................................212-869-7440
Orlando Multimedia 95/Feb 22-24,1995/
(SALT) Soc for Applied Learning Technology
Hyatt Orlando, Kissimmee, FL...........................800-457-6812

MIDWEST

Image World Chicago/April 26-28, 1994
/Knowledge Industry Publctns Inc
Chicago, IL..914-328-9157
Mead Paper/Shawn Palmer
Courthouse Plz NE, Dayton, OH..............513-495-3308
Midwest Graphics®/Apr 7-9, 1994/Patrick Graphic Arts
Show Co Inc LaFramboise
Cobo Conf & Exhibit Ctr, Detroit, MI 703-264-7200
Soc of Manufacturing Engineers/Carolyn Best/351
One SME Dr/PO Box 930, Dearborn, MI............313-271-1500

WEST

"Seybold San Francisco" Sept 14-16, 1994/Seybold
Moscone Center, San Francisco, CA..................310-457-8500
11th Annual Showbiz Expo West/June 11-13, 1994/
Live Time
Los Angeles Convention Ctr, Los Angeles, CA 213-668-1811
12th Annual Showbiz Expo West/Jan 8-10,
1995/Live Time
Los Angeles Convention Ctr, Los Angeles, CA 213-668-1811
136th Tech Conf & World Media Expo/Oct 12-15,
1994/SMPTE
Los Angeles Convention Ctr, Los Angeles, CA 914-761-1100
28th Adv Techn&Electr Imag/Feb 10, 1995/SMPTE
Weston/St Francis Hotel, San Frncisco, CA 415-397-7000
Baird Productions/Doug Baird
578 Vermont St, San Francisco, CA...................415-626-9494
COMDEX Fall/Nov 14-18, 1994/The Interface Group
Las Vegas Conv Ctr/Sands Expo, Las Vegas, NV .617-449-6600
Digital Hollywood/Feb-March 1995/Victor Harwood
Beverly Hills Hilton, Beverly Hills, CA 212-226-4141
Digital Video San Francisco/Oct 11-13, '94/
Victor Harwood
Moscone Center, San Francisco, CA...................212-226-4141
Frank Relations/Nancy Frank
1342 Stevenson, San Francisco, CA...................415-626-5742
Gutenberg Expositions/David Jacobson
PO Box 11712, Santa Ana, CA 714-921-3120
IKONIC/Robert May
188 The Embarcadero 7th fl, San Francisco, CA..415-864-3200
Intermedia Conference/March-April, 1995/
Reed Exhibition
San Jose Convention Ctr, San Jose, CA 203-964-0000
Laser & Graphics Conf/March 1995/Patrice Dunn
San Diego, CA 619-758-9460
Live Marketing/William J Steinmetz
96 Hawthorne Ave, Los Altos, CA.....................415-941-8188

Multimedia Expo West/Oct 11-13, 1994/
Victor Harwood
Moscone Center 747 Howard St,
San Francisco, CA 212-226-4141
Online/CD-Rom '94 Oct 24-46, 1994/Online Inc
SF Hyatt Regency, San Francisco, CA 203-761-1466
Seybold San Francisco/Sept 13-16, 1994/Seybold
Moscone Ctr, San Francisco, CA 415-578-6900
SIGGRAPH '94 Show/July 24-29, 1994
Orange County Convention Ctr, Orlando, CA......312-321-6830
Wescon/Sept 27-29, 1994/Donna Call
Anaheim Convention Center, Anaheim, CA 800-877-2668

CANADA

Canada Online Ltd/Early Fall 1994/Edel Ching
Toronto, ON..416-777-2020
ECM-A Exhibit & Conference Management Ltd/
SIIM '94 Intl Computer & Office Exh/June 1-4
Pl Bonaventure, Montreal, QB 416-497-9562
MACWORLD Expo Toronto/Sept 20-23 1994/
Mitch Hall Associates
Toronto Convention Center, Toronto, ON 617-361-8000

INTERNATIONAL

Virtual Reality Conference Intl/April13,14,
1994/Meckler Conference Management
Olympia Conference Center, London, UK 800-635-5537

Professional Development

NEW YORK CITY

Apple Market Center/NY/Tara Griffin
135 E 57th St, New York, NY...........................212-339-3700
Bklyn Coll/Ctr for Computer Music/Dr Charles Dodge
2900 Bedford Ave & Ave H, Brooklyn, NY..........718-951-5000
City College of NY/Art Dept/Annie Shaver-Crandell
138th St & Convent Ave, New York, NY..............212-650-7421
Columbia Univ/Div of Continuing Educ/Paul McNeil
206 Lewisohn Hall, New York, NY 212-854-2820
Fashion Inst of Technology/Terry Blum
227 W 27th St #C220, New York, NY 212-760-7938
Grundy, Jane
150 W 82nd St, New York, NY..........................212-787-8454
Meyerowitz + Co, Michael
175 Fifth Avenue #609, New York, NY 212-353-1561
New School for Social Research/Mark Schulman
2 W 13th St 12th fl, New York, NY....................212-229-5600
NYIT/Metropolitan Center/Martin Bressler
1855 Broadway/Computer Grphcs,
New York, NY...212-261-1500
NYU/Tisch School of the Arts/Red Burns
721 Broadway/Interactive Tele, New York, NY ...212-998-1820
Pratt Institute/Manhattan/Karen Miletski
295 Lafayette St, New York, NY........................212-925-8481

EAST

Art Institute of Philadelphia/Michael Santaspirt
1622 Chestnut St/Visual Comm, Philadelphia, PA215-567-7080
Boston Univ/Comp Graphics Lab/Glenn Dresnahan
111 Cummington St, Boston, MA.......................617-353-2780
Bucks County Community College/Craig Johnson
Comp Graphics Ctr, Newtown, PA 215-968-8191
Buffalo State College/Stephen Saracino
Upton Hall 212/1300 Elmwood, Buffalo, NY......716-878-6032
Carnegie Mellon Univ/Architecture/John Eberhard
5000 Fords Ave, Pittsburgh, PA.........................412-268-2355
Carnegie Mellon Univ/Art/Bryan Rogers
Coll of Fine Arts Bldg, Pittsburgh, PA.................412-268-2409
Carnegie Mellon Univ/Design/Stephen Stadelmeier
Margaret Morrison Bldg #110, Pittsburgh, PA......412-268-2828
Center for Creative Imaging
51 Mechanic St, Camden, ME 800-428-7400
Central Conn State College/Clifford Pelletier
Comp Sci/1615 Stanley St, New Britain, CT 203-827-7000
Computer Applications Learning Ctr/Ed Roffman
100 Hanover Ave, Morristown, NJ 201-539-6050
DiGena, Louis
356 First St, Hoboken, NJ................................201-963-5339
Fairleigh Dickinson Univ/Madison/Harvey Flaxman
285 Madison Ave, Madison, NJ........................201-593-8500
Gambino, Donald
12 Carolyn Way, Purdys, NY.............................914-277-1837
George Washington University/Prof S Molina
Art&Vis Comms/2300 I St NW, Washington, DC 202-994-7455

RESOURCES & INFORMATION
trade shows/conferences/
professional development

Glassboro State College/William Travis
Art Dept, Glassboro, NJ...................................609-863-7081
Gloucester County Coll/Lib Arts/John Henzy
Tanyard Rd/Deptford Twnshp, Sewell, NJ............609-468-5000
Graphic Arts Sales Foundation/Judy Warren
113 E Evans St/Matlack Bldg, West Chester, PA..610-431-9780
Graphic Arts Suppliers Assn/Jennifer Cochran
1900 Arch St, Philadelphia, PA.........................215-564-3484
Graphic Arts Technical Foundation/Charles Lucas
4615 Forbes Ave, Pittsburgh, PA.......................412-621-6941
Graphic Comms Intl Union/James Norton
1900 L St NW, Washington, DC........................202-462-1400
Green Mountain College/Steve Ingram
16 College St/A-V Dept, Poultney, VT 802-287-9313
Guild of Natl Science Illustrators/Leslie Becker
PO Box 652/Ben Franklin Sta, Washington, DC ..301-309-1514
High Resolution Inc/Peter Koons
PO Box 397, Camden, ME 207-236-3777
Hofstra University/Don Booth
Fine Arts/Clkns Hall #218,107 H U,
Hempstead, NY 516-463-5474
Howard University/Floyd Dr Colman
6th St NW/Dept of Fine Arts, Washington, DC....202-806-7047
ImageSoft Inc/Phil Schwartz
2 Haven Ave, Port Washington, NY...................516-767-2233
Indiana Univ of Penna/Dr Anthony DeFurio
115 Sprowls Hall/Art Dept, Indiana, PA.............412-357-2530
Kean College/Lennie Pierro
Fine Arts/Morris Ave, Union, NJ 908-527-2000
Law Holographics/Linda E Law
234 Main st, Huntington Station, NY 516-673-3138
Marist College/Dr O Sharma
Math & Computer Science/290 North Rd,
Poughkeepsie, NY...914-575-3000
Mass Coll of Art/Comptr Arts Ctr/Hubert Hohn
621 Huntington Ave, Boston, MA 617-232-1555
MINDesign/Mihai Nadin
35 Butts Rock Rd, Little Compton, RI 401-635-1675
MIT Research Lab/Electronics #36-763/David Zeltzer
50 Vassar St, Cambridge, MA 617-253-5995
Mohawk Valley Community College/Ronald Labuz
1101 Sherman Dr/Graphics Dept, Utica, NY 315-792-5400
Montclair State College/John Luttrop
Fine Arts Dept, Upper Montclair, NJ 201-655-4000
Montgomery Coll/Applied Technologies/George Payne
51 Mannakee St, Rockville, MD.........................301-279-5142
Moore College of Art/Nancy Stock-Allen
20th & The Pkwy/Graphic Design,
Philadelphia, PA 215-568-4515
New England Schl of Art & Design/Jean Hammond
28 Newbury St/Grphc Dsgn, Boston, MA 617-536-0383
New Overbrook Press/Charles Altschul
356 Riverbank Rd, Stamford, CT 203-329-7251
Northeastern Univ/Univ College/Maryann Frye
Graphic Arts/360 Huntington Av, Boston, MA...617-373-2000
Notre Dame College/Sister LaMarr
2321 Elm St/Art Dept, Manchester, NH 603-669-4298
Princeton Univ/CIT/Jacqueline Brown
87 Prospect Ave, Princeton, NJ.........................609-258-6061
REF Associates/Dr Robert Fidoten
118 Greyfriar Dr, Pittsburgh, PA.......................412-963-8785
Rensselaer Polytechnic Inst/
Design Manufacturing Inst/Mary Johnson
110 Eighth St Bldg CII Rm 7015, Troy, NY..........518-276-6751

Resources & Information

professional development

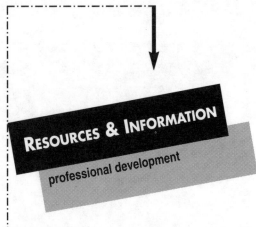

Rochester Inst of Techn/Graduate Prog/
Joanne Dr Szabla
Fine Arts Dept/1 Lomb Meml Dr, Rochester, NY...716-475-2641
Rochester Inst of Techn/Natl Inst for Deaf/Bill Castle
1 Lomb Meml Dr/Applied Art, Rochester, NY.......716-475-6418
Rochester of Inst Techn/Printing/Frank Romano
PO Box 9887, Rochester, NY.............................716-475-7023
Rutgers University/Roberta Dr Tarbell
311 N 5th St/Art Dept, Camden, NJ 609-225-1766
Rutgers University/Art History/Rona Goffen
71 Hamilton St/Voorhees Hall,
New Brunswick, NJ..908-932-7041
School of Computer Science/Andrew Witkin
Carnegie Mellon University, Pittsburgh, PA..........412-268-6244
Seton Hall Univ/Comms Dept/Kenneth Dr Hoffman
400 S Orange Ave, S Orange, NJ 201-761-9474
Skidmore College/Doretta Miller
815 N Broadway/Art Dept, Saratoga
Springs, NY 518-584-5000
Smith College/Susan Heideman
Hillyer Hall 112/Art Dept, Northampton, MA 413-584-2700
SUNY/Buffalo/Art Dept/Tyrone Georgiu
2917 Main St/202 Fine Arts Ctr, Buffalo, NY......716-645-6878
Syracuse Univ/Art Media Studies/Onen Shapiro
Shaffer Art Bldg #102, Syracuse, NY..................315-443-1870
Trenton State Coll/Art Dept/Howard Dr Goldstein
Hillwood Lakes/CN 4700, Trenton, NJ 609-771-2652
Univ of Maryland/Baltimore/David Yeager
5401 Wilkens Ave #111/Visual A,
Baltimore, MD 410-455-2150
Univ of Massachusetts Art Dept/Hanlyn Davies
368 Fine Arts Ctr, Amherst, MA..........................413-545-1902
Univ of the Arts/Brian Feeney
320 S Broad St/A-V Dept, Philadelphia, PA.........215-875-4800
Voelkl, Michael
1994 Five Mile Line Rd, Penfield, NY..................716-381-4205
William Patterson Coll/Art Dept/Alan Lazarus
Ben Shan Hall/300 Pompton Rd, Wayne, NJ 201-595-2401

SOUTH

Anderson College/Susan Wooten
316 Boulevard/Art Dept, Anderson, SC 803-231-2000
AppleTree Technologies Inc/Larry Hall
3020 Mercer Univ Dr, Atlanta, GA.....................404-457-4500
Art Inst of Atlanta/Mike Lawsky
3376 Peachtree Rd NE/A-V Dept, Atlanta, GA 404-266-2662
Art Inst of Ft Lauderdale/John Morn
1799 SE 17th St Causeway/M&V,
Ft Lauderdale, FL 305-463-3000
Art Institute of Dallas/Wanda Lundy
2 N Park/8080 Park Ln, Dallas, TX....................214-692-8080
Baylor University/Fine Arts/John McClanahan
PO Box 97263, Waco, TX....................................817-755-1867
Coll of Arts & Sciences-CPR107/Rollin Richmond
Univ of S Florida, Tampa, FL 813-974-2804
Computer Imaging Resources/Pat Johnson
401 69th St, Miami Beach, FL.............................305-868-1206
George Mason University/Scott Breivold
4400 University Dr/A-V Dept, Fairfax, VA...........703-993-2209
James Madison University/Dr Philip James
Art Dept, Harrisonburg, VA.................................703-568-6216
John C Calhoun State Commty Coll/Art Bond
PO Box 2216/Art Dept, Decatur, AL...................205-306-2500
Louisiana State University/Michael Dougherty
Int Design #402 New Dsgn Bldg,
Baton Rouge, LA 504-388-8461

Loyola College/Silvia Pengilly
6363 St Charles Ave, New Orleans, LA 504-865-2595
Memphis College of Art/Chris Haywood
1930 Poplar Ave, Memphis, TN..........................901-726-4085
Memphis State Univ/Art/Dr Robert Lewis
Jones Hall #201, Memphis, TN 901-678-2216
New World School of the Arts
300 NE Second Ave, Miami, FL 305-237-3135
Our Lady of the Lake Univ/Jody Cariolano
Art Dept/411 SW 24th St, San Antonio, TX.........210-434-6711
Ringling School of Art & Design/Maria Palazzi
2700 N Tamiami Trl/Cmptr Grphc, Sarasota, FL..813-359-7574
Texas Tech University/Melody Weiler
PO Box 42081/Art Dept, Lubbock, TX 806-742-3825
Texas Womans Univ/Adv Art/Dr Betty Copeland
PO Box 22995, Denton, TX 817-898-2530
Univ of Florida/Dr Lee Mullally
Instr & Curric/258 Norman Hall, Gainesville, FL ..904-392-0761
Univ of N Texas/Scott Sullivan
Art Dept/PO Box 5098, Denton, TX....................817-565-2000
Univ of N Texas/Exp Music & Int/Dr Cindy McTee
PO Box 13887, Denton, TX 817-565-3764
Univ of S Alabama/Larry Simpson
Dept of Art #172/Vis Arts Bldg, Mobile, AL.........205-460-6336
Univ of SW Louisiana/Edward Cazayoux
Architecture PO Box 43850, Lafayette, LA..........318-231-6225
Univ of Tenn/Arch/William Lauer
1715 Volunteer Blvd, Knoxville, TN 615-974-3261
Univ of Virginia/School of Arch/Kenneth Schwartz
Campbell Hall, Charlottesville, VA 804-924-3715
University of Texas/Rick Lawn
Dept of Music, Austin, TX..................................512-471-7764
Virginia Commonwealth Univ/John DeMao
Comp Grphcs/325 N Harrison St, Richmond, VA 804-367-1709
Virginia State Univ/Dept of Art & Dsgn/
Valery Bates-Brown
PO Box 9247, Petersburg, VA.............................804-524-5543
Visualization Lab/Coll of Architecture /Bill Jenks
Texas A&M University #216, College Station, TX .409-845-3465

MIDWEST

ACA Coll of Design/Dan Rodgers
Comp Grphcs/2528 Kemper Ln, Cincinnati, OH..513-751-1206
American Acad of Art & Grphic Comms/Mark Jarcinski
332 S Michigan Ave #300, Chicago, IL...............312-939-3883
Atonelli Inst of Art & Photog/Bob Wilson
124 E 7th St, Cincinnati, OH 513-241-4338
Bowling Green State Univ/Thomas Hilty
School of Art, Bowling Green, OH.......................419-372-2015
Chicago Schl of Holography/Loren Billings
1134 W Washington Blvd, Chicago, IL.................312-226-1007
College of St Mary/Tom Schoosser
Comp Grphcs/1901 S 72nd, Omaha, NE 402-399-2621
Columbia Coll/Academic Computing/Geof Goldbogen
600 S Michigan Ave, Chicago, IL........................312-663-1600
Columbus Coll of Art & Design/Carl Grant/Ron Saks
107 N 9th St/Computer Grphcs, Columbus, OH..614-224-9101
Cranbrook Academy Of Art
Box 801, Bloomfield Hills, MI.............................810-645-3300
Creative Center for Advanced D/Anthony Main
201 SE Main St, Minneapolis, MN 612-649-4641
Grand Valley State/Alex Nesterenko
Communications, Allendale, MI 616-895-3668
Illinois State Univ/Coll of Fine Arts/John Hayes
Micro Computer Lab, Normal, IL.........................309-438-7934
Illinois State Univ/Coll of Fine Arts/Ron Mottram
Dept of Art/5620, Normal, IL...............................309-438-8321
Illinois Wesleyan Univ/Miles Bair
Art Dept/302 E Graham, Bloomington, IL............309-556-3134
Indiana State Univ/Wayne Enstice
Art Dept/Fine Arts Bldg #108, Terre Haute, IN 812-237-3697
Iowa State Univ/Nancy Polster
Art & Design, Ames, IA.......................................515-294-3928
Kansas City Art Inst/Jack Lew
Design/4415 Warwick Blvd, Kansas City, MO......816-561-4852
Lansing Community College/Nancy Lombardi
Media & Art/PO Box 40010, Lansing, MI...........517-483-1957
Laserworks/Garret Peaslee
6401 Quail Run, Kalamazoo, MI 616-375-8996
Marycrest University/Alan Garfield
1607 W 12th St/Comp Grphcs, Davenport, IA ...319-326-9532
Mead Paper/Shawn Palmer
Courthouse Plz NE, Dayton, OH 513-495-3308
Michigan State University/Carrie Heeter/Comms Tech Lab
253 Communications Arts Bldg, E Lansing, MI 517-353-5497

Milliken University/James Schietinger
Art Dept/1184 W Main St, Decatur, IL.................217-424-6227
Milwaukee Inst of Art & Design/Robert Solem
273 E Erie St, Milwaukee, WI.............................414-276-7889
Minneapolis Coll of Art & Design/Russ Mrcvek
2501 Stevens Ave S/Comp Grphcs,
Minneapolis, MN..612-874-3700
Minneapolis College of Art & Design/
Thomas A DeBiasso
Media Arts/2501 Stevens Ave S,
Minneapolis, MN..612-874-3638
Minneapolis Tech College/Rena Gray/Dianne Michels
1415 Hennepin Ave S, Minneapolis, MN............612-370-9400
NE Illinois Univ/Laverne Ornelas
Art Dept/5500 N St Louis, Chicago, IL................312-583-4050
Northern Illinois Univ/Richard M Carp
School of Art/Jack Arends Hall, DeKalb, IL 815-753-7851
Northern State University/Art/Mark Shekore
S Kline & 12th Ave, Aberdeen, SD......................605-622-2514
Northwestern Univ/Comp Music Studio/Gary Kendall
711 Elgin/Acad Studies & Comp, Evanston, IL.....708-491-3178
Oberlin College/Susan Kane
Art Dept, Oberlin, OH 216-775-8181
Ohio State Univ/Adv Comptng Ctr for Art&Dsgn/Charles Csuri
1224 Kinnear Rd, Columbus, OH.........................614-292-3416
Ohio State Univ/Comp Aided Arch Dsgn/Jose Oubrerie
189 Brown Hall/190 W 17th Ave, Columbus, OH614-292-2918
Ohio State Univ/Ind Design/James Kaufman
380 Hopkins Hall, Columbus, OH 614-292-6746
Ohio State Univ/Schl of Music/Thomas Wells
1866 College Rd, Columbus, OH.........................614-292-7837
Prairie State Coll/Grphc Comms/James Moore
202 S Halstead, Chicago Heights, IL...................708-756-3110
Ray College of Design/Steve Farrell
Comp Grphcs/401 N Wabash #610, Chicago, IL312-280-3500
S Illinois Univ/Dave White
Commcl Graphic Design, Carbondale, IL..............618-536-6682
School of the Art Inst/Claudia Cumbie-Jones
112 S Michigan/Data Bank, Chicago, IL..............312-345-3550
South Dakota State Univ/Tim Steele
Visual Arts/102 Folberg Hall, Brookings, SD 605-688-4103
St Louis Community Coll at Merramec/Diane Carson
Comms/11333 Big Bend Rd, Kirkwood, MO 314-984-7532
Univ of Ill at Champaign/Theodore Zernich
Dept of Art/408 E Peabody, Champaign, IL 217-333-7713
Univ of Ill at Chicago/Thomas Dr DeFanti
Elec Vis Lab/EECS Dept/MC154, Chicago, IL 312-996-3002
Univ of Kansas/Robert Brawley
#300 Art & Design, Lawrence, KS 913-864-4401
Univ of Minnesota/Earl Morris
1985 Buford Ave/Dsgn, St Paul, MN..................612-624-9700
Univ of Missouri/Ruth Dr Brent
135 Stanley Hall/Env Dsgn, Columbia, MO 314-882-7224
Univ of Wisconsin/T C Farley
Art Dept, Oshkosh, WI 414-424-2235
Univ of Wisconsin/Eau Claire/Scott Robertson
Art Dept, Eau Claire, WI.......................................715-836-3277
Univ of Wisconsin/Madison/Truman Lowe
455 N Park St/Art Dept, Madison, WI..................608-262-1660
University of Notre Dame/John Sherman
Dep of Art, Art Hist & Des, Notre Dame, IN 219-631-5000
Wayne State University/Jeffrey Abt
Art & Art Hist/150 Commty Arts, Detroit, MI.........313-577-2985

WEST

Acacia Systems
681 Miramar Ave, Long Beach, CA......................310-437-7690
Academy of Art College/Dan Gonzales
Comp Ed Ctr/79 New Montgomery,
San Francisco, CA 415-274-2245
Accelerated Computer Training/Sharon Panone
3255 Wilshire Blvd #903, Los Angeles, CA 213-388-0551
American River College/Greg Gregory
4700 College Oak Dr, Sacramento, CA...............916-484-8666
Antelope Valley College/Dr Dennis White
Fine Arts West/3041 Ave K, Lancaster, CA..........805-943-3241
Applied Graphic Technologies
601 Rodier Dr, Glendale, CA...............................310-245-4111
Aproprose/L Walford
1511 Sawtelle Blvd #376, Los Angeles, CA 310-451-3948
Arizona State Univ/West/Carl Chapman
PO Box 37100, Phoenix, AZ...............................602-543-8252
Arizona State University/Julie F Codell
School of Art, Tempe, AZ....................................602-965-3468

Art Ctr College of Design/Rob Hennigar
Comp Grphcs/1700 Lida St, Pasadena, CA........818-396-2359
Art Institute of S Calif/John Walker
2222 Laguna Canyon Rd, Laguna Beach, CA......714-497-3309
Bay Area Video Coalition/Luke Hones
1111 17th St, San Francisco, CA......................415-861-3282
Brodsky, Michael
1327 S Carmona Ave, Los Angeles, CA..............310-338-3060
Brooks Institute/Advanced Imaging Group/Barbara Hawkins
801 Alston Rd, Montecito, CA............................805-966-3888
Business Information Solutions Inc/Mary Lambert
4500 Campus Dr #284, Newport Beach, CA......714-966-1180
Calif Coll of Arts & Crafts/David Meckel
1700 17th St, San Francisco, CA......................415-703-9500
Calif Institute of the Arts/Thomas Lawson
Art Dept/24700 McBean Pkwy, Valencia, CA.....805-255-1050
Calif State Polytechnic Univ/Barry Wasserman
Dept of Arch/3801 W Temple Ave, Pomona , CA909-869-2682
Calif State Univ/Chico/Orlando Dr Madrigal
Dept of Comp Sci, Chico, CA............................916-898-6442
Calif State Univ/Long Beach/Jim Van Eiermen
Vis Comm/1250 Bellflower Blvd, Long Beach, CA310-985-4111
Calif State Univ/Northridge/Dorothy Miller
Comp Sci/18111 Nordhoff St, Northridge, CA....818-885-3398
Calif State Univ/Sacramento/Anna Louise Radimsky
Comp Sci Dept/6000 J St, Sacramento, CA........916-278-6834
Calif State Univ/Sacramento/Lita Dr Whitesel
Art Dept/6000 J St, Sacramento, CA..................916-278-6166
Calif State/LA/Techn & Art Depts/Charles Borman
5151 State University Dr, Los Angeles, CA.........213-343-4550
Captured Images/Bill Grimm
31316 Via Colenas #117, Westlake Village, CA.818-707-9491
Center for Electronic Art/Harold Hedelman
950 Battery St #3D, San Francisco, CA..............415-956-6500
Chapman Univ/Comp Sci Dept/Gary Ramet
333 N Classell St, Orange, CA..........................714-997-6729
Colorado Institute of Art/Lee Park
Vis Comm/200 E 9th Ave, Denver, CO 303-837-0825
Competitive Edge
17542 Briarwood St, Fountain Valley, CA...........714-962-9921
Computer Arts Institute/Richard Howard
310 Townsend #230, San Francisco, CA............415-546-5242
Computer Education Intl/Ryan Granard
230 N Maryland #300, Glendale, CA 818-242-0828
Computer Evolution/Don Roark
2500 Abbot Kinney Blvd #15, Venice, CA..........310-821-6184
Computer Intl
10517 W Pico Blvd, Los Angeles, CA................310-474-6409
Cyber Network Svcs Inc/Eva Way Konigsberg
514 Bryant St, San Francisco, CA......................415-974-7000
Cypress Coll/Fine Arts/Arden Alger
9200 Valley View St, Cypress, CA 714-826-2220
DeAnza Coll/Cmptr Grphc Design/Michael Cole
21250 Stevens Creek Blvd, Cupertino, CA 408-864-8832
Denver Commty Coll/Downtown Aurora Campus/Patricia Lehman
PO Box 173363/Campus Box #850, Denver, CO303-556-2600
Eastern Washington University/Richard Twedt
Gallery of Art MS-102/Art Dept, Cheney, WA.....509-359-2493
Fenster, Diane
140 Berendos Ave, Pacifica, CA.......................415-355-5007
Fullerton College/Todd/Graphics Lab Glen
321 E Chapman Ave, Fullerton, CA....................714-992-7351
Goodman, Barbara
PO Box 585, Tustin, CA 714-997-2626
Great Wave Software/Roberta Sosby
5353 Scotts Valley Dr, Scotts Valley, CA.............408-438-1990
Gunnar Swanson Design Office
739 Indiana Ave, Venice, CA 310-399-5191
Hada, Gail
23290 Clearpool, Harbor City, CA....................310-539-5114
Halley Design/Lynda Halley
PO Box 3685, Chatsworth, CA 818-407-1642
Hands On Software Training/Chris Allen
9868 Scranton Rd #110, San Diego, CA............619-627-9857
Honolulu Commty College/Marcia Roberts-Deutsch
Art Dept/874 Dillingham Blvd, Honolulu , HI......808-845-9211
Imaging Supplies & Eqpt/Richard Brock
200 Carob St , Compton, CA............................800-526-3727
Info Systems Computer Ctr
19950 Mariner Ave, Torrance, CA.....................310-214-4200
Integrated Solutions/Eliot Kaplan
712 Hillcrest Dr, Topanga, CA..........................310-455-3950

Joule Group, The/Jaye Towne
10428 Apache River Ave, Fountain Valley, CA....714-968-8902
LA Trade Tech Coll/Art/Grphc Comm/Richard Brown
400 W Washington Blvd, Los Angeles, CA 213-744-9500
Learning Tree/Comp Graphics Dept/Albert Nades
20920 Knapp St, Chatsworth, CA......................818-882-5599
Learnsoft Inc/Mark Jongeward
5880 Oberlin Dr #600, San Diego, CA...............619-546-1400
Level 6 Computing/Kevin Garrett
22647 Ventura Blvd #201, Woodland Hills, CA..818-888-0675
Little Gems Computer Consulting Svcs/Jason Levine
23114 Canzonet St, Woodland Hills, CA 818-346-1684
Media Culture/Frank Wiley
215 Fair Oaks, San Francisco, CA.....................415-824-3993
Metropolitan Skills Ctr/Spence McIntyre
2801 W Sixth St, Los Angeles, CA.....................213-386-7269
MicroAge/Santa Monica/Mike Ikegami
2020 Santa Monica Blvd, Santa Monica, CA 310-828-4911
Mind Over Macintosh/Bruce Kaplan
600 Corporate Pointe #1170, Culver City, CA 310-216-4000
Modesto Jr Coll/Art Dept/Robert Gauvreau
435 College Ave, Modesto, CA..........................209-575-6081
Moorpark Jr College/John Grzywacs-Gray
7075 Campus Rd, Moorpark, CA........................805-522-5427
Mt Hood Community College/Eric Sankey
Vis Art/26000 SE Stark, Gresham, OR 503-667-6422
N Arizona Univ/Charles Hiers
School of Art & Dsgn/Box 6020, Flagstaff, AZ.....602-523-9011
NASA Ames Research Ctr/Steve Bryson
MST045-1, Moffett Field, CA..............................415-604-5000
New Horizons Cmptr Learning Ctr
1231 E Dyer Rd #140, Santa Ana, CA................714-556-1220
New Horizons Comp Learning Ctr/Mike Brenda
1231 E Dyer Rd #140, Santa Ana, CA................714-556-1220
NM Highlands Univ/Steve Wallace
Liberal & Fine Arts, Las Vegas, NM...................505-454-3593
Orange Coast Coll/Fine Arts/Edward Baker
2701 Fairview Rd, Costa Mesa, CA 714-432-5629
Otis College of Art & Design/John T Thaxton
2401 Wilshire Blvd, Los Angeles, CA 213-251-0550
Pacific NW Coll of Art/Michael Ambrosino
Comp Lab/1219 SW Park Ave, Portland, OR......503-226-4391
Palomar College/Graphic Comm/Neil Bruington
1140 W Mission Rd, San Marcos, CA..................619-744-1150
Personal Training Systems/Devin Lynch
173 Jefferson Dr, Menlo Park, CA 800-832-2499
Phoenix College/Roman Reyes
Art Dept/1202 W Thomas Rd, Phoenix, AZ........602-264-2492
Platt College/Computer Graphics/Mark Jaress
7470 N Figueroa St, Los Angeles, CA 213-258-8050
Primary Color/Romero Flores
16841 Milliken Ave, Irvine, CA.........................714-660-7080
Quickstart Technology Inc/Mitch Argon
1500 Quail St 6th Fl, Newport Beach, CA 714-894-1448
Rancho Santiago College/Tom Hill
Art & Telecomms/17th & Bristol, Santa Ana , CA.714-564-5600
Randy and Cindy Tobin/PC Consultant
1309 Riverside Dr, Burbank, CA........................818-955-5830
San Diego Supercomputer Ctr/John Healy
PO Box 85608, San Diego, CA 619-534-5000
San Francisco Art Inst
800 Chestnut St, San Francisco, CA..................415-771-7020
San Francisco State Univ
Art Dept/1600 Holloway Ave, San Francisco, CA415-338-1111
Sharon McCormick Holography Studio
PO Box 38, White Salmon, WA.........................509-493-1334
Specialty Graphic/Lou Butera
16644 Johnson Dr , City of Industry, CA 818-855-1232
Synergy Computer Training/Ed Woodhull
225 S Lake Ave #401, Pasadena, CA.................818-356-0380
UCLA Extension/Visual Arts Div/Ruth Iskin
10995 Le Conte Ave #414, Los Angeles, CA 310-206-8876
UCSB Extension/Jean Feigenbaum
6550 Hollister Ave, Goleta, CA..........................805-893-4143
Ultimac/Gigi Michaels
3444 Cloudcroft Dr, Malibu, CA 310-459-4099
Univ of CA-Extension Grphcs Prgrms/Judy Rose
740 Front St #155, Santa Cruz, CA....................408-427-6620
Univ of Calif/San Diego/Kim MacConnel
Vis Arts 0327/9500 Gilman Dr, La Jolla, CA 619-534-5784
Univ of California/Berkeley/Prof Anne Healy
Art Dept/232 Kroeber Hall, Berkeley, CA...........510-642-6342
Univ of Colorado/Academic Media Svcs/David Stirts
Graphics/Campus Box 379, Boulder, CO...........303-492-2672

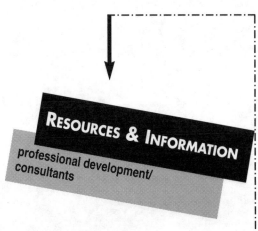
Univ of Denver/School of Art/Cal Sparks
2121 E Asbury/Art Dept, Denver, CO 303-871-2000
Univ of Utah/Comp Sci Ctr/Tom Henderson
3190 Merrill Eng Bldg/MEB, Salt Lake City, UT ...801-581-8224
University of Colorado/Jerry Kunkel
Dept of Fine Arts #CB318 , Boulder, CO.............303-492-6504
University of Oregon/Jerry Finrow
Fine&Appl Art/Arch 1206 U OR, Eugene, OR 503-346-3609
University of Washington/Doug Zuberbuhler
Dept of Arch/JO-20, Seattle, WA.......................206-543-4180
Washington State University/Rafi Samizay
School of Architecture, Pullman, WA..................509-335-5539
Weber State College/Eric Jacobson
Academic Computing, Ogden, UT 801-626-6866
West Coast University/Tony Longson
Comptr Grphcs/440 Shatto Pl, Los Angeles, CA ..213-386-7782
West Coast University/Vladimir Lerner
Computer Sci/440 Shatto Pl, Los Angeles, CA 213-487-4433
Western Montana College/Eva Mastandrea
Art Dept/710 S Atlantic, Dillon, MT 406-683-7312
YLEM
Box 749, Orinda, CA 510-482-2483

C A N A D A

Cochrane & Cassidy Design/Darrell Cassidy
2700 Simpson Rd #240, Richmond, BC..............604-276-0838
College du Vieux Montreal/Danielle Voisard
Design/255 E Ontario, Montreal, QB 514-982-3437
Durham Coll/Applied Arts & Technology/Bill Swann
Graphic Design/PO Box 385, Oshawa, ON.........416-576-0210
Emily Carr College of Art & Design/Greg Bellerby
1399 Johnson St, Vancouver, BC.......................604-687-2345
Imagicians/Bonni Evans
PO Box 1005, Manotick, ON 613-692-4306
Kwantlen College/Graphic Design/Frank Ludtke
PO Box 9030, Surrey, BC.................................604-599-2100
Mohawk Coll/Applied Arts & Techn/Stephen Dunn
TV Brdcstng/PO Box 2034, Hamilton, ON 416-575-1212
Nova Scotia Coll of Art & Design
Multimedia Ctr/5163 Duke, Halifax, NS.............902-422-7381
Ontario College of Art/Peter Mah
100 McCaul St, Toronto, ON............................416-977-5311
Sheridan Coll/Comp Grphcs Lab/Jim Sayers
1430 Trafalgar Rd, Oakville, ON.......................416-845-9430
Univ Coll of the Cariboo/Digital Art & Design/David DiFrancesco
900 College Dr/PO Box 3010, Kamloops, BC 604-374-0123
University of Calgary
Computer Science Dept, Calgary, AB.................403-220-6009

Consultants

N E W   Y O R K   C I T Y

Browne, Jeanne/Editorial Services
110 West End Ave #8F, New York, NY.............212-721-0652
Coppola, Richard
50-16 213th St, Bayside, NY...........................718-428-3636
Desktop Publishing/Melody Reed
353 W 57th St #2006, New York, NY.................212-246-6354
Gay Young Agency/Gay Young
700 Washington St #3 Upper, New York, NY.....212-691-3124
Harris Musicology/Matthew Harris
24 W 69th St #5A, New York, NY....................212-362-7087
HD/CG New York/Minako Sugiura
34-12 36th St/Landmark Front, Astoria, NY 718-361-1118

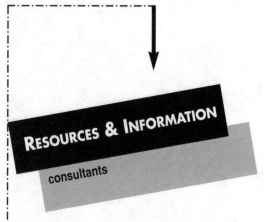

J & M Studio/Consulting/Jacqueline Skiles
236 W 27th St, New York, NY 212-675-7932
Juzwik, Amy Stahl
43 E 60th St, New York, NY............................212-759-5475
Kallish Group, The/Peter Kallish
78 W 85th St #7B, New York, NY......................212-496-6879
Kilkelly Associates/Jim Kilkelly
40 W 73rd St #2R, New York, NY......................212-873-6820
Latent Image Development Corp/David Geshwind
2 Lincoln Sq, New York, NY............................212-873-5487
Machattan/Edward E Stern
254 Park Ave S #GLB, New York, NY.................212-545-7171
Mirenburg & Co/Barry L Mirenburg
301 E 38th St, New York, NY...........................212-573-9200
Mythic Graphics/Bonnie Robin Cohen
230 Riverside Drive #8H, New York, NY.............212-866-6059
Preda, Dan
598 Ninth Ave #2A, New York, NY....................212-265-2145
Solomon, Mark
1826 Second Ave #105, New York, NY...............212-439-7264
Studio Graphic Systems/Susan M Blake
750 Eighth Ave #503, New York, NY.................212-575-4705
Synthetic Imaging Inc/Sam J Merrell
201 E 25TH St #11C, New York, NY...................212-684-6311
zpwk/Ikar Kozak
161 S First St, Brooklyn, NY............................718-963-2925

EAST

Artlandish Graphic Design/Bevi Chagnon
7417 Holly Ave, Takoma Park, MD.....................301-585-8805
Baka Inc/Peter Krakow
200 Pleasant Grove Rd, Ithaca, NY....................607-257-2070
Beehive Graphics/Carla Jaffe
PO Box 411, Greens Farms, CT 203-454-2303
Bolognese & Assocs Inc, Leon
135 Connecticut, Freeport, NY.........................516-379-3405
Cableputer Inc/Paul Michael Henry
236 North Rd/Box 2111, Greenport, NY..............516-477-1226
Centre Grafik/Arthur Bromley
900 W Valley Rd #802, Wayne, PA 215-688-2949
Citron Assocs, Lisabeth
110 Bi County Blvd #124, Farmingdale, NY 516-420-4200
Cognetics Corp
51 Everett Dr #103B/POB 386,
Princeton Junction, NJ..................................609-799-5005
Davis Inc/L Mills Davis
2704 Ontario Rd NW, Washington, DC.............202-667-6400
DeltaTech Corp/Robert Turner
8700 Georgia Ave, Silver Spring, MD 301-588-2200
Digital Constructs/Robert Moran
759 N Park Ave, W Redding, CT 203-452-1116
DMR Group/Bruce Beck
57 River St, Wellesley Hills, MA 617-487-9000
Egeland Wood & Zuber/Elizabeth Wood
176 Freemans Bridge Rd, Scotia, NY.................518-374-3131
Forsight Inc/Ronald Gregory
POB 0427/1700 Rockvll Pike#150,
Rockville, MD 301-816-4900
Gambino, Donald
12 Carolyn Way, Purdys, NY 914-277-1837
Graphic Arts Service Bureau/Rowland Brandwein
159 Coburn Woods, Nashua, NH......................603-880-4844
IBM TJ Watson Research Ctr/Clifford A (Dr) Pickover
Yorktown Heights, NY..................................914-945-3000
ImageNet/Craig Shrader
23 Penwood Rd, Basking Ridge, NJ...................908-647-0353

ImageTech Communications Inc/Jeff Seideman
1 Kendall Sq #2200, Cambridge, MA................617-621-7111
Kottwitz & Assoc/Randal L Kottwitz
27 Peaslee Rd, Merrimack, NH.........................603-889-4808
Mac in Design/Jeff Loechner
127 Post Rd E, Westport, CT............................203-221-1545
MINDesign/Mihai Nadin
35 Butts Rock Rd, Little Compton, RI.................401-635-1675
New Edge Technologies/Nick Michael
PO Box 3016, Peterborough, NH......................603-547-2263
PM Associates/Park Gerald
71 Hundreds Rd, Wellesley, MA 617-235-7975
REF Associates/Dr Robert Fidoten
118 Greyfriar Dr, Pittsburgh, PA........................412-963-8785
Stellarvisions/Stella Gassaway
4041 Ridge Ave/Bldg 17 #105,
Philadelphia, PA..215-848-9272
Steuer, Sharon
205 Valley Rd, Bethany, CT.............................203-393-3981
Videos Multimedia Inc/Bill Mutsler
South River Rd/Rte 130 Bldg C, Cranbury, NJ....609-395-1120
Villanova, Joseph
2810 Altantic Ave, Stottville, NY......................518-828-0141
Voelkl, Michael
1994 Five Mile Line Rd, Penfield, NY................716-381-4205
Wurman, Richard Saul
180 Narragansett Ave, Newport, RI...................401-848-2299

SOUTH

AppleTree Technologies Inc/Larry Hall
3020 Mercer Univ Dr, Atlanta, GA....................404-457-4500
Caluger & Assocs/J Wayne Caluger
237 French Landing, Nashville, TN....................615-255-2792
CompuWorks Inc/Philip C Ellis
9130 Wiles Rd #124, Coral Springs, FL 305-345-5450
Gestalt Systems Inc./Harry Lee
610 Herndon Pkwy #900, Herndon, VA 703-471-6842
Graphics Group/Vivian Darby
2820 Taylor St, Dallas, TX 214-749-2222
Impact Ideas Inc/Geoff Amthor
4280 Twin Rivers Dr, Gainsville, GA..................404-967-9700
MacInstitute/Mark Miller
3200 N 29th Ave, Hollywood, FL......................800-328-4349
Micron/Green/Millard Pate
1240 NW 21st Ave, Gainesville, FL...................904-376-1529
Parker, Nick
Intergraph Corp MS CR2-901 , Huntsville, AL 205-730-6208
State of the ART Dept/Bobby O'Rourke
631 E 62nd St, Hialeah, FL.............................305-364-0100
Thomas Mrktng Comms, Steve
409 East Blvd, Charlotte, NC...........................704-332-4624
Tolan, Stephanie
14232 N Dallas Parkway #1404, Dallas, TX.......214-392-2774

MIDWEST

17th St Studios/Gary Olsen
PO Box 855, Dubuque, IA...............................319-589-5017
Andersen Consulting /Liz Landon
33 W Monroe St, Chicago, IL...........................312-507-4205
BGN Unltd/Bryan Gerard Nelson
2420 31st Ave S, Minneapolis, MN 612-724-3430
Butler Graphics Inc/Jack Butler
294 Town Center, Troy, MI.............................313-528-2808
Color Image/Jim Trotter
12342 Conway Rd, St Louis, MO......................314-878-0777
Connecting Images/Terry Hickey
2469 University Ave N, St Paul, MN..................612-644-1977
Data Service Company/Robert Chudek
3110 Evelyn St, St Paul, MN...........................612-636-9469
Digital Pictures Inc/Tom Grotting
1212 N Second S, Minneapolis, MN..................612-371-4515
Fontastik Inc/Cliff Ping
911 Main St #1717, Kansas City, MO 816-474-4366
Hughes Design/Edward Hughes
2122 Orrington Ave, Evanston, IL.....................708-869-2330
JD Media Consultants/Dan Brennan
1309 E Dawes Ave, Wheaton, IL......................708-668-0559
Kadison-Shapiro, Sari
1432 Elmwood Ave, Evanston, IL.....................708-866-8080
KV Graphics/Bob Cavey
710 Canterbury Cir, Chanhassen, MN................612-949-2902
N2 Communications Inc/Randall Powell
817 W 2nd St, Wichita, KS..............................316-265-1630

● **SANDBOX DIGITAL PLAYGROUND**/
Sandy Ostroff **(P 84)**
203 N Wabash #1602, Chicago, IL 312-372-1170
The Roland Co/Will Sullivan
420 Summit Ave 3rd Fl, St Paul, MN 612-222-8716
Threshold Software Inc/Richard Wilson
29 S Park Ave, Hinsdale, IL..............................708-887-0480
Wace Resource Center/Rocco Matera
200 E Ohio St, Chicago , IL..............................312-915-0695
Xenas Communications/Greg Peck
354 Gest St, Cincinnati, OH.............................513-621-2729

WEST

Abaci Gallery of Computer Art/Daria Barclay
312 NW Tenth Ave, Portland, OR......................503-228-8642
Absolute Advantage Inc/Bob Gossom
10444 Canoga Ave #12, Chatsworth, CA.........818-718-2027
Alpha-Lex
CA 310-453-9000
Animation Farm/Don Smith
21850 Poppy Lane, Nuevo, CA 909-928-1577
Axcess
PO Box 5016, Woodland Hills, CA....................818-880-5532
Biedny, David
PO Box 151498, San Rafael, CA.......................415-721-0638
Brainiac/Jeffrey Schneider
290 Stuyvesant Dr, San Anselmo, CA................415-453-7853
C Design/Pat Coleman
915 Oak Ln #4, Menlo Park, CA.......................415-321-2094
Captured Images/Bill Grimm
31316 Via Colenas #117, Westlake Village, CA.818-707-9491
Chalsie Systems & Design/Barry Chalsie
1641 E Maple Ave #E, El Segundo, CA..............310-322-6433
Chaparral Software/Russell Kohn
9201 W Olympic Blvd #201, Beverly Hills, CA ...310-273-4904
Charles H Williams Consulting/Charles H Williams
977 Ashbridge Ln, Harbor City, CA...................310-539-9439
Christensen, Steve
615 Coeur D'Alene Ave, Venice, CA 213-991-1889
Closed Circuit Products/Jon Williams
6395 Gunpark Dr #PHJ, Boulder, CO................800-999-3130
Commotion/Bob Robbins
3131 Cahuenga Blvd W, Los Angeles, CA..........213-882-6637
Compas Technologies
2510-G Las Posas Rd #424, Camarillo, CA.........805-484-9205
Computed Designs
PO Box 802, Temple City, CA...........................213-263-5590
Computer Education Intl/Ryan Granard
230 N Maryland #300, Glendale, CA 818-242-0828
Computer Graphics Group/Joel Spaunberg
23181 Via Celeste, Trabuco Canyon, CA 714-858-8978
Computer Graphics Group/William Brown
1325 W Imola #121, Napa, CA........................707-257-7975
Computer Graphics Group/Jim Boren
1450 Manhattan Beach Blvd #B,
Manhattan Beach, CA.................................310-372-3228
Computer Intl
10517 W Pico Blvd, Los Angeles, CA...............310-474-6409
CompuWrite/Dona Meilach
2018 Salient Way, Carlsbad, CA.......................619-436-4395
Contracted Computer Training Inc/Azmina Kanji
4640 Admiralty Way #310,
Marina Del Ray, CA...................................310-827-0303
Coyne Co/Richard Coyne
25108 Marguerite Pkwy #B-200,
Mission Viejo, CA......................................714-855-4689
Critical Connection/Betty A Toole
PO Box 452, Sausalito, CA..............................415-388-3549
Crucible/Brian Raney
1717 N Seabright Ave #1, Santa Cruz, CA 408-423-4600
Design Source/Doug Hines
1121 Broadway #G-1, Boulder, CO................303-447-8604
Desktop Graphics & Design/Marion Wilson
17921 #H Sky Park Cir, Irvine, CA....................714-261-1881
Diehl, Michael
3251 Primera Ave, Los Angeles, CA.................213-851-3111
Digital Resources/Mitch Krayton
14545 Friar St #106, Van Nuys, CA.................818-901-4848
Dreamers Guild, The/Robert McNally
98100 Owensmouth Ave, Chatsworth, CA.........818-349-7339
Duffy, Fred
10535 Wilshire Blvd #405, Los Angeles, CA 310-475-8400
Expressive Images Ltd/Jim Plowden
8211 Dumbarton Ave, Los Angeles, CA 310-645-2558

Fenster, Diane
140 Berendos Ave, Pacifica, CA 415-355-5007
Frank Antonides Design/Frank Antonides
1395 Linda Vista, Pasadena, CA..................818-795-3310
Future Image Rep,The /Alexis J Gerard
1020 Parrott Dr, Burlingame, CA......................415-579-0493
Gader, Bertram
2760 Hollyridge Dr, Los Angeles, CA 213-856-9201
Gee Whiz Enterprises/Jackie Gee
PO Box 2069, Artesia, CA 310-924-4847
Genius Inc/Byron Wagner
4179 Sunnyslope Ave, Sherman Oaks, CA 818-905-8866
Goodman Consulting Systems/Philippe Goodman
1615 Hilts Ave #3, Los Angeles, CA..................310-470-2998
Goodman, Barbara
PO Box 585, Tustin, CA 714-997-2626
Gunnar Swanson Design Office
739 Indiana Ave, Venice, CA 310-399-5191
IBIS Graphics/Ralph Bentley
3443 SE Grant Ct, Portland, OR........................503-231-1781
IDesign/David Robinson
750 N Diamond Bar Blvd #208,
Diamond Bar, CA 909-396-1323
Info Systems Computer Ctr
19950 Mariner Ave, Torrance, CA....................310-214-4200
Integrated Solutions/Eliot Kaplan
712 Hillcrest Dr, Topanga, CA...........................310-455-3950
Intercommunication
620 Newport Ctr Dr, Newport Beach, CA...........714-644-7520
Jeff Barnett Consulting
3734 Monon St Unit 2, Los Angeles, CA.............213-664-9169
KHL Consulting/Kathy Hecht-Lopez
3871 Hatton St, San Diego, CA 619-576-4140
KnoWare/Ron Lawrence
11288 Ventura Blvd #702, Studio City, CA.........818-769-7589
Level 6 Computing/Kevin Garrett
22647 Ventura Blvd #201, Woodland Hills, CA..818-888-0675
Little Gems Computer Consulting Svcs/Jason Levine
23114 Canzonet St, Woodland Hills, CA 818-346-1684
MacArtist/Fritz Richard
119 E Alton Ave #D, Santa Ana, CA 714-262-1546
Macintosh Systems Consulting/Jess Georgevich
80 Crest Dr, Manhattan Beach, CA 310-372-7947
MicroAge/Santa Monica/Mike Ikegami
2020 Santa Monica Blvd, Santa Monica, CA 310-828-4911
Mind Over Macintosh/Bruce Kaplan
600 Corporate Pointe #1170, Culver City, CA 310-216-4000
Neotech Interactive/Greg Hill
2663 Centinela Ave #504, Santa Monica, CA 310-392-2711
On Target Marketing/Ken Drake
5699 Kanan Rd #328, Agoura Hills, CA 818-707-9453
Pacific Interface Inc/Laurin Herr
5703 College Ave #4, Oakland, CA 510-652-5300
PANALOG Computer Guides/Dan Nash
4450 Lakeside Dr #200, Burbank, CA................818-559-4277
Photovault
1045 17th St, San Francisco, CA 415-552-9682
Pre-Press Associates/Doug Peltonen
1722 32nd Ave, Seattle, WA 619-259-8230
Primary Color/Romero Flores
16841 Milliken Ave, Irvine, CA.........................714-660-7080
Prince, Patric
160 W Jaxine Dr, Altadena, CA 818-797-7674
QuillTech/David Gibbons
16458 Bolsa Chica Rd #410, Huntington
Beach, CA 714-761-9040
Randy and Cindy Tobin/PC Consultant
1309 Riverside Dr, Burbank, CA 818-955-5830
Software Assist US/Richard Wilson
76 Perine Pl, San Francisco, CA 415-921-3417
Specialty Graphic/Lou Butera
16644 Johnson Dr , City of Industry, CA 818-855-1232
Stieglitz, Carol
Santa Monica, CA 310-395-5978
SuperVision/Alan Jerram
PO Box 4867, Chatsworth, CA 805-520-3956
Swaim, Howard
605 S Adams St, Glendale, CA.........................818-500-0337
Synergy Computer Training/Ed Woodhull
225 S Lake Ave #401, Pasadena, CA 818-356-0380
Tolman, Brian
11620 Wilshire Blvd #370, Los Angeles, CA 310-312-1080
Ultimac/Gigi Michaels
3444 Cloudcroft Dr, Malibu, CA 310-459-4099

Uni-Mesa Co/Gary Koch
7 Swift Ct, Newport Beach, CA.........................714-646-1400
Varner, Vicky Jo
950 N Kings Rd #225, W Hollywood, CA..........213-656-3433

RESOURCES & INFORMATION

consultants

A

07-TV Commercial Production Co, 177
10th St Video Productions, 171, 176, 182
1185 Design, 148
11th Annual Showbiz Expo West, 136, 208
2 Dimensions Inc, 151
21st Century Media, 127
22 Product, 182
28th Adv Techn&Electr Imag, 160, 164, 171, 176, 182, 186
3D Magic/FTI, 182
3M Image Reprographic Systems, 198
4 Designerly Types, 151
4-Front Video Design Inc, 141, 162
4Media Company, 171, 176, 187
4th Wave Inc, 159, 163, 166
525 Post Production, 176, 187
615 Music Productions, 184
7th Level, 194
911 Gallery, 178

A

A1 Printers Inc, 191
A2Z Creative, 148
Aartvark Photography, 128
Aartvark Studios, 145
AB Communications, 158, 166, 168
AB Graphics, 133, 143
Abaci Gallery of Computer Art, 127, 148, 208
Abbey Photographers, 123, 143
Abbott, Waring, 122
ABC News, 194
Abekas, 198
Abolafia, Oscar, 122
ABRA, 130
Abracadabra Slide Production, 192
Abraham Animation, 160, 163
Abrams Design, Kym, 147
Abramson, Elaine, 145
ABS Productions Ltd, 177
Absolute Advantage Inc, 153
Absolute Music, 185
Abt, Patricia A, 123
ACA Coll of Design, 122
AccuData Inc, 154, 156
Ace Group, The, 188
Acer America Corp, 154
Acius Inc, 156
ACM/SIGGRAPH, 1995, 163, 200
Acme Holography, 128
Acme Recording Studios, 172, 184
Acmetek, 176, 182, 187
Acosta, Ralph, 137
Act III Publishing, 152, 158, 172
Action Nicholson Color Inc, 191
Action Video, 171, 176, 182
Active 8, 158, 166
Active Video Inc, 172
Activision, 194
Acuris Inc, 153
Ad-Visual Communications, 160, 163, 167
Adage Graphics, 192
ADAM Software, 167

Adaptec, 198
Adco Productions, 173
Addis, Kory, 123
Addison Wesley, 194
ADIC, 196
Admark Inc, 147, 153, 174
Admaster, 158, 189
Adobe, 127, 156, 194, 200, 201
Adobe Systems, 156, 194
Adpro Inc, 160, 170, 181
Adron Graphics Inc, 191
Adv Computer Imaging, 191
Advance Concepts Inc, 153, 173
Advanced Audio Visual, 174
Advanced Concepts, 137
Advanced Digital Imaging, 182
Advanced Gravis Computer Technologies Ltd, 199
Advanced IDAS, 160
Advanced Laser Graphics, 190
Advanced Micro Devices Inc, 154, 156
Advanced Retrieval Systems, 198
Advanced Storage Concepts Inc, 170, 198
Advanced Systems/IDG Communications, 154
Advanced Video Communications, 160, 170, 174
Advanced Video Services, 172, 179
Advent AudioVisual Design, 153
Advertising Club of Cincinnati, 143, 158, 172
Adwar Video, 172
Aerial Image Video Services, 168, 170
Aero Graphics, 188
Aerodrome Picture Inc, 148
Aesthetic Engineering, 184
AFCG Inc, 141, 162
After Science, 160, 164
Agelopas, Mike, 123
Agency Images, 130, 189
Agfa Compugraphic, 194
Agfa, 155
AGFA/Div of Miles, 196, 198
Agfa Division, 194
AGS, 160, 174, 181, 191, 192
AGS & R Communications, 191
AGS&R Communications, 160, 174, 181
Aguilar, Bob, 137
Ahhhhh Graphics, 137, 148
Ahrens Photo, Bob, 136, 147, 156
Aich, Clara, 122
AIE Studios, 173
AIGA/Minnesota, 185, 204
Aimtech (Advanced Interactive, 200
Ainsworth Group, The, 194
Akavia, Eden, 122, 162
Aker, 125
Akins Design, Charles, 145
Akis, Emanuel, 123
Aktulun Design, 145, 159
Alacrity Systems Inc, 155
Aladdin Systems Inc, 156
Alan C Ross Productions, 160, 166, 167
Alben & Faris, 148, 160, 168
Albers, Louis, 131
Alco Computer Graphic, 133, 143

Alcorn, Kathryn, 143
Alde Publishing, 194
Aldridge Reps Inc, 151
Aldus Corp, 148, 153, 156, 203
Aldus Corp/Aldus Magazine, 147
Alexander, Martha, 145
Alexander, 151
AlexanderDesign Inc, 160, 163, 167
Alias Research, 135
Alias Research Inc, 141
All American TV, 179
Allen, Dave, 136
Allen, Don, 125
Allen, Jim, 128
Allen Communication, 160, 166, 167
Allen Design, Mark, 137
Alley Newspaper, Steve, 191
Alliance Peripheral Systems, 154, 196
Allied Arts of Seattle Inc, 170, 175
Allied Film and Video, 173, 175, 181, 185
Allied Graphic Arts, 141
Allison, Linda, 137
Alper Studio, A J, 131, 141
Alpha Channel, 191
Alpha Video & Audio, 185
Alpha-Lex, 208
Alphagraphics Printshops of the Future, 192
Alsberg, Peter, 133
AlSoft Inc, 156
Altered Image, 158, 166
Altered+Images+Inc, 130
Alternative Design, 147
Alternative Type & Graphics, 191
Altman, Ben, 126
Altronix Systems, 196, 197, 199
Altsys Corp, 156
Alvarado, Leon, 135
Alvarez Group, 148
Alvuso Corp, 190
Amazing Media, 160, 164, 166
Amber Productions Inc, 147
Amendola, Steve, 131, 162
Amenta Jr, Joseph L, 126
Americ Disc Inc, 171
American Acad of Art & Grphic Comms, 151
American Assn for the Advancement of Science, 203
American Assn of Adv Agencies, 128
American Booksellers Association Inc, 160, 175, 185
American Center for Design, 188-192
American Color, 189-191
American Computer Imaging, 153, 200, 201
American Express, 194
American Film Technology, 171, 182
American Holographic Inc, 170
American Holographic Society, 203
American Institute of Graphic Arts, 191
American Media Productions, 173, 180
American Production Services, 171, 176, 182
American Research Corp, 154, 156
American River College, 133
American Translation Services, 192
Amgraf Inc, 156
Amicus Communications, 153
Amigo Business Computer, 158, 170, 172
Amitech Corp, 188
Amphion Music, 184
AMS Productions Inc, 173, 180
Amtronics Inc, 131, 141
Anagraphics Inc, 189
Analog Devices Inc, 197, 198
Analog Video, 163, 175, 181
Andersen, Rolf, 131, 141
Andersen Consulting , 168
Andersen-Bruce, Sally, 123
Anderson, Darrel, 137
Anderson, Lisa, 148
Anderson, Sara, 137
Anderson, Terry, 137
Anderson Audio Light, 185
Anderson College, 147
Anderson Perlstein, Ltd, 130
Anderson Video, 171, 176, 187
Andraleria, 137
Andre's Photo Lab, 130, 192
Andrea Mistretta, 134, 178
Andresen Typographics, 192
Andrew Faulkner Design, 137
Andrew Rodney Photography, 127
Andrews, Jackie, 153, 197, 198

Anello Technical Services, 178
Angel Studios, 176, 182
Angelsea Productions, 172
Ani-Majic Productions Inc, 173, 180
Anigraf/x, 137
Animated Arts, 179
Animated Design, 167-169
Animated Design, 179
Animatics Inc, 183
Animation Farm, 182, 208
Animation, 177
Animation Station Ltd, 175, 181
Ankeny, Martha Langley, 165, 168
Ann-imation Video, 183
Anne Phillips Productions, 184
Annenberg Ctr at Eisenhower/TV Svcs, 160, 176, 187
Anning & Associates, 153
Another Girl Rep, 152
Antelope Valley College, 147
Antigravity Workshop, 138, 182
Antler & Baldwin Design Group, 141
Anzographics Computer Typographers, 191
Apartment 3D, 136, 147
Apex Systems Inc, 169
Apocalyptik Sparkz Design, 141
Appalshop Inc, 178
Apple Computer Inc, 196-198
Apple Market Center/NY, 165, 168, 205
AppleTree Technologies Inc, 145, 206, 208
Application Techniques Inc, 155
Applications, 191
Applied Graphic Technologies, 206
Applied Graphics, 131, 141
Applied Graphics Technologies, 141
Applied Imagination Inc, 158, 165, 166
Applied Integration, 160, 167, 169
Applied Memory Technology Inc, 198
Applied Optical Media Corp, 152, 158, 170, 194
Applied Technical Video Inc, 176, 182, 187
Applied Video Communications, 173, 180
April Greiman Inc, 148, 176
Aprose, 192
Aquidneck Graphics Inc, 190
Aquila Technologies Group Inc, 192
ArborText, 156
ARCCA, 177
Archambault Group, The, 143
Archetype Inc, 155
Arcimage, 180
Arday Illus, 125
Arenella Inc, 129
Arens & Associates, 127
Ario Inc, 151
Arion, Katherine, 138
Aristosoft, 156, 187
Arizona Four Color Corp, 130, 189
Arizona State Univ/West, 172
Arkay Technologies, 158, 198
Arkwright, 158
Armock, Amy, 187
Armour Productions, 176, 182, 187
Arneson, Peter, 160
Arni Katz, 125, 129
Arnold Black Productions, 184
Aronoff, Susan, 131
Art & Design, 152
Art & Visual Communications, 165
Art (to the Nth) Lab, 191
Art Bunch, The, 130
Art Ctr College of Design, 145
Art Direction, 192
Art Directors and Artists Club of Sacramento, 204
Art Directors Club of Denver, 171, 176, 182
Art for Advertising Inc, 129, 130
Art for Medicine, 138, 148
Art House Graphic Design, 145
Art Images, 135
Art in Media Productions, 153, 173, 180
Art in Progress, 130
Art Inst of Atlanta, 184, 204, 205
Art Institute of S Calif, 196
Art on Fire, 133, 165, 168
Art Resources & Technologies, 131, 141
Art Science & Tech Inst, 128
Art Source Design, 145
Art-Pro Graphics, 141
Artattack Studios, 143
Artbeats, 156, 201
Artco, 151, 190
Artco Graphics Inc, 190

ArtCom, 172, 178, 179
Artform Communications, 153
Arthur Andersen & Co, 169
Artichoke Arts, 131, 152, 158
Artichoke Productions, 176, 182, 187
ARTiculate Imaging, 127, 130
Artificial Reality, 163, 198, 199
Artisan Professional Freelance, 152
Artist Graphics, 197
Artistic Perceptions, 131
Artistic Typesetting & Graphic, 192
Artlab, 130, 138, 166
Artlandish Graphic Design, 143, 208
Artmarx, 138, 148
Artray Ltd, 177, 183
Arts in Action, 148, 160, 166
Artspace, 194
Artype Publications, 153
As Soon As Possible Inc, 191
ASA Darkrooms, 127, 138, 148
Asal, Ahamad, 148
ASC Systems, 167, 197, 198
ASCAP, 185, 204
Asche & Spencer Music, 185
Ascroft, Robert, 133, 143
ASDG Inc, 156, 201
ASG Inc, 188
Ashe/Bowie Productions, 170, 173
Ashford, Janet, 138, 148
Ashley, Bruce, 127
Ashton, Larry, 138
ASI Image Studios, 154, 191
Asmus, David, 136
ASP Computer Products Inc, 154
Aspex Inc, 197
Assn for Dev of Computer, 204
Assn of Professional Design Firms, 160, 181, 202
Associate Producers, 175, 181, 185
Associated Graphic Services Ltd, 190
Associated Images, 131, 141, 152
Associated Media Productions, 152
Associated Production Music, 187
Associated Publications & Graphics, 192
Associates Creative Inc, 160, 163, 167
Association for Multi-Image, 196
Astavision Post Prod, 148, 176, 182
Astro Audio-Visual, 173
ASV Corp, 159, 163, 180
Asymmetrix Corp, 166
AT&T Graphics Software Labs, 194
Atari Corp, 197
Atechno-Marketing Inc, 167
Atelier Graphics, 138, 148
Atelier Infographe, 177
Atex Commercial & Color, 154, 155
Atlantic Group, 190
Atlantic Television System (CHCH-TV), 177
Atlantic Video, 170, 172, 179
Atomic Art, 136
Atomic Media Group, 130, 182, 187
Atomique Film International, 141, 178
Atonelli Inst of Art & Photog, 184
Audio Engineering Society, 172, 184
Audio Visual Systems, 153
Audio-Visual Associates, 160, 175, 185
Auras Design, 143, 190
Aust, Hale, 128
Auto-trol Technology Corp, 156
AutoDesk Inc/Multimedia Div, 199, 200
AutoDesk Retail Products, 190
Autographix, 152
Autologic Inc, 154, 203
Automated Broadcast Consultants Inc, 173, 180
AV COM Productions, 191
AV Design, 158, 172, 179
AV Workshop Inc, 171
Avalanche Development Co, 156
Avanti Computer Systems, 157
Avenue Edit, 175, 181, 185
Avery Dennison Corp, 154, 156
Avery Illustration, Design & Adv, 138
AVI Visual Productions, 171, 189
AVID Inc, 173
Avid Productions, 176, 182, 187
Avid Productions Inc, 123
Avid Technology Inc, 197, 201
Avila, Winona, 138
AVP Inc, 175
AVPE Systems, 154
AVS Inc, 153, 160, 175
Avsense Productions, 185

AVT, 172, 179
Avtex Research Corp, 160, 166, 167
Award Publications, 191
Axcess, 208
Axelrod, Dale, 138
Axiom Design Systems, 189
Axiom Inc, 125
Axion, Pierce, 141, 168
Axmann, Doug, 133
AXS/Optical Technology Resource, 126, 163, 196
Aztek, 154, 156

B
B Graphic, 138
B.N. Architect, 148, 169
Baby Blues, 138
Bachmann Photography, 125, 153
Back Stage, 175
Bagel Design Studio, Richard, 148
Bahry, Sharon, 138
Bailey Inc, Robert, 148
Bailey Productions, 160, 164, 167
Baird, Darryl, 125
Baird Productions, 179, 205
BAIRDesign, 143
Bajuk, Mark, 167
Baka Inc, 154, 155, 208
Baker, Brenna, Madison, 147, 153
Baker, Don, 138
Baker, Kolea, 138, 152
Baker, Michael, 151
Baker Design Assocs, 148
Baker MediaSource, 158, 163, 168
Baker Photography, 125
Bakhtiar Photo, Sherry, 123
Bakst, Edward, 141, 178
Baldino Design, Patt, 141
Baldridge Studios Inc, 153, 173, 180
Baldwin, Doug, 127
Baldwin, Scott, 133
Balfour Walker Photoggraphy, 127
Baltimore Color Plate, 188
Balzer, 189
Bamundo, David, 131
Banchero, J, 145, 178
Bancroft, Monty, 123
Bang Co, The, 151
Bangert, Colette & Charles, 147
Banghaus Music Productions, 185
Banjo Video & Film Productions, 173, 180, 184
Bantam Books Inc, 194
Baquero, George, 129, 133
Barba, Jac, 122
Barbeau, Dan, 133
Barbero, Teri, 122
Barco Inc, 154
Barfuss Creative Services, 147
Barking Art Productions, 182
Barlow Productions Inc, 126, 175, 185
Barnard, Doug, 138
Barnum Design, 145
Barr, Kevin J, 133
Barrett, Anne S, 131, 162
Barry, Judith, 171
Barta, Les, 138
Bartalos, Michael, 138
Bartels Associates, Ceci, 152
Barth, Henrietta, 147
Bartholomew, Sandra Steen, 135
Bartholomew Enterprises, 126
Bartlett, Michael, 138, 164
Bartley Collection Ltd, 133
Base Arts, 127
Bass Litho Color, 188
Batelman, Kenneth, 131
Bateman, John H, 125
Bates, Betty, 138
Bates, Carolyn, 123
Bates, Karla M, 138
Batsry, Irit, 141
Baudville Inc, 156
Bauman Communications Inc, 153, 175, 185
Bautista, David R, 133
Baxley, Diane, 143
Bay Area Video Coalition, 161, 176, 207
Baylor University/Fine Arts, 160, 181, 185
Bazooka Graphics/Div Swell Pictures, 160, 181, 185
BBP Graphic Design Inc, 141

BC Printing, 191
BCA/Desktop Designs, 156
BCD Ink Ltd, 141
BDG Production Studio, 126, 130, 136
Beach, Lou, 138
Beachwood Productions, 173, 180, 184
Beacon Software, 160, 163, 170
Beals, Steven K, 127
Bear River Associates Inc, 169
Beatty TeleVisual Productions, 170, 175
Beauchamp, Jaime, 131
Bechtold, John, 122
Bechtold Studios, 161, 182
Becker, Lisa, 145
Becker, Peter, 158
Becker, Scott, 136
Beckerman, Carol, 138, 148
Beckman, Melissa, 129, 131
Beckman Instruments Inc, 179, 182, 197
Bedford Photo Graphic, 123
Bedrock Design, 158, 166, 168
Bee, Paula, 148
Bee Show!, Johnee, 138
Beebe, Rod, 122
Beechler, Greg, 122
Beehive Graphics, 141
Before & After, 138, 148
Bego, Dolores, 131
Behr, Sheri Lynn, 122, 131
Beidler, Barbara, 122
Belding, Pam, 136, 178
Bellamy, Gordon, 131
Bellamy, Mike, 133, 143
Belman, Vickie, 131, 141
Benante, Catherine, 133
Benchmark Display Graphics, 153
Bender + Bender, 126
Bender, Brenda, 151
Bender, Jon, 138, 161, 164, 168
Bengal Graphics Inc, 188
Benjamin Nowak, 129, 165, 167
Berendsen & Associates Inc, 136
Bergdorf Productions, 153
Bergeron, Joe, 133
Berglund, Lawrence, 129
Bergman Graphics Limited, 194
Berinstein, Martin, 123
Berkhout, Rudie, 128
Berkley, Miriam, 122
Berkwit Studio, 122
Berle Cherney, 122
Berlin, Jeff, 138, 148, 161
Berlin Productions Inc, 133
BERMAC Communications Inc, 159, 165, 166
Berman, Bruce, 125
Bernhard Studio, Ivy, 151
Bernsau, W Marc, 123
Bernstein & Andriulli, 151
Bernstein, Linda A, 131, 179
Berry & Homer Inc, 133, 190
Berry, Rick, 133, 143
BES Teleproductions, 173, 180
Bessen Tully & Lee, 141

Besser, Doug, 126
Best Digital, 130, 138
Bestinfo Inc, 155
BetaCorp Technologies Inc, 196, 198, 200, 201
Betelgeuse Productions, 170, 171
Betz, Charles, 122
BGL Technology Corp/LaserLeaders, 198
BGM Color Labs, 130
BGN Unltd, 147, 178, 208
Bidco Manufacturing Corp, 155
Biedny, David, 138, 208
Bielenberg Design, John, 148
Big City Productions, 185
Big Deal Graphics, 138
Big Fat TV, 171
Big Hand, 135
Big Nasty Redhead Inc, 136
Big Noise Software, 138
Big Time Productions, 184
Big Zig Video, 171, 176, 182
Biggs, Ken, 127
Bigtwin, 131
Bilby, Glade, 125
Bill & Virgies Publications, 145
Bill Feigenbaum Designs, 171, 200
Billian, Cathy, 141
Bimstein, Phillip Kent, 187
Biner Design, 147
Bio Design, 138
Bioscan Inc, 156
Birenbaum, Molly, 151
Birkey, Randal, 136
Birmingham, Barbara, 131, 178
Birmy Graphics Corp, 154
Bklyn Coll/Ctr for Computer Music, 126
Black & White Dog Studio, 141
Black Box Corp, 198
Black Creative Professionals, 141
Black Lamb Studios, 141
Black Lamb Studios Ltd, 122
Black Point Group, 138, 148
Blackbourn Inc, 188
Blackhawk Color Corp, 129, 188
Blackhawk Data Corp, 156
Blackman Studios Inc,, 122
Blae, Ken, 133
Blair Inc, 153, 159
Blair Photography, Richard, 127, 148
Blake + Barancik Design, 143
Blake, Juliet, 135
Blake Design, 141
Blakeslee Group, 123, 143, 152
Blanchard-Healy Video, 172, 196
Blank, Pat, 128
Blanks Color Imaging, 188
Blate, Samuel R, 123
Blaustein, Alan, 127
Blauweiss, Stephen, 131, 141, 158
Blavatt, Kathleen, 138, 148
Blazing Graphics Inc, 190
Blazing Pages, 192
Blell, Dianne, 122
Blessen, Karen, 135

Blik Design, Tyler, 148
Blink Inc, 122
Bliss, Anna Campbell, 138
Blue Earth Picture Inc, 175
Blue Jacket Studios, 151
Blue Rose Studio, 153
Blue Sky Productions, 175, 178, 181
Blue Sky Research, 156
Blueberry Software, 156
Blumberg Communications Inc, 185
Blume, Neil, 179, 182
Bluming, Joel, 131
Blumrich, Christoph, 133
Bob Clarke Illustrations, 133
Bob Gerardi Music, 184
Bob Olhsson Digital Audio, 167, 187, 201
Bober, Marlene, 143
Boca Laser Inc, 191
Bodily, Michael, 138
Bodin, Fredrik, 123
Bodkin Design Group, 143
Boelts Brothers Design, 148
Boger, Claire, 135, 165
BOGH AV Productions, 152, 172
Boileau, Lowell, 136
Bokland Custom Visuals, 190
Bolduc, Damon, 123
Bolger Publications Creative Printing, 191
Bollinger, Rebeca, 138
Bolognese & Assocs Inc, 143, 208
Bonauro, Tom, 138, 148
Bond, Kendra, 123
Book 1, 143, 152
Bookmakers Ltd, 138, 152, 182
Boone, Tim, 133
Borderud, Mark, 122
Bordnick, Barbara, 122
Borges, Ruth, 148
Boris, Daniel, 143
Borkoski Photography, Matthew, 123
Borland International, 156
Bornstein, Myer, 123
Bornstein, Stephen, 135, 145
Borruso, John, 138
Boss Film Studios, 148, 182
Boston Business Graphics Inc, 190
Boston Computer Society, 168
Boston Photographers, 123
Boston Univ/Comp Graphics Lab, 169
Bottiglieri, Dessolina, 141, 162
Boucek, Al, 129
Boudreau, Janice, 123
Bourbaki Inc, 179
Bourke, Dennis, 122
Bovey & Assocs, 160, 163, 200
Bow-Haus Inc, 130, 189, 192
Bowling Green State Univ, 197
Boyd, Cathy, 141, 152
Boyington Film Productions, 176, 182
Brad & Maggie Palm, 130
Brad R Callahan & Co, 185
Bradfield, Rod, 136
Bradford, Peter, 141
Bradford-Cout & Jansen, 147
Bradley, Shane, 177
Bradley Johnson Productions, 185
Bradshaw, Stephen, 138
Brady Design Consultants, John, 143
Brainiac, 141, 158

Brand Design, 143
Bransby Productions, 159, 163, 173
Brauer, Bruce Erik, 131
Braun, Khyal, 133
Braun Photography, 126
Brave Young Artists, The, 167, 169, 204
Bravin Learning Technologies, 164, 166, 167
Breene Kerr Production, 176
Breiger, Elaine, 131
Bremmer, Mark, 138
Breskin, Michael, 122
Breth, Jill Marie, 143
Brett, Clifton, 122
Brewer, Benita, 136
Brianstorm Unlimited, Mark, 131, 141
Brice, Jeff, 138
Bridge Communication Group, 147, 160
Bridgestone Multimedia, 156, 200
Bridgette Inc, 154
Bright & Associates Inc, 148
Bright Star Technology Inc, 189
Brilliant Image Inc, 152, 189
Brilliant Media, 161, 164, 168
Briones Engineering, 129
Britt, Ben, 125
Britt, Ron, 178
Britton, Arlen, 136, 147
Broad Street Productions, 152, 170, 171
Broadcast Productions Media Image, 177
Broadcast Resource Group, 173
Broadcast Video Inc, 173
Broadway Video Graphics, 131, 141
Brocke Graphic Design, Robert, 138, 148
Brodsky, Michael, 166, 207
Brody, Bill, 138
Brody, Ellen, 131
Brogren, 148
Bromley Chapin Designs, 160, 163, 181
Bronstein/Berman/Wills, 122
Brooke & Company, 151
Brookhouse, Winthrop, 127, 164
Brooks, Charles, 125
Brooks Communication Design, 148, 164, 167
Brooks Institute, 154, 198
Brown, Gloria, 148
Brown, Jensen & Garloff, 185
Brown, Jim, 123
Brown, Rob, 135
Brown, Ron, 127
Brown Graphics, 188
Browne, Jeanne, 143, 152, 165
Brownwood, Bruce, 138
Bruce Hayes Productions, 161, 176, 187
Bruml, Kathy, 151
Bruner, Rick Ernest, 135
Brunn, K David, 161
Bruno, Patricia J, 123
Bruton, Jon, 126
Bryan-Brown Photo, Marc, 122
BT&D Audio, 153
BTS/Broadcast TV Systems, 124
Buck & Barney Kane, Sid, 151
Buck, Sid & Kane, Barney, 131
Buck Consultants, 168
Bucks County Community College, 138
Buffalo State College, 125
Buhl Studios, 122
Buhler, Ray Varn, 138, 148
Burdick, Gary, 124, 143

Bureau, 128, 151
Bureau of Elec Publishing, 194
Burian Imagesetting, 192
Burke Design Group, 135
Burke, 127
Burkey, J W, 125, 135
Burnett Group, 158, 165, 168
Burns & Assocs Inc, 127, 148
Burt, Pat, 127
Burwinkel, David, 133
Busby Productions Inc, 170, 175
Buschner Studios, 124
Bush, Diana J, 136
Bush, Edward, 128
Bush Associates, Diane, 124
Busher, Dick, 127
Business & Professional Software Inc, 152, 201
Business Information Solutions Inc, 189
Business Presentation Services, 124, 158, 163
Business Publishing, 133, 143
Butler Creative Resources, 152
Butler Graphics Inc, 191, 208
Button Interactive, 161, 167, 176
Buz Design Group, 149
BXB Inc, 178
Bybee Digital Studios, Gerald, 127
Byron Preiss Multimedia, 194
Byte By Byte Corp, 158, 170-171

C

C Design, 161, 179, 208
C Graphics, 167
CA/GRAFX, 190
Cabana Seguin Design Inc, 151
Cabarga, Leslie, 133, 155
Cabimat, 141, 200
Cableputer Inc, 141, 171, 179
CADalyst, 155
Caere Corporation, 156, 197
Caesar Photo Design Inc, 127
Caesar Video Graphics, 171
Cahan & Assocs, 149
Cahners Publishing Co, 127, 138, 149
CalComp Inc, 154, 156
Calderhead & Phin, 141
Calera Recognition Systems Inc, 154, 156
Calfo, 141
Calibre Digital Design, 183
Calico, 161, 176, 182
Calif Coll of Arts & Crafts, 176, 182, 187
California Image Associates, 176, 182, 187
California Video Center, 176
Call, Ed, 149, 179
Callaghan, Charles, 124
Callaway Editions, 131, 188
Calliope, 185
Callis, Chris, 122
Caluger & Assocs, 173, 180, 208
Calvert, Carl, 203
Calvin Lew Design, 149
Cambridge Media, 168, 201
Cambridge Television Productions, 172, 179
Cameron & Co, 153, 160, 185
Campbell Consulting Ltd, 169
Campbell Studio, Thomas, 127
Campos, Mike, 161
Campus, Peter, 122
Canada Online Ltd/Early Fall 1994, 183
Cannata Communications Corp, 163, 173, 180
Cannedy, Carl, 125
Cannon + Eger, 169
Canon USA Inc, 154, 155
Cantor, Phil, 124
Cantrell Design, David, 133, 143
Cantwell, James, 141
CAP Comms (CKCO-TV), 177
Capital Presentations, 158, 172
Capital Vectors, 171, 179
Capitol Color Imaging Inc, 130, 189, 192
Capps Studio, 130
Capstone Studios, 138, 149
Captain NY/Ed Sullivan Theater, 171
Captured Images, 192, 207, 208
Carden, Vince, 138, 149, 153
Carey, Richard A, 168
Carey Color, 191
Caribiner Group, 171
Carl M Rodia & Assoc, 168
Carleton Productions, Inc., 177
Carlson Marketing Communications, 153, 160, 175

Carnegie Mellon Univ/Architecture, 125, 129
Carol Chiani Digital Workshop, 152, 158
Caroline Hightower, 203
Carpe DM Design, 149
Carrara Design, 143
Carrera Design, Joann, 145
Carrier, Alan, 138
Carroll, James, 122
Carsello Design, 136, 147, 153
Carson Design, David, 149
Carter, Greg, 135
Carter, Mary, 138, 149
Cartesia, 155
Carver Design, Stephen, 138, 149
Caryn Professional Modeling School & Agency, 185
Casdin-Silver, Harriet, 128
Casper, Daniel S, 133
Cassette Productions Unlimited, 187
Castle, Ed, 124
Castleman, Valerie, 131
Cat Walk, 173, 180
Catalog Design & Production Inc, 138, 149
Catalyst Productions, 176, 182, 187
Caton, Chip, 151
CAV Productions Inc, 162, 166, 167
Cbm Type, 192
CBS Inc, 171
CBS News, 162
CCI COmmunications Inc, 172, 179
CCS Princeton Inc, 188
CD Consultants Inc, 168
CD Folios/CD Publishing, 130, 196, 197
CD Rom Inc, 169
CD Rom Rights, 169
CD Technology, 196, 198
CD-CINC, 161, 189, 196
CDS Design, 127
CDTD, 168
CEC Audio Visual, 173
Ceisel, Beth, 147
CELCO (Constantine Engineering Labs Co), 154, 155
Celefex, 171, 179
Center City Film & Video, 170, 172, 180
Center for Advanced Whimsy, The, 131, 141, 158
Center for Art & Design, 137, 148
Center for Creative Imaging, 205
Center For Disease Control, 145
Center for Electronic Art, 161, 207
Center, 164, 181
Center Stage Communications, 168
Central Apex Engraving Co, 188
Central Conn State College, 192
Central Image Video & Photography, 176, 182, 187
Centre de Montage Electronique-TM, 177
Centre Grafik, 133, 143, 208
Century III at Universal Studios, 170, 173, 180
Cerulli, Andrew, 141
CFC Applied Holographics, 129
CG Graphic Arts Supply Inc, 155
Chada, Ritu, 131
Chalkmark Graphics, 122, 131
Chalsie Systems & Design, 125
Chan, Ron, 138
Chandler, Roger, 138
Chang, Dr Rodney, 138
Chang, George, 138
Channel 3 Productions, 178
Channel 3 Video, 152, 172
Chaparral Software, 168, 169, 208
Chapman Univ/Comp Sci Dept, 176
Charles, Bill, 151
Charles H Williams Consulting, 127, 157, 201
Charlex Inc, 170, 171, 179
Charmed Productions Inc, 175
Chartmakers, 152
Chartmasters, 153
Chase Design, Margo, 138, 149
Chattum Design Group, 131, 141
Chausse, Norbert, 133
Cheap Computer Graphics, 179
Chelsea Television Studios, 171
Chenn, Steve, 125
Chernishov, Anatoly & Irene, 133, 143
Cherry, Jim, 138
Cherry Optical Holography, 129
Cheshire Designs, 159, 163
Chesney Communications, 176

Chess Productions, 153
Chey, Dong Cheol, 124
Chezem Studio, 135
Chiavenna Design, 145
Chiba, Kaz, 122
Chicago Schl of Holography, 129, 206
Childrens Book Council Inc, 198
Chinon America Inc, 154, 196
Chipsoft Inc, 156
Chipurnoi, Minda, 133, 163, 166
Chisholm, 197
Chisholm Rich & Assoc, 125
Chislovsky Design Inc, Carol, 151
Chmelewski, Kathleen, 147
Chollick, Phyllis, 131, 141
Chotas, James, 122
Chris Jones & Associates, 185
Christensen, Steve, 208
Christensen Media, 161, 169
Christenson, Paul, 130
Christopher Designs, 133, 144
Chroma Studios Inc, 130
Chromagem Inc, 129
Chromagen Digital Imaging, 129, 190
Chromakers, 190
Chromavision Corp, 170, 171, 184
Chui, John, 138, 149
Chung, Harry, 131
Chung, Mei K, 131
Church, Diane, 151, 156
Cicchetti, John, 129, 133
Ciemny, Ray, 133
Cies/Sexton Photo Lab, 130, 149
Cimarron Productions, 161, 167, 169
Cimity Arts/Special Visual Effects, 161, 164, 182
Cine Service, 185
Cinecraft Audio Visual Services, 172
Cinecraft Productions Inc, 153, 160, 175
Cinema Concepts, 159, 163, 173
Cinema East, 173
Cinema Engineering Co, 182, 187
Cinema Network (Cinenet), 161, 176
Cinema Research Corp, 176
Cinema Video Center, 175
Cinemasound Video Productions, 173, 180
Cinetel Productions, 173, 180
Cinetron Computer Systems, 173, 180, 198
Cineworks Productions, 172
Ciné-Med Inc, 158, 163, 172
Cinque, Michael, 127
CIP, 152, 172
Circle Video Productions, 170, 173, 184
Cirone, Bettina, 122
CiS Graphics, 199
CIS-Hollywood, 171, 176, 192
Citizen's Graphic Arts, 192
Citron Assocs, 144, 208
City Animation Co, 170, 175, 185
City College of NY/Art Dept, 184
City College of NY, 131
CKS Partners, 164
Clarinet Communications, 194
Claris Corp, 157, 194, 201
Clark Design, James, 149
Clark University, 128, 172
Class Line Graphics, 131, 141
Classic Animation, 181
Classic Video Inc, 175, 181, 185
Clear Light Studio, 125
Clicks Inc, 153, 191
Cliggett Design Group, 133, 144
Closed Circuit Products, 138, 149, 161
CMI Business Communications, 175, 185, 191
CN Graphics Inc, 191
Coast, The, 185
Coates, Peter, 141
Cobb Inc, 191
Cobi Productions, 122
Cocchiarella Design, 126
Cochenille Computer, 156, 201
Coddbarrett Associates, 155
Coffman, Claudia, 144, 158
Cognetics Corp, 201, 208
Cohen, Adam, 131, 141
Cohen, Hagit, 138, 166
Cohen Design Inc, Hayes, 131, 141, 189
Cohn, Larry, 138
Colby, Tracy, 138
Colby Systems Corp, 196, 198
Coleman Lipuma Segal & Morrill, 141
Coll Engnr, 183

Coll of Arts & Sciences-CPR 107, 122
Collector, Stephen, 127
College du Vieux Montreal, 165
College of New Rochelle, 144
College of St Mary, 138
Collignon, Daniele, 151
Collins & Collins, 124, 133, 144
Collyer, Frank, 133
Colony Multimedia, 158, 166
Color Age, 154, 155, 201
Color Associates Inc, 188
Color Graphics Inc, 191
Color Group/Reubens Color, 190
Color Image Inc, 189
Color Image, 126, 130, 147, 188, 208
Color Logic, 129, 188
Color Masters Inc, 189
Color Perfect Inc, 191
Color Place, The, 129, 153, 191
Color Professionals Inc, 188
Color Service, 192
Color Systems, 188
Color Trend Corp, 188
Color Wheel Inc, 129
Color-Imetry Inc, 188
Colorado Design Associates, 152
Colorado Institute of Art, 188
ColorBus Software, 189
ColorCopy Inc, 191
Colorcraft Inc, 188, 191
Colordynamics, 153, 173, 180
Colorgraphics, 130, 189
Colorpointe Design, 147
Colorprep Inc, 154, 157
ColorScan Services Inc, 188
ColorType, 189, 192
Colotone Inc, 188
Columbia Coll/Academic Computing, 163, 206
Columbia Lithograph Inc, 192
Columbia Software Inc, 157
Columbia Univ/Div of Continuing Educ, 124
Combined Services, 153, 160, 175
ComCorp, 147
COMDEX Fall/Nov 14-18, 1994, 129
Comitini, Peter, 131
Command Post & Transfer, 177, 184, 188
Combine, 147
Commercial Art Service, 191
Commercial Projection Service, 153
Commodore Business Machines, 196, 198
Commonwealth Creative Group, 152, 190
Commotion, 176, 182, 208
Commty TV of S Calif (KCET), 176
Communica, 147
Communicate, 161, 164, 192
Communicating Arts Group of, 204
Communication Bridges, 161, 166-167
Communication Design, 169
Communication Research Inc, 168
Communication Technologies, 170
Communications Concepts Co, 161, 164, 176
Communications Design, 138, 149
Communications Plus, 158, 171, 179
Communications Services, 175, 185
Communications, 172, 196
Communications Video, 161, 176, 182
CommuniCreations, 187
Communigrafix, 136
Comp Ed Ctr, 192
Comp-U-Type, 188, 191
Compact Publishing, 194
Compaq Computer Corp, 135
Compas Technologies, 208
Compatible Systems Engineering Inc, 156
Competitive Edge, 207
Complete Post, 182, 187
Complete Post Inc, 149, 176
Compographics, 192
Composition Type, 192
Compton's Learning, 194
Compton's New Media, 194
Compu, 153
CompuColor Inc, 188
Compugraphia, 131, 155, 162
Compumation, 155
Computed Animation Technology, 163, 180, 199
Computed Designs, 208
Computer Aided Technology Inc, 160, 165, 167
Computer Associates International Inc, 155
Computer Chrome Inc, 191

Computer Consultants, 158, 180
Computer Education Intl, 166, 207
Computer Friends, 196, 201
Computer Generated Imagery Inc, 172, 180
Computer Graphic Center, 191
Computer Graphic Group, 191
Computer Graphic Resources, 171
Computer Graphics Ctr ARC Bldg,, 122, 131
Computer Graphics Dept, 176
Computer Graphics Design, 133, 144
Computer Graphics Group, 161, 208
Computer Graphics Resources, 158, 163, 172
Computer Graphics Services, 192
Computer Graphics World, 152
Computer Group, 196
Computer Imagery, 125, 129, 135
Computer Images, 133
Computer Imaging, 192
Computer Imaging Resources, 154
Computer Presentations Inc, 156
Computer Prompting & Captioning Co, 198
Computer Putty, 164, 200
Computer Reseller News, 122
Computer Studio, The, 163, 173, 180
Computer Support Corp, 156
Computer Typesetting & Graphics, 192
Computerized Bizmart Systems, 194
Computerized Video Service, 159, 170, 173
Computing Technology Industry Assoc, 191
CompuWorks Inc, 154-155
Comware Inc, 178, 200-201
Conant, Chrissy, 133, 144
Concentrics Co, 169
Concept & Design, 141
Concept Publishing Systems, 156
Concrete Couple Productions, 141, 179
Condon Norman Design, 147
Conejo Valley Mac Users Group, 1994, 168
Connacher, Nat, 144, 169
Connecticut Art Directors Club, 160, 165, 208
Conner Holographic Services, 129
Conrad & Associates, Jon, 138
Conrad, Larry, 126
Constructive Communications, 151
Conter Software, 194
Continental Cablevision/Madison Hts, 175, 178, 185
Continental Cablevision of Cook Cnty, 175
Continental Productions, Continental TV Network, 176, 182, 187
Contracted Computer Training Inc, 185
Cook, Jamie, 125
Cook Design, Robert, 145
Cooke, Charles, 141, 166
Cooper, David, 128
Cooper, Ken, 128
CoOperative Printing Solutions Inc, 156
Cope, Doug, 138
Copeland, Burns, 126
Coppola, Richard, 158, 178, 207
Copy Print Co, 190
Copy Spot Inc, 192
Copytone, 152, 188, 189
Corbett Communications, 185
Corbin Design, 147
Corbitt, John, 135
Cordaro, Mary, 169
Cordero, Felix L, 128
Corel Systems, 155, 157

Corelli Jacobs Recording, 184
Corkum, Paul, 124
Cornell/Abood , 167, 180
Cornerstone Associates, 198
Cornerstone Technology, 154
Corplex Inc, 175, 181, 185
Corporate Communication Group, 178
Corporate Communications, 158
Corporate Graphics Inc, 153, 174, 180
Corporate Media Services, 159, 163, 174
Corrente, Linda, 141
Corrette, Nicholas Moses, 133
Cosmos Comm, 122
Cossette, Diane I, 126
Costanzo, Jim, 168
Cotton Communications, Jim, 188
Counts, Clinton, 133
Cowen Design, 146
Coy, Los Angeles, 149
Coyne Co, 127
CPI Inc, 157
CPN Televison Studios, 159, 163
CPS Technologies Inc, 155
CR 2 Studio Inc, 124
CR Waldman Graphics Communications, 190
Craig, Andrew B, 168
Craig, John, 136
Cramer Productions, 158, 163, 172
Cranbrook Academy Of Art, 206
Crane, Gary, 135, 146
Crawford Communication, 170, 174, 180, 184
Creative Alliance, 161, 176
Creative Ape, 138
Creative Center for Advanced D, 189
Creative Computing, 149
Creative Concepts, 126, 153, 175
Creative Dept Inc, 135
Creative Edge, 174, 180
Creative Freelancers, 131, 141, 151
Creative Images, 160, 163, 175
Creative License-Film & Video, 182
Creative Media Concepts, 122
Creative Media Development, 161, 176, 182
Creative Media Services, 157
Creative Multimedia Corp, 194
Creative Photo, 191
Creative Post & Transfer, 174, 180
Creative Productions Inc, 152
Creative Productions/WEAR-TV, 174
Creative Resource, The, 152
Creative Resource Center Inc, 147
Creative Text + Page, 147
Creative Typesetting Inc, 190
Creative Video Design & Prod, 144, 172, 180
Creative Video, 159
Creative Video Inc, 174, 180
Creative Vision, 176
Creative Visual Associates, 177
Creative Visuals Inc, 153, 185, 191
Creative Ways Inc, 158, 171, 179
Critical Connection, 122, 152, 178
Crocker, Jim, 130
Crocker, Will, 125
Crocker Inc, 144
Croma-Video, Inc., 172
Cronan, Michael Patrick, 149, 161, 167
Cronopius, 130
Crosier, Dave, 127
Cross, Lloyd, 129, 171, 201

Crossley, Dorothy, 124
Crosspoint Productions, 176, 182, 187
Crowder Design, N David, 144
Crowley & Associates, 176, 187
Crown & Shield Software Inc, 155, 196
Crown Color Corp, 188
CrownGraphics Inc, 188
Croydon, Michael, 129
Crucible, 153, 161, 208
Crushing Enterprises, 184
Crutchfield, William, 138, 149
Cruz, Frank, 127
Crystal Graphics, 175, 181, 185
Crystal River Engineering, 199
CSA Archive, 126
CSR Productions Inc, 152, 172
CSS Laboratories, 155
CST Entertainment Imaging Inc, 176
Cudlitz, Stuart, 138
Cuevas, George, 135, 146
Culbertson Design, Jo, 149
Cully, Mike, 136, 163
Culver & Assocs, 147
Cumberland Graphics, 191
Cummins, Karla, 141
Cunningham, Jerry, 124
Cunningham, Peter, 122, 158
Curious Pictures, 158, 162, 170
Current Designs Corp, 196, 198
Curry Design Inc, 149
Curtin Design, Paul, 138, 149
Curtis Inc, 175, 181, 185
Curved Space, 153, 157
Cushwa, Tom, 131
Custom Color, 192
Custom Color Corp, 130, 153, 191
Custom Composition, 192
Custom Recording Studios, 185, 196
Customizer, The, 146
Cutler Productions, Howard M, 144
Cutler Studio Inc, 122
Cutler-Graves, 130
CWA Inc, 149
Cyan Inc, 166, 167, 194
Cyber Network Svcs Inc, 166, 167, 207
CyberChrome Inc, 155
Cyberiad Project, 180
Cyberkids, 163
Cyberplex, 136, 147, 160
Cycle Sat, 170, 175, 185
Cynosure Films, 141, 178
Cypress Coll/Fine Arts, 174, 180, 184
Czechowicz, Lesley, 169

D
D'Andrea Productions, 162
D'Art Studio Inc, 144
D'Elia-Wittkofski Productions, 170, 172, 184
D'Hamer, M, 127
D'Pix Inc, 156
Da Silva Animation, 162
Dadabase Design, 133, 144
DADC-Sony, 170
Daedalus Systems, 131
Dahl + Curry, 191
Dahlquist, Roland, 138
Dahm, Bob, 133
Dai, 154

Daily Planet Ltd, 170, 175, 181
Daisley, Dawn, 168
DAK Productions, 172
Dakini Designs, 138, 149
Dakota Photographics Inc, 188
Dakota Teleproductions, 170, 175, 185
Dale Photo, Larry, 126
Daley, Joann, 138
Dallas Photo Imaging, 129
Dallas Society of Visual Communications, 184
Daly, Perla Linda, 144
Daman Studio, Todd, 127, 138
Dammer, Mike, 136
Dan, Noah, 144
Dan Krech Productions Inc, 177
Dana Industries, 125
Dana White Productions, 161, 164
Dangel, Corey, 138
Danger Studios, 185
Daniels, Charles, 127
Daniels, Mark, 124
Daniels Printing Co, 129, 194
Danny Pietrodangelo, 146, 174, 181
Dar Electronic Research & Development, 160,
    165, 170
Darino Films, 158, 162, 171
DAROX Interactive, 161, 164, 167
Darrell Brand/Moving Images Inc, 175, 185
Darshan Associates, 125
Data Motion Arts, 158, 171, 179
Data Service Company, 154, 156, 208
Data Slide, 191
Data Translation, 154, 202
Datalogics Inc, 156
Dataquest Inc, 169
Datastream, 191
Datavision Technologies Corp, 169
Dataware Technologies Inc, 155
Dataware Technology Inc, 196, 198, 200
Dave Mueller, 160, 181, 185
Dave Restuccia Productions, 175
Dave Trescot, 156, 200
Davick, Linda, 135
David A Roth Music, 185
David Blum Animation, 164
David Cundy Inc, 144
David Curry Design, 141
David Horowitz Music Assoc, 184
David Slavin Design, 149
David Tremblay Productions, 185, 191
Davidian, Peter, 122
Davidson & Co, 129, 191
Davidson, Peter, 124, 133, 144
Davis & Assoc, Bruce, 149
Davis, Harold, 122, 166
Davis, Jack, 138, 149
Davis, Mills, 144
Davis, Robert, 138
Davis, Sally, 138
Davis, Stephen, 135
Davis Inc, 131
Davison, Bill, 133
Dawber & Co, 153
Dawson, Hank, 138
Dawson, Will, 135
Day, Rob, 133
Dayal, Antar, 138
Dayna Communications Inc, 157
Dazzleland Digital, 127, 130

DBF Media Company, 172
DC Heath & Co, 194
DCP Communications Grp Ltd, 158, 163, 172
DDB Needham Worldwide, 141
De Cerchio, Joe, 133, 144
De Goede & Others Inc, 147
de Kerangal, Pierre, 158
Deak, David, 131
Dean, Glenn, 131
Dean Digital Imaging, 124
DeAnza Coll/Cmptr Grphc Design, 178
Debela, Acha, 135
Debold, Bill, 125
DeBolt Photography, Dale, 126
Deborah Herath, 154, 156
Deborah Miller, 196, 200
Deborah Wolfe Ltd, 133
DeBrey Design, 147
Deck Design, Barry, 131, 141
Decotech, 141
Dedell Inc, Jacqueline, 151
Dee Typographers Inc, 192
Deem, Rebecca, 129
Deen, Georganne, 138
Deep River Publishing, 194
Defreitas, Frank, 128
DeGraf/Wahrman, 182
Deico Electronics, 155
Delago, Ken, 133, 144
Delbert, Christian, 124
Dell'Aquila, Mei Ying, 138
Dellinger, Joseph, 131
Delmar Communications Co, 174
DelMontagne, 204
Delrina Technology Inc, 157
Delta Tao Software, 161, 171
DeltaPoint Inc, 157
DeltaTech Corp, 131, 141
Demark Keller & Gardner, Inc, 131, 141
Denbo Multimedia, 122, 162, 168
Deneba Software, 156, 200
Denham, Karl, 138
DeNicola, Robert, 131
Denon America, 196, 198
Denon Digital Industries, 170
Denver Commty Coll/Downtown Aurora Campus,
    144, 158, 163
Depixion, 161, 164, 182
Depthography Inc, 122
DeRiggi, Danamarie, 144
DeRose, Andrea Legg, 146
Dervis, Aris, 122
DeScribe, 157
Desert Rat Design, 138
DeSeta, Maxine, 131
Design & Type Inc, 192
Design at Work, 133, 144
Design Center, 147
Design Co, 147
Design Collective, 149
Design Comp Inc, 144
Design/Efx, 163
Design Form, 169
Design Graphics, 149
Design Heads, 131, 141
Design II, 147
Design Imaging, 144, 188
Design Imaging, 188
design M design W, 144
Design Manufacturing Inst, 149
Design Moves, 136, 147
Design Office of George Wong, 147
Design Plus, 141
Design Point, 144
Design Provisions, 141
Design Source, 149, 208
Design Space, 141
Design Studios Teleproductions, 175, 185
Design Team, 144
Designed to Print & Assoc, 141
Designers Atelier, 129, 188, 189
Designers Desktop, 149
DESIGNS +, 149, 161, 166
Designs for Business, 149
DesignSoft, 156
Designwerks, 147
Deskprint, 192, 193
Desktop Design, 147
Desktop Graphic Solutions, 153
Desktop Graphics & Design, 192, 208
Desktop Plus, 156

Desktop Publishing Inc, 192
Desktop Publishing, 152, 207
Desktop Solutions, 169
Desktop Studio, 192
Destiny Technology Corp, 157
Development, 160, 165, 170
Development Communications, 159, 174, 184
Devereaux, Carole, 149
Deverich, Michael, 149, 169
Devon Video Productions, 172
Dégallier Animation, 163
DFI Inc, 198
DFW Photography Inc, 125
DHD Post Image, 184
DI Group, 158, 163, 172, 180
di Liberto, Lisa, 141
Di Re, John, 141
Diadul, Robert, 122
Dial Music Productions, 187
Diamant Noir, 122, 179
Diamond Video Productions Inc, 174, 180, 184
Diaquest Inc, 198, 200, 202
Diaz, Jorge, 122
Diaz, Jose, 131
DiChello, Joseph, 124
Dickson, Ellie, 131
Dicomed, 196, 198
Diebold, George, 124
Diehl, Michael, 149, 208
Dietmeier, Homer J, 169
Dietrich, David, 125
Diette, Paul, 129
Dietz, Mike, 138
Diffraction Ltd, 128
DiGena, Louis, 124, 205
DiGiacomo, Carole, 144
DigiChrome Imaging, 130
Digicliques, 153, 191
Digidesign Inc, 182, 187
Digipress, 196
Digit Eyes, 127, 138
Digital Art, 138
Digital Arts, 182
Digital Constructs, 133, 208
Digital Creations, 181
Digital Design Inc, 130
Digital Design Simulations, 138, 149, 182
Digital Equipment Corp, 198
Digital Exchange, 189
Digital F/X Inc, 197, 201
Digital Foto Arts, 127
Digital Hollywood/Feb-March 1995, 170
Digital Image Design Inc, 167
Digital Image, 129, 191
Digital Imaging & Design, 130
Digital Imaging Group, 125
Digital Ink, 126, 136, 147
Digital Knowledge Corp, 126
Digital Magic, 177, 184
Digital Medical Images, 135
Digital Multi-Media Post Inc, 174, 180, 184
Digital Photo Studio, 122
Digital Photo Works, 129, 131, 141
Digital Photographic Imaging, 191
Digital Pictures Inc, 198, 208
Digital Pond, 192
Digital Post & Graphics, 171, 176, 182
Digital Publishing, 194
Digital Resources, 127, 157, 201
Digital Stock Professional, 127, 157, 201
Digital Technology Intl, 157
Digital United, 170
Digital Valley Design, 144, 166, 169
Digital Video Arts, 158, 169
Digital Video NY/April 19-21 1994, 127, 197
Digital Vision, 197
Digital Wave, 180
Digital Wisdom Inc, 194
Digital Wrist, 130
Digital Zone, 149
Dignum Computer Graphics, 177, 184
Dillon Jr, Emile, 124
DiMarco, Paula, 131
Dimension 3 Inc, 128
Dimension Inc, 188
Dimension Teleproductions, 179
Dimensional Foods Corp, 128
Dimensional Imaging Cons Inc, 129
Direct Holographics, 128
Direct Images, 127, 179, 201
Directors Guild Of America, 204

Disc Manufacturing Inc, 171
Discis Knowledge Research Inc, 195
Disk Productions, 184
Diversified Graphics, 136
Diversified Video Ind Inc/Sierra Video, 176, 182
Division Ltd, 199
DiVitale Photo Inc, 125
Dixieland Productions Inc, 174
Dixon & Associates, 161, 166, 167
Dixon, David, 135, 146
Dixon Design, 141
DJM Films & Tape, 171
DLH Studios, 153, 182, 187
DM Infosearch, 169
DMC & Co, 133, 144
DMR Group, 185
Documentation Development Inc, 158, 165, 166
Dodge, Jeff, 122
Dodson, Liz, 136
Doerner Graphics, 138
Dogmatic, Irene, 141
Doherty, James, 138, 164
Dojc, Yuri, 128
Dole, Jody (Mr), 122
Dollar & Associates, 161, 176
Dolphin Multi Media, 153, 164, 182
Dome Productions, 177
Domin, Jacqueline, 133, 158, 178
Don Krech Productions, 177
Don Pierce Productions, 182
Donath, Emeric, 130
Donath Communications, Ellen, 152
Doorbell Productions, 172
Dorf Studios, 122
Doris Seitz-Boothe, 159, 165, 169
Doros Motion, 179
Dorothea Taylor-Palmer, 135, 146
Dots Enuff!, 190
Dots Per Inch/Meyer Associates, 191
Dots, 188
Double Click Design, 133, 144
Double D Associates Inc, 147, 153, 160
Double Vision Productions, 161, 182, 187
Doublespace, 142
Doudoroff, Martin, 178
Doug Mackeroy, 197, 198
Dougherty, Suzanne, 138
Douglas Graphics, 191
Douglas, 171, 183, 187
DouPonce, Kirk, 147
Dourmashkin Productions, 161, 176, 182
Dow, Carter, 127
Dowlen Artworks, 130, 138, 182
Downtown Design, 135, 146
DP&C, 192
Dr T's Music Software Inc, 184
Drabick, Matt, 178
Drake, Patti, 133
Dream Maker Software, 157
Dream Merchant Graphics, 138, 149
Dream Quest Images, 171, 182, 187
Dreamers Guild, The, 133, 144, 172
Dreamtime Systems, 125, 129
Drebelbis, Marsha, 146
Drenttel Doyle Partners, 142
Dressler, Brian, 125
Dressler, Marjory, 122, 131, 142
Dressler, Rick, 127
Drew, Ned, 135
DrewPictures Inc, 182
DRI, 194
Drivas, Joseph, 122
Drucker Associates, 169
Drucker Motion Pictures, 179
Drummond, Deborah, 133
DSC Laboratories, 177
DSM Producers, 184
Dub Masters/SP, 171, 176, 187
Dubl-Click, 157
Dubreuil, Michel, 151
Duda Design Inc, 164, 166-167
Duffus, Bill, 182
Duffy, Fred, 208
Duffy Design Group, The, 147
Duffy Designs, 146
Dufour, 136, 147
Duggal Color Projects Inc, 189
Duin, Ed, 149
Duke, William, 127
Dumptruck Studios, 138
Dunn & Rice Inc, 144

Dunning, Chris, 129, 133
Dupont Imaging Systems, 194
DuPuis Design, 149, 192
Duraset Typesetting, 192
Durbin Associates, 155
Durham Coll/Applied Arts & Technology, 171
Dwight Yaeger Typographer, 191
Dycam Inc, 197
Dyens, Georges M, 129
Dyma Engineering Inc, 153
Dyna Pac, 127, 149
Dynacom Inc, 160, 165, 167
Dynaimages, 175, 181
DynaMedia Design & Graphics Inc, 192
Dynamic Graphics Inc, 147, 156, 203
Dynamic Perspectives, 138, 182
Dynaware USA Inc, 157, 200
Dysart Creative, 144

E
E Fitz Art, 133, 144
E&J Advertising, 192
E Lambert Inc, 188
E M 2 Design, 146
E-Conspiracy, 138
E3 Inc, 142, 168
Eagle, Steve, 178
Eagle Eye Publishers Inc, 196
Eagle Graphics, 146
Eaglevision Inc, 158, 172
Eales, Ray, 178
East Design, 194
Eastern Washington University, 146
Eastman Kodak, 194
Ebersol, Rob, 135
Ebook Inc, 194
ECI Video, 174, 180
Eclipse, 126, 147, 178
ECM-A Exhibit & Conference Management Ltd, 155
EDCO Services Inc, 156
Edcom Productions, 185
EDEFX, 171, 174, 179, 180, 184
Edge Graphics, 138
Edge Multimedia, 170, 181, 185
Edgerton, Brian, 124, 133, 144
Edit Decisions Inc, 171
Edit One Video Productions, 172
Edit Suite, 175, 185
Editech Post Productions Inc, 175, 181, 185
Editel Design, 170, 181, 185
Editel/Los Angeles, 161, 182, 187
Editing, 171
Editvision, 170-171
Editworks, 159, 180, 184
EDR Media, 160, 175, 181
Edson, Jennifer, 158
Educorp, 194
Educorp/Technical Support Dept, 196
Edwards, Andrew, 131
Edwards Design Inc, Sean Michael, 142
Edwin Schlossberg Inc, 158, 165, 166
Effective Communication Arts, 163, 165, 172
Effective Graphics Inc, 189
Effective/Visual Imagery, 161, 182, 187
Effects House/The Optical House, 170
EFX Communications Inc, 174
Egas, Eric, 133
Egeland Wood & Zuber, 158, 208
Ehrentreu, Devora, 142
Ehrlich Multimedia, 158, 165, 166
Eicon Technology Corp, 157
Eisenberg, Sheryl, 142, 168
El Dorado Graphics, 189, 191
El-Darwish, Mahmoud, 124, 133, 144
Eldred, Dale, 147
Electra Nova Studios, 184
Electric Art Company, 130, 136
Electric Arts, 174
Electric Canvas, 161, 167
Electric Easel, 138
Electric Image Center, 152
Electric Image Inc, 138
Electric Page, The, 192
Electric Pages, 149, 164, 182
Electric Pencil, 192
Electric Pencil Studio, 135
Electric Zebra, 179
Electrim, 154, 155
Electro/Grafiks, 124, 152, 172

Electrohome Ltd, 177
Electronic Artists Group, 165, 168
Electronic Easel, 191
Electronic Frontier Foundation, 127, 138
Electronic Imaging Inc, 191
Electronic Media, 140
Electronic Publishing Ctr, 191
Electronic Publishing Ctr, 122, 131, 142
Electronic Publishing Network, 193
Electronic Publishing, 153, 193
Elias Associates, 184
Elite Color Inc, 130, 188, 189
Elite Post of Nashville, 163, 180
Elixir Technologies Corp, 138
Elledge, Paul, 126
Elliot, Tom, 125
Elliott, 151
Ellison Graphics Inc, 188
Elmsoft Inc, 169, 197
Elsberry Production Services, 185
Elsner, John P, 185
Elson, Matt, 138, 182
EMA Video Productions, 164, 176
EMC Productions, 185
Emerald City Animation/Post Prod, 153, 181, 185
Emerging Technology Consultants Inc, 169
Emerson, 142
Emily Carr College of Art & Design, 149
Empire Video, 170, 171, 179
EMR Systems Comm, 129, 188
EMS Communications, 161
Encompassed Graphics, 135, 146
Encore Video, 171, 176, 187
Endo, Stan, 138
Energy Productions' Timescape, 176, 193
Engineered Software, 174, 180
Engle & Murphy, 149
Ennis, Phillip H, 124
Ensemble Productions, 159, 178, 180
Entercor Studio, 189
Enteron Group, 189
Enterprise Information Service, 191
Envision Inc, 164, 175, 181
EP Graphics/Burbank, 161, 182, 187
Epix Inc, 197
Epure, Serban, 131, 158, 178
Equilibrium Technologies, 158, 171
Erin Morrissey, 143
ESPN Inc, 158, 172
Etgin Co, 157
Etkin, Rick, 128
ETS Graphics Inc, 189
Euphonics, 138
Evans, Virginia, 133
Eve Design, 131
Evenson Design Group, 149
Everett Studios, 190
Everex, 196, 198
Everyday Music Co, 185
Evon, Susan, 161, 167
Ewasko Studios, 125
Ewert, Pat, 160
Ewert, Steve, 126
Ewert's Photo & Audio Visual, 153
Executive Information Base, 162, 171
Executive Technologies Inc, 196, 200
Exographic, 147
Exos 8, 199
Exoterica Corp, 157

EXP Graphics, 138, 179
Expanded Video Inc, 158, 163, 172
Experimental Intermedia, 188
ExperTech Corp, 158, 165
Express Slide, 191
Expressive Images Ltd, 190
Extended Systems, 155, 157
Eyebeam, 129, 188, 191
EZTV Video Ctr, 170, 174, 180

F
Facit Inc, 154
Facts On File Inc, 194
Fairchild Publications, 131
Fairleigh Dickinson Univ/Madison, 169, 172, 205
Faith Inc, 141, 151
Falcone & Assoc, 144
Falk Design Group, Robert, 147
Familian, David, 127
Fanning, Rich, 138
Farace Photography & Comm, 127, 153
Faragher, Patsy, 138
Farallon Computing, 196
Farmer, Roscoe, 127
Farr Photography, 124, 201
Farrell, Richard, 133
Fastforward Communications Inc, 124, 144, 163
Fat Baby Productions, 158, 162, 168
Fatone, Bob, 124
Faxgraphix, 191
FCL/Colorspace, 129, 131, 189
FCL, 171
FEA Laser Service, 190
Fearey, Patricia, 161, 164, 167
Fearless Designs, 135, 146
Fearless Eye, 136, 164
Feder, Eudice, 138
Feidler, Anita, 124
Feigus, Jan, 133
Feiling, David, 124
Feinberg, Susan, 131
Feinen, Jeff, 144
Feldman, Simon, 122
Feltenstein, Keith, 131
Felzman Photography, Joe, 127
Fenelon, Daniel, 133
Fenster, Diane, 138, 207, 209
Ferguson, Heleman, 133
Ferguson, Melanie, 131, 162
Feroe Holographic Consulting , 129
Ferranti Educational Systems, 165
Ferrari Color Digital Imaging, 130, 193
Ferreira, Al, 124
Ferri-Grant, Carson, 142, 158, 162
Ferster, Gary, 138
Feuereisen, Fernando, 131
Fidelity Color Inc, 188
Figura, Paul, 125
Filippucci, Sandra, 131
Film & Tape Works Inc, 170, 175, 181
Film Craft Video, 164, 175, 181
Film Div, 172
Filmack Studios, 160, 164, 181
FilmCore, 176, 187
Filmet Color Labs, 190
Filmworks, The, 189, 193, 201
Final Proof Inc, 191

Findley, John, 135
Fine Art, 128, 151
Fine Arts, 129
Fine Arts Video, 178
Fine Print Design Co, 131
Finfirst, 138
Finishing Group, 175
Fink, Mike, 139, 149
Finot Group, The, 157
Finzi, Jerry, 122
Firstcom/Music House/Chappel, 184
Firstdesk Systems Inc, 155
Fisch, Amy, 128, 131
Fischer Photography, Ken & Carl, 122
Fisher & Day Design, 149
Fisher, Elaine, 124
Fisher, Reed, 139
Fisher Yates Communication, 152
Fishman, Miriam, 122, 131
Fitz, Tracy, 171, 179, 184
Flaherty, David, 131
Flaherty, Michael, 149
Flatow, Carl, 122, 128
Flax, Carol, 139
Fleisher, Audrey, 131, 152, 168
Fleishman, Carole, 142, 178
Flink Design Inc, HGans, 144
Flom, Eric, 139, 179
Flood, David Williams, 133, 144, 180
Flora & Co, 154
Florida Film & Tape, 174, 184
Florida Production Group Inc, 174, 180
Floyd Design, 146, 159, 165
Flying Color Graphics Inc, 189
Flying Colors Inc, 185
Flying Colours, 176, 182
Flying Foto Factory, 180
Flying Turtle Productions, 144, 158, 165
Flynn, Bryce, 124
Focal Point 3D Audio, 199
Focus Enhancements, 197
Foge, Kurt, 135
Follett Software Co, 196, 198
Folsom Research Inc, 198
Font Company, The, 157
Font World, 155
Fontastik Inc, 147, 192, 208
Fontographics, 193
Forcade & Associates, 147, 160, 181
Ford Myers & Co, 133, 144
Foremost Comm's, 124, 172
Foresight Resources Corp, 156
Form & Function, 161, 164
Forma, 146
Fornari, Arthur, 128
Forsight Inc, 158, 163, 208
Forward Design, 158, 166, 168
Foster, Kim A, 146
Foster Digital Imaging, 139, 164
Fotocraft, 193
Four/Four Productions, 184
Fox & Perla Ltd, 172, 184
Fox, John, 127, 149
Fox, Mark, 139, 149
Fox Colour, 189
Fox Hollow Enterprises, 124
FPG International, 122
Fractal Design Corp, 157
Frame One Inc, 160, 164, 175

Frame Runner, 171
Frame-by-Frame Media Services, 176, 182, 200
Frampton, Bill, 131
Francois, Emmett W, 124
Frank, Brooks, 127
Frank Antonides Design, 126, 136
Frank Relations, 142
Franklin, Charly, 127
Franklin Video Inc, 174
Frassamito & Assoc, John, 167, 168
Frasser, Oscar, 122
Frazier, Craig, 149
Frazier, Jillian, 133
Fred Wolf Films, 182
Fredrick Paul Productions Ltd, 170, 175, 181
Free Spirit Software, 194
Freed, Gary, 122
Freed, Hermine, 131
Freedom Sound/Tallman Music, 184
Freeman, Nancy J, 135
Freeman Associates Inc, 169
FreeSoft Co, 155
French & Prtnrs, Paul, 153
French, Graham, 128, 130
Freund, Art, 129
Friday Saturday Sunday Inc, 142
Friedman Consortium, Harold, 142, 162, 171
Fritz, Tom, 126
From Art to Design Inc, 139, 149
Froman, Loralie, 139
Frumkin, Peter, 131
Fry III, George B, 127
FSA Video Productions Inc, 172
Fujitsu Computer Product of America, 157, 198
Fujitsu Imaging Systems, 194
Fukuda, Fujie, 135
Full Circle Production, 149, 176, 179
Full Spectrum Productions, 161, 164, 187
Fuller Dyal & Stamper Inc, 146
Fullerton College/Todd, 133, 144
Fumi Color Engineers, 129
Funk Software, 155
Fusion Communications, 158, 165, 166
Future Image Rep,The , 168
Future Perfect, 139
Future Video Products Inc, 196
FWB Inc, 155, 196, 198
FX Plus Design, 161, 176, 182

G

G & G Designs, 162, 164, 167
G&G Laser Typesetting, 190
G K Hall & Co, 194
G.R.A.P.E. Video Recording, 176, 182, 187
G R Waldman Graphics, 190
G W Hannaway, 167, 201
G-Nine Productions, 149, 179
Gable Design Group Inc, 149
Gader, Bertram, 209
Gadsden-Grant Graphic, 142, 179
Gage, Hal, 127, 139, 149
Gaijin Studios, 135
Galante, Dennis, 122
Galeano, Margaret, 144
Galic Maus Ventures, 198
Galileo, 135, 159
Gallagher, Matthew, 133, 144
Gallucci Studio Inc, 122

Gambino, Donald, 144, 205, 208
Gamma One Inc, 188
Ganley Design, 133
Gannett Production Services, 175
Gannett Production Services, 174, 180, 184
Gannett Productions, 171, 176, 187
Ganson, John, 124
Garber, Helen Kolikow, 127
Gardiner, Jeremy, 122, 131
Gardner Design, 147
Garman Audio/Video, 153, 174, 185
Garrin Prods, Paul, 178
Garrison, Gary, 125, 146
Gartel, Laurence, 122, 131, 142
Garti, Anne Marie, 168
Gasowski, Igor, 139, 161
Gastown Post & Transfer, 177
Gates, Jeff, 124, 133, 144
Gauldin, 149
Gavin, Bill, 133, 144
Gaviota Graphics, 139, 164
Gay Young Agency, 131
Gazelle Technologies, 194
GBH Productions, 172, 180
GCC Technologies Inc, 154
Gee Design, Earl, 149
Gee Whiz Enterprises, 125
Gellman, Rachel, 131
Gelman, Candice, 152
Gencarelli, Elizabeth, 131
General Graphics Co, 190
General Parametrics Corp, 161, 164, 166
General Television Network, 170, 175, 185
Generated Image, 190
Geniac, Ruth, 135, 146
Genigraphics, 158
Genigraphics Inc, 152
Genius Inc, 133
Gensurowsky, Yvonne, 149
Gentile Studio, 142
Gentry, David, 149, 161, 182
George & Company, 149, 161, 164
George, Cathy, 139
George Coates Performance Works, 167
George Mason University, 190
George Washington University, 152, 165, 205
Georgia Inst of Technology, 159, 163, 180
Georgia Pacific Television, 174, 180
Geovision Inc, 196, 198
Gerald Alters Inc, 184
Gerardi Studios, 124
Gerber, Mark & Stephanie, 133
Gerrard Associates, 149, 164, 176
Gerry McIntyre Photography, 127
Gersch, Wolfgang, 139, 168
Gershoni, Gil, 142
Gestalt Systems Inc., 141
Gibson Research Corp, 157
Gibson Studio Inc, 193
Gidley, EF, 124
Gil Bellin Productions, 158
Gilbert, Thom, 124
Gillian, 149
Gilmore Assoc Inc, Robert, 172
Gilo, Dave, 126
Ginsberg, Robin, 127
Ginsburg Productions, 176, 183
Girard Video, 158, 163, 172
GIST, 190
Gittler, Barbara, 122
Gladstone, Dale, 131
Glaser Media Group, 161, 164, 167
Glasgow & Assoc, Dale, 135
Glass Design, Michael, 147
Glassboro State College, 126
Glazer & Kalayjian Inc, 142
Glendale Studios, 176
Glenn & Associates, 161
Glenn, Mary Jane, 133
Glenn Roland Films, 176, 183, 187
Glessner, Marc, 133, 144
Glitch Graphics, 131, 142
Global One Design & Innovation, 139, 183
Global Vision, 183
Gloffke, Barry, 122
Gloucester County Coll/Lib Arts, 194
GMG International, 163, 174, 180
Gmucs, Rebecca, 131
Goavec Photography, 127
Godfrey, Ally, 151
Goehring, Steven, 139

Golan Productions, 160, 164, 181
Gold, Gary D, 124
Goldberg, Ken, 139
Goldberg, Lenore, 122
Golden, Helen, 139
Golden, Kenneth Sean, 131
Golden Dome Prdctns/WNDU-TV, 160, 181
Golden Dome Productions, 185
Golden Gaters Productions, 171, 176
Goldman, David, 122
Goldsholl: Film Group, 160, 178, 181
Goldstein, Edith, 136
Goldstein, Howard, 139, 149
Golici, Ana, 133
Goll, Charles R, 122
Gomberg, Susan, 151
Gomes, Lori, 169
Gomes, Steve, 171, 183, 187
Gomez, Rick, 125
Gonzales, Mariano, 166
Gonzauras, 193
Good, Zane, 139
Goode, Michael, 142
Goodfellow, Stephen, 136
Goodman, Alan, 144
Goodman, Barbara, 207, 209
Goodman, John, 136
Goodman Consulting Systems, 166, 209
Goodman/Orlick Design, 142, 178
Goossens, Jan, 122
Gordon, Joel, 122
Gordon Associates, Barbara, 131, 151
Gorelick Design, Alan, 144
Gorewitz, Shalom, 142, 178
Gorglione Holography, Nancy, 129
Gosfield, Josh, 131
Goss, John C, 149, 179, 187
Gould, Tom & Stephanie, 149
Gove, Geoffrey, 122
Goyer, Mireille H, 139
GP Color Inc, 130, 139, 193
GR Barron & Co, 185
GR Graphics Audio Visual, 161, 183
Graboff, Paul, 144
Grace & Wild Studios, 153, 181, 185
Grace, Alexa, 131
Grace, Laurie, 131
Graffito Inc, 144
Grafica, 149, 193
Grafica Multimedia Inc, 161, 164, 167
Grafiks Comm, 146
GraFX Creative Imaging, 130, 136
GRAFX, 161, 166, 167
Gragame-Harding Productions, 161, 168
Graham, Thomas, 131, 142
Graham-Henry, Diane, 126
Grahame, Donald, 139
Gramercy Broadcast Center, 171
Granberg, Al, 133
Grand Valley State, 1994, 152, 190
Graphic Access, 130
Graphic Actualization, 146
Graphic Alchemy, 136
Graphic Art & Design, 190
Graphic Art Marketing Information, 204
Graphic Art Resource Assoc, 158, 168
Graphic Artists Guild, 192
Graphic Arts Mrktng Info Svc, 204
Graphic Arts Sales Foundation, 190, 208
Graphic Arts Suppliers Assn, 131, 142
Graphic Color Image, 188
Graphic Comms Intl Union, 190
Graphic Consortium, 129, 144, 152
Graphic Data, 122
Graphic Design: USA, 139, 179
Graphic Design Services, 189, 193
Graphic Designers, 190
Graphic Enterprises, 156
Graphic Enterprises Inc, 188, 194
Graphic Evidence, 161, 164, 176
Graphic Express, 193
Graphic Expression, The, 142
Graphic Hearts, 190
Graphic Images, 133, 144
Graphic Media Inc, 154, 164, 187
Graphic Presentation Services, 193
Graphic Services, 156, 189
Graphic Systems, 129
Graphic Traffic, 193
Graphic Zone, The, 135, 146, 191
Graphics 150, 172, 190

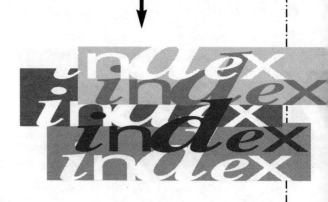

Graphics 4 Typography Inc, 193
Graphics at the Speed of Light, 135
Graphics by Nostradamus, 142
Graphics for Industry, 129, 142
Graphics Group, 188, 191, 208
Graphics Illustrated, 194
Graphics in Medicine, 131
Graphics Plus, 189
Graphics, 131
Graphics Trade Services Inc, 188
Graphics Unlimited Inc, 192
Graphics Unlimited, 144
Graphics West, 139
Graphics Workshop, 190
Graphique, 190
Graphis Type, 193
Graphisoft, 144
Graphix Zone, 161, 166, 183
Graphsoft, 155
Graphtec Inc, 188, 191, 194
Graphtec Inc, 188
Graphx, 146
Grass Valley Group, 198, 202
Graves, Tom, 127
Gravity Design, 139
Gravlee, Bob, 146
Gray Matter Advertising, 171, 176, 183
Grayphics Type & Design, 193
Grayson Media, 175, 181
Great Projections Inc , 154
Great Wave Software, 175, 181, 185
Greater Media Cable Adv, 158, 163, 166
Green, Howie, 144
Green, John, 133
Green Grphc Dsgn & Adv, Mel, 133, 144
Green Mountain College, 122
Greenberg Associates, R, 122, 151
Greenboam & Casey, 144
Greene Productions, Jim, 127, 130, 149
Greenfield Belser Ltd, 133, 144
Greenlight Corporation, 177
Gregg, Mutsumi, 149
Gregg, Rene, 122
Greiman Inc, 149, 176
Greke Film/Video/Animation, 170, 174, 180
Greland Graphics, Gerald, 142
Grey Matter Design, 139, 149
Greystoke, 152
Grid, Steve Chang, 142
Grien, Anita, 131, 151
Griffith, Sam, 126
Grigg, Roger Allen, 125
Griggs, Peter, 149
Grolier Electronic Publishing Inc, 194
Gross, Jonathon, 122
Grossman, Myron, 139
Grossman, Rhoda, 139
Grossman, Wendy, 131
Group 1 Design, 146
Group, 144
Group Video Productions, 161, 171, 176
Groupe Andre Perry, 177, 184, 188
GRS Inc, 175, 181, 185
Gruel, George, 139
Grumman Data Systems, 167
Grundy, Jane, 122, 205
GS Imaging Services, 188
GSI Ltd, 175
GT Group, 171
GTCO Corp, 197
GTP Design Studios, 136, 147
GTS Graphics, 193
Gudynas, Peter, 133
Guenzi Agents Inc, Carol, 152
Guerin, Francois, 139
Guild of Bookworkers, 174, 185
Gumucio Creative Comm, 147
Gundlach, Elisabeth, 129, 134
Gunnar Swanson Design Office, 149, 207, 209
Gussin, Jane E, 131
Gutenberg Expositions, 158, 172, 190
Gyration Inc, 199

H

H K Portfolio, 151
H&S Reliance Ltd, 189
H-GUN/Digital Div, 136, 147, 181
Habal, Jan, 128
Haber-Schaim, Tamar, 134
Hada, Gail, 139, 149, 207

Haedrich, Todd, 134
Haefele, Steve, 134
Hafeman Design Group, 147
Hagner, Dirk, 149
Hahn, Alexander, 142
Hahn, Bob, 124
Hahn Graphics Inc, 188
Haislip, Kevin, 127
Halcyon Software, 154, 157
Haleen, Brentano, 139, 169
Half Moon Video Productions, 172, 180
Half-Inch Video, 176, 183, 187
Hall, Jeffrie, 139
Hall, Lane, 136
Hall, Susan, 135, 146
Hall Kelley Organization, 149
Hallas Photo Lab, 130, 193
Halley Design, 139, 149, 207
Halsband, Michael, 122
Halsey Creative Svcs Inc, Dan, 126
Hamann, Brad, 131, 142
Hamann, Horst, 122
Hamelton, Meredith, 131
Hamill, Larry, 126
Hamilton, Bruce & Susan, 139, 164
Hamilton, Jeffrey Muir, 127
Hamilton Jr, Robert, 135, 165
Hamm & Assocs, 183
Hamm, Julian, 142
Hammarlund, Vern, 126
Hammer, Bonnie, 139
Hammes, Alan, 131, 200
Hampstead Computer Graphics, 155
Han, Calla & Peter, 122
Hand, Ray, 125
Hand to Mouse Arts, 136
Handler Group Inc, 142
Hands On Software Training, 147
Hanley, Katherine, 136
Hanna Productions, 179
Hannaway & Assoc, G W, 193
Hanson Assocs Inc, 144
Hanson Graphics of Memphis, 191
Hard Drives International, 196-198
Harholdt, Peter, 124
Harlan, Steve, 135, 146
Harlan Type, 192
Harman, Richard, 136, 147
Harmon's Audio-Visual Services, 153
Harmonic Ranch, 184
Harmony Press, 190
Harnett, Gena, 144
Harrington & Assocs, 149
Harris, Ellen, 134, 144
Harris, Martin, 136
Harris Design, George, 142, 158, 167
Harris Musicology, 127
Hartley, Jill, 142
Hartley Metzmer Huenink Comm, 170, 181, 185
Hartley Metzner Hunick Comms Inc, 160, 165, 167
Hartwig Music, 185
Harvest Communications, 175, 181, 185
Harvest Productions, 194
Hastings, Cynthia, 134
Hastings, Pattie Belle, 146
Hatcher, Mary Josie, 146
Hatfield, Lee, 134
Hathaway, Andrew J, 135
Hathaway Point Design, 144
Hatt, Shelley, 141
Hattersley, Lissa, 135
Haukom Associates, 161, 166
Hauser, Karl X, 139
Have A Happy, 149
Haveman, Josepha, 149
Hawaii Production Center, 176
Haxton, David, 162
Haydock Associates, 161
Hayes, Kevin, 130, 136
Haynes, Barry, 127, 201
Hays, Sorrel, 142
Haywood & Sullivan, 144
HB Type & Graphics, 193
HBO Studio Productions, 129, 131, 171
HD/CG New York, 171, 207
HDM Design, 146
Head Spin Studio, 136
Headline Typography, 193
Heal, Patricia, 122
Healy, Anne, 142
Hearn, Walter, 144

Heavy Meta, 142
Hebert, Jean-Pierre, 139
Hedstrom/Blessing Inc, 147
Heeter, Carrie, 136
Heiberg Studio, 122
Heifner Communications, 136
Heisler, Gregory, 122
Held, Cynthia, 152
Helical Productions, 176, 183, 187
Hellawell, Dennis, 172
Heller, Brian, 170
Heller, Donald, 144
Hellerman, William, 144
Hellman Animates, 178, 181, 185
Hellman Associates, 136, 147, 181
Helmar, Dennis, 124
Hemmeon, Bruce, 124
Henderling, Lisa, 139
Henderson, C William, 149
Henderson Typography, 192
Henneman-Hopp Design, 153
Henninger Video Inc, 167, 180, 185
Henriquez Studio, 136
Henry, Paul, 165
Heppert, David, 124
Herbert, Jonathan, 131, 200
Herigstad, Dale, 149, 161, 183
Herman, Ben, 146
Herman, Michael P, 141
Hermine Design Group, 134, 144
Herndon Jr, Tom, 129
Herring, David, 135, 146
Herschel Commercial Inc, 185
Hersh, Seth, 158
Hershey Multimedia, 161, 166
Hershman, Lynn, 149
Hess, Mark, 134
Hess, Robert, 129, 134
Hesselberg, Brenda, 136
Heun, Christine, 131, 142
Hewitt, Scott, 124
Hewlett Packard Co, 194, 196, 198
Heyl, Fran, 122, 151
Hi-res Slide Graphics, 189
Hi-Rez Graphics, 193
Hi-Tech Productions, 175, 181, 185
Hickman, Louis, 124
Hicks, Robin, 139
Hidy, Lance, 134
Hierro Studio Inc, 134, 144
Higgins, Michael Francis, 144
Higgs, Simon, 139
High, Philip, 135, 153
High Priority Consulting, 131, 142, 162
High Resolution Inc, 190, 205
High Technology Solutions, 154, 155
High-Res Solutions Inc, 122, 165, 166
Highton, Scott, 127
Hill-Cresson, Pat, 134
Hillier, Karen, 135
Hillis, Craig F, 139, 149
Hillstrom Stock Photo, Ray, 130
Himmelstein, Shelley, 142
HiRes Graphics, 134
Hitachi Denshi America LTD, 154, 155
Hitachi Sales Corp of America, 161, 196, 198
Hitchcock, Paul, 125
Hixson, George, 125
HK Graphics, 188

Hobbs, Pamela, 139
Hoberman, Perry, 158, 162, 168, 178
Hodgson, David, 122
Hoekstra Graphics, Grant, 147
Hoerr, Fred, 139
Hoey, Peter, 134
Hoffer, Jane, 122
Hoffert Communications, 188
Hoffman, Patricia, 139, 164
Hoffman Communications, 185
Hofstra University, 158, 172, 184
Hoit, Wayne, 178
Hoke, Ken, 134
Holcomb, Mike, 139
Holewski, Jeff, 134
Holland & Co, Mary, 152
Holland Litho Service, 189
Holland Printing, 192
Hollenbeck, Cliff, 127
Hollywood Interactive, 161, 167
Hollywood South Producers Centre, 176, 183, 187
Hollywood Vaults Inc, 193
Holmberg, Irmeli, 151
Holmes, Michael, 135, 188
Holmes Agency, 124, 134, 144
Holmes Animation, 163
Hologram Research Inc, 128
Holographic Applications Inc, 128
Holographic Images Inc, 129
Holographic Studios, 128, 129
Holographics Inc, 128
Holographics North Inc, 128
Holography Collection, 128
Holography Institute, 129
Holomat, 129
Holter, Catherine, 158
Holzer, Chris, 134
Homer & Associates Inc, 149
Hong, Won-Hua, 131, 162
Honkanen, William, 131, 142
Honolulu Commty College, 189
Honowitz, Ed, 127
Hoon, Samir, 142
Hoover, Duane, 122
Hopkins, Paul, 127
Horizon Entertainment Inc, 153, 174, 180
Horizon Images, 131, 142, 162
Horn, Jenny, 142, 162
Horn, 171
Hornall Anderson Design Works, 149
Hornbacher, Sara, 122, 131, 152
Horowitz, Jason, 125
Horowitz, Ryszard, 122
Hot Graphics & Printing Inc, 191
Hot Source Media, 163, 174, 180
Hot-tech Multimedia Inc, 132, 142, 162
Hothouse Designs Inc, 134, 144
Houck, Julie, 124
House Film Design, 161, 164, 179
House Graphics, 149, 154, 193
House of Graphics, 147, 192
Houston Photolab, 191
Hovde, Nob, 122
How, 149, 179
Howard Berman & Steve Bronstein, 122
Howard University, 136, 147
Howe, Philip, 139
Howtek Inc, 155
HP Productions, 159, 174

HSC SellMEDIA Creative Svcs, 161
HSD Microcomputer US Inc, 196, 198
Huang, Eric, 122
Huba, John, 122
Hudson, Ross, 142
Hudson Media, 178
Huggins, Rucker & Cleo, 139, 149
Hughes Design, 147, 208
Hui, Sandy, 149
Hull, Cathy, 132
Human Interface Technology Lab, 166
Human Performance Institute, 164, 167, 169
Hume, Kelly, 149
Hummingbird Productions, 185
Humphrey Jr, John J, 135
Hundertmark, Karen, 134, 144
Hunt, Robert, 139
Hunt, Steven, 127
Hunt Productions Inc, 185
Hunter, Marian, 178
Hunter, Nadine, 139
Hunter Publishing, 203
Hunza Graphics, 193
Husom, David, 126
Hutton, Lisa, 139
HVS Video/HVS Cable Adv, 175, 181, 185
Hyperglot Software Co Inc, 194
HyperMedia Group Inc, The, 161, 164, 167
Hypermedia Productions, 162, 166, 167
Hyperspeed Technologies, 196, 197, 199
Hyperview Systems Corp, 158, 165, 166
Hyphen Inc, 155

I

I/O Productions Inc, 176, 183, 187
IBIS Graphics, 154, 209
IBM/Educational Systems, 196, 197
IBM Marketing Svcs, 196, 197
IBM Media, 159, 170, 174
IBM Multimedia, 169, 198, 200
IBM Multimedia Solutions, 167
IBM TJ Watson Research Ctr, 191
Ice Tea Productions, 171
Icon Associates, 146
Icon Computer Graphic, 177
ICON Graphics, 134, 144
Icon West, 193
Iconcepts, 144, 158, 165
Iconics Graphic Design, 134
Iconink, 132
Iconos, 147, 160, 169
ICONS, 134, 144
ICT Productions, 158, 166
ID Studios, 142, 179
Idea Works, 158, 172
Ideaction Inc, 165
Identity Center, 147
IDesign, 142, 158, 166
IDG Communications/Amigaworld, 198
Ikeda, Tomoya, 139
Ikegami, 154, 197
Ikon Communications Inc, 149
IKONIC, 164, 167, 205
Ikonographics, 171
Ilford Photo, 198
Illl Inst of Technology, 147
Illinois State Univ/Coll of Fine Arts, 139, 149, 176
ILM (Industrial Light & Magic), 171, 183, 200

Image Advantage, 149
Image Architects, 154
Image Assoc, 170, 181, 185
Image Assocs, 146
Image Assocs Inc, 146, 153, 163
Image Axis Inc, 129, 142, 188
Image Bank, The, 122, 126, 127
Image Base Videotex Design Inc, 162, 166, 167
Image Center, 146, 159, 191
Image Club Graphics, 157
Image Communications, 193
Image Concepts Technologies Inc, 161, 164, 183
Image Foundry, 124, 188
Image Group, 158, 170, 172
Image Group Canada Ltd, 177
Image Maker, The, 149, 161, 164
Image Network, 129
Image Press, 188
Image Processing Software, 156
Image Producers, 175, 181, 185
Image Production, 158, 163, 172
Image Productions, 126, 136, 147
Image Recordings, 172
Image Resolutions, 190
Image Resources Inc, 146, 174
Image Source, 193
Image Systems Corp, 154
Image Technix, 177
Image Technology, 154, 156, 200
Image Transform Ltd, 194
Image Video Teleproductions, 175, 185
Image Ware, 177
Image Works, 165, 172, 180
Image World San Jose, 149, 154, 164
Image World/Video Expo NY 1994, 205
Imageland, 137, 147
Imageland, 149
Imagen, 135
ImageNet, 122
Images, 146
Images for Medicine Inc, 152
Images Inc, 196
Imageset Design, 190
Imagesetter, 190
Imagesetter Inc, 192
ImageSetters, 193
ImageSoft Inc, 156
Imagetects, 161, 200
Imageworks, 123, 152, 153, 190, 201
Imagic, 127, 130, 146, 176, 180
Imagicians, 162, 166, 167, 207
Imagics Design Group, 149
Imagination Effects, 175, 181, 186
Imagine That, 126, 139
Imaging, Inc., 126
Imaging Automation Inc, 155
Imaging Consortium Inc, 129, 189
Imaging, 130
Imaging Supplies & Eqpt, 168, 201
ImaginThat!, 169
Imapro Corp, 196, 199
Imergy, 158, 165, 168
Immaculate Concepts, 144, 166
Immersive Technologies Inc, 161, 167, 168, 199
Impact Communications Group, 161, 164, 167
Impact Corporate Communications, 161, 176, 183
Impact Ideas Inc, 124, 152
Impact Studios, 124
Impossible Images Inc, 125, 129

Imspace Systems, 168
In Sync, 185
In You Wendo Design, 135, 146
In-House Video Editing, 172, 180
Independent Graphics, 149, 154, 166
Independent Media Corp, 152, 178
Independent Media Producers, 154
Index Stock International Inc, 122
Indian Rock Image Setting, 193
Indiana State Univ, 166
Industrial Video, 160, 181, 185
Infinity USA Inc, 188
Info Systems Computer Ctr, 207, 209
Info Systems, 155, 196
InfoAccess, 141
Infomania, 193
Informatics Group Inc, 169
Information/Context, 144, 165, 167
Information International Inc, 198, 190
Information Navigation Inc, 160, 181, 202
Informed Solutions Inc, 144
Infoview Corp, 165
Ingham Photo Inc, Stephen, 127
Ingram Design, 147
Initiatives, 165, 166, 168
Ink Tank, 142, 179
InkSpot Designs, 146
Inkspot Printing Media Service, 191
Inkwell Inc, 142
Inland Audio Visual Co, 161, 176, 187
Inland Imaging Service/Via Type Corp, 193
Inman, Rebecca, 146
Inner Media Inc, 155
Inner Visions Group, 139
InnerStellar Productions, 139, 149, 161
Innervision Studios, 170, 175, 185
Innovative Data Design, 157
INSET Systems Inc, 124
Insight Productions, 159, 163, 170
InSight Systems Inc, 156
Inspired Solutions, 190
Instant Replay Video & Film, 175, 181, 185
Institute of Business Designers, 169
Institute of Electrical & Electronic Engineering, 204
Instruction Systems, 158, 165
Int Design #402 New Dsgn Bldg, 206
Intechnica Int'l Inc, 194
Integrated Graphic Media, 193
Integrated Solutions, 194
Intellichoice Inc, 161
Intelligenceware, 167, 168
Intelligent Light, 172, 180
Intelligent Resources Integrated, 199
Inter Video, 176, 187
Inter-Media Art Center, 173, 180
Inter-Tel Image Ltd, 177
Interactive Arts, 159, 163, 174
Interactive Audio, 161, 187
Interactive Computer Entertainment, 152
Interactive Design Inc, 161, 164, 167
Interactive Design, 158, 166
Interactive Illusions, 169
Interactive Innovations, 169
Interactive Media Technologies, 161, 166, 167
Interactive Multimedia 94/Aug 17-19, 1994, 161, 164
Interactive Music Co, The, 194
Interactive Personalities, 160, 164
Interactive Records, 187
Interactive Services Association, 164
InterCAP Graphics Systems, 155
Intercommunication, 209
Intercon Assocs, 154, 155
Interface Arts, 142, 178
Interface Comms, 146
Interface Video Systems, 170, 173, 180
Intergraph Corporation, 154, 156
Interleaf Inc, 155
Interlight International, 159
Intermedia/San Jose, CA/ March-April, 1995, 205
International Advertising Association Inc, 149, 183, 187
International Corporate Video Inc, 149, 183, 187
International Digital Fonts, 157
International Imaging, 146, 188, 191
International Interactive Comms Society, 194
InterNetwork Inc, 161, 166, 167
Interoptica Publishing, 194
Intersep, 130, 189, 193
Interset, 193
Intersound Inc, 176, 183, 187

InterVision, 149, 179
Intex Solutions Inc, 155, 200
Intl Assn of Busn Communicators, 190
Intl Design by Electronics Assn, 155, 173
Intran Electronic Publ Svcs, 192
Intrepid Productions, 161, 164, 166
Inventive Images, 175
Inverse Media, 134, 144, 173
INVIEW Corp, 161, 169
Invisions Inc, 164, 183, 203
IOC/Westbrook Technologies, 155
Iowa State Univ, 160, 170, 185
Ipa/The Editing House, 170, 175, 181
IPG Digital Imaging Ctr, 193
IPP LithoColor, 189
Irwin, Virginia, 134
IS Grupe Inc, 196, 200
Iseki, Keiko, 132
Iselin, Josephine, 139
Isgett, Neil, 125
Isis Inc, 161, 164
Island Graphics Corp, 157
Island Video Productions, 153, 174
Isley Design, Alexander, 142
Itami, Michi, 132
ITC, 159, 163, 174
Iterated Systems Inc, 177
Ivanoff, Deborah, 139, 149
Iverson Photomicrography, Bruce, 124
Ivey-Seright International, 130
IVL Post, 153, 175, 181
Iwasaki, Karen, 149
Ixion Inc, 164, 167
Iylem: Artists Using Sci & Technology, 129, 149
Izquierdo, Abe, 126
Izui, Richard, 126

J

J & D Publishing, 194
J & M Studio/Consulting, 132, 142, 208
J Dyer Inc, 163
J Gibson & Co, 144
J Parker Ashley & Associates, 175, 178, 181, 185
Jackson, Crystal, 178, 180
Jackson, Troi, 132
Jackson Design, 135, 146
Jackson Studios, 147
Jacobs Digital Arts, Todd, 126, 130
JAG Broadcast Video, 161, 179
Jager DiPaola Kemp Design, 134, 144
Jaime Martovamo, 124, 144, 163
James, Randy, 129
James, Rick, 134, 144
James Design Studio, Inc, 146, 159
James Fasso, 149, 183, 187
James Madison University, 158, 165-166
Jamison, Chipp, 125
Jane Lily Design, 139, 149, 183
Jane's Information Group, 194
Jane Sallis & Associates, 161, 166-167
Janesko, Lou, 129, 134
Jareaux, Robin, 134
Jariya, Peat, 146, 156
Jarpotawanich, Ampon, 122
Jarvis & Assoc, Nathan Y, 147
Jasin, Mark, 139, 149
Jasmine Multimedia Publishing, 194
JCH Group Ltd, 189
JD Media Consultants, 137, 147, 208
Jed Schwartz Productions, 180
Jeff Barnett Consulting, 209
Jeff Nichols, 155
Jeffrey Gardner, 131, 141
Jensen Communications Group Inc, 142
Jet Litho, 130, 137
Jet Propulsion Laboratory, 167
Jett & Associates, Clare, 151
Jex Fx, 129
JFM Graphic Design, 149
JHL & Assocs, 161, 166, 167
JHT Multimedia, 159, 174, 180
Jiempreecha, Wichar, 132, 162, 179
Jilling, Helmut, 126, 147
Jim Muse Presentation Technology, 161, 176
Jim Walters Color Separation, 189
Jim Zangmeister, 160, 164, 181
JIMANDI, 174, 178, 180
JMTV, 161, 171, 183
Joachim Studio Photography, Bruno, 125
Joanne Dr Szabla, 206

Jodee Kulp Graphic Arts Services, 192
Joe Bevilacqua Stills/Film/Video, 172
Joe's Type Shop, 193
Joel, Seth, 122
Joel Bennett Design, 146, 167-168
Joffe, Barbara, 139
John C Calhoun State Commty Coll, 178
John Flomer's Primal Cinema, 185
John McKee Productions, 173
John Ulliman, 140, 168
Johnson, Bud & Evelyne, 151
Johnson, Craig, 134
Johnson, Diane, 137
Johnson, James, 134, 169
Johnson, Jay, 139, 161, 179
Johnson, Paul, 124
Johnson, Stephen, 127
Johnson Design Inc, Dean, 137, 147
Johnson Galleries, Stephen, 146
Johnson Illustration, Paul, 137
Johnson Productions, Ric, 153
Jon Anthony Blumhagen, 136, 163
Jon Conrad, 138
Jon Garon Creative Inc, 178
Jonason, Dave, 139
Jonathan Helfand Music, 184, 197
Jones, Aaron, 127
Jones, Mark, 126
Jones, Maureen, 134
Jones, Ric, 134
Jones Design, Brent A, 149, 157
Jones-Rasikas, 160, 175, 185
Jordan, G Steve, 122
Josell Communications, 151, 184
Josephs, Carolyn, 142
Joule Group, The, 186
JPM Enterprises, 154-155, 184
JRA Interactive , 164, 166, 187
Judy Neis, 152
Juenger, Richard, 147
Juliana, Peter, 139
Julie Research Laboratories, 142, 196, 201
Jung, Mike, 126
Juntunen Video/Field & Post Production, 186
Jurewicz, Arlene, 128
Juzwik, Amy Stahl, 142, 208
JVC Co of America, 196, 197
JVC Disc America, 198

K

K&L Color Graphics, 188
K Landman Design, 142, 152, 158
K&S Photographics, 126, 130
K Squared Inc, 126, 137, 147
Kaake, Phillip, 127
Kaczmarkiewicz, Maryanne, 134, 144
Kadison-Shapiro, Sari, 137, 208
Kaetron Software Corp, 153
Kahl, David B, 134
Kahler, Charlotte, 122
Kaino, Glenn, 139
Kalamazoo Color Lab, 130, 192
Kalamazoo Label Co, 192
Kaliczak & Associates, 169
Kallish Group, The, 161, 164, 166
Kalnin Graphics, 190
Kaltman, Len, 126
Kamen Audio Productions, 184
Kana, Titus, 122
Kanar, Jodi, 147
Kane, Dennis, 152
KanImage Division, 122
Kannofsky, John, 168
Kansas City Art Inst, 146
KAO Optical Products, 170
Kapili, Jesse V, 200-201
Kaplan, Alan, 122
Kaprow, Alyce, 144
Karas, G Brian, 139
Kasper Studio, 160, 181, 186
Kass Communications, 142
Kastaris & Assoc, Harriet, 152
Katz, Gary S, 137
Katz, Marco, 184
Katzenberger, George, 127
Kauffman, Kim, 126
Kaufman, John, 129
Kauftheil/Rothchild, 132, 142
Kayenay Graphics, 189
Kayser, Alex, 158

KAZ, 122, 127
KCFW Productions, 176, 183, 187
Kean, Christopher, 126
Kean College, 125
Kedie Image Systems, 193
Keedy, 149
Keeley, Chris, 124
Keenan Assocs, 127
Kelemen, Stephen, 132
Keller & Cohen, 187
Keller, Steve, 139, 154, 179
Kelley Studios Inc, Tom, 127
Kellum, Ron, 142
Kelly, Beth, 122
Kelly, Sloan W, 144, 163
Kelly Michener Advertising, 144
Keltner, Stephen, 142
Kemper, Lewis, 127
Kendrick, Dennis, 132, 142
Kenngott, Barbara, 132
Kensinger Studio, 142
Kensington Microware, 155, 157, 197
Kent, David H, 152
Kent, Eleanor, 150
Kent, Nicholas, 134
Kent Kallberg Studios, 128
Kenwood Group, The, 161, 164, 166
Kern, Geof, 125
Kerns, Ben, 127
Keswickhamilton, 150
Ketchoyian Design, Suzanne, 142
Keynote Productions Inc, 186
Keystone Media Group, 173
KeyTronic, 155, 157
KGTV Channel 10, 171, 176, 187
KHL Consulting, 139, 150, 209
Kibby, 130, 189, 192
Kildow, William, 126
Kilkelly Associates, 122, 208
Killer Tracks, 187
Killian, Ted, 139
Killingsworth Presentations, 164, 183, 187
Kimball, Ron, 127
Kinard, Lou, 136
Kinesix Corporation, 168
Kinetic Presentations Inc, 156
King, Cindy, 144
King, Stephen, 139
King Graphics Inc, 130, 189, 194
King-Judge, Cynthia, 139, 150
Kingery Printing, 192
Kirk, Bev, 137, 147
Kirk, Roberta, 139
Kirkland, Douglas, 127
Kirwan Graphic Design, 146
Kirwin Communications, 176, 183
Kitses, John, 134
Kitzerow, Scott, 130, 137, 147
Klang Sound, 161, 187
Klein, Josh, 132, 179
Klein, Nikoai, 142
Klein, Renee, 132
Klein Design, 150, 193
Kleiser-Walczak Construction East, 180, 201
Kliger, David, 137
Kline, Daniel A, 129
Klineman, Peggy, 132, 142
Klinkowstein, Tom, 142
Klitsner Industrial Design, 139, 161, 164
Klopp, Karyl, 134, 144
Kluhspies, Christine, 144
Knight, Kevin, 122
Knight, William J, 122
KnoWare, 196, 201
Knowledge Garden Inc, 149, 154, 164, 205
Knowledge Industry Pubs, 159, 174, 185
Knowlton, Ken, 134
Knox, David, 139, 150
Knox Video, 197
KNTV Production Services, 171, 176, 183
Koala Acquisitions, 155, 157, 197
Kodak, 154, 194, 196-198
Kode, 142
Koehnline, James, 139
Kohanim, Parish, 125
Kohut, Paul, 137
Kollberg-Johnson Assoc Inc, 142
Kolnick Consulting, 169
Konkle, Kathy, 132
Koopman, Mary Ann, 122
Koosh, Dan, 127

Koplar Communications Ctr, 160, 175, 186
Kothari, Sanjay, 124
Kottwitz & Assoc, 122
Kowalski, Stephen, 137
Kozlowski Productions, 124
Kraft, Heather, 150
Krasner, Carin, 127
Krause, Dorothy Simpson, 134
Kreber Graphics Inc, 189, 192
Kress, Michael, 124
Kretzschmar, Art, 134
Krishna Copy Center, 193
Kristofik, Bob, 124
Krogstad Design, 147
Krongard, Steve, 122
Kroyer Films Inc, 183
Krueger Wright Design, 147
Krumme, Leah, 146
Kruza, Jay, 124
KTVU Retail Services, 171, 176, 187
Kubota & Bender, 144
Kubota Pacific Computers, 196
Kuehl, Allan, 129
Kuether Music Productions, 186
Kuntz, Diane, 150
Kursar, Ray, 132
Kurta Corp, 155
Kurzen, Aaron, 128
KV Graphics, 137, 147, 208
KWA Inc, 188
KXTX-TV, 170, 174, 185
Kyocera Electronics Inc, 198

L

L Graphix, 193
L Squared Studios, 123, 162
L-E-O Systems Inc, 160, 170, 186
La Driere Studios, 130
LA Trade Tech Coll/Art/Grphc Comm, 176, 183
LaBerge Graphic Design, 139, 150
Lackow, Andy, 134
Lacy, John, 126
LaFramboise, 205
Laguna Productions, 164, 183, 187
Lahr, Kimberly, 134
Laird Illustration & Design, 134, 144, 152
Lake Champlain Productions, 158, 163, 173
Lake Compuframes Inc, 130, 137, 147
Lamb & Co, 175, 181, 186
LaMotte, Les, 147, 153
Lamotte Studios Inc, 127
Lancaster Design, 150
Lance Studios, 142
Land, Fred, 127
Landman, Mark, 139
Landreth Studios, 127
Landwehrle, Don, 124
Lane, Edmund, 134
Lane, Morris, 123, 178
Lange Design, Jim, 137, 147
Langenstein, Michael, 142
Language & Graphics, 132
Language By Design, 169
Lanman Companies, 188
Lansing Community College, 144
LaRocco, Richard, 134
Larsen & Ludwig, 180
Larsen, Ernie, 124

Larsen Design Office, 147
Lasart Ltd, 129
Laser & Graphics Conf/March 1995, 129
Laser Edit East Inc, 170, 172, 179
Laser Express, 193
Laser Image Reprographics, 190
Laser Images Inc, 188
Laser Light Ltd, 128
Laser Magnetic Storage International Co, 196, 198
Laser Masters Inc, 156
Laser Media Inc, 179
Laser Pacific Corp , 161, 171, 187
Laser Pacific Media Corp, 161, 171, 176
Laser Storage Solutions, 169
Laser Tech Color Inc, 129, 188, 191
Laser Writing Inc, 193
Laser-Pacific, 167
Lasercom Inc, 192
LaserGo Inc, 157
Lasergraphics Inc, 154, 155
LaserMaster Corp, 154, 156
Laserscan Inc, 130, 189
Laserscan Systems Inc, 129
LaserTools Corp, 157, 194
Laserworks, 178
Lasry, Ronen, 141
LateNite, 154, 176
Latent Image Development Corp, 199-201, 208
LaTona, Kevin, 127, 150
Lattice Inc, 156, 192
Lauth Photography, 126
Lavin, Arnie, 134, 144, 180
LaVoie's Photography, 160, 175, 178
Law Cypress Distributing, 197
Law Holographics, 128, 199, 205
Lawhead, Elizabeth, 139
Lawler, Pat, 146
Lawson, Pamela, 127
Lax Syntax Design, 139, 150, 189
Laxer, Jack, 127, 161
Layton Intl, 159, 165, 167
Lazansky, Aaron, 133
Lazar, Jeff, 142, 178
Lazer Graphix, 193
Lazertouch, 193
Lazerus, 139
Le Baugh Software Corp, 156
Le Studio Du Centre-Ville, Inc, 177
Leaf Systems, 197, 198
Leaping Raster, 165
Learn-PC Video System, 186
Learning Tree/Comp Graphics Dept, 134, 144
LeBlanc, Terry, 134
Leckie, Laina, 134
Leduc, Lyle, 123
Lee DeForest Communications Ltd, 160, 170, 178
Lee Gagnon Music Production, 184
Lee Martin Productions, 159, 163, 180
Lee's Copy Network, 193
Leeds Design, Richard, 150
Leep Systems Inc, 199
Leete, William W, 134
Legs Akimbo, 137
Lehigh Interactive Multimedia, 158, 165, 166
Lehman, Cassandra, 139
Lehn & Assocs, John, 126
Lehner & Whyte Design, 134, 144

Leibowitz, David Scott, 124, 178
Leicht, Christina, 136
Leinwand, Freda, 123
Leitch Inc, 199
Lemelman, Martin, 134
Leng, Brian, 127
Lens Design, Jenny, 139, 150
Lentz & Associates, 153
Leonard Still Video, 124, 178
Leonhardt Group, 150
Lepine, Philip W, 134
LePrevost Corporation, 183
Les Animation Drouin, Inc, 184
Lesinski Photography, Martin, 123
Letraset USA, 124
LeVan, 134
Leveile Stawarz & Holl, 134, 144
Level 6 Computing, 123
Levin, Lon, 139, 150
Levin Design, Lisa, 150
Levine, Les, 142
Levy, Brian, 139
Levy, Jaime, 158
Lewczak, Scott, 136, 146
LeWinter, Renee, 134
Lewis Cohen & Co, 159, 163, 165
Lewis Creative Technologies, 191
Liao Inc, Sharmen, 139
Licata Associates Inc, 130
Lieberman, Fred, 124
Lieberman Productions, Jerry, 162, 172, 179
Lifetime Television, 172
Light & Power Productions, 152, 159, 173
Lightborne Communications, 164, 175, 181
Lightform Design, 150, 183
Lighthouse Productions Inc, 153, 175, 181, 186, 192
Lightpath Imageworks, 123, 201
Lightscape Graphics Software, 172
LightSource Images Inc, 147, 160, 165
Lightworks, 124
Lilie, Jim, 152
Lim, Deborah, 137, 164-165
Lincon, Denise, 124
Lindbloom, Bruce, 137
Lindgren & Smith, 132, 151
Lindroth, David, 134
Line & Tone, 189
Linhoff Corporate Color, 192
Linker, James Alan, 139
Linker Systems, 191
Linotext Group of Companies, 130
Linotype Hell, 154, 155
Lipman, Michael, 139, 164
Lipner, Robin, 132
Lipson, Stephen, 125
Liska & Assoc, 147
Litho Arts Inc, 189
LithoCraft Color Service Inc, 189
Lithographic Services Inc, 191
Lithographics, 193
Little & Co, 153
Little, Brown & Co, 194
Little Caesar Productions, 172
Little Gems Computer Consulting Svcs, 130
Live Marketing, 164, 167, 205
Live Time, 205
Lizarraga, Sergio, 139
LK Design Source, 147

Lloyd & Germain, 188
Location & Back Ltd, 186
Lochray, Tom, 137
Lockett, Carolyn L, 139
Lockwood, Scott, 127, 139, 150
Logical Art, 137
Logitech Inc, 199
Lombardo, William, 132
Long Island Video Entertainment, 173
Long Run Productions Ltd, 178
Longworth Communications, 174, 180
Look Inc, 184
Look Twice, 150, 168
Looking, 150
Lopez, Nola, 123
Loral Ads, 193
Lord, Rosalind, 132, 142, 178
Lori Castro, 155, 157
Lorick, Blake, 134
Los Angeles Creative Club, The, 150, 161, 164
Louey/Rubino Design Group, 143, 165, 168
Louisiana State University, 139, 150
Lovejoy, Margot, 142
Lovitt, Anita, 142
Lowery, Ron, 125, 146
Lowry, Rose, 134, 144
Lowry Graphics, David, 136, 146
Loyola College, 185, 206
LRP Video, 172
LSI Graphic Evidence, 139, 150
LSI Logic, 169, 199
LTS Productions, 175, 181, 186
Lubeck & Assocs, Larry, 130
Luce, Ben, 132
Luckett, Julie L, 125, 129, 136
Luckwitz, Matthew, 139
Ludtke, Jim, 139
Lui, David, 132
Lukas, Chris, 144
Lumeni Productions, 154, 164, 183
Lumigenic Media, 164, 166, 167
Luminarts, 150
LunaGrafix, 130, 137
Lunar Graphics, 191
Lund, John, 127, 130
Lundy Computer Graphics, 154
Lusk, Danielle, 150
Lussier, Robert, 134
Lutheran Laymens League, 147
Lux & Associates Inc, Frank, 152
LVI Video Productions, 186
LVT-Kodak Company, 198
Lynch, Alan, 151
Lynch, Jeffrey, 134
Lynch Graphic Design, 139, 150
Lynn, Jenny, 124, 134
Lynn Goodman, 196, 200
Lynx Digital Design, 130, 139, 164
Lyon Design, Catt, 147
Lyon Lamb Video Animation Systems Inc, 199
Lyons, Lisa, 144
Lyons, Steven, 139
Lyons Inc, 152

M

M & Co, 142
M & I Recording, 184
M A D, 150

M Design, 144
M Graphics, 147
M Plus M Inc, 142
M Scullin & Associates, 134
M2 Design, 142
Mably, Greg, 141
Mac in Design, 126, 147
MacDonald, Greg, 139
Macfactory, 191
Machattan, 169
MacInstitute, 137
Macintosh Product Registry, 139
Macko, Nancy, 168, 169
MacMedia Systems of Orlando, 136, 163
Macmillan New Media, 194
MacNeill and MacIntosh, 134, 144
MacNicol, Gregory, 139, 150, 179
MacProducts, 154, 196
MacShane Laser Art, 129
Mactemps, 156
MacWeek, 1994, 139
Madridejos, Fernando, 142
MAF Graphic Design, 147
Magazine, 134
Magic Graphics, 129, 190
Magic Venture Studios, 184
Magical Media, 159, 163, 169
Magiera, Rob, 139
Magika, 139, 150
Magna IV Color Imaging, 188
Magnetic Fax, 177
Magnetic Music Publishing Co, 184
Magnetic North, 177
Magni Systems Inc, 196-197
Magno Sound & Video, 172
Magnum Communications Inc, 153
Magnum Design, 150, 167
Magnus, Mark, 172
Mahoney, Bob, 124
MAI2 / The Graphics Room, 152, 163
Maikoetter, Mitch, 136
Maile, Richard W, 136
Main Frame Inc, 153
Main Point Productions, 152, 163
Main Street Multimedia, 161, 164
Main Street Video Productions Inc, 161, 171, 187
Mainsail Production Services, 164, 175, 181
Mainstay, 155, 157
Mainstream Communications, 160, 175, 181, 199
Maisel, David, 127
Majlessi, Heda, 150
Mallory Marketing Communications, 178, 186
Maloney, Jeff, 127
Mambo Music Ltd, 184
Mammoth Microproductions Inc, 196, 199
Man, 129, 199
Management Graphics Inc, 154
Mandelkorn, Richard, 124
Manelis, Jessica, 134
Manhattan, Maria, 132
Manhattan Transfer/Edit, 172, 179
Mani, Anand, 141
Mankus, Gary, 126
Mann, Heidi, 134
Mannes, Don, 132
Mannesmann Tally Corp, 198
Manning, Ed, 124
Manning, Lawrence, 127
Manzione, John, 134
Mar Design, William, 139, 150
Marc Multimedia Ltd, 159
Marcolina Design Inc, 134, 144
Marcus & Assocs, Aaron, 166, 168, 179
Marcus, Joel, 125
Margolin, Diane, 132
Mariah Productions Ltd, 152, 158, 172
Marie & Friends, Rita, 152
Marinstat, 193
Marist College, 186
Marketing Comms By Design, 147
Marketing Connection, The, 160, 164
Marketing Div, 158
Marketing Network, The, 153, 159
Marketing Resources, 193
Markley, Andy, 139
Marley, Stephen, 127
Marotte, Frank, 125
Marr, Julie, 123
Marsden Reproductions, 172, 179
Marseille, Anne, 144
Marsh, Finnegan, 169

Marshall Photo, Don, 126
Marshall Productions Inc, 159, 163, 174
Marsico, Dennis, 124
Marstek Inc, 155, 157
Martha Productions Inc, 152
Martin, Sean, 141
Martin/Bastian Communications, 186, 192
Martin Brinkerhoff Associates, 127, 139, 161
Martin/Greater Film Graphics Inc, 188
Martino, Jacquelyn, 132
Martins, Marcos, 141
Marx Production Ctr, 170, 181, 186
Mary Perillo Inc, 142
Marycrest University, 144
MAS Services, 186
Mascioni, Michael, 168
Mason & Associates Inc, Dave, 151
Mason Studio, John, 139
Masque Publishing Inc, 194
Mass Coll of Art/Comptr Arts Ctr, 155, 157
Mass Productions, 150, 179
Mast, 154
Master Communication, 171, 183, 187
Master's Touch Inc, The, 137
Master Video Productions Inc, 186
Master Videoworks, 171, 183, 187
Masters Presentation Graphics, 154, 193
Matador Design, 142
Match Frame Post Production, 174, 180
Math & Computer Science, 186
Matrix Communication Arts & Svcs, 193
Matrix Communications, 193
Matrix Photo Labs, 126, 130
Matrix Video Inc, 170, 175
Matrox Electronic Systems Ltd, 197
Matsuri Corporation, 142
Matthew, Matthew K, 144
Matthews, Scott, 137
Matthews/Griffith Music, 187
Mattingly, David B, 134
Mattingly, Matthew, 134, 163
Mattingly Design, George, 150
Mattioli, Angela, 139, 150
Mattlin, Brian, 158, 178
Maurio, 123
Mauro, George, 124, 144
Maus Haus, 161, 164
Max Grafix, 142
Max Loftins Quality Graphics Inc, 193
MAX, 139
Max Place, 146, 159, 174
Max PostScript, 192
Maxedon, Terry, 128, 132
Maximum Impact Design, 139, 150
Maxon, Paul, 127
Maya Technology, 142, 178
Mayer, Judith, 169
Maziacyzk, Claire, 134
MB Productions, 180
MBTV Inc, 180
MC2, 127, 130, 139
McBrien Productions Inc, 184
McCamant Productions Animation, 161, 164, 183
McCann, Michael K, 126
McCarroll Advertising & Design, 150
McCarthy, Tom, 123, 124
McClain, Craig, 127
McCord, Jeff, 139
McCormack, Geoffrey, 132
McCormack, Richard, 124
McCormick, Peter, 139
McCourt, Tim, 139
McCoy, Dan, 124, 134, 144
McDaniel, Jerry, 132, 142
McDonald, Keith, 125
McDonnell, Patrick, 141
McDonnell Douglas, 171, 183, 187
McDonough, Patricia, 123
McElroy, Darlene Olivia, 132
McGrath, Bonnie, 124
McGraw Hill, 194
McGroarty Advertising, 145
McGuire, Steve, 150
Mchale Videofilm, 176
McIntyre Advertising, 193
McKenzie, Crystal, 142
McKinley, Liz, 146
McLain, W Clay, 136
McLean Media, 161, 166, 167
McLeod, Lisa, 142
McManus, Robert, 139

McMillan Co Public Relations, 179
McNamara, Casey, 124
McNicholas, Florence, 142
McWilliams, Jack, 124
MDC Teleproductions, 171, 183, 187
Mead Paper, 192, 205, 206
Media Corp, 161, 167
Media Arts, 137, 206
Media Construction Inc, 159, 165, 166
Media Consultants, 175, 181, 186
Media Culture, 150, 203, 207
Media Cybernetics Inc, 155
Media Design, 139
Media Enterprises West, 176
Media Events Concepts Inc, 153, 174
Media Foundation, 141
Media Group Inc, 170, 181, 186
Media Group Television, 170, 175, 186
Media Lab Inc, 166, 168
Media Lab, 161, 164, 183
Media Learning Systems, 166, 167, 169
Media Loft Inc, 170, 175, 186
Media Management Systems Inc, 136, 163, 200
Media Mark, 140, 150
Media Mix, 173, 180
Media Power, 190
Media Process Group Inc, 160, 170, 175
Media Productions Inc, 186
Media Projects Inc, 178
Media Services-SoundMax, 187
Media Staff, 176, 183, 187
Media Team, 186
Media Technics, 169
Media Vision, 152, 159, 163, 173, 194
Media Vision Productions, 152, 173
Media Works, 152, 188, 190
Mediadesigns, 169
Mediainterface, Ltd., 128
Medialogic Inc, 161, 166, 167
Mediamix Limited, 178
Mediascape, 136
Mediaware, 168
Medina Software Inc, 156
Medio Multimedia, 194
Medior, 161, 166, 167, 193
Mednick Group, The, 150
Medvec, Emily, 124, 145
Meehan, Joseph, 124
Meeting Media Enterprises Ltd, 160, 164
Mega Music Corp, 184
Megavision, 197
Megcomm Film & Video Productions, 173, 180
Meggs, Philip, 146
Meisel Photographic Corp, 129
Mel Typesetting, 193
Melanson Assoc, Donya, 145
Melnick, Anil, 132
Melovision Productions Inc, 173, 180
Melton, Janice Munnings, 124
Memory Technology Texas, 154, 196
Memphis College of Art, 123
Mendola Ltd, 151
Mendoza Design, 145, 159, 165
Menten Music Inc, 186
Meola, Eric, 123
Mercer, Celia, 150
Merchandising Workshop, 142, 172
Meredith, Diane, 125
Meridian Data Inc, 169, 196
Merrill, John, 145
Merrill, Lizanne, 123, 132
Messex, Mike, 140
Meta 4 Multimedia, 159, 163, 168
Metafor Imaging, 130, 140
MetaLanguage, 161
Metamorphosis Conputer Concept, 142
Metatec Corp, 199
Meteor Photo & Imaging, 126, 130
Meteor Photo & Imaging Co, 129, 191
Metro ImageBase Inc, 157
Metro Productions Inc, 175, 181, 186
MetroCreative Graphics Inc, 155
MetroGrafik, 142, 190
Metrolight Studios, 176, 183
Metropolitan Entertainment, 164, 183, 187
Metropolitan Skills Ctr, 174
Metzger's Prepress, 189, 192
Meyer & Assocs, Aaron D, 126
Meyer, Bonnie, 145
Meyer Software, 159, 163
Meyerfeld, Alan, 140, 150

Meyerowitz + Co, Michael, 142, 205
Meyler, Dennis, 125
Mezzina/Brown Inc, 142
MFS Inc, 137
MGS Services/Viacom Div, 175, 192
MHX Designs, 172
Michael Dansicker Music, 184
Michael Furman Photography Ltd, 124
Michael Weiss Photo, 124
Michaels, Christion, 127
Michaels, Robert, 127, 130
Michaud, Brian, 124
Michels Photo, Bob, 127
Michigan Guild of Artists & Artisans, 196, 198
Micro Dynamics Ltd, 155
Micro Interactive, 158
Micro Visual Systems Inc, 159, 166
MicroAge/Santa Monica, 132
Microcomputer Publishing Center, 190, 203
Micrographics, 181
Micrographx, 156
Microlytics Inc, 154, 155
Micron/Green, 159, 191, 208
MicroNet Technology Inc, 155, 196
Micropage/Microtype, 188, 190
Microprint, 190
MicroPublishing News, 203
MicroSearch, 196-198
Microseeds Software, 155
Microsystems Engineering, 156
Microtech International Inc, 196-198
Microtek Lab Inc, 155, 198
MicroTimes, 153
Microvitec Inc, 156
Microware Systems Corp, 156, 200
MID-CO TV Systems Inc, 186
Midi Inc, 159, 165, 166
Midland Production Corp, 161, 164, 183
Midnight Oil Studios, 132, 134, 140, 142, 145, 150
Midtown Imaging, 192
Midwest Graphics, 192
Midwest Litho Ctr/3D Imaging, 130, 137
Midwest Teleproductions, 175, 186
Migliozzi, Kathy, 124
Migraph Inc, 155, 157
Mike Jones Film Corp, 181
Mikel Covey & Traci O'Very, 150, 164, 167
Mikros, Nikita, 132
Miles 33 International, 155
Mill City Records, 186
Millenium Group, The, 124
Miller, J T, 124
Miller, Jack Paul, 140, 161, 176
Miller, Jordan, 127
Miller, Marilyn, 123
Miller, Melissa, 134
Miller Design & Associates, 176, 183
Miller Mauro Group Inc, 152
Miller Multi Media, 153
Millet, Cecile, 137, 147
Milliken University, 125
Mills & Co, L Jane, 151
Mills, Elise, 132
Mills, Robert, 123
Mills-James Prdctns, 160, 181
Mills-James Prdctns/Teleproduction Ctr, 160
Mills-James Productions, 164
Milne, Bill, 123
Milwaukee Advertising Club, 125
Mind Links, 154, 155, 157
Mind Meld Reps, 152
Mind Over Macintosh, 166, 207, 209
Mind's Eye Graphics, 153, 174, 180
MINDesign, 169, 205, 208
Mindflex, 159, 163, 174
Minelli Design, 145
Minerbi, Joanne Sheri, 145, 166, 168
Minneapolis Coll of Art & Design, 137, 147, 206
Minneapolis Tech College, 168
Minnesota Film Board, 186
Minnix, Gary, 137
Mintz, Les, 151
Mira Film & Video, 176
Mirage Visuals, 152, 159, 188
Mirak Video Communications Services, 178
Mirenburg & Co, 190, 208
Mirror Technologies, 154, 198
Mison, Vesna, 132
Mistretta Illustration Studio, 134, 178
MIT Ctr for Education Computing, 165, 166, 168

MIT, 168
Mitchell, Diane, 123, 142
Mitchell, Mindy, 137, 147
Mitchell Photo, Josh, 127, 140
Mitnick Design Inc, Joel, 145
Mitsubishi Electronics America Inc, 197, 198
Mitsubishi Intl Corp, 154, 155, 198
MLH Communications Group, 132
MMI, 191
MMZ Graphics, 193
Moberley, Connie, 188
Mobil Image, 178
Mobile-Video Productions Inc, 173, 180
Mobius Artists Group, 145, 178
Mock Design Assoc, 150
Mockensturm, Steve, 137
ModaCAD, 168, 200
Modern Graphics, 193
Modern Imaging/WACE USA, 130
Modern Mass Media, 173
Modern Video Productions Inc, 159, 170, 184
Modern Videofilm Inc, 164, 167, 176
Modern World Media Inc, 173, 180
Modesto Jr Coll/Art Dept, 123
Moetus, Olaf, 137, 147
Mofoto Graphics, 126, 147
Mohawk Coll/Applied Arts & Techn, 162
Mok Designs, Clement, 150, 161
Molenhouse, Craig, 127
Molina, Jose R, 125
Monaci, Steven, 179
Monderer Design Inc, Stewart, 145
Mondo 2000, 161, 164, 169
Monotype Inc, 154, 156
Monotype Typography, 153, 156
Monroy, Bert, 140
Montage Publ Inc, 134
Montana Graphics, 132, 142
Montano, Daniel, 127, 150
Montclair State College, 146, 165, 169
Montgomery, George, 150
Montgomery Coll/Applied Technologies, 150
Moojedi, Kamran, 127, 140, 179
Moonlight Press Studio, 132
Moore, Tom, 132
Moore College of Art, 125
Moorpark Jr College, 162, 166, 167
Morales Design, Frank, 146
Moran, Michael, 134
Moran Studios, 150
Morawa, Amy Lynne, 132, 158
Moreau, Sylvain, 162
Morecraft, Ron, 134
Morgan Associates, Vicki, 132, 151
Morgan Communicating Art Ltd, 136
Morgan Guaranty Trust Company, 158
Morgan Inc., Scott, 127
Morita, Tatsuya, 123
Morla Design, 150
Morrell, Paul, 127
Morris, Burton, 134
Morris, Don, 136
Morris, Grey, 132
Morris, 142
Morrison, Joanna, 132
Morrissey Photography, 125
Morrow, Beret, 136
Morrow, Michael, 136, 146
Morrow, Skip, 134

Morse, Bill, 140
Morse, Michael, 123, 132
Mortensen Design, 150
Mortice Kern Systems Inc, 157
Morton Graphics, 190
Moser, 124, 145, 178
Moskovic, Stephen, 132
Motion City Films, 162, 166-167
Motionart Studios, 145, 180
Motivation Media Inc, 160, 164, 165
Motivation Media Inc, 170, 175, 181
Mouse Systems Corp, 155, 157
Movidea Inc, 145
Moving Pixels, 183, 187
Moxie Media Inc, 159, 163, 174
Moy, Willie, 126
MPL Film & Video, 170, 174, 185
MPTV Video Production Unit, 159, 165, 173
Mr David Savage, 136, 146
Mr Film, 162, 177, 183
MSI Marketing/Communications, 193
Mt Hood Community College, 158, 170, 172
Mulhern, Mike, 142
Mulkey, Roger, 127
Muller & Co, John, 147
Mullican, Matt, 142
Mulligan, Donald, 132, 142
Multi Image Group, 174
Multi Image Productions Inc, 162, 166, 167
Multi Video Group Ltd, 172
Multi-Ad Services Inc, 156
Multi-Ad Services, 156
Multi-Media Computing Corp, 169
Multicom Publishing, 194
MultiDimensional Images Ltd, 123
Multimedia & VideoDisc Monitor, 1994, 162, 164, 167
Multimedia Library, 162, 166, 168
Multimedia Products Corp, 194
Multimedia Professionals, 127, 140, 150
Multimedia Publishing Studio, 194
Multimedia Research, 169
MultiMedia Resources, 159, 165, 169
Multimedia Review, The Journal of, 203
Multimedia Technology Lab, 163
Multimedia Technology Labratory, 159
Multimedia Workshop, 159, 165, 166
Multivision, 159, 173, 174, 185
Multivision Video & Film, 159, 174, 185
Mundy, Ann, 169
Munro/Goodman, 152
Muresan, Jon, 126
Murphy, Alan, 183
Murphy, Charles, 137
Murphy, Michael, 125
Murray, Bill, 127
Murray Lienhart Rysner & Co, 147
Murrie, Lienhart, Rysner & Assoc, 147
Musgrave, Steve, 137
Music Advantage, 186
Music Staff Inc, 186
Music Works, 185
Musivision, 179
Muskie, Stephen O, 124, 145
Muskovitz, Aaron, 129
Muskovitz, Rosalyn, 147
Mustafa Bilal Photography, 127
MVI/POST, 174, 181, 185
Myers, Barry, 204

Myers, Jeff, 125, 146
Myers Graphics, David, 132
Myman, Barbara T, 140
Myrvik Productions, Ron, 142, 172
Mythic Graphics, 132, 142, 208
Mytilene Enterprises, 154

N

N Arizona Univ, 192, 208
NAB Multimdedia News, 173, 180
Nairn Gaphic Design, Doreen, 140
Nakamura, Ikuo, 123
Napoleon Video Graphics, 172
Nappi Inc, Maureen, 142
NASA Ames Research Ctr, 168, 207
Nasty Productions, 140
National Assn of Artists' Organization, 173
National Center for Super Computing, 204
National Color, 191
National Digital Corp, 156
National Education Training Group, 159, 165, 166
National Geographic Society, 194
National Photographic Labs, 191
National Teleprinting Inc, 193, 196, 198
National Video Industries, 172
Natl Assn of Broadcasters, 180
Natl Computer Graphics Assn, 147
Naudin, Suzanne, 145
NBC Magic, 150, 177, 183
NBC Telesales, 172, 179
NC Supercomputing Ctr, 165
NE Illinois Univ, 137
NEC Home Electronics Computer, 194
NEC Inc, 129, 188, 191
NEC Technologies (USA) Inc, 196-198
Necro Enema Amalgamated, 162
Needham Graphics Inc, 154, 155, 190
Nehman Kodner Design, 147
Nelson, Jeffrey, 147
Nelson, Pamela, 136
Nelson, Tom, 126
Nemtzow, Scott, 128
Neo Geo, 153, 174, 181
Neographic, Inc, 158
Neotech Interactive, 129
Neptune, Frantz, 145, 152, 163
Nesnadny & Schwartz Inc, 147
Nessim & Assocs, Barbara, 132, 142
Network 90, 169
Network Music Inc, 187
Neumann, William, 123, 142
Neumeier Design Team, 150
New Edge Technologies, 134, 145, 208
New England Computer Arts Assn, 159, 166
New Horizons Cmptr Learning Ctr, 207
New Media Group, 134, 145, 163
New Media, 129, 165, 167
New Microtime, 199
New Overbrook Press, 145, 205
New School for Social Research, 184
New York Holographic Labs, 128
New York Inc, 151
Newfound Music, 184
NewGen Systems Corp, 198
Newman & Associates, Carole, 152
Newman, Tom, 168
NeXT Inc, 196
Next Wave Productions, 152, 190

Next World/PCW Communications Inc, 140, 150
Nexus Productions, 172, 179
Nexvisions, 123, 142
NFL Films Video, 159, 163, 166, 180
Ng, Michael, 142
NG Visual Art, 128
Nicholas Assocs Design Consultants, 148
Nichols, Mary Ann, 142
Nicola Katharina Wewer-Elejalde, 178
Nielsen, Ron, 126, 148
Nielson Design Group, 148
Niemann, Andrew, 141, 165
Niffeneger, Bill, 137
Nikon Inc/EID, 154, 197
Nimbus Info Systems, 170, 194, 196
Nimbus Records, 194
NISCA Inc, 154, 156, 198
NJ Motion Pic & TV Comm, 170
NM Highlands Univ, 140, 150
Nolan, Roy, 162, 166, 167
Noneman & Noneman Design, 142
Norac Productions , 162, 164, 177
Noral Color Corp, 189, 192
Nordell Graphic Communications Inc, 189
Nordstrom, Jennifer, 145
Norkin Digital Art, 129
Norling Studios Inc, 130, 191
North Country Media Group, 162, 177, 183
North Forty Productions, 184
Northeastern Univ/Univ College, 129
Northern State University/Art, 154, 178
Northwest Media, 177, 183, 187
Northwest Mobile Television, 175
Northwest Teleproductions, 170, 175, 178, 181
Northwestern Mutual Life, 170, 181, 186
Northwestern Univ/Comp Music Studio, 134
Notovitz Design Inc, 142
Notre Dame College, 191
Nova Media, 137
Nova Scotia Coll of Art & Design, 207
Novachrome, 192
Novack, Dev, 132, 152, 168
Novak, Dennis, 123
NovaWorks, 142
Novell, 155
Novich, Bruce, 145
Novocom/GRFX Productions, 162, 164, 183
Novus Visual Communications Inc, 142
NPES, 194
NTERGAID, 196, 197
Nu Vox Animation & Design, 145, 163
NuLight Studio, 125
Number Nine Computer Corp, 154, 155
Number One Graphics Inc, 192
Number Seventeen, 143
NUS Training Corp, 159, 165
Nuthouse Studios, 192
Nutopia Digital Video, 164, 183
nView Corp, 197
NW Printcrafters, 192
NY Inst of Technology, 134, 145, 178
Nyberg, Tim, 137
Nygard & Associates, 160, 175, 186
NYIT/Metropolitan Center, 126

O

O'Connell, Francis C, 134
O'Connor, Buster, 146

O'Dell, Dale, 125
O Design Group Inc, 143
O'Farrill Music Ltd, 184
O&J Design, 143
O'Kelley, Lynn, 132
O.N.E. Color Communications, 189
O'Neil, Julie, 124
O'Shea, Kevin, 140
O'Toole, Terence, 127
O'Very/Covey, 150, 164, 167
Oasis Music Inc, 184
Oasis Post Production Services, 171, 178, 184, 188
Oasis Studios, 171, 177
Obata Design, 126, 137, 148
Oberlin College, 155
Oceana, 173
Ocron Inc, 198
Octree Corporation, 168
Odom, Laddie Scott, 160, 169, 200
Odyssey Communications Group, 159, 174, 181
OEC Graphics Inc, 189
Of Mice & Graphics Inc, 136
Ogdemli/Feldman Design, 140, 150, 162
Ogilvy Mather Direct, 158
Ogrudek Ltd, 123
Oh, Jeffrey, 134
Ohio State Univ/Adv Comptng Ctr for Art&Dsgn, 137, 148
Ohio State Univ/Comp Aided Arch Dsgn, 140
Oko & Mano Inc, 143
Olausen, Judy, 126
Oldach, Mark, 148
Oldroyd Ltd, Bob, 191
Olduvai Corp, 156
Olive, Tim, 125
Olsen, Dan, 124
Olsen Muscara Design, 143
Olson, Carl, 125, 146
Omaha Graphics Inc, 189
Omega Films, 174, 181, 185
Omicron, 169
Omni Media Productions Ltd, 154, 178
Omni Productions, 170, 181, 186
Omnimusic, 184, 187
Omniprint, 193
Omnivore, 150
Oms Business Services, 192
On Line Video, 175, 181, 186
On Tape Productions, 171, 177, 183
On Target Marketing, 193
On the Go Software, 197
On the Wave Visual Communications, 140, 168
On Video Inc, 153, 174
On Word Inc, 193
On-Q Productions, 162, 179
One on One Design, 150
One World Interactive, 164, 167, 168, 169, 171
Ong, Philip, 143
Onli, Turtel, 137, 165, 167
Online, 169, 196, 200
Ono Stuff, 162, 166, 168
Ono-Sendai Corp, 199
Ontario College of Art, 194
Opcode Systems Inc, 187
Open Computing, 203
Optical Access Intl, 155, 196, 198
Optical Media International, 196
Optical Publishing Assn, 190
Optimage Interactive Services LP, 196, 200
Optimax, 154
Optimedia Systems, 159, 166, 173
Optimum Group, 148, 160, 164
Optimus Inc, 164, 181, 186
Optisonics Productions, 145, 173
Optisys, 157
Optronics/Intergraph Div, 155
Orange Coast Coll/Fine Arts, 140
Orange Micro Inc, 155, 157, 197
Orcutt, Bill, 123
Orent Graphic Arts, 189
Original Cinema Inc, 172
Orlando Multimedia 95, 146
Osama Ltd, 132, 179
Osiecki, Lori, 140
Osiow, Andrew, 140, 164
Ostro Design, 145
Otis College of Art & Design, 145, 178, 180
Otterson Television Video Inc, 172, 179, 184
Otto, Jeff, 134

Otus, Erol, 140, 150
Our Designs Inc, 143
Our Lady of the Lake Univ, 145
Our Town TV, 173, 180
Outerspace, 128
Outwater Prods, Robert, 128
Ovation Graphics, 173, 180
Overdrive, 141, 151
OWL Software, 157
Oxberry/Div of Cybernetics Products, 124, 129, 170
Ozam Film & Tape Productions,, 175, 181, 186

P

P Beach Illustration, 134
P E Photron, 199
P T Pie Illustrations, 134
Pace, Julie, 140
Pace Color Separation Inc, 188
Pace Video Center, 177, 183, 187
Pacific Data Images, 150, 183, 193
Pacific Data Products, 155
Pacific Digital Image, 130, 189, 193
Pacific Focus Inc, 177
Pacific Interactive Design Corp, 162, 167, 183
Pacific Interface Inc, 169, 209
Pacific Media Associates, 162, 171, 177
Pacific Micro, 157
Pacific Motion, 150
Pacific NW Coll of Art, 177, 183, 187
Pacific Post Productions, 178, 188
Pacific Rim Video, 162, 171, 187
Pacific Title & Art, 193
Pacific Video Resources, 162, 164, 177
Packard, Wells, 143, 168
Page & Page Slidemarket, 124, 145, 152
Page, Lee, 123
Page One, 193
Page Printing, 193
Page Studio Graphics, 157
Page-Trim, Debra, 134
PageMaker User Group/NADTP, 190
Pagliaro Design, Joseph, 148
Paglietti & Terres Design Co, 150
Paintpot Productions, 140
Palermo, David, 140, 150, 179
Palette Studios, 124
Palm, Brad & Maggie, 130
Palmer, Laura Leigh, 134, 145, 163
Palmieri, Jorge, 124
Palms High Res Digital Imagery, 130
Palomar College/Graphic Comm, 167, 169, 209
Panasonic, 196-198
Pandemonium Design, 140
Panoptic Imaging, 134
Panorama Productions, 177
Pantuso, Mike, 132, 143
PaperDirect Inc, 155
Paradigm Communication Group, 160, 182, 186
Paradise Software, 132
Paragraphics, 132
Parallax Graphics, 143
Parian, Levon, 128
Park, Chun-Sin, 132
Park Avenue Teleproductions Inc, 159, 174, 185
Parker, Charles A, 124
Parker, Nick, 146, 208
Parks, Peggy, 128
Parsons Design, Glenn, 140, 150
Partners & Agostinelli Productions, 170, 172, 184
Partners by Design, 140
Partners III, 131
Paschal, Richard, 140
Pasinski Assocs, Irene, 145
Paskevich, John, 140
Passage Productions Inc, 189
Passin, Jim, 160
Passport Designs Inc, 187
Paston, Herbert, 134
Patchwork Productions Inc, 162, 164, 167
Pate & Assocs, Randy, 152
Paternoster, Nance, 140, 164
Paterson, Nancy, 151
Patrick Graphics, 191
Patrick LaFramboise Inc, 205
Patton Brothers Design, 140
Patton Sight & Sound, 169
Paul, David, 134
Paul, Dayan, 134, 159, 163
Paul, Edie, 140, 164

Paul Aho Artworks, 136
Paul Hertz Media Arts, 137, 201
Paul Mace Software, 140
PC Comix, 194
PC Graphics & Video, 146
PC Pix, 178
PC World/PCW Communications Inc, 173, 180, 184
PCM, 168
PCPC, 154, 155, 197
PCPI (Personal Computer Products Inc), 155, 157
Pearle, Eric, 123
Pearlman Productions, 159, 174
Pearlstone & Assoc, 125, 146
Peed Corp/PC Today, 188
Pegasus Productions, 160
Peirce, George E, 124, 145
Peji, Bennett, 140, 150
Pelavin, Daniel, 132
Peled, Einat, 132, 143
Pelikan Pictures, 130, 137, 148
Pellerin, Dana, 140
Pendergrast, James, 143
Penfield Productions Ltd, 173, 180
Pennington, Juliana, 150
PennWell Publishing Co/Computer Graphics World, 177, 183, 187
Penta Software Inc, 155
Pentagram Design, 143
Pentleton, Carol, 134, 145
Pepper, Missy, 152
PERC, 159
Perception Labs, 128, 177
Perceptive Solutions Inc, 154
Percey, Roland, 128
Percivalle, Rosanne, 132
Pereg, Larry, 123
Performing Artists Gallery, 136
Pergament Graphics, 132, 188
Perich & Partners Ltd, 148
Peripheral Land Inc, 155, 197-198
Perkins & Associates, 178
Perlman, David, 134
Pernambuco, Fernando, 123
Perry, Eric, 126
Perry, Rebecca, 140
Personal TEX Inc, 157
Personal Training Systems, 198
Perweiler & Assoc, 172
Peskor, Sharon, 145
Pete Bleyer Studio Inc, 130, 150
Peter X (+C) Ltd, 190
Peters Design, David, 140
Petersen Music, John, 184
Peterson & Co, 146
Peterson, Elsa, 123
Peterson, Grant, 123
Peterson, Marty, 140
Peterson Associates, 148
Peterson Photography, Bruce, 128
Petitto, Andrea, 134, 165
Petrillo, Jane, 134
Petroff, Tom, 126
Petronio, Frank, 124, 134, 145
Pettigrew, Stuart, 177, 183
Pettingill, Andre, 132
PFA Film & Video, 178
Pfleger, Mickey, 128
PG Music, 184
PH Brink International, 188
Phaedrus Productions, 128, 130, 189
Phaneuf, Arthur P, 124
Phar Lap Software Inc, 199
Phar-Mor Productions, 153, 182, 186
Phase 2 Digital Arts, 140, 150
Phase One Graphics, 124, 129, 134
Phelps, Greg, 125, 146
Phideaux Communications, 150
Philip, Nick, 140
Philips Consumer Electronics Co, 196
Philips Interactive Media Systems, 194
Phillips, Chet, 136
Philpott, Keith, 126
Phlange Design, 148
Phoenix Press, 193
Phoenix Technologies Ltd, 155
Phosphor Inc, 154, 166, 179
Photo Communications Corp, 190
Photo Concepts Inc, 130
Photo Concepts, 126
Photo Lab Inc, 192

Photo Lith, 189
Photo Offset Plate Service Inc, 188
Photo Specialties Inc, 189
Photocom Inc, 151
PhotoEffects, 130, 191
PhotoEngraving Inc, 188
Photographers/Aspen, 123
Photographic Specialties Inc, 130, 192
Photography Unlimited Inc, 130
Photogroup, 128
Photonics Graphics, 126, 137
Photos Ink, 126, 137
PhotoSep Inc, 190
PhotoSonics, 125, 174
Photosynthesis, 128
Phototime, 140
Phototype, 188, 189
Phototype Color Graphics, 188
Phototype Engraving Co, 189
Photovault, 193, 209
Physically Handicapped, 192
Picture Conversion Inc, 159, 191
Piece of Cake, 187
Pietrodangelo Prod Group Inc, 146, 174, 181
Pilgrim New Media, 194
Pina, C David, 140
Pink, Patty, 129
Pinkstaff, Marsha, 151
Pinnacle Effects, 177, 183
Pinnacle Post, 162, 164, 183
Pinnacle Publishing, 193
Pintura Graphics, 190
Pinzke, Nancy, 148
Pioneer New Media Technology, 170, 197, 198
Pip Printing, 193
Pipeline Associates Inc, 155
Pittard Sullivan Design, 183
Pittman, Dustin, 123
Pittman Hensley, 150, 164
Pitts, Tom, 128, 140
Pivotal Graphics, 168
PIX Imaging, 130, 193
Pixar, 157, 201
Pixcel Eyes, 124
Pixel Factory, 124
Pixel Graphics, 159, 163, 174
Pixel Ink, Consultants, 166, 169
Pixel Light Communication, 159, 163, 173
Pixel Media, 136, 146
Pixel Productions, 140
Pixel-The Computer Animation News People Inc, 203
Pixelink Corp, 155
Pixelworks Inc, 154
Pixmil Digital Imaging Service, 130, 193
Plager, Doug, 128
Planet Blue, 177
Planet Productions, 162, 165, 166
Platinum Design, 143
Platinum Productions, 160, 186
Platt College/Computer Graphics, 177
Pleasant, Ralph B, 128
Pleasure Retouching, Robert, 130
Plotkin Music Assoc, 184
plus design inc, 145, 166, 168
PM Associates, 162, 169
PMC Intl, 184
Pocock-Williams, Lyn, 178
Podevin, Jean-Francois, 140
Point-of-Purchase Advertising Institute, 197
Polaroid Resource Ctr, 152, 154, 198
Polatty, Bo, 124
Polhemus, 199
Pollack, Steven, 124
Pollman Marketing Arts Inc, 150
Polniaszek, John, 145
Ponzoni Photographic Inc, 125
Popa, Joseph, 126
Porett, Thomas, 134
Port-To-Print, 192
Portella, Dalton, 132
Porter, Jill, 143
Portfolio Artists Network, 162, 179
Porto, James, 123
Posch, Michael, 152, 163, 178
Posey, David, 136, 146
Possenti, Peter, 125
Post, Marilyn, 145
Post Edge Inc, 159, 163, 174
Post Effects, 160, 182, 186
Post Group At Disney/MGM Studios, 174, 185

Post Group, 162, 183, 187
Post Perfect Inc, 172, 179
Post Press, The, 145
Post/Producers Quarterly, 160, 182, 186
Postcraft Intl Inc, 157
Poster Studio, James, 123
Postique Inc, 175, 182, 186
Postmasters, 153
Potomac Graphic Industries Inc, 129, 188
Potts & Assocs, Carolyn, 152
Powell, Todd, 128
Power Presentations, 160, 164
Power Productions, 174, 181, 185
Powerhouse Productions, 125
Powers, Tom, 134
Pragma Design, 159, 163
Prairie Production Group, 170, 175, 186
Prairie State Coll/Grphc Comms, 132, 162
Pratt Institute/Manhattan, 140
PRAXIS Digital Solutions, 124, 134
Praxis Film Works Inc, 177
Pre-Press Associates, 157
Precision Color Inc, 188
Precision Color Plate Inc, 189
Precision Tapes Inc, 175, 182, 186
Precision Type, 155
Precision Visual Communications, 177, 183, 187
Preda, Dan, 132, 143, 208
PREFIT Corporation, 159, 163, 173
Premier Post, 158, 172, 179
Premore Inc, 171, 187
Prendergast, Michael, 134, 145
Preplink, 192
PrePress Computer Works, 190
PrePress Studio, 189, 193
Prepworks Inc, 188
Presentation Graphics, 190
Presentation Magazine, 203
Presentation Resource, 159, 188, 191
Presentation Services, 124, 153, 158, 163, 193
Presentation Source, 190
Presentation Technologies, 155, 157, 197
Presentation Works, 162, 200, 201
Presenting Solutions, 154
Preslicka, Greg, 137
Pressley Jacobs Design Inc, 148
PressLink, 191
Prestige Color, 188, 190
Prestige Production & Graphic, 137, 148, 178
Presto Studios, 165-168
Preston Productions, 152, 173
Price, Clayton J, 123
Price, Joan, 140
Priestly, Joanna, 150, 179
Primalux Video, 172
Prime Option Inc, 197, 198
Primo Angeli Inc, 150
Prince, Patric, 209
Princeton Univ/CIT, 132, 143
Princzko Productions, 172, 179
Print Craft Inc, 192
Print Expression, 128, 150
Printers Inc, 137, 148
Printing, 169, 202, 206
Printing Ind Assoc of the South, 192
Printronix, 155
Printware Inc, 154
PRINTZ, 150, 162, 179

Prism Enterprises Inc, 155
Prism Printing & Design, Inc., 145
Prism Studios, 126, 137, 148, 160, 164, 182
Pro Graphics Inc, 192
Pro Lab, 193
Pro Litho Plate Inc, 189
Pro Photo Productions, 128
Pro Type Computer Graphics, 193
Pro-Color, 130, 192
Producers Color Service Inc, 170, 182, 186
Producers East Media, 173
Producers Video Corp, 173, 180
Product Illustrations Inc, 137
Production Co, The, 162, 177, 183
Production Group, The, 174, 181
Production House, 170, 173, 180, 182, 186
Production Inc, 175, 181, 185
Production Masters Inc, 177, 183
Production Masters, 173, 180
Production West Inc, 177, 179, 187
Production Works, 173, 180
Products Group, 196, 198
Professional Litho Art Inc, 189
Professional Printing Svcs, 189, 193
Professional Publications, 188, 203
Proforma, 150
Progressive Image, 154
Project Four Communications, 186
ProLab Imaging & Graphics, 193
Promark Ltd, 157
Promedia Productions, 160, 182, 186
Promotovision Inc, 175
Prototype Graphics, 191
Provideo, 170, 175, 182
PSI Inc, 162, 165, 177
Psychic Dog Illustration, 140
PTN Publishing, 191
Publications Inc, 203, 205
Publish, 140
Pulling, Nathaniel H, 124
Pulse Communications, 170, 175, 182, 186
Pulse Imaging, 137
Punin, Nikolai, 132, 143
Purington, Camille & Mark, 150
Purup Pre-Press America, 154, 156
Puryear, Jack, 125
Pushpin Assoc, 151
Putcher, Terry, 151
Putnam Photo Lab, 190
Puza, Greg, 126, 137
PVS (Professional Video Services), 159, 163, 173
Pyramid Recording, 152, 172, 184
Pyramid Studios, 159
Pyramid Tele-Productions Corp, 174
Pyrate Animation, 184
Pyrate Group, The, 162, 165

Q

Q-Burn Records, 136
Q1 Productions, 152, 173
QMS, 154, 156
Quad Right, 190
Quadram, 154, 156
Quality Chrome, 191
Quality Color Process, 189
Quality Film & Video, 173, 180
Quality Graphics, 189
Quality Software Inc, 155

Qually & Co Inc, 148
Quanta Press Inc, 194
Quantum Computers, 196
Quantum Group, 160, 170, 182
Quantum Research Corp, 169
Quark Inc, 157
Quebecor Printing Prepress Group, 188, 189
Quebecor Printing, 191
Quest, 193
Quicksilver Assoc Inc, 126, 137, 153, 160, 164, 175, 182, 192
Quickstart Technology Inc, 193
QuillTech, 159, 163, 174
Quintar Co, 155, 157
Qume, 155
Quon Design Office, Mike, 143

R

R B Images, 193
R C Communications Inc, 190
R/G Video, 172
R/GA Print, 123, 132, 151
R P & A Inc, 162, 165, 167
Raabe, Dan, 123, 158, 162
Rabinovitch, William, 132
Radencich, Mike, 126
Radio Band of America, 184
Radius Group, 145
Radius Inc, 155, 197, 201
Radmar Inc, 153
Rae, William, 134, 143
Rafferty Communications, 137, 153, 160
Raging Fingers, 189, 193
Rahner, Andrea, 140
Rainbow Productions, 141
Rainbow Technology, 140
Rainbow Video Productions, 170, 175, 182
Ralcon Development Lab, 129
Ralston Design, 148
Ramba Design, 148
RamPage Electronic Imaging, 130, 137
Rampion Visual Productions, 159, 169, 180
Ran-Ger Technologies Inc, 154
Rancho Santiago College, 198
Randall, Robert, 128
Randall Lubert, 127, 139, 161
Random Access Media Works, 146, 153, 159
Randy and Cindy Tobin, 132
Raphael, Dick, 124
Raphael Film, 179
Raphaele, 130
Rapid Lasergraphics Photography &, 193
Rapoport, David, 123
Rapp Inc, Gerald & Cullen, 151
Rase Productions, Bill, 171, 183, 187
RasterOps Corp, 196-197
Rattan Design, Joseph, 146
Rauchman & Assocs, 136, 146
Ray, Al, 199
Ray, Elise, 130
Ray C Photographics, 126
Ray College of Design, 157, 200, 201
Rayson Films Inc, 189
RB Photographic, 128
RBL Publications, 146
RC Publications Inc, 175, 197
RCF Graphics, 137, 148
RDB Productions Inc, 170, 175, 186

RE Miller Communications, 126, 137, 186
Reactor, 170, 201
Reactor Worldwide Inc, 152
Reagan, Russell, 150
Real Productions, 182, 186, 199
Real Time Media, 162, 165, 168
Realta, 177, 183
Rearick, Kevin, 140, 150
Rebo Group, 172, 197
Recognition Concepts Inc, 197
Red Beam Inc, 129
Red Window Studio, 137, 147, 148, 159, 193, 197, 199, 208
Redmond Design, Patrick, 148
Reed Exhibition, 205
Reed Exhibtion Co, 205
Reed Sendecke, 148
Reel Good Productions, 186
Reelworks Animation Studio, 160, 182, 186
REF Associates, 159, 205, 208
Reflection Technology Inc, 197
Reflective Arts International, 194
Regester Photo Services Inc, 190
Regional Typographers Inc, 155, 190
Rehabilitation Inst of Chicago, 153, 160, 164, 170, 175
Reider Video Productions , 173
Reineck & Reineck, 150
Reiser, Beverly, 150
Reitz Data Communications, 160, 170, 182
Relax Technology, 196-198
Relf, Geoff Graphics Group, 150, 179
Renaissance Art Center, 158, 168, 190
Renaissance Communications, 145
Renaissance Video Corp, 175, 182, 186
Renard Represents, 151
Rencom, 181
Renda Gaphic Design, Molly, 146
Rensselaer Polytechnic Inst, 193
Reproductions Inc, 188
Research & Eng Council of the Graphic Arts Indus, 173
Ressler, Susan, 137
Results! Computers & Video, 183, 197
Results Video, 159, 163, 181
Reteaco Inc USA, 170
Reuter & Associates, 134, 145
Revelation Video, 186
ReVerb, 150, 183
REZ-N-8 Productions, 177, 183
Rezendes, Paul, 124
RFX Inc, 193
RGB Spectrum, 197
Rhoades, Dean, 125
Rhythm & Hues, 177, 183, 193
Richard Kidd Productions Inc, 159, 163, 174
Richard Salzman Represents, 152
Richards, Kenn, 134
Richards, Robin M Nance & Courtland, 136, 146
Richards, Tamarra, 123
Richards Group, The, 146
Richardson or Richardson, 150
Rick Barry, 132, 143
Rick Barry Desktop Design Studio, 132, 143
Rick Resnick Audio Visual Design, 177, 183, 187
Rick Ulfik Productions, 184
Rickabaugh Graphics, 148
Rico, Antonio, 130, 140
Ricoh Corp, 198

Riddick Advertising Art, 191
Ridgeway, Ronald, 132
Ridgley Curry & Assoc Inc, 150, 162, 167
Rifai, Christian, 143
Riggs, Robin, 128
Right Brain Inc, 156, 201
Riley, George, 125, 145
Rimmereid, Renell, 169
Rinaldi, John, 132
Ringier America, 191
Ringling School of Art & Design, 132, 143
Riordon Design Group Inc, 151
Rischawy, Jonathan, 134
Rising Star Graphics, 169
Riskind, Jay, 137
Risser Digital Studio, 130
River Design Inc, 145
River Image, 178
Rivera, Tony, 125, 129
Riverside Color Corp, 189
Rivet Films, Inc., 175, 182
RIX Softworks Inc, 157, 201
RJ Bauer Studio, 160, 164
RJG Design, 132, 143
RK Music Production, 184
Roaman, Brad, 123
Roanoke College, 136
Roaring Fork Productions, 186
Robbins, Edward, 168
Robert Gilmore Associates, 173
Robert L Biel Associates, 152, 173
Roberts, Mathieu, 123
Roberts, Terence, 125
Roberts Renditions, 153
Robilotto, Philip, 134, 159, 184
Robin Color Imaging, 192
Robinson, Dave, 128
Robinson, Lenor, 132, 143
RoboShop Design, 151
Roche, Alan, 123
Rochester Inst of Techn/Graduate Prog, 169, 206
Rock Solid Productions, 162, 177, 183
Rocky Mountain AV Productions, 177, 183, 187
Rocky Road Music Productions, 184
Roden, Bill, 128
Rodgers, Ted, 125
Rodney Alan Greenblatt, 131, 141, 158
Rodriguez, Claudio, 132, 143
Rodriguez, Gisela, 134
Rogala, Miroslaw, 137, 160, 164
Rogers Seidman Design, 143
Rogow, David, 178
Rohr, Dixon, 132
Roland, George S, 134
Roland Corp US, 196
Roland House Inc/Cosby, 191
Roman, Dianne, 134
Roman Assoc, Helen, 145
Romanello, Kim Marie, 143
Romeo, Richard, 136, 146
Romero Design, Javier, 143
Romero Design, Javier, 132
Romero Gaphics, Artie, 150
Romney, Michael, 132, 143
Romulus Productions Inc, 158
Ron Shook, 148, 178
Roolaart, Harry, 125
Ropkey Graphics, 189
Roscoe, Kurt, 148
Roscor Corp, 160, 170, 175
Rose, Lee, 136, 146
Rose Design, David, 150, 162
Rosebush Visions Corp, 132, 143, 178, 203
Rosen, Barry, 128
Rosen, Terry, 132
Rosenthal, Marshal M, 123
Rosenthal, Mel, 123
Rosenworld, 143
Ross, Bill, 128
Ross, Ken, 128
Ross Culbert & Lavery, 132, 143
Ross Ehlert Photolabs, 192
Ross Roy Communications, 165, 167, 175
Ross Typographic Service Inc, 193
Ross-Ehlert, 188
Ross-Gaffney Inc, 172
Roth, Wayne, 135
Rothaus, Valli, 129
Rothstein, Jeff, 123
Rouda, Saul, 177
Routch Design, 145

Rowin, Stanley, 125
Rowland, Lauren, 132, 143
Roxburgh, Ed, 140
Roy O Disney Production Ctr, 174, 185
Royal Graphics, 191
Royden, Elizabeth, 132
Rozasy, Frank, 140
Rua Nascimento Silva, 141
Rubenstein, Raeanne, 123
Rubin, Susan, 123, 143
Ruby Shoes Studio, 135, 145
Ruby-Spears Productions, 165, 177
Rubyan, Robert, 132, 143
Ruggeri, Lawrence, 125
Runyan Hinsche Assocs, 150
Rupert Design, Paul, 150
Rupp Art & Design, Katherine, 136, 169
Rusciano, Patty, 136
Russell, John, 123
Russell Communications Group, 150
Russell-Manning Productions, 186, 199
Rutgers Dept of Visual Arts, 145
Rutgers University/Art History, 150
Rutkovsky, Paul, 146
Rutt Video Inc, 172, 179
RW Video Inc, 170, 182, 186
Ryan Design, Thomas, 146
Ryane, Nathen, 140, 165

S

S A & A Inc, 126, 136, 146
S E Tice Consulting, 165, 169, 199
S I International, 151
S Illinois Univ, 154, 167, 183
S M Sheldrake Graphics, 150
S Presentation Design B4-1D-47, 159
Sachs, Jim, 140
Sackett Design, 150
Sacks, Ron, 137, 164, 182
Sacred Land Photography, 128
Safton, Carole, 132
Sahulka, Lawrence, 137
Saiki & Assoc, 143
Sakai, Kazuya, 136, 146
Salem Graphics, 189
Salient Software Inc, 157
Salsgiver Coveney Assoc I, 143
Salzman, Richard W, 152
Sam Flax, 190
Samata Associates, 148
Sammel, Chelsea, 140
San Diego Digital Post, 162, 171, 177
San Francisco State Univ, 207
San Jose State University, 162, 169
SANA, 126, 136, 146
Sanctuary Woods Multimedia, 165, 167
Sandbox Digital Playground, 126, 130, 137, 148, 208
Sanders & Co, 160, 182, 186
Sanders, Phil, 143, 162, 166
Sanders Agency, Liz, 152
Sanderson, Glenn, 126
Sandra Canniford, 152
Sandy Ostroff, 126, 130, 208
Sanford, Ron, 128
Sanson, Wells, Rogers, 136
Santo, Vincent, 129, 135, 145
Santoro Design, Joe, 143, 165, 168
Sapulich, Joe, 137
Sargent Design, Ann, 148
SAS Institute, 168, 200
Saturn Productions, 132
Sauer & Associates, 137, 165, 192
Sauers, Valerie, 143
Saunders, Michele, 151
Savage Art & Design, David, 136, 146
Savannah Color Separations Inc, 188
Savard, Sister Judith, 132
Sawyer Studios Inc, 132, 143
Sayett Technology, 169
Scabrini Design Inc, Janet, 145, 159, 180
Scallon, Ken, 132
Scan Magazine, 140
Scan-Graphics Inc, 150, 189, 193
Scantronix, 155, 157
Scarlett Letters, 193
Scene Three Inc, 174, 181, 185
ScenicSoft Inc, 157
Schaffer, Stephanie, 143
Scharf, David, 128

Schawk Inc, 189
Schell, Paul, 132, 143
Schiada-Smith, Laurie, 150
Schierlitz, Tom, 123
Schlowsky Photography and Computer Imagery, 125, 129, 135
Schmidt, Mechthild, 162, 168, 178
Schmidt Printing, 189
Schminke, Karin, 140
Schmitt & Co Prods Inc, 175
Schneemann, Carolee, 158, 165, 166
Schneider, Pat, 159, 173
Schneider, William, 137
Schneidman, Jared, 132
School of Architecture, 174, 207
School of Communications Arts Inc, 168, 206
School of Music, 187
School of the Art Inst, 137, 206
Schreiber, Dana, 135
Schubert, Christa, 150
Schulwolf, Frank, 146
Schuster, Robert, 132
Schwartz, Ariel, 159, 168, 169
Schwartz, Roberta, 143, 178
Schwartz Recording Inc, 184
Scientific Computing & Automation, 137
Scion Corporation, 154, 159, 178
Scitex America Corp, 154, 156
Scitex Graphic Arts Users Assn, 128
Scopinich, Robert, 140
Scott, Ron, 126, 201
Scott A MacNeill, 134, 144
Screaming Color Inc, 154, 156
Scriber, Cheryl, 136, 159
Scripps Research Institute, 140
Scruggs, Jim, 123
Scuderi, John, 145
Sea Studios Inc, 162, 167
Seaman Design, Robin, 150
Search West, 152
Seattle Ad Federation, 143
Sebastian, Benno, 150
Second Original Transparencies, 193
Second Sight Productions, 160, 182, 186
Sednaoui, Stephane, 123
SeeColor Corp, 157
Seeger, Stephen, 126
Seelig, Derek, 140
Segal Photography, Susan, 128
Segura Inc, 148
Sehmi, Gagan, 132
Seibold, J Otto, 132, 143
Seiffer, Alison, 132
Seigel/Inocencio, 140, 162, 168
Seitz, Arthur, 126
Seiwell, Yvonne, 146
Sela, Eliot, 132
Selbert Design, Clifford, 145
SelectGraphics, 189
Selectra, 197, 199, 201
Selkirk Communications Video Services, 174, 181, 185
Selman, Jan, 135
Semantic Music, 186
Semple, Rick, 128
Sense Interactive, 126, 136, 146
Sense of Design, 148
Sentry Color Labs Ltd, 190
Seong, Young-Shin, 132
September Productions Inc, 173
Seraphim Communications Inc, 186
Serious Robots, 174, 181
Seroka Group Inc, The, 159, 166
Serrano Company, 145, 152
Services, 171, 177
Sesto, Carl, 135
Seton Hall Univ/Comms Dept, 173
Severson, Michael R, 186
Severtson, Jeff, 132, 178
Sexton Design, 140, 150
SH Pierce & Co, 156
Shaff, Tom, 137
Shahidi, Behrooz, 123
Shane Productions Inc, Lunny, 178
Shaner, Tom, 128
Sharon McCormick Holography Studio, 207
Sharpe, Stuart, 165
Shaw, Barclay, 135
Shaw, Ned, 137
Shaw Video Communications, 175, 182, 186
Shawver Associates, 150

Sheffield Audio Video Productions, 184
Shelly, Jeffrey, 132
Shelton Communications, 169
Shelton Leigh Palmer & Co, 184
Sheridan Coll/Comp Grphcs Lab, 137
Sherman, Ron, 126
Sherwin, Cynthia A, 136
Sherwood, Melanie, 135
Shields, Charles, 150
Shima Seiki USA Inc, 154, 156
Shin, Young, 132
Shoffner, Terry, 141
Sholik, Stan, 128
Shooters Inc, 173
Shooting Stars Post Inc, 159, 163, 181
Short, Robbie, 136, 146
Shoulder-High Eye Productions, 148, 178
Show & Tell, 129, 143, 190
Showbiz Expo East, 141
Shtern, Adele, 132, 143
Shultz, David, 140, 150
Sicurella & Assoc, 184
Sid's Typographers, 193
Side Effects, 178, 184
Sidley Wright & Associates, 177, 183
Siegel, Dink, 132
Siegel, Jeffrey, 168
Siemer, Patrick, 137, 164, 178
SIGGRAPH '94 Show, 1994, 205
Sigma Designs Inc, 155, 157
Signet Sound Studio, 187
SIIM '94 Intl Computer & Office Exh, 150
Silbert, Barbara Bert, 132
Silicon Graphics Inc, 196, 200
Silva, Keith, 128
Silva, Raul, 137
Silver Shadow Images, 125
Silverleaf Design & Illustration, 145
Silverman, Gail G, 145
Simerman, Tony, 136, 163, 178
SimGraphics Engineering Corp, 197, 199, 201
Simmons, Jessica, 145
Simon, Peter Angelo, 123
Simone, Luisa, 143, 168
Simons, Stuart, 125
Simpson, Elizabeth, 151
Sims, John L, 186
Sincyr Creative Services, 150
Sinick, Gary, 128
Sinkler Or Sinkler, 160, 164, 192
Sinnott & Associates, 182
Siren Design Inc, 132, 143
Sirius Image, 159, 163, 181
Sirko Design, R, 148, 160
Site One, 129
Skidmore College, 155, 157, 193
Skilset Grafix, 169
Skjei Design Co, M, 148
Skogsbergh, Ulf, 123
Skolnik, Lewis, 123
SKW Computers Inc, 157
Sky Tree, 140, 150
Skylight Graphics, 145
Skyline Displays Inc, 192
Skylite Productions, 123
Skyview Film & Video, 170, 175, 186
Sladcek Studio, 126
Slavin, Daniel, 140, 150
SLD Communications, 169
Slide Center, 184, 190
Slide Factory, 140, 177, 187
Slide House, The, 190
Slide House, 193
Slide Sense, 190
Slide Services Inc, 192
Slide Shoppe, 153
Slide Step, 153, 191
Slidemasters, 191
Sloan, Richard C, 145, 178
Sloan, Rick, 140, 150
Smalley, David, 145, 178
Smallwood, Bud, 135
Smart Concepts Ltd, 159, 163, 181
Smart Design Inc, 143
Smart Set Inc, 192
Smetts Design, Bonnie, 150, 162, 166
Smith & Co, 131
Smith, Brian, 125
Smith, CJ, 140
Smith, David, 150
Smith, Dick, 125

Smith, Ellen, 135
Smith, Mark, 180
Smith, Marty, 140
Smith, Rusty, 125
Smith, Stephen C, 123
Smith, Steve, 140
Smith, Thayer, 125
Smith College, 143
Smith Design, Laura, 146
Smith/Taylor Productions, 174
Smith-Evers, Nancy, 135, 145
Smoke & Mirrors, 162
Smolenski, Peter, 135
Smool, Carl, 140, 150
Smyth, Kevin, 125
Smyth, Richard F, 137
Smythe & Co, 184
Snelson, Kenneth, 132
Snitzer, Herb, 125
Soc for Applied Learning Technology, 205
Soc of Manufacturing Engineers, 126
Society for Computer Simulation, 141, 184
SoftCraft Inc, 156
SoftQuad Inc, 157
Softsync Inc, 156, 200
Software Assist US, 154, 209
Software Complement, 156
Software Consulting Services, 156
Software Media & CD ROM, 170
Software Publishers Assn, 128
Sokolowski, Ted, 135
Solomon, Mark, 208
Solomon, Paul, 126
Solomon Video Productions, Rob, 143, 158, 165
Solution Technology Inc, 156
Solzer, Wolfgang, 189
Sonic Images Productions Inc, 159, 165, 166
Sonneville, Dane, 125, 135, 151
Sony Corp of America, 196, 198
Sony Electronic Publishing, 194
Sorensen, Chris, 123
Sorensen, Marcos, 140
Sorensen, Vibeke, 140
SOS, 153
SOS Productions, 170, 182, 186
Sound & Motion, 173, 184
Sound Decisions Inc, 186
Sound Image Inc, 135, 152, 173
Sound Photosynthesis, 168, 177, 187
Sound Resources, 186
Sound/Video Impressions, 170, 182, 186
Soundlight Productions, 153, 160, 182
Soundtrack/NY, 184
South Bay Film & Video, 177
South Dakota State Univ, 193
Southern Productions, 174, 181
Southwest Teleproductions Inc, 174, 181, 185
Southwest Television Production, 171, 177
Southwind Publishing Co, 203
Sowash, Randy, 137
Spaceball Technologies Inc, 199
Spaceshots Inc, 128
Spade, Sergio, 132
Spaeth, Dana, 123
Spangler Design Team, 148
Spano Design, 143, 158, 179
Sparkman & Assoc, Don, 145
Sparks, Joe, 162
Spatial Data Architects, 169

Speaker Support Group, 152, 159, 173
Spear, Charles, 135
Special Interest Group for Computers, 203
Special Interest Group for Software, 203
Specialty Graphic, 146
Specter Video, 152, 159, 173
Spectra Action, 126
Spectracomp, 190
Spectratone, 152, 190
Spectris, 150, 165
Spectrum, 188, 190
Spectrum Arts Ltd., 190
Spectrum Image Group, 130, 148, 192
Spectrum Inc, 130, 189
Spectrum Sight & Sound, 162
Spectrum South Inc, 174
Specular International, 156, 200
Speed Graphics, 188
Speer, Stephan, 132, 178, 179
Spellman, David, 128
Spellman Photomontage, Naomi, 123
Spicer Productions Inc, 159, 180, 184
Spiece, Jim, 148
Spiegelman & Assoc, 150
Spinnaker Software, 156
Spohn, David, 135
Sprintout Corp, 190
Sprintout New York, 190
Square One Desktop Publishing, 156, 192
SRG Design, 151
St John, Bob, 135, 145
St Louis Community Coll at Merramec, 194
St Onge, Cheryle, 125
Staada, Glenn, 135, 145
Staartjes, Hans, 126
Stabin, Victor, 132
Stabler, Barton, 135
Stafford, Rod, 135, 145
Staging Techniques, 177
Stahl, Nancy, 132
Stanard Inc, Michael, 148
Stanford Telecommunications Inc, 168
Stanford University, 167, 179
Stankiewicz, Steven, 132
Stapleton, Kevin, 128
Star Connection, 125
Star Graphic Art, 189
Star Video Services, 177
Starfleet Productions, 173, 184
Stargardt, Fred, 136
Stark Studio, 172, 179
Starlite Studios, 177, 187
Stars Production Services, 172
Stat Cat, 136
stat-media, 151, 162, 179
State of the ART Dept, 129, 151
Steele, Melissa, 128
Stehney, Regina, 130, 140
Stein, Marion, 135
Steinkamp, Jennifer, 140
Stellarvisions, 145, 159, 208
Stemrich, J David, 145, 151, 173
Step 2 Software, 157
Stephens, Anait Arutunoff, 129
StereoGraphics, 199
Sterling Group, 143
Steuer, Sharon, 129, 135, 208
Steve Blexrud Studio, 179, 186
Steve Bon Durant, 134, 144

Steve Butterfield, 157, 198
Steve Ford Music, 186
Steve Greenberg Photography, 125
Steve Hallett, 124, 129, 170
Steve Larson Design Associates, 169
Steve Michelson Productions, 177, 183, 187
Stevens, Alex, 140
Stevenson Photo Color Co, 189
Steward Digital Video, 173, 180
Steward Studio, 148
Stewart Design Group, 143
Stieglitz, Carol, 151, 209
Still Life & Kicking, 160, 164, 182
Stillman Design Assocs, 143
Stimulus, 177, 183, 187
Stockland Martel, 151
Stockler, Len, 129
Stockton, Cheryl, 123
Stokes Imaging Services, 181, 191
Stokes Retouching & Computer, 130
Stone, April, 140
Stone Design Group, 128
Storter Childs Printing Company, 191
Strata Studio, 157, 200
Strategic Mapping Inc, 155, 157
Stratton, Mary M, 140, 151, 162
Stratton, Robert, 168
Strauss Design, 137, 148
Straylight, 199
Streetworks Studio, 146
Stribiak & Assoc, 148, 170, 175
Stroukoff, Ann, 178
Struthers, Doug, 132, 136, 162, 163, 181
Struve-Dencher, Goesta, 131, 141, 179
STS Production, 162, 165, 177
Stuart, Preston, 145, 152, 169
Stubbs, Diane N, 140
Stuck, Jon D, 126, 169, 174
Studer, Gordon, 140
Studio, The, 162, 166
Studio 46, 145
Studio Advertising Art, 157
Studio Francesca, 143
Studio Graphic Systems, 140, 151
Studio M, 179
Studio Macbeth, 123, 129, 132
Studio MD, 140
Studio One, 136, 137, 148
Studio Post & Transfer, 162, 165, 171, 178, 184
Studio Productions Inc, 162, 165, 183
Studio Q, 123, 143
Studio Stitchers, 153
Studio Twenty Six Media, 135
StudioGraphics, 165
Studiolink, 173, 180, 184
Subia Corporation, 193
Subjective Technologies Inc, 162, 167
Substance Interactive Media, 195
Sugar, James, 128
Sugiyama Design, 148
Sukolsky-Brunelle Inc, 125, 190
Sullivan, Steve, 132
Sumeria Inc, 157, 195
Summa Logics Corp., 163, 174, 181
Summagraphics Houston Instrument, 197, 198
Summerfield Graphics, 189
Sun Group, 184
Sun Microsystems Inc, 196, 198
Sun Moon Star, 196, 198, 200

Sun Moon Star/Personal Computer Div, 196, 198, 200
Sunbelt Video, 163, 174, 181
Sund, Harald, 128
Sundance Technology Group, 196-198
Sunset Post Inc, 171, 177, 187
SUNY/Buffalo/Art Dept, 159, 163, 165
Superior Color Graphics, 192
Supermac Technology, 195, 197, 199
SuperScan Div, 189
Superslides, 193
Supersuite, 178
SuperVision, 145
Sutcliffe Music Inc, 184
Sutton, Eva, 133, 168
Sutton, Jeremy, 140
Swaim, Howard, 151, 209
Swaine, Mike, 140
Swan Engraving, 125, 188
Swann Design, 133, 143
Swanstock, 152
Swenson, Barbara, 125
Syd Mead Inc, 140, 183
Symantec Corp, 157
Symbolics, 168, 196
Symington, Gary, 135, 145
Symmetry Software Corp, 157
Synergy Art & Tech, 137, 164
Synergy Computer Training, 148
Synthesis, 174, 178, 181
Synthetic Imaging Inc, 194
Syracuse Univ/Art Media Studies, 165
SYSTAT Inc, 156
System Generation Assoc Inc, 199
Systems, 126, 148
Szabo, Michelle, 159, 163, 165, 169

T

T/Maker Co, 157
T P Design, 135, 146
Tachibana, Kenji, 128
Taft Printing/Deskprint Graphics, 193
Tagteam Film & Video, 186
Take 1 Productions Ltd, 170, 182, 186
Take One Productions, 174, 181, 185
Take Ten Teleproductions, 174, 185
Takessian Creative, Adam, 151
Takigawa Design, 151
Talaris Systems Inc, 198
Talent*Media Presents, 186
Tall Tree Systems, 198
Tamagni, Charles, 177
Tamura, David, 135
Tanenbaum Inc, Robert, 140
Tangent Design/Communications, 159, 163, 166
Tanglewood Productions, 187
Tapestry Productions, 172, 179
Tara Visual Corporation, 168
Tarleton, Gary, 128
Tarzan Communication Graphique, 151
Tassian, George Org, 148
Tate, Clarke, 137
Taya, Hisao, 178
Taylor & Browning Design Assoc, 151
Taylor, Dawn Marie, 179
Taylor, John B, 123
Taylor, Joseph, 137
Taylor Corporation, 137

Taylor Design, Robert W, 151
Taylor Photo, Dan, 128
Taylored Graphics, 156
Tcherevkoff, Michel, 123
TDS Inc, 180
TDS, 130, 192
TEAC America Inc, 198
Teach Services, 156
Team Xerox, 157
Tech Graphics, 135, 180
Tech Pool Studios, 175
Techniarts, 173
Technicad, 169
Technical Photographers, 152
Technidisc, 170
Technigraphics Inc, 188
Technologies, 165, 168
Technology Group, 156
TechPool Studios Inc, 182
Techtron Imaging Center, 130, 189
Techware, 156
Tedesco, Thomas, 145
Teich, David, 135, 145, 163
Tektronix Inc/Display Products, 197
Tele Edit Inc, 175, 186
Tele-4, 178
Tele-Measurements Inc, 190
Tele-Producers Inc, 186
Telemation Denver, 162, 167, 183
Telemation Productions, 160, 182, 186
Telephoto Communications Inc, 197, 201
Teleproductions Unlimited, 174
Telerobotics Intl Inc, 196, 199
TeleScene Inc, 177, 183, 187
Telesis Productions Inc, 170, 173, 180
Teletechniques, 172, 178, 179
Teletime Video Productions, 170, 180, 184
TeleVideo Productions, 170, 182, 186
Television Broadcast, 197
Telezign, 172, 179
Tell-A-Vision, 177
TEM Associates Inc, 154
Tempra Software, 156
Tepper, Marlene, 145, 159
Terezakis, Peter, 133
Terra Incognita Prods, 143, 178
Terre Haute Engraving Co Inc, 189
Terry Fryer Music, 186
Terry Waldo Enterprises, 184
Tetes Video, 178
TEX Interactive Svcs, 165, 168, 169
Texas Instruments Inc, 198
Texas Photocomp, 191
Texas Tech University, 163, 174, 185
Texas Womens Univ/Adv Art, 146
TGA Recording Co, 171, 182, 186
Tharnstrom Music, 186
The Big Picture, 190
The Blank Company, 140, 151
The Future Image Rep, 166, 203
The Image Bank, 126
The Music Picture, 184
The Real Design, 125
The Roland Co, 186, 208
The San Diego Convention & Visitors, 128, 151
The Window Book Co, 170
Thiel Visual Design, 148
Thin Air, 145, 152
Third Wave Computing, 154, 156
Third World TV Exchange, 178
THIRST, 148
Thirteen Fifty-one prdctns, 160, 186
Thirteen Fifty-one Productions, 182
Thirty Second St, 177, 183
Thomas, Paul, 148
Thomas, Tony, 126
Thomas A DeBiasso, 137, 206
Thomas/Bradley Design, 137
Thomas EFX Group, 182
Thomas Group, The, 148
Thomas Marketing Communications, 136
Thomas Mrktng Comms, 146, 208
Thomas Piwowar & Assocs, 165, 169
Thompson, Arthur, 135
Thompson, Jim, 136
Thompson Type Inc, 193
Thonen, Rod, 136, 163
Thornell, Ian, 133
Thornock, Christopher, 140
Thorpe, Peter, 133, 143
Three & Associates, 137, 148

Three G Graphics Inc, 155, 157
Three in a Box, 152
Three-D Graphics, 157
Three-D Visions, 157
Threinen, Cher, 140, 151
Threshold Software Inc, 177, 183, 187
Thunderware Inc, 155, 157
Thurlbeck, Ken, 128
Thurston Productions, 193
Tiani, Alex, 135, 159, 163
Tiberi Graphics, 148
Tier One Communications, 178
Tillander, Michelle, 136
Tillis, Harvey S, 126, 148
Time Arts Inc, 190
Timescape Image Library, 128
Timeworks Inc, 156
TiNi Alloy Co, 199
Tishman, Jill Rosean, 140
Tivadar, Bote, 141
TM Century Inc, 159, 170
TMP Video Communications, 174, 185
TMS Inc, 169, 201
Toby's Tunes Inc, 186
Today's Graphics Inc, 190
Today Video, 172, 179
Todd Enterprises Inc, 196-198
Todd Johnson Productions, 128
Tolan, Stephanie, 146, 208
Tolman, Brian, 209
Tom Buhl Imaging Ctr, 193
Tom Nicholson Associates, 158, 162, 165
Tompkins, Bill, 123
Tomzak, Paul, 143
Tonal Values Inc, 152
Toney, Allen, 136
Tonkin, Thomas, 133
Tony Giordano, 129, 188
Top Communications, 162, 166, 171
TOPIX, 177, 178, 183, 184
Topix LA Inc, 177, 183
Toppan Printing Co, 129
Torinus, Sigi, 140
Toronto Image Works, 194
Torrance, Scott, 128
Total Video 3, 171, 175, 186
Total Video Co, 177, 183, 187
Total Vision Inc, 151, 162, 179
Total Visuals, 190
Touchton, Ken, 126
Tourtellott, Mark, 135, 163, 165
Tower Graphics, 143, 158
Towers Perrin, 148
Towler, Matthew, 135
Toy Specialists, 170
Tozzi, Graig, 133, 163
TPS Electronics, 155
Tracer Design, 140, 165, 183
TRACES, 173, 180
Track Record Studios Inc, 186
Track Seventeen Productions, 186
Tracking Station, 186
Tracy, Donna, 140, 151, 165
Trade Litho Inc, 189
Trailer Photo, Martin, 128
Trans FX, 130
Trans Media Creative Assoc, 153
Trans-Delta Productions, 165, 167, 185
Trans-Ocean Video Inc, 172, 179
Transfer Zone, 171, 175
Transparencies Inc, 130
Transparent Media, 159, 165, 169
Traugot-Weldman, Marsha, 168
Trend Multimedia, 135, 159, 163
Trenton State Coll/Art Dept, 184
Tri-Comm Productions, 153, 159
Triad/Artbytes, 148, 164, 167
Triad Media, 152, 162, 173, 177, 183, 202
Triad Media Services, 162, 177, 183
Tribotti Designs, 151
Trici Venola & Co, 140
Trico Graphics, 189
Trident Microsystems Inc, 196
Trillion Post Production, 175
Trillium Computer Resources, 196
Trimble Production Studios, 153, 174, 185
TriMetrix Inc, 157
Trinity Photo Graphic, 126, 146
TriTel Productions Inc, 192
Trofimova, Marianna, 133
Trollbeck, Jakob, 143

Tru-Color Inc, 189
Truckenbrod, Joan, 148
TruColor Labs, 188, 190
Truevision, 126
Truline Graphics, 193
Truly Computer Graphics, 136, 163
Trunk Line, 193
Truvel Corporation, 154, 156
TSA Design Group, 140, 151, 154
Tsutsumi, Adi, 123
Tucker, Mark, 126
Tugboat, 123
Tukaiz Litho, 189
Tullio & Ruppert, 186
Tulloss Design, 148
Tumble Interactive Media, 133, 158, 201
Tung, Claudia, 133
Turbitt, Kevin, 145
Turner & DeVries, 128, 140, 151
Turner, Dave, 136
Turner, Pam, 159
Turner Photography, 125
Turpin Design Assocs, 146
Tuttle, Jean, 141
Tuveson, Christine, 140, 151
TV Svcs, 160, 176, 187
TV-5 WNEM, 171, 175, 186
TVA/Television Associates Inc, 162, 171, 177
TW Design, 146
Twelve Point Rule Ltd, 143
Twenty-Nine Point Five Music, 186
Two Pie Are Music, 184
Two Twelve Assocs, 158, 163, 168
TX Digital Imaging & Retouching, 130
Tyava Productions, 125, 135, 190
Tyler, Wayne R, 140
Type Case Inc, 191
Type Directors Club of New York, 193
Type Inc, 194
Type Plus, 194
Typecasters Ltd, 160
Typeline, 190
Typesetting Room, The, 189, 194
Typetronics, 192
Typeworks Plus, 190
Typografiks Inc, 191
Typogram, 190
Typographic House, 190

U

U&Lc, 125
Ultimac, 157
ULTITECH Inc, 163, 165, 173
Ultra Design Group, 146
Ultratype Inc, 194
Uman, Michael, 133, 163
Umansky, Steven, 135, 156
UMAX Technologies Inc, 155
UMed Ct, 164, 167, 169
Uni Disc Inc, 195
Uni-Graphix, 126, 136, 146
Uni-Mesa Co, 157
Unisys, 146, 201
Unisys Corporation, 146
United States Video, 160, 167, 174
Unitel Video/Hollywood, 171, 177, 183
Unitel Video, 172
Univ Coll of the Cariboo, 165, 183
Univ of Calif/San Diego, 151, 177, 207
Univ of California/Berkeley, 126, 206
Univ of Ill at Chicago, 168, 206
Univ of Kansas, 170, 174
Univ of the Arts, 192
Universal Images, 153, 160, 175
University of Calgary, 207
University of Colorado, 145
University of North Colorado, 187
University of Notre Dame, 137, 206
University of Oregon, 163
University of Texas, 135
University of Washington, 179
Upton, Thomas, 128
Urban Communications, 176, 186
Urban Taylor & Assocs, 146
Urbane USA Inc, 123
US Animation, 183
US Lynx Inc, 155
USA Teleproductions, 171, 182, 186
Utopix Infographie, 178

V

V Graph Inc, 159, 166, 169
Valentine Productions at Omega, 187
Valery Bates-Brown, 206
Valesco, Frances, 140
Valla, Victor R, 135, 145
Valle, Robin, 140
Valley Printers, 194
Valley Production Center, 177, 187
Van Campen, Tim, 145
Van der Veer Photo Effects, 183
Van Dyke Columbia Printing, 145
Van Nostrand Reinhold, 203
Vance, Bill, 129
Vander Houwen, Greg, 128
Vanderbos, Joseph, 140
Vanderbyl Design, 151
VanderLans, Rudy, 151
VanderSchuit Studios Inc, Carl, 128
Vandervoort, Gene, 140
Vanderwarker, Peter, 125
Vanguard Productions Inc, 171, 182, 186
Vanguard Video Productions, 154, 177, 179
Varis Photomedia, 128
Varitel Video, 171, 177, 183
Varitype Inc, 154, 156
Varner, Vicky Jo, 209
Vasulka, Steina & Woody, 151
Vaughn Communications Inc, 186
Vaughn Wedeen Creative, 151
Vegas Valley Prods/KVVU Bdcst Corp, 162, 177, 187
Veldenzer, Alan, 123
Vella, Ray, 135
Ventura, Michael, 125
Ventura Publisher Users Group, 174
Vera, Clara Claudia, 126, 137
Verba Communications Network, 176
Verbum Magazine, 140, 154, 183
Veritel, 189, 194
Veritel Select, 130
Vermont Microsystems Inc, 197, 200
Vesna, Victoria, 140, 167
VGC Corporation, 156
VGI Productions/Video Genesis Inc, 160, 182, 186
Via Type, 194
Viacom New Media, 123
Victor Harwood, 205
VIDCAM Inc, 160, 171, 176
Vidcam Productions Inc, 171, 182, 186
Vidconn Productions Inc, 173, 180
Videk Corp, 196, 197
Video Ad Services, 173
Video Arts, 176, 192
Video Arts, 165, 177, 183
Video Assist, 179
Video Co, The, 174, 181
Video Communication Services, 173
Video Communications SE, 174, 181
Video Copy International, 173
Video Data Services of Montana, 162, 165, 177
Video Disc System Sound & Video, 172
Video Duplication Inc, 186
Video Editing Centers, 173
Video Editing Services Inc, 160, 174, 181
Video Electronics Standards Assn, 174
Video Features, 176, 186
Video Genesis, 171, 182, 186
Video I-D Inc, 176
Video Image Associates, 165
Video Image Productions, 163, 174, 181
Video Impressions, 171, 176, 186
Video Media Productions, 183
Video Nova Inc, 160, 167
Video One Inc, 171, 177, 187
Video One, 174, 181, 185
Video One Teleproductions, 170, 174, 181
Video Park, 174, 181, 185
Video Post & Graphics Inc, 153, 182, 186
Video Post & Transfer, 174, 181, 185
Video Post, 171, 182, 186
Video Pro of Wyoming, 171, 179
Video Production Assocs, 174
Video Production Inc, 173
Video Professionals, 179
Video Promotions Intl, 153, 160, 170
Video Resource, 173, 180
Video Resources, 162, 165, 169
Video Services Unlimited, 173, 180

Video Snapshots, 197
Video Solutions for Business, 169
Video Systems Magazine, 173
Video Transform, 177
Video/Visuals Inc, 156, 159
Video West Productions, 177, 183, 187
Video Wisconsin, 171, 182, 186
Video Works, 158, 172, 179
Video Works Productions, 173
Video Workshop, 173, 174, 181
Videocenter of New Jersey, 173, 180
Videofax Company, 199, 201, 202
Videofonics Inc, 160, 181, 185
Videograf, 158, 165, 179
VideoGraphicArts, 140
Videographics Corp, 176
Videologic Inc, 196-197
Videomation, 162, 165, 166
Videomax Productions Ltd, 178
Videomedia Inc, 197
Videomix/Aquarius, 184
Videoplex, The, 177
Videos Multimedia Inc, 153, 159, 208
Videosmith Inc/Post & Graphics, 173
Videosmith of NJ, 173
Videotechniques Inc, 173
Videotex Systems Inc, 156, 202
Videotexting, 160, 165, 174
Videotroupe, 180
VideoTutor Inc, 156
Videx Corp, 174
Vidsat Communications, 174
Viele Inc, Maureen, 145
Viesti, Joe, 123
Vietor, Noel, 140
View Studio, 177, 179
Viewpoint Computer Animation, 180
Vigon, Jay, 151, 171, 177
Vila, Doris, 128
Vilinsky/Snyder Music, 184
Villager Graphics, 192
Villanova, Joseph, 135, 145, 208
Vincent, Wayne, 136
Viner, Kevin, 152
Viper Optics, 151, 162, 196
Virginia Commonwealth Univ, 163
Virtual Images, 177, 183
Virtual Reality Conference Intl, 199
Virtual Technologies, 199
Virtus Corp, 168, 200, 201
Visicom Design Group, 194
Vision Associates, 178
Vision Communications Inc, 176
Vision Database Systems Inc, 174, 181, 185
Vision Graphics, 190
Vision Productions/KSAZ-TV, 165, 183, 187
Visionary Art Resources, 151, 165, 166
Visioneering International, 160, 170
Visionetics International Corp, 199
Visions, 125, 186
Visions of Naples, 169
Visions Plus, 171, 177, 183
Visitel, 157
Visitors Bureau, 128, 151
Vista Color Corp, 188
Vista Color Lab, 130
Visual Access Technology Inc, 146
Visual Communications Group, 152
Visual Effects Lab, 183

Visual Graphic Communications, 135, 145, 151
Visual Impressions, 159, 163
Visual In-Seitz, 190
Visual Information Development Inc, 135, 159, 180
Visual Motivation Inc, 160, 182, 192
Visual Promotions, 162, 165, 183
Visual Services Inc, 153, 159, 169-170
Visual Solutions, 153
VisualEdge, 165, 167, 169
Visualization Lab/Coll of Architecture, 168
Vital Images Inc, 201
Vitali, Julius, 135
Vivid Graphics & Typesetting, 194
Vivid Group, 199-201
VMI Inc, 135, 206, 208
Vogelei & Assocs, 181
Volan Design Associates, 151, 162
Volotta Interactive Video, 162, 168
Volunteer Lawyers for the Arts, 133, 143
Voris, Rebecca J, 140
Vortex Interactive, 159, 170, 173
Voyager Co, 195
Voyager, 155, 201
VPL Productions, 160, 176, 186
VREAM, 199
VSA Partners Inc, 148
VSC, 172, 179, 184
VSE Corp, 177

W

Wace Resource Center, 156, 208
Wace USA/Graphic Warehouse, 130, 189
Wacom Technology Corp, 155
Wade, Renee, 135
Wadlund & Associates Inc, 176
Wages Design, 146
Wagner, Daniel, 123
Wagner, Marilyn M, 125
Wagner Studio, 123
Wahlstrom, Richard, 128
Wait Design, 151
Walcott-Ayers Group, The, 151
Walker, Donna, 133
Walker, Kevin, 136
Walker, Todd, 140
Walker Graphics/A WACE USA Com, 130
WalkerVision Interarts, 162, 177, 187
Walkoe Design, Don, 148
Wallace/Church Assocs, 133, 143
Wallace Engraving Co, 188
Wallin, Michael, 158, 163, 165
Walls, David, 168
Walter, Melissa Jo, 136, 146
Walton & Assocs, 148
Walton Electronics Inc, 174
Waltzer Digital Service, Carl, 123, 129, 190
Wander Communications, 172, 179, 190
Wang Studio Inc, Tony, 123, 143
Warman, Brian, 137, 148
Warnell, Ted, 141
Warner New Media, 195
Warren, William James, 128
Wasatch Computer Technology, 157
Waselle Production Artists Inc, 130
Washburn Graphics, 194
Washington House Photography, 126, 130
Washington State University, 125, 135, 145
Waters Design Assoc Inc, 143

Waters Design, 133, 143, 163
Waterworks Productions, 160
Watt/Peterson Inc, 189, 192
Watts, Mark, 135
Wave Inc, 159, 173, 180
Wavefront Technologies, 168, 200, 201
Wax, Wendy, 133
Wayne Boyer Studio, 182
Wayne State University, 194
Wayzata Technology, 156
WBMG Inc, 143
WCVB TV Design, 145
WDBJ-TV Inc, 174
Weaver Graphics, 156
Web Co, 191
Web Tech Inc, 192
Weber, John, 137
Weber Designs Inc, Elaine, 146
Weber State College, 126
Weems, Al, 125
Weidlein, Peter, 125
Weigand Report, 123
Weiland, Juliette, 125
Weimer Typesetting, 192
Weinberg & Clark, 128
Weinberg, Jan, 151
Weinman, Lynda Animation, 177
Weinrebe, Steve, 125
Weiser, Barry, 123
Weisman, David, 140, 166, 168
Weiss Associates, 148
Weiss Group, The, 151
Weisser, Carl, 133, 143
Weissman, Walter, 123, 143
Weithers, Arlington, 136, 146, 160
Weller Inst for the Cure, 151
Werbickas, Joseph, 169
Wescan Color Corp, 189
Wescon, 162, 183, 187
West Coast University, 141
West Design, Suzanne, 151
West Design Studio, Harlan, 151
West End Post, 174, 181
Westcom Creative Group, 162, 183, 187
Western Front, 151
Western Images, 177, 187
Western Imaging, 189, 194
Western Montana College, 194
Westgate Graphic Design Inc, 148
Westheimer, Bill, 123
Westinghouse CSS, 135, 145
Westlight Inc, 128
Weston Productions, 125
Wexler, Glen, 128
Wexler, Jerome, 125
WFRV-TV, 176
WGBH Design, 173
Wheeler Arts, 156, 194
Whitaker, Corinne, 141
Whitby, Michelle, 123, 133
White, Lee, 128
White Hawk Pictures, 174, 181, 185
White Pine Software, 126, 146
Whitney-Edwards Design, 145
WICI, 177, 183, 187
Wickham & Assoc Inc, 145
Widen Colourgraphics Ltd, 189
Widstrand, Russ, 128
Wiggin Design Inc, 145

Wiggins, Mick, 133
Wijtvliet, Ine, 143
Wilcher Design, 151
Wilczynski, Alina, 135
Wiley & Flynn, 126
Wiley, David, 152
Wiley, Matthew, 128
Wiley, Paul, 133, 143
Willens + Michigan Corp, 137, 148
Willi Communications, 153
William Patterson Coll/Art Dept, 128, 177, 189
Williams & Ziller Design, 151
Williams, Monica, 135
Williams, Wayne, 128
Williams Marketing Services, 153
Williamson Janoff Aerographics, 123, 133, 143
Williamson Photography, 123
Willman Design, 148
Willming Reams Animation, 181
Willow Graphics, 137, 148
Willow Peripherals, 154, 197
Willson Creative Group, 130, 137, 148
Willson Creative Group, 130, 148
Wilson, Barry, 135, 145, 188
Wilson, Lin, 137
Wilson, Mark, 145
Wilson, Nancy Patton, 141
Wilson Engraving Co Inc, 188
Winchell Company, The, 190
Winchester, Shawn, 179
Windows Watcher, 172, 179
Windsor Digital, 143
Windstar Studios Inc, 162, 183, 188
Windy City Communications, 148, 166
Wink, David, 136
Winkie Interactive, 169
Winners Circle Systems, 155, 157
Winsted Corp, 192
Winston, Matthew, 135
Winters Productions, 162, 171, 177
Wise, Jennifer, 179
Wisenbaugh, Jean, 133
Witkin, Christian, 123
Witte, Mary Stieglitz, 137
Wizardly Teleproductions Inc, 186
Wizywig, 194
WKBN, 176
WLNE-TV, 173
WNDU-TV, 160, 181, 185
WNET, 172
WOFL-Productions, 174
Wohlmut Media Services, 162, 167
Wojnar Drake, 125
Wokuluk, Jon, 141, 151
Wolf, Anita, 145
Wolf, Dixon, 141
Wolf Design, Stephen, 135, 145, 159
Wolf-Hubbard, Marcie, 135, 145
Wolfe Ltd, Deborah, 151
Wolff, Jennifer Snow, 168
Wolfram Research Inc, 168
Woloshin Inc, 184
Women in the Arts, 146
Wong, Leslie, 123
Woo, Don, 141
Wood, Joan, 129
Wooden, John, 126
Woodholly Productions, 177, 183, 188
Woods + Woods, 151

Woodward Associates, 162, 166
WordPerfect Corp, 154, 157
Words & Pictures, 153, 188, 191
Words Worth, 151
WordStar International Inc, 157
Wordwrap, 190
Works, The, 141
Workstation Technologies Inc, 196
World Computer Graphics Association, 191
Worrell, Rand, 179
Worth, Dennis, 147
Worthington, Nancy, 141
Worthwhile Films, 153, 176
WP Plus, 194
Wrap-Around Inc, 190
Wrap-Arounds Inc, 188
Wray, Lisa, 135
WRE/Colortech Prepress, 130, 188, 191
Wright, Michael Ragsdale, 141
Wright, Richard, 135
Wright, Ted, 137
Wright Lithography, 194
Wrinkle, James, 141
WTEN, 160, 175, 185
Wu, Brian, 143
Wu Multimedia, Brian, 158, 168
Wuilleumier Inc, Will, 135, 153, 178
Wurman, Richard Saul, 168, 208
Wurster, George, 153
Wy'east Color, 194
Wy'east Color Inc, 194
WYD Design Inc, 145
Wynn, Dan, 133
Wynne Ragland, 136, 163, 200
Wyse Technology, 155, 157

X

XAOS, 154
XAOS Tools, 157
Xenas Communications, 182, 186, 208
Xenon Micromedia Productions Inc, 162
Xerox Corp, 192
Xerox Imaging Systems, 154, 156
Xerox Imaging Systems Inc, 154
Xerox Media Center, 173, 180, 184
Xiphias, 195
Xitron Inc, 154, 156
Xpress Graphics, 192
XRS Corp, 155
Xtensory Inc, 199

Y

Yamaha Corp of America, 188
Yankus, Marc, 133
Yarmolinsky, Miriam, 135
Yaworski, Don, 126
Yazzolino, Brad, 141
Yeager, Carol, 129
YLEM, 204, 207
York Graphic Services, 145
York Photo Imagery, 128
Young, Andrea, 178
Young, Emily, 141
Young Assoc, Robert, 129, 135, 145
Young McKenna & Assocs Inc, 126
Youngblood, Michael S, 137
Yourke, Oliver, 133, 143
Yurth Video Production Services, 171, 177, 188

Z

Z-AX-IS, 136, 147, 160
Z-AXIS, 177, 183, 188
Zada, Nida, 137, 148
Zakari, Chantal, 137, 148
Zale Studios, 137
Zamiar Photography, 126
Zander Productions, Mark, 172
Zang Design, Ulla, 158, 163, 165
Zarley, Tim, 152
Zedcor, 157
Zeisse, Brook, 143
Zen Over Zero, 179
Zender & Assocs Inc, 148, 160, 167
Zenographics, 157
ZER, 129
Zero One, 190
ZFX, 181
Zgodzinski, Rose, 141
Ziff-Davis/MacUser, 141, 162, 183
Zima, Al, 135

Zimmer Inc, 192
Zimmerman, Robert, 136
Zimmerman, Sharon, 129
Zink Communications, 158, 166
Zinn, Elizabeth, 151
Zlowodzka, Joanna, 133, 143
Zoot, Ira, 137, 148
Zoran Corp, 198, 201
Zuber-Mallison, Carol, 136
Zubkoff Photography, Earl, 125
ZZYZX Visual Systems, 130, 194

ACCORNERO, FRANCO ——— 28
ACE GROUP, THE ——— 101
AGFA ——— 109
ANTHONY, MITCHELL ——— 74
ARTCO ——— 50-53
BAKER, KOLEA ——— 54,55
BAKER, DON ——— 55
BELDING, PAM ——— 77
BERMAN, HOWARD ——— 8,9
BERNSTEIN & ANDRIULLI REPRESENTATIVES = 64
BIG PIXEL, THE ——— 70
BLACKMAN, BARRY ——— 14,15
BLANK, JERRY ——— 59
BRICE, JEFF ——— 54
BRONSTEIN, STEVE ——— 8,9
BUCHMAN, DOUGLAS ——— 70
BURKEY, J.W. ——— 35
CARROLL, DON ——— 18,19
CHROMA STUDIOS, INC. ——— 92
CONRAD & ASSOCIATES, JON ——— 68
COPYTONE ——— 103
COVEY, MIKEL ——— 10
CUSTOM COLOR CORPORATION = 102
DAMAN STUDO ——— 25
DEAN DIGITAL IMAGING ——— 38
DIGICHROME IMAGING ——— 92
DIGITAL ART ——— 40,41
DIGITAL IMAGE ——— 90
DIGITAL STOCK ——— 99
DREAMTIME SYSTEMS ——— 29
FARR PHOTOGRAPHY ——— 36,37
FEARLESS DESIGNS, INC. ——— 58
GLASGOW & ASSOCIATES, DALE ——— 67
GROSSMAN, MYRON ——— 75
GROSSMAN, WENDY ——— 49
GUDYNAS, PETER ——— 24
HERBERT, JONATHAN ——— 63
H-GUN LABS ——— 43
HOWE, PHILIP ——— 32,33
HUNT, STEVEN ——— 22
ICON GRAPHICS INC. ——— 76
IKEGAMI ——— 108
JACKSON, LANCE ——— 61
LAX SYNTEX DESIGN ——— 61
LETRASET ——— 113
LVT A KODAK COMPANY ——— 117
LYNCH, ALAN / ARENA ARTISTS ——— 24
MacNEILL, SCOTT ——— 69

MAD WORKS ——— 74
MATSURI CORPORATION ——— 72
McELROY, OLIVIA ——— 64
MC2 ——— 48
MERSCHER, HEIDI ——— 48
META 4 DIGITAL DESIGN ——— 95
MICROCOLOR ——— 97
MILLET, CÉCILE ——— 73
MORRELL, PAUL ——— 17
NEITZEL, JOHN ——— 34
NEW MEDIA PRODUCTIONS, INC. = 96
OLDROYD DIGITAL ——— 100
OSTROFF, SANDY ——— 84
OUTERSPACE ——— 91
O'VERY COVEY, TRACI ——— 10
PACE, JULIE ——— 79
PALMS HIGH RES DIGITAL IMAGERY, THE = 85
PELIKAN PICTURES ——— 88,89
PETERSON, BRUCE ——— 11
PHOTOEFFECTS ——— 87
PIXAR ——— 116
PUNIN, NIKOLAI ——— 60
RAPHËLE/DIGITAL TRANSPARENCIES,INC. ——— 82,83
RENARD REPRESENTS ——— 63
R/GA PRINT ——— 44,45
ROMERO, JAVIER ——— 57
SANDBOX DIGITAL PLAYGROUND ——— 84
SCHLOWSKY, BOB & LOIS ——— 20,21
SILICON GRAPHICS ——— 106,107, BACK COVER
SKYLITE PHOTO PRODUCTIONS, INC. 34
SMITH, MARTY ——— 65
SPECULAR INTERNATIONAL = 114,115
STOKES RETOUCHING, LEE ——— 86
STRATA INC. ——— 110,111

STUDIO MacBETH ——— 50,51
STRUTHERS, DOUG ——— 52,53
TAYLOR - PALMER, DOROTHEA ——— 62
TEICH, DAVID ——— 71
TRACER DESIGN, INC. ——— 42
TUCKER, MARK ——— 23
TULL, JEFF ——— 58
UPTON, THOMAS ——— 12,13
VAN NOSTRAND REINHOLD ——— 120
VARIS PHOTOMEDIA ——— 26,27
WACOM ——— 112
WEISS, MICHAEL ——— 16
WESTLIGHT ——— 94
WILEY, PAUL ——— 56
WOJNAR DRAKE PHOTOGRAPHY, INC. ——— 30,31
ZAP ART ——— 24
ZUBER-MALLISON, CAROL ——— 66